T0192127

Theoretical and Mathematical Physics

The series founded in 1975 and formerly (until 2005) entitled *Texts and Monographs in Physics* (TMP) publishes high-level monographs in theoretical and mathematical physics. The change of title to *Theoretical and Mathematical Physics* (TMP) signals that the series is a suitable publication platform for both the mathematical and the theoretical physicist. The wider scope of the series is reflected by the composition of the editorial board, comprising both physicists and mathematicians.

The books, written in a didactic style and containing a certain amount of elementary background material, bridge the gap between advanced textbooks and research monographs. They can thus serve as basis for advanced studies, not only for lectures and seminars at graduate level, but also for scientists entering a field of research.

Roger Balian

From Microphysics to Macrophysics

Methods and Applications of Statistical Physics

Volume I

Translated by D. ter Haar and J.F. Gregg

With 39 Figures

Roger Balian

CEA, Direction des Sciences de la Matière, Service de Physique Théorique de Saclay
F-91191 Gif-sur-Yvette, France and
Ecole Polytechnique, F-91128 Palaiseau, France

Translators:

Dirk ter Haar

P.O. Box 10, Petworth, West Sussex
GU28 0RY, England

John F. Gregg

Magdalen College
Oxford OX1 4AU, England

Title of the original French edition: *Du microscopique au macroscopique,*
Cours de l'Ecole Polytechnique
©1982 Edition Marketing S.à.r.l., Paris; published by Ellipses

2nd Printing of the Hardcover Edition with ISBN 978-3-540-534266-8

Library of Congress Control Number: 2006933305

ISBN 978-3-540-45469-4 Springer Berlin Heidelberg New York

Typesetting: Data conversion by Springer-Verlag
Cover design: eStudio Calamar, Girona, Spain
Production: LE-TEX Jelonek, Schmidt & Vöckler GbR, Leipzig

Printed on acid-free paper SPIN: 11870784 55/3100/YL 5 4 3 2 1 0

Preface

Although it has changed considerably in both coverage and length, this book originated from lecture courses at the Ecole Polytechnique. It is useful to remind non-French readers of the special place this institution occupies in our education system, as it has few features in common with institutes with a similar name in other parts of the world. In fact, its programme corresponds to the intermediate years at a university, while the level of the students is particularly high owing to their strict selection through entrance examinations. The courses put a stress on giving foundations with a balance between the various natural and mathematical sciences, without neglecting general cultural aspects; specialization and technological instruction follow after the students have left the Ecole. The students form a very mixed population, not yet having made their choice of career. Many of them become high-level engineers, covering all branches of industry, some devote themselves to pure or applied research, others become managers or civil servants, and one can find former students of the Ecole amongst generals, the clergy, teachers, and even artists and Presidents of France.

Several features of the present volume, and in particular its contents, correspond to this variety and to the needs of such an audience. Statistical physics, in the broadest meaning of the term, with its many related disciplines, is an essential element of modern scientific culture. We have given a comprehensive presentation of such topics at the advanced undergraduate or beginning graduate level. The book, however, has to a large extent moved away from the original lecture courses; it is not only intended for students, but should also be of interest to a wider public, including research workers and engineers, both beginning and experienced. A prerequisite for its use is an elementary knowledge of quantum mechanics and general physics, but otherwise it is completely self-contained.

Rather than giving a systematic account of useful facts for specialists in some field or other, we have aimed at assisting the reader to acquire a broad and solid scientific background knowledge. We have therefore chosen to discuss amongst the applications of statistical physics those of the greatest educational interest and to show especially how rich and varied these applications are. This is the reason why, going far beyond the traditional topics of statistical mechanics – thermal effects, kinetic theory, phase transitions, radiation laws – we have dwelt on microscopic explanations of the mechanical, magnetic, electrostatic, electrodynamic, ... properties of the various states of matter. Examples from other disciplines, such as astrophysics, cosmology, chemistry, nuclear physics, the quantum theory of measurement, or even biology, enable us to illustrate the broad scope of statistical physics and to show its universal nature. Out of a concern for culture, and also in trying to keep engineers and scientists away from too narrow a specialization, we have also included introductions to various physical problems arising in important technological fields, ranging from the nuclear industry to lighting by incandescent lamps, or from solar energy to the use of semiconductors for electronic devices.

Throughout this abundance we have constantly tried to retain a unity of thought. We have therefore stressed the underlying concepts rather than the technical aspects of the various methods of statistical physics. Indeed, one can see everywhere in the book under various guises two main guiding principles: on the one hand, the interpretation of entropy as a measure of disorder or of lack of information and, on the other hand, a stress on symmetry and conservation laws. At a time when excessive specialization tends to hide the unity of science, we have deemed it instructive to present unifying points of view, showing, for instance, that the laws of electrodynamics, of fluid dynamics, and of chemical kinetics all go back to the same underlying, basic ideas.

The French tradition, both in secondary education and in the entrance examinations to the Ecole Polytechnique, has to some extent given pride of place to mathematics. We have tried to benefit from this training by putting our treatment on a strict logical basis and giving our arguments a structured, often deductive, character. Mathematical rigour has, however, been tempered by a wish to present and to explain many facts at an introductory level, to avoid formalistic stiffness, and to discuss the validity of models. We have inserted special sections to present the less elementary mathematical tools used.

A first edition of this book was published in French in 1982. When the idea of publishing an English translation started to take shape, it seemed desirable to adapt the text to a broader, more international audience. The first changes in this direction brought about others, which in turn suggested a large number of improvements, both simplifications and more thorough discussions. Meanwhile it took some time to find a translator. Further lecture courses, especially one given at Yale in 1986, led to further modifications. One way or another, one thing led to another and finally there was little left of the original text, and a manuscript which is for more than eighty per cent new was finally translated; the present book has, in fact, only the general spirit and skeleton in common with its predecessor.

The actual presentation of this book aims at making it easier to use by readers ranging from beginners to experienced researchers. Apart from the main text, many applications are incorporated as exercises at the end of each chapter and as problems in the last chapter of the second volume; these are accompanied by more or less detailed solutions, depending on the difficulty.

At the end of each of the two volumes we give tables with useful data and formulae. Parts of the text are printed in small type; these contain proofs, mathematical sections, or discussions of subjects which are important but lie outwith the scope of the book. For cultural purposes we have also included historical and even philosophical notes: the most fundamental concepts are, in fact, difficult to become familiar with and it is helpful to see how they have progressively developed. Finally, other passages in small type discuss subtle, but important, points which are often skipped in the literature. Many chapters are fairly independent. We have also tried clearly to distinguish those topics which are treated with full rigour and detail from those which are only introduced to whet the curiosity. The contents and organization of the book are described in the introduction.

I am indebted to John Gregg and Dirk ter Haar for the translation. The former could only start on this labour, and I am particularly grateful to the main translator, Dirk ter Haar, for his patience (often sorely tried by me) and for the care bestowed on trying to present my ideas faithfully. I have come to appreciate how difficult it is to find exact equivalents for the subtleties of the French language, and to discover some of the subtleties of the English language. He has also accomplished the immense task of producing a text, including all the mathematical formulae, which could be used directly to produce the book, and which, as far as I can see, should contain hardly any misprints.

The Service de Physique Théorique de Saclay, which is part of the Commissariat à l'Energie Atomique, Direction des Sciences de la Matière, and in which I have spent the major part of my scientific research career, has always been like a family to me and has been a constant source of inspiration. I am grateful to all my colleagues who through many discussions have helped me to elaborate many of the ideas presented here in final form. They are too numerous to be thanked individually. I wish to express my gratitude to Jules Horowitz for his suggestions about the teaching of thermodynamics. As indicated in the preface to the first edition, I am indebted to the teaching staff who worked with me at the Ecole Polytechnique for various contributions brought in during a pleasant collaboration; to those mentioned there, I should add Laurent Baulieu, Jean-Paul Blaizot, Marie-Noëlle Bussac, Dominique Grésillon, Jean-François Minster, Patrick Mora, Richard Schaeffer, Heinz Schulz, Dominique Vautherin, Michel Voos, and Libero Zuppiroli, who to various degrees have helped to improve this book. I also express my thanks to Marie-Noëlle Bussac, Albert Lumbroso, and Marcel Véneroni, who helped me in the tedious task of reading the proofs and made useful comments, and to Dominique Bouly, who drew the figures. Finally, Lauris and the other members of my family should be praised for having patiently endured the innumerable evenings and weekends at home that I devoted to this book.

Paris, April 1991 *Roger Balian*

Preface to the Original French Edition

The teaching of statistical mechanics at the Ecole Polytechnique used for a long time to be confined to some basic facts of kinetic theory. It was only around 1969 that Ionel Solomon started to develop it. Nowadays it is the second of the three physics "modules", courses aimed at all students and lasting one term. The first module is an introduction to quantum mechanics, while the last one uses the ideas and methods of the first two for treating more specific problems in solid state physics or the interaction of matter with radiation. The students then make their own choice of optional courses in which they may again meet with statistical mechanics in one form or another.

There are many reasons for this development in the teaching of physics. Enormous progress has been made in statistical physics research in the last hundred years and it is now the moment not only to reflect this in the teaching of future generations of physicists, but also to acquaint a larger audience, such as students at the Ecole Polytechnique, with the most useful and interesting concepts, methods, and results of statistical physics. The spectacular success of microscopic physics should not conceal from the students the importance of macroscopic physics, a field which remains very much alive and kicking. In that it enables us to relate the one to the other, statistical physics has become an essential part of our understanding of Nature; hence the desirability of teaching it at as basic a level as possible. It alone helps to unravel the meaning of thermodynamic concepts, thanks to the light it sheds on the nature of irreversibility, on the connections between information and entropy, and on the origin of the qualitative differences between microscopic and macroscopic phenomena. Despite being a many-faceted and expanding discipline with ill-defined boundaries, statistical physics in its modern form has an irreplaceable position in the teaching of physics; it unifies traditionally separate sciences such as thermodynamics, electromagnetism, chemistry, and mechanics. Last and not least its numerous applications cover a wide range of macroscopic phenomena and, with continuous improvements in the mathematical methods available, its quantitative predictions become increasingly accurate. The growth of micro-electronics and of physical metallurgy indicates that in future one may hope to "design" materials with specific properties starting from first principles. Statistical physics is thus on the way to becoming one of the most useful of the engineering sciences, sufficient justification for the growth of its study at the Ecole Polytechnique.

This book has evolved from courses given between 1973 and 1982 in the above spirit. The contents and teaching methods have developed considerably during that period; some subjects were occasionally omitted or were introduced as optional extras, intended only for a section of the class. Most of the major threads of statistical mechanics were reviewed, either in the course itself, or in associated problems. Nevertheless, on account of their difficulty, it has been possible to treat some important topics, such as irreversible processes or phase transitions, only partially, and to mention some of them, like superconductivity, only in passing. The published text contains all the material covered, suitably supplemented and arranged. It has been organized as a basic text, explaining first the principles and methods of statistical mechanics and then using them to explain the properties of various systems and states of matter. The work is systematic in its design, but tutorial in its approach; it is intended both as an introductory text to statistical physics and thermodynamics and as a reference book to be used for further applications.

Even though it goes far beyond the actual lecture programme, this is the text circulated to the students. Its style being half way between a didactic manual and a reference book, it is intended to lead the student progressively away from course work to more individual study on chosen topics, involving a degree of literature research. Typographically, it is designed to ease this transition and to help the first-time reader by highlighting important parts through italics, by framing the most important formulae, by numbering and marking sections to enable selective study, by putting items, supplementary to the main course, and historical notes in small type, and by giving summaries at the end of each chapter so that the students can check whether they have assimilated the basic ideas. However, the very structure of the book departs from the order followed in the lecture course, which, in fact, has changed from year to year; this is the reason why some exercises involve concepts introduced in later chapters.

Classes at the Ecole Polytechnique tend to be mixed, different students having different goals, and some compromises have been necessary. It is useful to take advantage of the mathematical leanings of the students, as they like an approach proceeding from the general to the particular, but it is equally essential that they are taught the opposite approach, the only one leading to scientific progress. The first chapter echoes this sentiment in using a specific example in order to introduce inductively some general ideas; it is studied at the Ecole as course work in parallel with the ensuing chapters, which provide a solid deductive presentation of the basis of equilibrium statistical mechanics. Courses at the Ecole Polytechnique are intended to be complemented later on by specialized further studies. When we discuss applications we have therefore laid emphasis on the more fundamental aspects and we have primarily selected problems which can be completely solved by students. However, we have also sought to satisfy the curiosity of those interested in more difficult questions with major scientific or technological implications, which are only qualitatively discussed. Conscious of the coherence of the book as a whole, we have tried to maintain a balance between rigour and simplicity, theory and fact, general methods and specific techniques. Finally, we have tried to keep the introductory approach of the book in line with modern ideas. These are based upon quantum statistical mechanics, richer in applications and conceptually simpler than its classical counterpart, which is commonly the first topic taught, upon the entropy as a measure of information missing because of the probabilistic nature of our description, and upon conservation laws.

Capable of being read at various different levels, and answering to a variety of needs, this course should be useful also outside the Ecole Polytechnique. Given its introductory nature and its many different purposes, it is not intended as a substitute for the more advanced and comprehensive established texts. Nevertheless, the latter are usually not easy reading for beginners on account of their complexity or because they are aimed at particular applications and techniques, or because they are aimed at an English-speaking audience. The ever increasing part played by statistical physics in the scientific background essential for engineers, researchers, and teachers necessitates its dissemination among as large an audience as possible. It is hoped that the present book will contribute to this end. It could be used as early as the end of undergraduate studies at a university, although parts are at the graduate level. It is equally well geared to the needs of engineering students who require a scientific foundation course as a passport to more specialized studies. It should also help all potential users of statistical physics to learn the ideas and skills involved. Finally, it is hoped that it will interest readers who wish to explore an insufficiently known field in which immense scientific advances have been made, and to become aware of the modern understanding of properties of matter at the macroscopic level.

Physics teaching at the Ecole Polytechnique is a team effort. This book owes much to those who year after year worked with me on the statistical mechanics module: Henri Alloul, Jean Badier, Louis Behr, Maurice Bernard, Michel Bloch, Edouard Brézin, Jean-Noël Chazalviel, Henri Doucet, Georges Durand, Bernard Equer, Edouard Fabre, Vincent Gillet, Claudine Hermann, Jean Iliopoulos, Claude Itzykson, Daniel Kaplan, Michel Lafon, Georges Lampel, Jean Lascoux, Pierre Laurès, Guy Laval, Roland Omnès, René Pellat, Yves Pomeau, Yves Quéré, Pierre Rivet, Bernard Sapoval, Jacques Schmitt, Roland Sénéor, Ionel Solomon, Jean-Claude Tolédano, and Gérard Toulouse, as well as our colleagues Marcel Fétizon, Henri-Pierre Gervais, and Jean-Claude Guy from the Chemistry Department. I have had the greatest pleasure in working with them in a warm and friendly environment, and I think they will excuse me if I do not describe their individual contributions down the years. Their enthusiasm has certainly rubbed off onto the students with whom they have been in contact. Several of them have given excellent lectures on special topics for which there has regrettably not been room in this book; others have raised the curiosity of students with the help of ingenious and instructive experiments demonstrated in the lecture theatre or classroom. This book has profited from the attention of numerous members of the teaching staff who have corrected mistakes, simplified the presentation, and thought up many of the exercises to be found at the end of the chapters. Some have had the thankless task of redrafting and correcting examination problems; the most recent of those have been incorporated in the second volume. To all of them I express my heartfelt thanks. I am especially indebted to Ionel Solomon: it is thanks to his energy and dedication that the form and content of the course managed to evolve sufficiently rapidly to keep the students in contact with live issues. On the practical side, the typing was done by Mmes Blanchard, Bouly, Briant, Distinguin, Grognet, and Lécuyer from the Ecole's printing workshop, efficiently managed by M. Deyme. I am indebted to them for their competent and patient handling of a job which was hampered by the complexity of the manuscript and by numerous alterations in the text. Indeed, it is their typescript which, with some adjustments by the publisher, was reproduced for the finished work. The demanding and essential task of proofreading was performed by Madeleine Porneuf from our group at Saclay. I also thank the staffs of the Commissariat à l'Energie Atomique and of the Ecole Polytechnique, in particular, MM. Grison, Giraud, Servières, and Teillac for having facilitated publication. Finally, I must not forget the many students who have helped to improve my lectures by their criticism, questions, and careful reading, and from whose interest I have derived much encouragement.

Roger Balian

Contents of Volume I

Introduction 1

1. Paramagnetism of Ionic Solids 15
1.1 Micro-states and Macro-state 16
1.1.1 Paramagnetism 16
1.1.2 The Model 18
1.1.3 The Probabilistic Nature of a Macroscopic State .. 20
1.1.4 Calculation of Expectation Values 21
1.2 Microscopic Interpretation of Thermal Equilibrium 22
1.2.1 Maximum Disorder 22
1.2.2 The Rôle of Energy Conservation 23
1.2.3 The Degree of Disorder 25
1.2.4 Thermal Contact and Partition of Energy 28
1.2.5 Absence of Correlations Between Magnetic Moments at Thermal Equilibrium 31
1.2.6 Spin Thermostat 32
1.3 Identification of the Thermodynamic Quantities 33
1.3.1 Relative Temperature 33
1.3.2 The Boltzmann-Gibbs Distribution 34
1.3.3 Entropy and Absolute Temperature 35
1.4 Experimental Checks 37
1.4.1 Equations of State 37
1.4.2 Curie's Law 38
1.4.3 Saturation 38
1.4.4 Thermal Effects; Characteristic Temperature 39
1.4.5 Magnetic Resonance 41
1.4.6 Discussion of the Model 43
Summary 46
Exercises 46
1a Negative Temperatures 46
1b Brillouin Curves 47
1c Langevin Paramagnetism 48

2. Probabilistic Description of Systems 49
2.1 Elements of the Quantum Formalism 50
2.1.1 Hilbert Spaces 50

2.1.2 Operators 52
2.1.3 Observables 56
2.1.4 Pure States, Preparations and Measurements 57
2.1.5 Evolution, Transformations, Invariances 58
2.1.6 Shortcomings of the State Vector Concept 62
2.2 Quantum Systems: Density Operators 63
2.2.1 Pure States and Statistical Mixtures 63
2.2.2 The Expectation Value of an Observable 64
2.2.3 Characteristic Properties of Density Operators 66
2.2.4 The Density Operator of a Subsystem 68
2.2.5 Measurements in Quantum Mechanics 70
2.2.6 Evolution with Time 73
2.2.7 Summary: Reformulation of Quantum Mechanics 75
2.3 Classical Systems: Densities in Phase 78
2.3.1 Phase Space and Micro-states 78
2.3.2 Classical Macro-states 79
2.3.3 Evolution with Time 81
2.3.4 The Classical Limit of Quantum Statistics 84
2.3.5 Reduced Densities 89
2.3.6 Uncertain Particle Number 93
Summary 93
Exercises 94
2a Density Operators for a Spin-$\frac{1}{2}$ 94
2b Density Fluctuations 98

3. Information Theory and Statistical Entropy 101
3.1 Information and Probability Theory 102
3.1.1 Statistical Entropy: The Measure of Missing Information 102
3.1.2 Properties of the Statistical Entropy 104
3.1.3 The Statistical Entropy Derived from the Additivity Postulate 107
3.1.4 Continuous Probabilities 110
3.2 The Statistical Entropy of a Quantum State 111
3.2.1 Expression as Function of the Density Operator 111
3.2.2 Inequalities Satisfied by the Statistical Entropy 113
3.2.3 Changes in the Statistical Entropy 118
3.2.4 Information and Quantum Measurements 119
3.3 The Statistical Entropy of a Classical State 122
3.3.1 Classical Statistical Entropy 122
3.3.2 Properties 123
3.4 Historical Notes About the Entropy Concept 123
3.4.1 Entropy in Thermodynamics 124
3.4.2 Entropy in Kinetic Theory 125
3.4.3 The Period of Controversies 127

3.4.4 Entropy and Quantum Mechanics 131
3.4.5 Entropy and Information 132
3.4.6 Recent Trends 134
Summary 136
Exercises 136
3a Information and Entropy: Orders of Magnitude 136
3b Subsystem of a Pure State 137
3c Relevant Entropy 137
3d Inequalities Concerning the Entropy 139
3e Entropy of Quantum Oscillators 140

4. The Boltzmann-Gibbs Distribution 141
4.1 Principles for Choosing the Density Operators 142
4.1.1 Equal Probabilities 142
4.1.2 Information About the System 142
4.1.3 Maximum of the Statistical Entropy 144
4.1.4 Macroscopic Equilibrium and Conservation Laws 146
4.1.5 Approach to Equilibrium 148
4.1.6 Metastable Equilibria 150
4.2 Equilibrium Distributions 152
4.2.1 Lagrangian Multipliers 152
4.2.2 Variational Method 156
4.2.3 Partition Functions 158
4.2.4 Equilibrium Entropy 160
4.2.5 Factorization of Partition Functions 161
4.2.6 Summary: Technique of Studying Systems at Equilibrium 163
4.3 Canonical Ensembles 165
4.3.1 The Canonical Ensemble 165
4.3.2 The Grand Canonical Ensemble 167
4.3.3 Other Examples of Ensembles 169
4.3.4 Equilibrium Distributions in Classical Statistical Mechanics 171
Summary 172
Exercises 173
4a Relation Between Fluctuation and Response 173
4b Adsorption 174
4c Free Energy of a Paramagnetic Solid 175
4d Absence of Magnetism in Classical Mechanics 176
4e Galilean Invariance 177
4f Oscillators in Canonical Equilibrium 177
4g The Ehrenfests' Urn Model 178
4h Loaded Die 179

5. Thermodynamics Revisited 181
5.1 The Zeroth Law 183
5.1.1 Relative Temperatures in Thermodynamics 183
5.1.2 Thermal Contact in Statistical Physics 183
5.1.3 Thermometers and Thermostats 186
5.1.4 Extensions: Open Systems 187
5.2 The First Law 188
5.2.1 Internal Energy 188
5.2.2 Heat 190
5.2.3 Work and Forces 191
5.2.4 Exchanges Between Systems and Sources 195
5.3 The Second Law 197
5.3.1 Energy Downgrading in Thermodynamics 197
5.3.2 Entropy and Absolute Temperature 198
5.3.3 Irreversibility 201
5.4 The Third Law or Nernst's Law 203
5.4.1 Macroscopic Statement 203
5.4.2 Statistical Entropy at the Absolute Zero 203
5.5 The Thermodynamic Limit 205
5.5.1 Extensive and Intensive Variables 205
5.5.2 Extensivity from the Microscopic Point of View 206
5.5.3 Equivalence of Ensembles 207
5.6 Thermodynamic Potentials 210
5.6.1 Entropy and Massieu Functions 210
5.6.2 Free Energy 213
5.6.3 Chemical Potentials 215
5.6.4 Grand Potential 218
5.6.5 Pressure; The Gibbs-Duhem Relation 219
5.6.6 Summary: Tables of the Thermodynamic Potentials 219
5.7 Finite Systems 221
5.7.1 Statistical Fluctuations 221
5.7.2 Finite Part of an Infinite System 224
5.7.3 Observation of Statistical Distributions 227
Summary 230
Exercises 231
5a Elasticity of a Fibre 231
5b Saddle-point or Steepest Descent Method 236
5c Electric Shot Noise 238
5d Energy Fluctuations and Heat Capacity 238
5e Kappler's Experiment 239

6. On the Proper Use of Equilibrium Thermodynamics 241
6.1 Return to the Foundations of Thermostatics 243
6.1.1 The Object of Thermodynamics 243

6.1.2 The Maximum Entropy Principle ... 243
6.1.3 Connection with Statistical Physics ... 246
6.1.4 Entropy and Disorder ... 248
6.2 Thermodynamic Identities ... 249
6.2.1 Intensive Quantities ... 249
6.2.2 Conditions for Equilibrium Between Two Systems . 251
6.2.3 Gibbs-Duhem Relations ... 251
6.2.4 Mixed Second Derivatives Identity ... 252
6.3 Changes of Variables ... 253
6.3.1 Legendre Transformations ... 253
6.3.2 Massieu Functions and Thermodynamic Potentials ... 256
6.3.3 Equilibrium in the Presence of Sources ... 256
6.3.4 Exchanges Between Systems in the Presence of Sources ... 259
6.3.5 Calculational Techniques ... 260
6.4 Stability and Phase Transitions ... 264
6.4.1 Concavity of the Entropy ... 264
6.4.2 Thermodynamic Inequalities ... 265
6.4.3 The Le Chatelier Principle ... 266
6.4.4 The Le Chatelier-Braun Principle ... 267
6.4.5 Critical Points ... 268
6.4.6 Phase Separation ... 270
6.5 Low Temperatures ... 273
6.5.1 Nernst's Law ... 273
6.5.2 Vanishing of Some Response Coefficients ... 273
6.6 Other Examples ... 274
6.6.1 Thermal Engines ... 274
6.6.2 Osmosis ... 278
6.6.3 Thermochemistry ... 279
6.6.4 Elasticity ... 283
6.6.5 Magnetic and Dielectric Substances ... 284
Summary ... 295
Exercises ... 296
6a Equation of State, Specific Heats, and Chemical Potential ... 296
6b Isotherms, Adiabats, and Absolute Temperature .. 298
6c Interfaces and Capillarity ... 299
6d Critical Behaviour in the Landau Model ... 301
6e Equilibrium of Self-Gravitating Objects ... 304

7. The Perfect Gas ... 307
7.1 The Perfect Gas Model ... 308
7.1.1 Structureless Particles ... 308
7.1.2 Non-interacting Particles ... 310

7.1.3 Classical Particles ... 311
7.2 The Maxwell Distribution ... 313
7.2.1 The Canonical Phase Density ... 313
7.2.2 The Momentum Probability Distribution ... 315
7.2.3 Applications of the Maxwell Distribution ... 316
7.2.4 Grand Canonical Equilibrium of a Gas ... 319
7.2.5 Microcanonical Equilibrium ... 321
7.3 Thermostatics of the Perfect Gas ... 322
7.3.1 Thermostatic Pressure ... 322
7.3.2 Thermal Properties ... 323
7.3.3 The Rôle of an Applied Field ... 325
7.4 Elements of Kinetic Theory ... 326
7.4.1 Effusion ... 326
7.4.2 Kinetic Pressure ... 328
7.4.3 Kinetic Interpretation of the Temperature ... 330
7.4.4 Transport in the Knudsen Regime ... 331
7.4.5 Local Equilibrium and Mean Free Path ... 331
7.4.6 Heat Conductivity and Viscosity ... 333
Summary ... 337
Exercises ... 337
7a Barometric Equation ... 337
7b Isotope Separation by Ultracentrifuging ... 338
7c Relativistic Gas ... 342
7d Doppler Profile of a Spectral Line ... 343
7e Liquid Nitrogen Trap ... 344
7f Adiabatic Expansion ... 345
7g Isotope Separation by Gas Diffusion ... 347

8. Molecular Properties of Gases ... 349
8.1 General Properties ... 350
8.1.1 The Internal Partition Function ... 350
8.1.2 The Equation of State ... 353
8.1.3 Thermal Properties ... 354
8.1.4 The Chemical Potential ... 355
8.1.5 The Entropy ... 356
8.2 Gas Mixtures ... 357
8.2.1 Non-reacting Mixtures ... 357
8.2.2 Chemical Equilibria in the Gas Phase ... 359
8.3 Monatomic Gases ... 363
8.3.1 Rare Gases; Frozen-In Degrees of Freedom ... 363
8.3.2 Other Monatomic Gases ... 364
8.4 Diatomic and Polyatomic Gases ... 366
8.4.1 The Born-Oppenheimer Approximation ... 366
8.4.2 The Energy Equipartition Theorem ... 370
8.4.3 Classical Treatment of Polyatomic Gases ... 372

8.4.4 Quantum Treatment of Diatomic Gases 375
8.4.5 The Case of Hydrogen 379
Summary 382
Exercises 383
8a Entropy and Work in Isotope Separation 383
8b Excitation of Atoms by Heating 385
8c Ionization of a Plasma: Saha Equation 386
8d Adiabatic Transformation and Specific Heats 387
8e Hemoglobin 387

9. Condensation of Gases and Phase Transitions 391
9.1 Model and Formalism 392
9.1.1 Interactions Between Molecules 393
9.1.2 The Grand Potential of a Classical Fluid 394
9.2 Deviations from the Perfect Gas Laws 396
9.2.1 The Maxwell Distribution 396
9.2.2 Perturbation Methods 397
9.2.3 The van der Waals Equation 398
9.2.4 The Virial Series 401
9.2.5 The Joule-Thomson Expansion 406
9.3 Liquefaction 408
9.3.1 The Effective Potential Method 408
9.3.2 The Gas-Liquid Transition 412
9.3.3 Coexistence of Gas and Liquid Phases 420
Summary 426
Exercises 427
9a Ferromagnetism 427
9b Spins with Infinite-Range Interactions 433
9c Linear Chain of Spins 435
9d Stable and Metastable Magnetic Phases 437
9e Liquid-Vapour Interface 440
9f Dilute Solutions 442
9g Density of States of an Extensive System and Phase Separation 446

Subject Index 449

Units and Physical Constants 462

A Few Useful Formulae 464

Contents of Volume II

10. Quantum Gases Without Interactions
10.1 The Indistinguishability of Quantum Particles
10.2 Fock Bases
10.3 Equilibrium of Quantum Gases
10.4 Fermi-Dirac Statistics
10.5 Bose-Einstein Statistics
Exercises

11. Elements of Solid State Theory
11.1 Crystal Order
11.2 Single-Electron Levels in a Periodic Potential
11.3 Electrons in Metals, Insulators and Semiconductors
11.4 Phonons
Exercises

12. Liquid Helium
12.1 Peculiar Properties of Helium
12.2 Helium Three
12.3 Helium Four and Bose Condensation
Exercises

13. Equilibrium and Transport of Radiation
13.1 Quantizing the Electromagnetic Field
13.2 Equilibrium of Radiation in an Enclosure
13.3 Exchanges of Radiative Energy
Exercises

14. Non-equilibrium Thermodynamics
14.1 Conservation Laws
14.2 Response Coefficients
14.3 Microscopic Approach
14.4 Applications
Exercises

15. Kinetic Equations
15.1 Lorentz Model
15.2 Applications and Extensions
15.3 The Boltzmann Equation
15.4 Microscopic Reversibility vs Macroscopic Irreversibility
Exercises

16. Problems
16.1 Paramagnetism of Spin Pairs
16.2 Elasticity of a Polymer Chain
16.3 Crystal Surfaces
16.4 Order in an Alloy
16.5 Ferroelectricity
16.6 Rotation of Molecules in a Gas
16.7 Isotherms and Phase Transition of a Lattice Gas
16.8 Phase Diagram of Bromine
16.9 White Dwarfs
16.10 Crystallization of a Stellar Plasma
16.11 Landau Diamagnetism
16.12 Electron-Induced Phase Transitions in Crystals
16.13 Liquid-Solid Transition in Helium Three
16.14 Phonons and Rotons in Liquid Helium
16.15 Heat Losses Through Windows
16.16 Incandescent Lamps
16.17 Neutron Physics in Nuclear Reactors
16.18 Electron Gas with a Variable Density
16.19 Snoek Effect and Martensitic Steels

Conclusion: The Impact of Statistical Physics

Introduction

"– Est-ce la physique que vous voulez apprendre?
– Qu'est-ce qu'elle chante, cette physique?
– La physique est celle qui explique les principes des choses naturelles et les propriétés du corps; qui discourt de la nature des éléments, des métaux, des minéraux, des pierres, des plantes et des animaux, et nous enseigne les causes de tous les météores, l'arc-en-ciel, les feux volants, les comètes, les éclairs, le tonnerre, la foudre, la pluie, la neige, la grêle, les vents et les tourbillons.
– Il y a trop de tintamarre là-dedans, trop de brouillamini."

Molière, Le Bourgeois Gentilhomme

The Unification of Macroscopic Sciences

It is striking that the progress of Science has so far confirmed most scientists in their belief that there exist *simple* principles which should enable us to understand the various aspects of Nature, even though she appears to be so complex. In fact, the search for a *unique foundation* as the basis of most dissimilar phenomena, together with a quest for facts going beyond the well trodden paths, have been amongst the most powerful driving forces behind scientific progress. Strangely enough, aesthetic criteria often have led to the construction of new theories which describe Nature more efficiently. For example, the Ptolemaean system was changed to the Copernican one in order that one could more simply explain the apparent motions of the planets; later Kepler's laws gave way to Newton's universal gravitation and much later to Einstein's relativistic mechanics, each stage introducing a *more general* and at the same time a *more accurate* description of celestial dynamics.

In the second half of the nineteenth century it was commonly – but, of course, wrongly – thought that science was about to be completed. The unification through induction had made it possible to regroup all known phenomena into a few broad, coherent disciplines: so-called *rational mechanics*, containing the statics and dynamics of point particles and rigid bodies, *mechanics of continuous media*, consisting of elasticity, fluid dynamics, and acoustics, *electromagnetism*, from electrostatics to the study of fields and currents in various kinds of substances, geometric and wave *optics, thermodynamics*, which studies thermal properties, pressure, and phase changes, and *chemistry*, which is concerned with equilibria, reactions, and thermochemistry.

Each of these disciplines is based on a few *fundamental principles* – such as the Hamiltonian, Navier-Stokes, or Maxwell equations, or the Laws of Thermodynamics – which appear as a kind of axioms discovered through successive inductions and justified by the extent of their consequences. Here the term "fundamental" refers not so much to the importance of the principles, as to their character of "foundations". These principles make it possible, in theory if not in practice, quantitatively to predict the behaviour of various complicated physical systems. The additional ingredients are the *specific properties* of the substances considered, with the data being available either as tables or graphs, or as empirical laws – for, say, the viscosity, the resistivity, the magnetic susceptibility, the refractive index, the specific heat, the equation of state, the mass action law – distilled from experiments.

One hardly needs stress the rôle of *mathematics* as a means for unifying science. It has made possible an accurate and succinct formulation of the basic principles and a derivation of their observable consequences. In fact, the larger the domain covered and the more general the principles, the more difficult it is to derive the consequences. One should note that the periodic reconsideration of physical theories may need invoking new mathematics, and that often these new tools lack rigour to start with, as witness the theories of differential and integral calculus, Fourier transforms, distributions, or functional integration. From a philosophical point of view, it is remarkable how efficient mathematics is in describing the real world: from the ancient Greeks to Einstein, scientists have felt that "God is a geometer".

The growing dependence of science on mathematics should, however, not hide the essential rôle played by experiments. They are the basis of induction and the only means for distinguishing between rival theories and checking new hypotheses. They provide exact data, indispensible both for the progress of fundamental science and for its applications, and open the way to discoveries thanks to a search for possible failures of the existing theories.

Note finally that science moves simultaneously to a greater unification and to an ever increasing specialization of the activities of the scientists. Even though the vision of an all embracing synthesis is a necessary part of all progress, it goes hand in hand with a deepening of knowledge in any particular domain. As soon as one realized that light is an electromagnetic wave, optics ceased to be an autonomous science, and one had to understand it as a branch of electromagnetism; however, this is no reason for using the Maxwell equations for designing a photographic lens!

Microscopic and Macroscopic Science

The scientific revolution of the first third of the twentieth century rendered the above-mentioned classification of science obsolete. Although it still exists to some extent in the organization of laboratories or of lecture courses, in active science itself it has faded away. Even the frontiers between physics, mechanics, and chemistry are fluid.

Nowadays, the major distinctions hinge upon the scale of the phenomena studied. Up to the end of the nineteenth century science was primarily concerned with the description and understanding of macroscopic phenomena, on our own scale. Certainly, interest in such matters has not abated since then, and one witnesses actually a renewed interest in research on our scale; indeed, much remains to be discovered and one finds all the time new applications to technology and to other sciences. However, the overall emphasis has shifted to microscopic physics; its discovery has completely modified accepted views about Nature; a similar evolution has taken place in biology.

Initially, microscopic research concentrated on phenomena on the scale of one Ångström (1 Å = 10^{-10} m) – the study of *atoms, molecules, and condensed matter* in its various forms, such as liquids, amorphous and crystalline solids, or mesophases. These investigations have more or less rapidly evolved towards a coherent, logical discipline, based upon a small number of principles, to wit, the postulates of quantum mechanics, together with the properties of the particles involved at this level – electrons, atomic nuclei, and photons – which are simply characterized by their mass, charge, spin, and electromagnetic interactions.

Nuclear physics, which started at the end of the nineteenth century through radioactivity, refers to the scale of one fermi (1 fm = 10^{-15} m). Its main development took place during the last sixty years; it describes the properties of *nuclei* regarded as assemblies of more elementary particles, protons and neutrons, bound together by the so-called strong interaction. More recently, *particle physics* has compelled us to go further down to even more microscopic scales where the laws of physics become more and more unified. At the scale $L \simeq 10^{-17}$ m, corresponding to energies $\hbar c/L$ of 100 GeV, one observes that a single mechanism is the basis of both electromagnetism and the so-called weak interaction, which is responsible for the β-radioactivity of nuclei and the instability of the neutron. An even more microscopic scale (10^{-25} to 10^{-30} m) is involved in the grand unified theories which try to include electromagnetic, weak, and strong interactions in a single scheme, and which describe protons, neutrons, or mesons as composite particles formed from elementary constituents, the quarks, bound together by gluons. Finally, including the gravitational forces would make it necessary to go down to the Planck length obtained by a dimensional argument from the gravitational constant G, $\hbar$, and c as $\sqrt{G\hbar/c^3} \simeq 10^{-35}$ m; at this level the theory can only be speculative.

The emergence of new branches of physics at ever more microscopic scales poses a difficult problem: how can one connect these scales? For instance, the aim of many papers in nuclear physics is to explain the properties of nuclei, starting from their components, the protons and neutrons, or even, going further back, from the more elementary constituents, the quarks and gluons.

In the present book we shall mainly study the passage from the Ångström, which we shall call the microscopic, scale to our own, the so-called macroscopic scale. There are obvious differences between these two scales. Not only the objects studied, atoms or molecules, on the one hand, pieces of matter, on the other hand, but also the phenomena and concepts are completely different. The macroscopic and microscopic laws seem unrelated, as they are in *qualitative* contrast to one another. Microscopic physics is *discrete*, governed by *probabilistic* quantum mechanics and by a *few simple* laws, whereas the laws on our scale are *continuous, deterministic, and manifold.* Moreover, microscopic dynamics is invariant under time reversal, whereas we observe daily *irreversible phenomena.* Many common macroscopic quantities, such as the temperature or the electric resistivity, and many phenomena such as the existence of different, solid and liquid, phases of the same material, have no obvious microscopic counterpart. We hit here a major difficulty. What are the relations between microscopic physics and the various macroscopic sciences? Are they distinct areas, governed by independent laws, as are, for instance, electromagnetism and thermodynamics? Or rather, does microscopic physics enjoy a privileged position in the unification of science in that it is possible to explain macroscopic effects solely from the interactions between particles, and from quantum mechanics? In other words, are classical mechanics, macroscopic electromagnetism, thermodynamics, and chemistry, truly fundamental, or can they be derived from a few simple laws of microscopic physics?

Nowadays everybody knows that the second statement is the correct one. In fact, even before the birth of microscopic physics, the atomic hypothesis, inherited from Democritus, had been recognized as supplying a coherent explanatory system for various macroscopic properties of gases.

Statistical Physics and Its Historical Development

One can consider Daniel Bernoulli (Groningen 1700–Basel 1782) as the forefather of statistical mechanics. Its two essential ingredients, namely simple laws governing a large number of constituent particles and a probabilistic description, are already at the heart of his theory from 1727, where he explained the pressure of a gas as the consequence of collisions of hypothetical molecules with the walls. After Lavoisier had established the conservation laws for the chemical elements and for their mass, the chemists of the end of the eighteenth century also foresaw the logical necessity of molecular physics as the basis of their own science. Similarly, Maxwell's and Boltzmann's kinetic theory continued the work of some of their predecessors in the first half of the nineteenth century, who assumed without much trouble that a gas consisted of molecules. The *kinetic theory of gases*, based upon purely mechanistic concepts, assumes that the molecules collide elastically with one another; it reduces heat to kinetic energy. It started around 1856–60 with the work of August Krönig, Rudolf Clausius, and first and foremost, James Clerk Maxwell (Edinburg 1831–Cambridge 1879); concepts such as the mean free path, velocity distributions, and ergodicity were introduced at that time. Ludwig Boltzmann (Vienna 1844–Duino near Trieste 1906) made a decisive contribution by introducing probabilities, finding the energy distribution in thermal equilibrium, and interpreting the entropy and the Second Law on a microscopic level (1877). The second progenitor of statistical mechanics, Josiah Willard Gibbs (New Haven, Connecticut 1839–1903) introduced statistical ensembles which describe thermodynamic equilibrium and showed how they are connected with the thermodynamic potentials for any classical system; he helped to clarify Boltzmann's approach thanks to his elegant mathematical formulation.

However, up to 1900, the scientific community ignored Gibbs's ideas and rejected those of Boltzmann. It is interesting to understand why this was the case, in order better to grasp the importance of statistical physics and its rôle amongst the other scientific disciplines. At that time it seemed epistemologically inadmissible to base a fundamental science on *statistics* while the theory of the then prevalent macroscopic sciences – such as thermodynamics or electromagnetism – only involved exactly known quantities. Moreover, the dominant trends forced scientists and philosophers to reject molecular approaches as long as there was no direct microscopic evidence in their favour – even though atomism had largely been accepted at the end of the eighteenth century. On the contrary, the experimental background, from fluid dynamics to electromagnetic fields, supported *continuous* ideas about Nature. It is thus hardly surprising to find eminent thinkers, such as Ernst Mach, famous for his contributions to aerodynamics, the thermodynamicist Pierre Duhem, and the physical chemist Wilhelm Ostwald, among the fiercest opponents of Boltzmann. Even chemists such as Marcelin Berthelot fought against atomistic and molecular approaches; chemistry teaching in France continued until 1950 to be based upon macroscopic principles such as the laws of "definite" or "multiple proportions", and it still talked about the "atomic hypothesis". The violence of the attacks against Boltzmann may have contributed, together with an unsurmountable and painful illness, to induce his suicide in 1906 – two years before Jean Perrin's experiments directly proved the existence of atoms.

It is not surprising that statistical mechanics had so many difficulties to become accepted, even after atomic physics had been solidly established, as it was not easy to show that all of macroscopic science followed from the latter. In fact, microscopic phenomena are far from everyday experience, since the systems that we observe on our scale contain a huge number of particles. Even though in some cases, such as chemical reactions in the gas phase, our observations may reflect rather directly what happens on a microscopic scale, this is not so in general. Above a certain degree of complexity, the *collective* properties of a system have nothing in common with the *individual* properties of its constituents, and their derivation, even though theoretically possible, in practice meets with well-nigh unsurmountable difficulties. For instance, a one carat diamond must be regarded as a molecule containing 10^{22} carbon atoms which are bound together. Although in principle its properties are supposed to be described by the Schrödinger equation of its nuclei and electrons and their Coulomb interactions, one cannot conceive the possibility to solve an equation with that many variables. Moreover, it is remarkable that most properties of solids, liquids, gases, or plasmas, that all of chemistry, and even biology itself, are virtually contained in this simple Schrödinger equation for nuclei and electrons with their Coulomb interactions. The diversity of the world which surrounds us testifies to the futility of even thinking about the possibility of simply extending to macroscopic substances the methods of quantum mechanics, even though these have been extraordinarily successful in the study of atoms and small molecules. The bridge between the two scales must be provided by a new science, *statistical physics.* In essence theoretical, it is directly in contact with other, experimental or theoretical, sciences dealing with various materials and with various kinds of properties: the physics of liquids, solids, plasmas, condensed matter, or amorphous substances – and electromagnetism in matter, mechanics of continuous media, thermodynamics, or physical chemistry.

The stakes are enormous. On the one hand, one is trying to *derive the principles of macroscopic sciences*, demonstrating an underlying hidden conceptual unity. This proof is a major step in the unification process: with the exception of nuclear and particle physics, nowadays all of science appears to be based upon atomic scale phenomena, and the qualitative differences between various scales have been explained. On the other hand, we expect that statistical mechanics should enable us to understand, to predict, and to *calculate* from the microscopic structure of a substance its manifold properties, which otherwise are only accessible experimentally. In this way one may *justify empirical laws* such as those of Gay-Lussac, Hooke, Ohm, or Joule. Even better, as it progresses, statistical physics, thanks to exchanges with experiments, helps the scientist or the engineer to produce substances or systems with properties which are defined beforehand, such as electronic components or special purpose glasses. In astrophysics also, statistical mechanics occupies a privileged position as, for instance, it is the only means for studying stellar interiors or evolution.

Statistical physics is an evolving science, never ceasing to improve and to change the subjects of its studies ever since it first appeared in the shape of the kinetic theory of gases. Around the end of the nineteenth century Lord Rayleigh (Essex 1842–1919) and Max Planck (Kiel 1858–Göttingen 1947) began to apply statistical mechanics to a new system, the black body, in order to explain the spectral distribution of *radiation*; a breakthrough by Max Planck in 1900 contributed to the creation of quantum mechanics. Indeed, the theoretical explanation in 1905–1907 by Albert Einstein (Ulm 1879–Princeton 1955) of Planck's formula through the quantization of radiation provided the basis of both quantum mechanics and *quantum statistical mechanics.*

The main subsequent progress of the latter came from *solid state* theory. Einstein himself already in 1907 and Petrus Debye (Maastricht 1884–Ithaca, New York 1966) in 1912 explained the behaviour of the specific heats of solids. In the same period, following experiments by Pierre Curie (Paris 1859–1906), theories, by scientists like Pierre Weiss and Paul Langevin, and later P.Debye, Léon Brillouin, and Louis Néel, appeared, explaining the various types of magnetic behaviour. Already at the end of the nineteenth century Paul Drude (Brunswick 1863–Berlin 1906) started the electron theory of metals, following H.A.Lorentz and J.J.Thomson. During our century one has been able to witness continuously an extraordinary blossoming of statistical mechanical applications to a wide range of subjects: adsorption, electrical, thermal, optical, magnetic, and mechanical properties of metals, alloys or insulators, metallurgy, electrolytes, quantum fluids, plasmas, electronics, and so on. In the twenties and thirties these developments went hand in hand with applications of quantum mechanics and many names are common to both fields: Irving Langmuir, Max Born, Paul Ehrenfest, Arnold Sommerfeld, Fritz London, Eugene Wigner, Rudolph Peierls, Werner Heisenberg, Hans Bethe, Satyendranath Bose, Wolfgang Pauli, Enrico Fermi, Paul A.M.Dirac. In particular, Lev Davidovich Landau (Baku 1908–Moscow 1968) made major contributions to magnetism, to the theory of phase transitions, to plasma theory where he showed how thermal motion leads to damping of the oscillations, and to the theory of quantum liquids where he explained the superfluidity of helium at low temperatures.

More recently, the understanding of the properties of semiconductors gave rise to the invention of the transistor (John Bardeen, Walter Brattain, and William Shockley, 1948). In the fifties statistical mechanics appeared in a new guise, as the "many-body problem", and the various techniques developed then made it possible to treat situations where the interactions between quantum particles play a major rôle. One of its applications, the BCS theory (John Bardeen, Leon N.Cooper, and John R.Schrieffer, 1957) gave an explanation of the superconductivity of various metals at very low temperatures, a phenomenon which had remained an almost complete mystery since its discovery by Heike Kamerlingh Onnes in 1911. The physics of the various condensed states of matter and nuclear physics continue to make contributions to the many-body problem.

The seventies have been the scene of the theory of critical phenomena, which has explained the awkward and time-worn puzzle of the behaviour of thermodynamic quantities in the vicinity of the critical point in a phase transition, or the properties of long polymer chains in solution. Many problems still remain open, for instance, in the fields of poorly ordered materials, intermediate between liquids and solids, such as quasi-crystals, glasses, amorphous or soft matter, of biological substances, of chaotic dynamics, or of phenomena which are self-similar under a change in the space or time scales. All those topics are the subjects of very active research. There also remains much to be done in the field of non-equilibrium processes; in fact, it was only in 1931 that one of the major laws of irreversible thermodynamics was found by Lars Onsager (Oslo 1911–Miami 1976), and kinetic theory was extended to cover liquids only at the end of the thirties (John G.Kirkwood, Joseph E. Mayer, Jacques Yvon).

Among the various guises in which statistical physics presents itself one finds a few aspects which unite this discipline. A first line of thought is a systematic exploitation of *symmetry, invariance, and conservation* laws which are known on the microscopic scale. The information thus obtained is not plentiful, but it is precious as it gives us cheaply a few definite conclusions. For instance, it pays to draw up a simple *balance* of conserved quantities, such as the energy or the particle number. Let us note, however, that going from one scale to another can introduce subtle changes. For instance, energy conservation is considered on our scale from a new standpoint, as the heat concept only emerges on this scale. Symmetry properties which are valid in microscopic physics, such as time reversal, can also be broken when one passes to the macroscopic scale, though leaving a few traces.

The major difficulty that statistical physics must overcome is the treatment of the *large number* of elementary constituents, measured by Avogadro's number, 6×10^{23} mol^{-1}, or by the ratio of the lengths which characterize the two scales: 1 cm $= 10^8$ Å. This makes it necessary to use mathematical techniques which are as diverse as they are elaborate. They range from exact or approximate solutions of more or less realistic models, where one often exploits the large size of the system by considering it to be infinite, to numerical calculations or computer experiments where one must confine oneself to rather small systems. A common feature of all these techniques is the use of *statistics.*

The Rôle of Statistics in Physics

Whereas in the theory of games one can easily identify probabilities with relative frequencies, in physics the probability concept has an unavoidable anthropocentric character.[1] The probabilities are not an intrinsic property of an object; they are a mathematical tool which enables us to make *consistent predictions*, starting from earlier acquired information. Even if we wish to describe an individual system, we must consider it as being part of a *statistical ensemble*, constructed from similar systems, to which our probabilistic information and our conclusions refer. This view of probabilities will always be at the back of our mind in what follows.

From this point of view every scientific law is statistical by nature. Indeed, scientific knowledge in any field relies on experiments, or at least on observations. The data that they provide invariably possess some degree of randomness, and the conclusions drawn from these data are coloured by the same uncertainties. Scientific truths are never absolute and progress is made by formulating laws which are less and less likely to be wrong.

Apart from this experimental aspect, statistics is involved in physics for various reasons. *Quantum* mechanics, the theory underlying all contemporary physics, is probabilistic *by nature.* According to its current interpretation, it is as much a theory about physical objects themselves as one about

[1] R.T.Cox, Am. J. Phys. **14**, 1 (1946); B. de Finetti, *Theory of Probability*, Wiley, New York, 1974.

our knowledge of these objects: its statistical nature reflects the intrinsic incompleteness of our predictions. Attempts to find another, non-probabilistic interpretation – one which would completely eliminate the rôle of the observer – have always failed. Quantum mechanics thus makes the statistical nature of microscopic physics unavoidable.

Nevertheless, other statistical aspects persist, even in the classical limit of quantum mechanics, to wit, classical dynamics, although *a priori* the latter is deterministic by nature.

First of all, it is possible that a system under investigation is subject to *unknown forces.* Such is the case, for example, for Brownian motion (1826): a small particle – a pollen grain – in a liquid is observed through a microscope and is found to move under the effect of random blows from neigbouring molecules in the liquid. Its trajectory is extremely complicated and cannot be predicted, but one can calculate quantities such as the root mean square displacement and show that it varies as the square root of the time (Einstein 1905, Smoluchowski 1906, Langevin 1908).

It is also possible that the *initial conditions* are incompletely known, as in experiments on unpolarized beams of non-zero spin particles. In a more subtle way, most physical systems, even though their motion is governed by completely deterministic differential equations, follow a *chaotic dynamics*: the least uncertainty about the initial conditions, or the smallest perturbation, gets amplified with time in such a way that a near-certainty becomes a probability. The general occurrence of this phenomenon – which, for instance, makes any long-term meteorological prediction impossible – has only been realized in the seventies. It occurs, as soon as a dynamic system possesses three degrees of freedom, but it is much more pronounced for complex systems.

The *large number of degrees of freedom* of a macroscopic system make it necessary for us to use statistics. This number is so huge that a complete description of the system is inconceivable. The largest computers available are capable of simulating the motion of a few thousand particles whereas one mm^3 of a gas contains 3×10^{16} molecules. It is sufficient to imagine the space required to write 10^{16} numbers, even restricting them to two significant figures, to see that it is completely impossible to record the positions of the molecules in a gas. Even if it were possible, such a cumbersome mass of information would be totally unusable and hence quite pointless.

In fact, most physically interesting quantities, experimentally measurable and of practical application, are macroscopic: volume, pressure, temperature, specific heat, viscosity, refractive index, magnetic susceptibility, resistivity, etc. If we wish to derive their values from a knowledge of microscopic properties, we must identify them with *statistical averages* over all particles, since the individual characteristics of the particles are both *inaccessible* and *irrelevant.* Starting from the microscopic constituents to explain macroscopic properties thus *makes it necessary* to use probabilistic ideas and methods, even if the laws governing the microscopic elements are exactly known, and even if the underlying microscopic theory is deterministic. Thus, as for quantum mechanics, but for completely different reasons, the predictions of statistical mechanical are *probabilistic by nature* and involve *averages*.

Nevertheless, the very fact that the systems studied are very large – and this is what initially impelled us to use statistical methods – has a redeeming consequence. The "law of large numbers" can, as shown in probability theory, lead to precise predictions whose fulfilment is a *near certainty*, and in these circumstances the system concerned is to all intents and purposes completely deterministic. For instance, playing head or tail a very large number, N, of times, the chance of finding heads equals $\frac{1}{2}$ with a relative uncertainty equal to $1/\sqrt{N}$. This is a general feature of statistical mechanics, where the large number involved is the number of atoms in the system: thus macroscopic physical variables have mean square deviations which are negligible in comparison with their mean values and *exact predictions* are possible. For instance, in a fluid of N particles which are randomly distributed in a large volume Ω, the number n of molecules in a volume v is a random variable. Its expectation value is vN/Ω, and its relative statistical fluctuation equals $1/\sqrt{\langle n \rangle}$ (Exerc.2b). For 1 mm^3 of gas, this quantity equals 0.6×10^{-8}; we can thus regard n/v as being exactly determined, and equal to the density N/Ω, its statistical fluctuations being negligible compared with the experimental errors. This kind of argument will enable us to reconcile the statistical nature of microscopic physics with macroscopic determinism.

Contents of the Book

The first chapter is devoted to a straightforward, detailed treatment of a simple statistical physics problem – paramagnetism of a particular class of ionic solids. The aim of this chapter is to introduce through an *actual example*, studied *inductively*, the most important methods and ideas of statistical physics which we shall use later on.

Chapters 2 to 5 follow an inverse path, which is *formal* in order to be as *general* as possible, and *deductive* in order to be simpler. They form a unity which starts at the microscopic description of systems and ends up with establishing their general equilibrium macroscopic properties. Deliberately we use *quantum mechanics*, conceptually simpler than classical statistical mechanics which we find by taking a *limit*. First of all, we recall in Chap.2 the laws of quantum mechanics; we introduce the density operator which plays the rôle of a probability law in quantum physics, and the density in phase, its classical counterpart. To a large extent this chapter serves as a reference and much of it is not needed for understanding the remainder of the book. Having thus set up the formalism of statistical physics we devote Chap.3 to a quantitative evaluation of our uncertainty about a system described by a probability law or, what amounts to the same, of the disorder existing in that system. To do this, we base ourselves upon *information* theory and upon the concept of a *statistical entropy* which plays a central rôle everywhere in this book. In particular, Chap.4 uses this concept to assign to a system in thermodynamic equilibrium an unbiased probability law. In this way we justify the introduction of the Boltzmann-Gibbs distributions which microscopically represent the equilibrium states, and we elaborate the techniques which can be used for efficiently making statistical predictions about macroscopic observables.

Having thus surveyed the road leading from the microscopic description of a system to the evaluation of its macroscopic properties, we are in a position to focus our attention on the two main goals of statistical physics.

On the one hand, statistical mechanics must enable us to derive from microscopic physics the basic postulates of the various macroscopic sciences. Keeping the treatment general, we perform this derivation in Chap.5 for the concepts and Laws of equilibrium thermodynamics. The two main aspects, conservation and degradation, appear on our scale as a reflection of two more fundamental microscopic properties, the invariance laws and the tendency towards disorder. Thus, statistical physics helps us to a better understanding of thermodynamics, exhibiting, in particular, the central rôle of entropy as a measure of the amount of disorder at the microscopic level. This idea is the starting-point for a modern presentation of equilibrium thermodynamics, given in Chap.6, whereas Chap.5 states the Laws of Thermodynamics in their traditional form. We include in Chap.6, which can be read in isolation, a summary of advanced thermodynamics, collecting together the main techniques of practical value, and a survey of general applications ranging from phase transitions to chemical equilibria and from thermal engines to dielectric or magnetic substances.

On the other hand, statistical physics aims to understand and evaluate specific macroscopic properties of the most diverse substances. Chapters 7 to 13 give examples of such applications. In contrast to Chaps.2 to 5, they are relatively independent of one another, even though we have tried to keep throughout a common line of thought. The emphasis is placed on explaining equilibrium phenomena; we calculate properties such as thermodynamic functions, equations of state, and specific heats of sundry substances. The first example studied, in Chap.7, is that of a gas in the limit where neither quantum mechanics nor interactions between the particles play a crucial rôle. In Chap.8 we show how the structure of the molecules in the gas is reflected, on the macroscopic scale, in the chemical and thermal properties of the latter, and we thus explain the laws of chemical equilibrium. In Chap.9 we study some effects of the interactions between molecules; they explain the deviations from the perfect gas law observed in compressed gases, and especially the vapour-liquid transition, that we choose as a prototype of the important phase transition phenomenon.

Chapters 10 to 13 deal with systems in which quantum effects have far-reaching macroscopic consequences. The required general formalism is laid out in Chap.10, and it is after that applied to specific problems. For the sake of simplicity we are especially interested in non-interacting particles, stressing the manifold macroscopic consequences of their indistinguishability. We give in Chap.11 an introduction to the vast field of solid-state physics, showing how the existence of these materials relies on quantum statistics and explaining properties that may be due either to their crystal structure, or to the lattice excitations, or to the electrons. We show here how one can approximately account for interactions between constitutive particles by describing the system as a collection of weakly interacting entities, the quasi-particles. The particular case of semiconductors has been singled out for a more detailed treatment on account of their practical importance. We also discuss the equilibrium properties of metals and insulators and show how the laws and concepts of macroscopic electrostatics in matter follow from microscopic physics. Chapter 12 reviews at an elementary level some of the curious aspects of quantum liquids – the isotopes ^{3}He and ^{4}He – and of superfluidity. Chapter 13 starts by quantizing the electromagnetic field and introducing the photon concept, then discusses the radiation laws, both as regards equilibrium, and as regards energy transfer.

The last two chapters are dealing with some aspects of the dynamics of non-equilibrium processes, an enormous field in which rigorous theory is difficult but which has important practical implications. Chapter 14 is devoted to the general laws of macroscopic thermodynamics of systems near equilibrium. The use of these laws is illustrated by a few examples of transport phenomena: diffusion, electrical or thermal conduction, cross effects, and fluid dynamics. As in Chap.5 for the case of equilibrium, but less exhaustively, we show how general macroscopic laws, such as the Navier-Stokes equations, follow from microscopic statistical physics. Here again our two main lines of attack are the symmetry and conservation laws, on the one hand, and the interpretation of the increase in the entropy as a loss of information, on the other hand. However, in order better to comprehend the relation between microscopic and macroscopic scales, we introduce an intermediate, the so-called mesoscopic, description which is useful to solve the irreversibility problem and to make the significance of the macroscopic entropy clearer. Finally, amongst the numerous phenomena pertaining to non-equilibrium statistical mechanics, we choose to present in Chap.15 the kinetics of various gases where the interactions can be taken into account in the form of collisions. This subject covers macroscopic dynamics and heat transfer in classical gases, neutron diffusion, and conduction in semiconductors or metals. We had already touched upon this in Chaps.7, 11, and 13 using balance techniques which are somewhat simpler, and rather efficient in the study of transport phenomena. The more detailed treatment of Chap.15 makes it possible to discuss for the systems studied the dynamics of the approach to equilibrium and the problem of irreversibility.

The Conclusion selects for discussion a number of concepts and methods emerging from statistical physics, which are sufficiently general to be useful in other fields.

Each chapter is completed by a few exercises. Some of them are meant to help understanding, some others to apply the methods of statistical physics to various subjects, ranging from astrophysics to technology. Still others introduce topics which go beyond the framework of the main text; they complement the latter and deal briefly with more advanced topics, such as Landau's theory of phase transitions and interfaces, or response theory. The problems at the end of the second volume serve the same purpose, but within a wider perspective; we have left most of them in the same form as they were given at various examinations. Detailed solutions accompany the more difficult exercises and problems; for the others we have given more or less detailed hints.

The text itself contains parts of varying difficulty. To ease access, passages which are not needed for a first reading, such as proofs, added material, discussions of subtle questions, or historical notes, have been printed in small type.

As this is an introductory book, it has few bibliographic references; these are given in the form of footnotes and they often contain extensive bibliographies which, we hope, will compensate for their scarcity. Our viewpoint is often complementary to those of the classical texts on statistical physics; the reader is referred to them for a comparison with other points of view, and for finding the most important references. Amongst those we may mention the intuitive and elementary treatments by R.P.Feynman in the first volume of the *Feynman Lecture Notes on Physics* (Addison-Wesley 1963) and by F.Reif in the fifth volume of the *Berkeley Physics Course* (McGraw-Hill 1965), as well as in R.Kubo's course of problems with comments (North-Holland 1964). Among the more advanced texts on statistical physics we mention those by L.D.Landau and E.M.Lifshitz, in the fifth, ninth, and tenth volumes of their *Course of Theoretical Physics* (Pergamon), D.ter Haar (Rinehart 1954), K.Huang (Wiley 1963), J.E. and M.G.Mayer (Wiley 1977), and S.-K. Ma (World Scientific 1985). The presentations of thermodynamics often suffer from the fact that the viewpoints of physicists, chemists, and engineers differ; unified approaches can be found for equilibrium thermodynamics in the text by H.B.Callen (Wiley 1975) and for non-equilibrium thermodynamics in the introductory book by I.Prigogine (Wiley, Interscience 1967) and in the more advanced book by S.R.de Groot and P.Mazur (North-Holland 1962).

A table of units and physical constants, and a list of useful mathematical formulae are printed for convenience at the end of each volume, just after the index. This enables us to omit such details in the exercises and problems; in particular, the table of units should help the reader systematically to use dimensional analysis to check his results.

1. Paramagnetism of Ionic Solids

"Dans la Physique, comme dans toute autre science, les commencements sont épineux; les premières idées ont peine à s'établir; la nouveauté des termes, autant que celle des objets, fatigue l'esprit par l'attention qu'elle demande."

Abbé Nollet, Leçons de Physique Expérimentale, 1775

"Pour former un plan de travail sur un objet quelconque, il est nécessaire de l'examiner sous ses principaux rapports, afin de pouvoir classer les différentes parties. Comment établir leur liaison successive, si l'on n'a pas saisi l'ensemble?"

Jean-Joseph Mounier, opening the debate of the National Assembly on the Declaration of Rights, the preamble to the Constitution, 9 July 1789

"... nous ne saurions assez faire apercevoir la relation intime de ces mêmes principes avec les vérités élémentaires dont ils émanent; vérités également simples et immuables, et qu'il suffit de montrer pour les reconnaître. Tout ce que l'on peut exiger, c'est qu'on le fasse d'une manière simple, claire, et à portée de tout le monde. Or, c'est précisément ce que j'ai tâché de faire."

A.Gouges-Cartou, ibidem, August 1789

This introductory chapter, which is meant to be studied in parallel with the next three chapters, has a dual purpose. On the one hand, by treating in full detail a simple example, we show how to deal with problems in statistical physics: starting from a bare minimum of assumptions, we construct a model for the real physical situation and we solve it; we then compare the results that we have obtained with experiment in order to discuss the validity of the model; if we are not satisfied with the outcome of this discussion we reconsider the model and improve it until a satisfactory agreement with experiment is obtained.

On the other hand, this simple example serves to introduce many of the concepts which the remainder of the book will develop and study in more depth. Our procedure – which we shall see later to be a general one – therefore is the following.

- We give a *microscopic description* of the physical system, involving three basic ingredients, to wit, the principles of *quantum mechanics*, a suitable *model* for the effect studied, and the use of *probability distributions* in which the macroscopic data appear as parameters (§ 1.1).
- We identify the state of *thermodynamic equilibrium* at the microscopic level as corresponding to the *least certain* – or the most disordered – probability distribution. This principle is supported by microscopic dynamics, energy being conserved in the evolution towards disorder. At the same time we elucidate the microscopic meaning of the thermal contact between two samples (§ 1.2).
- We introduce efficient calculation formalisms, which rely on the large size of the system and which are based upon the use of *partition functions* and the *statistical entropy*; in this way we find the macroscopic consequences of the microscopic statistical approach (§§ 1.2.3 and 1.3.2).
- We identify the *temperature* and the *entropy* corresponding to our probability distribution and we use our example to show that the *laws of thermodynamics* follow and how a *probabilistic* behaviour at the microscopic level leads to a *deterministic* behaviour at the macroscopic level (§ 1.3).
- Finally, we *derive physical laws* which are specific for the material studied and which can be compared with experiments. In this way, phenomenological, macroscopic laws appear as simple consequences of a microscopic theory based upon a small number of assumptions; they can be used to explain and predict properties which may have applications of practical interest (§ 1.4).

1.1 Micro-states and Macro-state

1.1.1 Paramagnetism

Experimental magnetism shows a wealth of phenomena.

In the simplest cases, when a substance is placed in a magnetic field $\boldsymbol{B}$, it acquires a magnetization parallel to the field, which vanishes when the field vanishes. We define the magnetization as the ratio of the magnetic moment $\boldsymbol{M}$ to the volume Ω. One can then distinguish two cases, depending on whether the magnetic susceptibility,

$$\chi = \lim_{B\to 0} \frac{1}{\Omega}\,\frac{M}{B}, \tag{1.1}$$

where M is the value of the magnetization in the direction of the field, is positive – *paramagnetism* – or negative – *diamagnetism.*

In some other substances, we find magnetic order even when there is no applied field present. The simplest example is that of *ferromagnetism*, where magnetization appears spontaneously if the temperature is sufficiently low

– for instance, $T < 1043$ K for iron and $T < 631$ K for nickel. We may also mention the case of *antiferromagnetic* crystals in which neighbouring ionic magnetic moments orient themselves alternatively parallel and antiparallel to a given direction, in such a way that the net magnetization is zero.

There are various kinds of substances which show paramagnetism; their common feature is that at the microscopic level they have constituents with a magnetic moment. For instance, the paramagnetism of metals (§ 11.3.1) is connected with the spin magnetic moments of the conduction electrons. Sodium vapour and, more generally, vapours containing atoms or molecules with odd numbers of electrons are paramagnetic; in this case the microscopic magnetic moments are connected with the orbital and spin angular momenta of the electrons in the atoms or molecules. We shall in the present chapter study yet another kind of paramagnetic substances, namely, insulating ionic solids in which the so-called paramagnetic ions, such as Cu^{2+}, Mn^{2+}, or Gd^{3+}, have a non-zero magnetic moment. We shall see that for an explanation of the magnetic properties of such salts we need consider only those ions.

The origin of paramagnetism can be understood qualitatively, if we note that in an external field $\boldsymbol{B}$, the N magnetic moments $\boldsymbol{\mu}_1, \ldots, \boldsymbol{\mu}_N$ in the sample have an energy $-\left(\boldsymbol{B} \cdot \left(\sum_{i=1}^{N} \boldsymbol{\mu}_i\right)\right)$. The magnetic moments $\boldsymbol{\mu}_i$ thus tend to line up parallel to $\boldsymbol{B}$ and in the same direction, in order to minimize the magnetic energy; this leads to a positive susceptibility. However, if the sample is *heated*, that is, if energy is supplied to it, the magnetic energy $-\left(\boldsymbol{B} \cdot \left(\sum_{i=1}^{N} \boldsymbol{\mu}_i\right)\right)$ must increase, the magnetic moments $\boldsymbol{\mu}_i$ become *disordered*, and the magnetization decreases. Hence, one expects a decrease in the susceptibility when the temperature increases, as is confirmed by experiment. For instance, the full drawn curve in Fig.1.1 shows the magnetic susceptibility χ_m per unit mass of a powder of $\underline{Cu}SO_4.K_2SO_4.6H_2O$ in which the paramagnetic ion is Cu^{2+}. The data are fitted with surprising accuracy – the measurement error is less than 1 % – by an inverse temperature law, $\chi \propto 1/T$. This empirical rule, known as *Curie's law*, is found to hold for a wide range of paramagnetic substances, although it tends to break down in both the low and the high temperature regions.

In the present chapter we develop a quantitative microscopic theory of paramagnetism. On the macroscopic scale the state of a sample is characterized by several parameters, relations between which are established by experiments. Some of these parameters, such as the magnetic field B, the magnetization M/Ω, or the energy U, are readily identified in a microscopic description. However, parameters such as the temperature or the entropy have so far only been defined phenomenologically, in the framework of macroscopic thermodynamics. In order to explain properties such as Curie's law, one of our tasks will be to understand the meaning of temperature on the microscopic level.

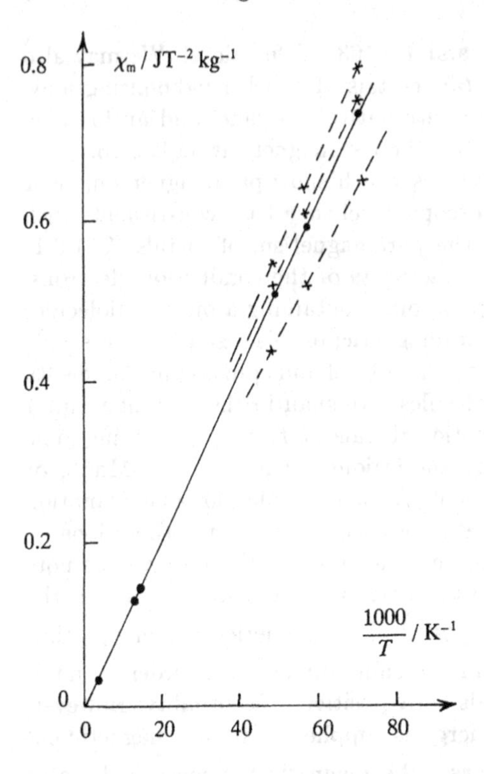

Fig. 1.1. Magnetic susceptibility of $CuSO_4.K_2SO_4.6H_2O$ as a function of the reciprocal temperature. The meaning of the dashed lines will be discussed in § 1.4.6. Laboratory *thermometers* based on measurements of the susceptibility of a known paramagnetic material placed within the experimental cell are commonly used

1.1.2 The Model

A microscopic theory that is derived completely from first principles would describe a piece of matter as a collection of elementary constituents, the nuclei and electrons, interacting through electromagnetic forces – which for most problems are the only relevant ones. Such an approach, however, is too complex to be viable. Moreover, even if supercomputers could deal with so many particles, their output would be unintelligible and useless, because the understanding of a phenomenon requires simplification and elimination of a huge amount of irrelevant data. We are thus reduced to finding a model which has a sufficient number of features in common with the real system so that we can draw useful conclusions from it, but which is simple enough so that we can treat it mathematically.

In order that a sample can show paramagnetic behaviour, it is essential that it contains microscopic magnetic moments. We know that such moments are associated with the orbital motion and the spins of the electrons. We also expect that in paramagnetic salts the permanent moments, localized in the vicinity of the paramagnetic ions, are the result of the interactions between electrons and nuclei. We wish to avoid, at least to begin with, any sophistications associated with the structure of the solid; we shall therefore postulate that there be a magnetic moment $\boldsymbol{\mu}_i$ associated with each paramagnetic ion and we shall not inquire after its origin. In order to simplify the problem, we assume, without any further justification, that each $\boldsymbol{\mu}_i$ behaves like the magnetic moment associated with an electron spin, that is,

$$\widehat{\boldsymbol{\mu}}_i = -\frac{e}{m}\widehat{\boldsymbol{S}}_i = -\frac{e\hbar}{2m}\widehat{\boldsymbol{\sigma}}_i,$$

where the caret indicates quantum-mechanical operators, where the angular momentum $\widehat{\boldsymbol{S}}_i$ is that of a spin $\frac{1}{2}$ represented by Pauli matrices, $\widehat{\boldsymbol{\sigma}}$, and where $-e$ is the electron charge. We shall take the z-axis along the direction of the applied magnetic field. The z-component of $\widehat{\boldsymbol{\mu}}_i$ is an operator $\widehat{\mu}_i$ with two equal and opposite eigenvalues $-\sigma_i\mu_{\rm B}$, where $\mu_{\rm B}$ is the Bohr magneton, that is, the absolute magnitude of the electron spin magnetic moment, $\mu_{\rm B} = e\hbar/2m = 9.27 \times 10^{-24}$ J T^{-1}, and where $\sigma_i = \pm 1$ denotes the z-component of $\widehat{\boldsymbol{\sigma}}_i$.

We thus portray our system as a collection of N magnetic moments $\boldsymbol{\mu}_i$, each associated with a paramagnetic ion and having an energy $-(\boldsymbol{B}\cdot\boldsymbol{\mu}_i)$ in the applied magnetic field $\boldsymbol{B}$. The magnetic part of the Hamiltonian is thus the operator $-\left(\boldsymbol{B}\cdot\left(\sum_{i=1}^N \widehat{\boldsymbol{\mu}}_i\right)\right)$. Each of its eigenstates is characterized by a set of N *quantum numbers* σ_1, ..., σ_N which independently take the values ± 1. The corresponding *eigenvalue of the magnetic energy* is

$$E(\sigma_1,\ldots,\sigma_N) = \mu_{\rm B}B\sum_{i=1}^N \sigma_i. \tag{1.2}$$

Our model thus reduces to a set of N identical two-level systems. This is the simplest possible model of statistical mechanics. We deliberately ignore all other degrees of freedom of the crystal. In particular, the moments are treated as being localized. Clearly, this last assumption is invalid in the case of paramagnetism in metals, since in that case the relevant magnetic moments are associated with itinerant electron spins: a different theory is required to take into account the rôle played by the electronic kinetic energy (Exerc.10b).

We neglect not only all interactions between the magnetic moments and the other degrees of freedom of the system, but also their interactions with one another through the magnetic fields they generate. This is justified for the substances that we are considering, such as $\underline{\rm Cu}SO_4.K_2SO_4.6H_2O$ where the paramagnetic Cu^{2+} ions are fairly dilute and hence well separated so that their magnetic interactions are cut down by their large distances apart.

Note that the indices $i = 1,\ldots,N$ refer to the *sites* of the paramagnetic ions in the crystal. These sites are *distinguishable*, even though the particles concerned may not be; we therefore do not need worry about the Pauli principle: the eigenket $|\sigma_1,\ldots,\sigma_N\rangle$ of the Hamiltonian represents a state which implicitly has the symmetry properties appropriate to indistinguishable particles, the electrons and the nuclei.

Note also that *quantum mechanics* is essential for the understanding of paramagnetism. There would be no magnetism in nature, if electrons and nuclei behaved according to classical mechanics. In fact, the electron spin and its associated magnetic energy $\pm\mu_B B$ are quantum properties. The treatment of orbital magnetism is more complicated, but it also requires quantum mechanics (Exerc.4d). The magnetic energy associated with the motion of a particle of charge q in a field $\boldsymbol{B} = \text{curl}\,\boldsymbol{A}$ is obtained by replacing in the Hamiltonian the momentum $\widehat{\boldsymbol{p}}$ by $\widehat{\boldsymbol{p}} - q\widehat{\boldsymbol{A}}$. In the case of an ion localized near the origin it is convenient to choose the gauge where $\widehat{\boldsymbol{A}} = \frac{1}{2}[\boldsymbol{B} \times \widehat{\boldsymbol{r}}]$. The Hamiltonian for each electron then contains the term $(e/2m)\big(\boldsymbol{B}\cdot[\widehat{\boldsymbol{r}}\times\widehat{\boldsymbol{p}}]\big) = \mu_B\big(\boldsymbol{B}\cdot\widehat{\boldsymbol{L}}\big)/\hbar$, where $\widehat{\boldsymbol{L}}$ is the orbital angular momentum; this term has the same form $-\big(\boldsymbol{B}\cdot\widehat{\boldsymbol{\mu}}\big)$ as the spin contribution. The Hamiltonian includes an extra term $e^2\widehat{\boldsymbol{A}}^2/2m$ which usually can be neglected. Orbital paramagnetism may then occur, if the angular momentum $\boldsymbol{L}$ is *quenched.*

1.1.3 The Probabilistic Nature of a Macroscopic State

The states usually encountered in quantum mechanics courses, which are represented by kets in the Hilbert space of the system concerned, should experimentally be generated by fixing the values of a *complete set of observables*. For instance, in our model, in order completely to prepare one of the eigenkets $|\sigma_1, \ldots, \sigma_N\rangle$ of the Hamiltonian, it would be necessary to orient each of the N magnetic moments $\boldsymbol{\mu}_i$ parallel or antiparallel to $\boldsymbol{B}$. Henceforth we shall refer to such states, characterized on the microscopic scale by specification of *all* their quantum numbers, as *pure states* or *micro-states.*

However, on a macroscopic scale the concept of a state is a different one. It relies on the specification of only a small number of macroscopic variables. In fact, experiments deal with *incompletely prepared states.* The number of degrees of freedom of the system is very large, and only a few of them can be observed. We can, for instance, measure the total magnetization of a paramagnetic sample or prepare such a sample that it has a given total magnetization, in each case to a fair degree of accuracy, but the values of the individual ionic magnetic moments cannot be determined. The only thing we know about the ket $|\sigma_1, \ldots, \sigma_N\rangle$ that describes the system is thus the sum $\sigma_1 + \sigma_2 + \ldots + \sigma_N$. The state of the system after a macroscopic experiment, where only a few observables are actually measured, is imperfectly known, as there are many micro-states, all corresponding to the same values of the measured observables.

As is usual when insufficient information is available, one resorts to a statistical treatment. A given macroscopic state is therefore, in terms of the microscopic structure of the system, represented by a probability distribution, which we call a *macro-state.* We assign to each of the micro-states, compatible with the available data, a probability of finding the system in it and the macro-state is represented by the set of these probabilities.

Predictions about macroscopic systems therefore are not deterministic in nature, but rather are best guesses in a statistical sense. Rather than describing a particular system, one treats a *statistical ensemble* of identical systems, all prepared in the given experimental conditions, but differing from one another on the microscopic scale due to the uncertainties arising from the incomplete nature of the preparations. This kind of theoretical approach corresponds to the problem as it occurs in practice: a physical law is established by repeating the same experiment a large number of times and the microscopic variables are not the same for each experiment. Only the probability law for these microscopic variables remains the same for all samples, for given macroscopic conditions. Predictions in physics are thus always statistical in nature, as only macroscopic quantities can be controlled, and as science does not deal with individual objects.

In order to avoid difficulties connected with the structure of the Hilbert space of the micro-states – a problem to be dealt with in Chap.2 – we assume here that the only micro-states allowed are the kets $|\sigma_1, \ldots, \sigma_N\rangle$. We do not take their linear combinations into account and also forget about the x- and y-components of the spins. A macro-state is thus characterized by the probability $p(\sigma_1, \ldots, \sigma_N)$ that the various magnetic moments are parallel or antiparallel to the applied magnetic field.

1.1.4 Calculation of Expectation Values

The various observables with which we are concerned in the present chapter are functions $A(\sigma_1, \ldots, \sigma_N)$ of the random variables $\sigma_1, \ldots, \sigma_N$ each of which can take the values ± 1. Their expectation values $\langle A\rangle$ follow immediately from the probability law p, which *contains all the information* about the system:

$$\langle A\rangle = \sum p(\sigma_1, \ldots, \sigma_N)\, A(\sigma_1, \ldots, \sigma_N); \tag{1.3}$$

the sum is over the 2^N configurations $\sigma_1 = \pm 1, \ldots, \sigma_N = \pm 1$.

For instance, the expectation value of the magnetic moment $\widehat{\mu}_i = -\mu_{\mathrm{B}}\widehat{\sigma}_i$ of the ion on the i-th site is

$$\langle \mu_i\rangle = -\mu_{\mathrm{B}} \sum_{\sigma_i} p_i(\sigma_i)\, \sigma_i; \tag{1.4}$$

we have introduced the reduced probability law $p_i(\sigma_i)$ for the i-th ion:

$$p_i(\sigma_i) = \sum{}' \, p(\sigma_1, \ldots, \sigma_N), \tag{1.5}$$

where the sum is over the 2^{N-1} configurations $\sigma_1 = \pm 1, \ldots, \sigma_{i-1} = \pm 1, \sigma_{i+1} = \pm 1, \ldots, \sigma_N = \pm 1$. Hence we get for the expectation value of the z-component of the total magnetic moment

$$\langle M\rangle = \left\langle \sum_{i=1}^{N} \mu_i \right\rangle = -\mu_{\mathrm{B}} \sum_{i=1}^{N} \sum_{\sigma_i = \pm 1} p_i(\sigma_i)\, \sigma_i, \tag{1.6}$$

and for the expectation value of the energy

$$U = \sum p(\sigma_1, \ldots, \sigma_N)\, E(\sigma_1, \ldots, \sigma_N) = -\langle M\rangle\, B, \tag{1.7}$$

which we can identify with the macroscopic *internal magnetic energy.*

Similarly we can derive all other physical quantities from the probability law $p(\sigma_1,\ldots,\sigma_N)$. For instance, the correlations between the i-th and the j-th ions are characterized by the reduced probability for two ions,

$$p_{ij}(\sigma_i,\sigma_j) \;=\; \sum_{ij}{}' \, p(\sigma_1,\ldots,\sigma_N), \tag{1.8}$$

where the summation is over all spins bar i and j. In particular, we find

$$\langle \mu_i \mu_j \rangle - \langle \mu_i \rangle \, \langle \mu_j \rangle \;=\; \mu_{\rm B}^2 \sum_{\sigma_i,\sigma_j} \left[p_{ij}(\sigma_i,\sigma_j) - p_i(\sigma_i) p_j(\sigma_j) \right] \sigma_i \sigma_j;$$

absence of correlations is expressed by $p_{ij}(\sigma,\sigma') = p_i(\sigma)p_j(\sigma')$.

1.2 Microscopic Interpretation of Thermal Equilibrium

1.2.1 Maximum Disorder

Once we have been given the probabilities $p(\sigma_1,\ldots,\sigma_N)$, we can calculate the properties of the macro-state. The problem which we are then faced with is thus how to *associate a probability to each possible configuration.*

Let us, to begin with, assume that there is no applied magnetic field in the system. In that case there is no reason to assume that the magnetic moment of an ion is aligned in one direction in preference to another. We are thus led to assume that the various possibilities are *equally probable*, as in a game of dice. There is no possibility to distinguish between the various micro-states so that they must all be considered to be equiprobable. As their total number is 2^N, we have therefore

$$p(\sigma_1,\ldots,\sigma_N) \;=\; 2^{-N}. \tag{1.9}$$

We have thus introduced the natural postulate that, if we have no information about the system, the probability law which we must associate with it must reflect that our *uncertainty is maximal.* In less subjective terms we can say that the macro-state of the system is the *most disordered of all states.* In Chap.3 we shall put the relation between uncertainty and disorder on a quantitative footing.

As we could have anticipated, we find that the expectation value (1.4) of the magnetic moment is zero for each ion. We see similarly that there is no correlation between the various magnetic moments, since

$$p_{ij}(\sigma_i,\sigma_j) \;=\; \tfrac{1}{4} \;=\; p_i(\sigma_i)\, p_j(\sigma_j).$$

The statistical fluctuation of the total magnetic moment is

$$\left[\langle M^2\rangle - \langle M\rangle^2\right]^{1/2} = \left[\sum_{i,j} \left(\langle \mu_i\mu_j\rangle - \langle \mu_i\rangle \langle \mu_j\rangle\right)\right]^{1/2}$$
$$= \left[\sum_i \langle \mu_i^2\rangle\right]^{1/2} = \mu_B \sqrt{N}. \qquad (1.10)$$

When N is large, we must compare this quantity with the maximum total magnetic moment $\mu_B N$ which is found when the moments of all ions are aligned with $\boldsymbol{B}$. Its relative value, $1/\sqrt{N}$, is small. We have here an example of the "law of large numbers": despite the complete disorder which prevails in the system, a *macroscopic quantity* such as the total magnetic moment is defined with a *small relative error* due to the large number of the elementary magnetic moments. From the maximum uncertainty about the micro-states a near certainty about this macroscopic quantity has emerged.

As an exercise one can evaluate the probability $p(M)\,dM$ for the total magnetic moment of the N spins to have a value M with a margin of dM. Using the technique described later on, one can show that, if N is large, this distribution, $p(M)$, is Gaussian (see Eq.(1.16) and § 1.2.4).

1.2.2 The Rôle of Energy Conservation

The systems in which we are interested in practice are, however, not completely random. Some of their parameters have been measured, so that they are *partially prepared.* In order to associate with a system a probability law which takes into account the experimental data, we can analyze how such a system evolves with time and how much it remembers of the way it has been prepared. To do this, we shall use intuitive rather than rigorous arguments.

When defining our model we neglected the interactions between the magnetic moments in expression (1.2) for the energy levels. Nonetheless, interactions, even infinitesimal ones, make it possible over a more or less long period for transitions to occur between micro-states which are characterized by different quantum numbers $\sigma_1, \ldots, \sigma_N$ but which in the field $\boldsymbol{B}$ have the same, or nearly the same energy. In particular, an interaction of the form $(\boldsymbol{\sigma}_i \cdot \boldsymbol{\sigma}_j)$ between the i-th and j-th magnetic moments can, if they are antiparallel, lead to flip-flops, the two spins flipping simultaneously from the $+-$ to the $-+$ configuration, and *vice versa.* If one waits long enough, this uncontrollable mechanism will lead to an *equalization of the probabilities for the micro-states of the same energy* even if at the start some of them were preferred. The state thus *evolves towards disorder, while conserving the energy* it orginally had (Exerc.2a).

The process which we have just described can be observed on a macroscopic scale: when various parts of a system are prepared separately and put in contact, one observes that they exchange energy until a permanent regime emerges – *thermal equilibrium.* The evolution towards thermal equilibrium can microscopically be interpreted as an evolution towards the most disordered macro-state possible with an energy equal to the initial energy. In our model, this evolution will tend, in particular, to equalize the mean values of the magnetic moments of the ions within the sample. The thermal equilibrium macro-state, characterized by giving the energy U of the system, is thus the one for which *all accessible micro-states* with energies close to U *are equally probable.* This choice of probabilities is also the natural one for a statistical ensemble of systems characterized by *only knowing* U. Just as in § 1.2.1 the least biased distribution corresponds to equiprobability.

Strictly speaking, as the spectrum is discrete, the energy takes on only discrete values. However, due to the large size of the system, the levels practically form a continuum: in fact, when $B = 1$ T, which corresponds to a rather large field, $\mu_B B$ only equals 10^{-23} J so that the distances between levels are always negligible as compared to the experimental errors ΔU. Moreover, the interaction terms in the Hamiltonian tend to blur the spectrum (1.2) by an amount which also should be taken into account in ΔU. We shall therefore avoid the complications arising from the discrete nature of the spectrum by assuming that the energy of the system is equal to U within a margin ΔU, where $\Delta U \ll U$. In thermal equilibrium the micro-states, the quantum numbers of which $\sigma_1, \ldots, \sigma_N$ satisfy the inequalities

$$U \leq \mu_B B \sum_{i=1}^{N} \sigma_i < U + \Delta U, \tag{1.11}$$

are thus taken to be all equally probable. Let W be the number of these micro-states which, as we shall see in a moment, is always huge, even when $\Delta U \ll U$. The macro-state describing the system at thermal equilibrium for a given value U of the energy is thus characterized by probabilities which are equal to

$$p(\sigma_1, \ldots, \sigma_N) = \frac{1}{W} \tag{1.12}$$

for σ_i satisfying (1.11) and which are zero for other micro-states. Such a statistical ensemble, in which the possible micro-states, with equal probabilities, are the ones with an energy U, within a margin ΔU, is traditionally called a *microcanonical ensemble.* This name has a historical origin without much logical justification and it reminds one of the fact that the energy is well defined with small fluctuations $\Delta U/U$.

Note that whereas on the macroscopic level a thermal equilibrium state is uniquely determined by a few characteristic variables, the concept of an equilibrium macro-state has a *statistical* character at the microscopic level. The orientations of the magnetic moments are random and badly known; they fluctuate from one sample to another and with time. Their probability distribution is the quantity which remains unchanged and it corresponds to the most disordered situation possible.

1.2.3 The Degree of Disorder

In order completely to determine $p(\sigma_1, \ldots, \sigma_N)$ we still must find the number W of micro-states involved. This number is always huge. For instance, one can easily check that for 1 kg of $\underline{Cu}SO_4.K_2SO_4.6H_2O$ in a field of 1 T there are altogether $10^{10^{24}}$ magnetic energy levels distributed over 25 J. The number W is connected with the random nature of the probability law: the more the number W of possible events increases, the less we know about the system. *Disorder increases with* W. In Chap.3 we shall introduce the statistical entropy S which is a measure of the disorder and we shall see that it is proportional to the logarithm of W.

To evaluate W we must know the values and multiplicities of the energy levels (1.2). The simplicity of this last expression would in the present case make it possible to find W directly using combinatorial arguments. However, we prefer to use here, as an exercise, a more general technique which will be useful later on.

We introduce the "*canonical partition function*"

$$\boxed{Z(\beta) \equiv \sum \mathrm{e}^{-\beta E}}\ , \tag{1.13}$$

where the sum is over all energy levels, each one weighted with its multiplicity. We shall evaluate $Z(\beta)$ directly and afterwards use it to find the values and multiplicities of the levels. To evaluate Z we use the fact that the energies (1.2) are sums of N independent terms:

$$\begin{aligned} Z(\beta) &= \sum_{\sigma_1,\ldots,\sigma_N} \mathrm{e}^{-\beta\mu_\mathrm{B} B \sum_i \sigma_i} = \sum_{\sigma_1,\ldots,\sigma_N} \mathrm{e}^{-\beta\mu_\mathrm{B} B\sigma_1} \ldots \mathrm{e}^{-\beta\mu_\mathrm{B} B\sigma_N} \\ &= \prod_i \left[\sum_{\sigma_i} \mathrm{e}^{-\beta\mu_\mathrm{B} B\sigma_i}\right] = \left(\mathrm{e}^{\beta\mu_\mathrm{B} B} + \mathrm{e}^{-\beta\mu_\mathrm{B} B}\right)^N . \end{aligned} \tag{1.13'}$$

By expanding $Z(\beta)$ as a sum of exponentials we can now find the energy spectrum:

$$Z(\beta) = \sum_{q=0}^{N} \frac{N!}{q!(N-q)!}\, \mathrm{e}^{-\beta[-N\mu_\mathrm{B} B + 2q\mu_\mathrm{B} B]}.$$

The energy levels thus have the values $(-N+2q)\mu_\mathrm{B} B$, where q is an integer lying between 0 and N, and have a spacing equal to $2\mu_\mathrm{B} B$. Their multiplicity equals $N!/q!(N-q)!$, a result which we could have found directly from (1.2) by noting that the set of micro-states corresponding to a given energy level is obtained by letting q spins point downwards and $N-q$ spins point upwards.

The number W of eigenstates of the Hamiltonian with energies between U and $U + \Delta U$ is thus approximately equal to

$$W \simeq \frac{\Delta U}{2\mu_\mathrm{B} B}\, \frac{N!}{q!(N-q)!}, \tag{1.14}$$

where we have taken the multiplicities into account and where q is connected with the energy U through the relation

$$\frac{-N+2q}{N} \simeq \frac{U}{N\mu_B B} = -\frac{\langle M\rangle}{N\mu_B} \equiv \varrho. \tag{1.15}$$

Expression (1.14) holds, provided the multiplicity $N!/q!(N-q)!$ does not change much over the range ΔU, that is, provided $2\mu_B B \ll \Delta U \ll N\mu_B B$. The quantity ϱ, defined by (1.15), lies between -1 and $+1$; according to (1.2), (1.6), and (1.7), it can be interpreted as the mean value of any of the spins σ_i. Provided U does not lie near either the top or the bottom of the spectrum, we can use Stirling's formula,

$$N! \sim N^N \mathrm{e}^{-N}\sqrt{2\pi N},$$

to evaluate (1.14). To dominant order in N we find

$$W \simeq A\,\Delta U\,\mathrm{e}^{S/k}, \tag{1.16}$$

where we have introduced the quantities

$$S(U,B,N) \equiv kN\left[\frac{1}{2}(1+\varrho)\ln\frac{2}{1+\varrho} + \frac{1}{2}(1-\varrho)\ln\frac{2}{1-\varrho}\right], \tag{1.17}$$

$$A \equiv \frac{1}{\mu_B B\sqrt{2\pi N(1-\varrho^2)}}. \tag{1.17'}$$

In (1.16) and (1.17) there appears a constant k which at the moment is arbitrary, but which will later be determined when we identify S with the entropy. Note that S *is proportional to* N and that W is large as an exponential of N.

The definition (1.15) of ϱ and its interpretation in terms of (1.5),

$$\varrho = \langle\sigma_i\rangle = p_i(+1) - p_i(-1),$$

imply that we can express the *probability distribution for the i-th spin* in terms of the internal energy U as follows:

$$p_i(\pm 1) = \frac{1\pm\varrho}{2} = \frac{1}{2}\left(1 \pm \frac{U}{N\mu_B B}\right). \tag{1.18}$$

We can then rewrite (1.17) as

$$\boxed{S \sim k\ln W \sim -kN\sum_\sigma p_i(\sigma)\ln p_i(\sigma)}\;. \tag{1.19}$$

Equation (1.19) anticipates the general expressions for the statistical entropy which will be introduced in Chap.3. In particular, it will be shown there that in the case of W equally probable configurations the statistical entropy, which *measures the degree of disorder*, is just equal to $k \ln W$. In the case of our example we can understand this by considering Eqs.(1.16) and (1.17) for W. In fact, the degree of disorder must, on the one hand, be an increasing function of W and, on the other hand, be proportional to the number N of paramagnetic ions; this is true for $k \ln W$. We shall also see in § 1.2.5 that the magnetic moments are statistically independent in the limit as $N \to \infty$. The total degree of disorder of the sample should therefore be equal to N times the *degree of disorder of one of the magnetic moments.* The latter quantity can thus be identified with the right-hand side of (1.19). We shall find the same expression in Chap.3 from a completely general argument.

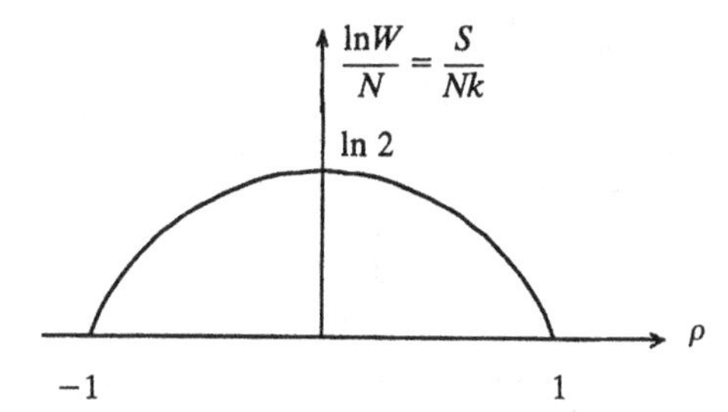

Fig. 1.2. The entropy as function of the energy $U = N\mu_B B\varrho$, or of the magnetization $\langle M \rangle = -N\mu_B\varrho$

The change of S with energy, shown in Fig.1.2, confirms its interpretation as a measure of the disorder or of our uncertainty. When U is close to the ground state energy, nearly all the magnetic moments are aligned in the same direction as $\boldsymbol{B}$; W is rather small and the system is nearly perfectly ordered. When U increases, W grows very fast – exponentially with N: the number of possible micro-states becomes larger and larger while $\langle M \rangle$ decreases and the number of magnetic moments antiparallel to the field increases. When $U = 0$ and $\langle M \rangle = 0$, we find in the large N limit that W is close to 2^N, that is, to the total number of micro-states: the disorder is maximal and the situation differs little from the one we studied in § 1.2.1. It makes, indeed, practically no difference whether we let the energy fluctuate freely around its expectation value $U = 0$, or constrain it to lie between 0 and ΔU, since we saw in the former case that the fluctuations in M, which are proportional to $\sqrt{N}$, were relatively small. When U continues to grow, the magnetic moments progressively become more ordered in a direction opposite to that of the field, and the disorder again decreases.

Later on we shall identify the degree of disorder S with the entropy of the thermodynamicists. Figure 1.2 thus shows the change in the *entropy of the paramagnetic ions as function of the internal magnetic energy* for a given magnetic field.

1.2.4 Thermal Contact and Partition of Energy

The technique used in the previous section to determine the probability distribution of the magnetic moments may seem to be uselessly complicated. We shall apply it here to a less simple situation. Consider two samples a and b, similar to the ones which we have studied, but of different sizes and subject to different magnetic fields (Fig.1.3). The macro-state of each of them is characterized by three variables, namely, the energy, the magnetic field, and the number of paramagnetic ions; we shall use (1.14), or, more explicitly, (1.16) and (1.17), to express the number of micro-states as a function of these variables:

$$\begin{aligned} W_{\rm a} &= W(U_{\rm a}, B_{\rm a}, N_{\rm a}), \\ W_{\rm b} &= W(U_{\rm b}, B_{\rm b}, N_{\rm b}). \end{aligned}$$

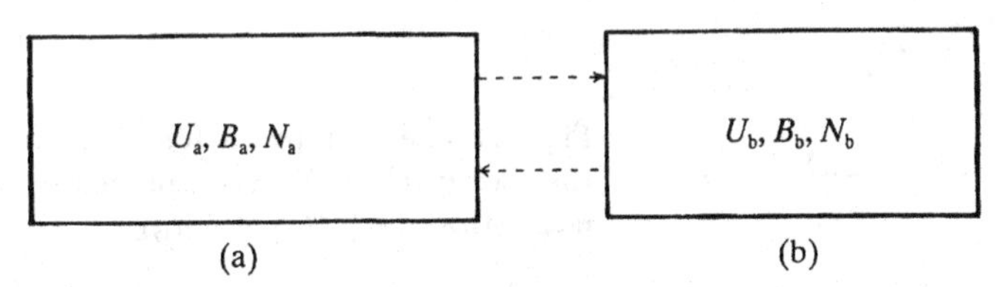

Fig. 1.3a, b. Thermal contact between two paramagnetic samples

When the samples are isolated from one another, these six variables are independent. We now put them into contact, that is, we assume that their magnetic dipoles interact very weakly with one another; this makes it possible for a and b to exchange energy. As the combined system is isolated, $U_{\rm a}$ and $U_{\rm b}$ can vary, but the total energy $U = U_{\rm a} + U_{\rm b}$ remains fixed – within a margin ΔU – and so do $B_{\rm a}$, $B_{\rm b}$, $N_{\rm a}$, and $N_{\rm b}$. On the other hand, in thermal equilibrium, the expectation value of the energy of a reaches a stationary value $U_{\rm a}$ – and so does $U_{\rm b}$ – around which it fluctuates thanks to energy exchanges with b. We now want to find out *how, on average, the energy U will be split between the two systems* when thermal equilibrium has been established. This problem of the so-called *energy partition* is of major importance: it will enable us to answer, starting from microscopic physics and using statistical calculations, a question which on a macroscopic level only the Second Law of thermodynamics could resolve, and thus to *prove* this *Second Law*, in the particular case considered here. Moreover, we shall understand the microscopic meaning of the equalization of the temperatures of two systems brought into thermal contact.

As in the case of a single sample we assume that the system is, in thermal equilibrium and at the microscopic level, described by a *microcanonical distribution*, that is, we assume that all accessible micro-states with energies between U and $U + \Delta U$,

$$U \leq E_a + E_b = \mu_B B_a \sum_{i=1}^{N_a} \sigma_i + \mu_B B_b \sum_{j=1}^{N_b} \sigma_j < U + \Delta U, \tag{1.20}$$

are equally probable. In the macro-state defined in this way, *the energy E_a is a random variable* and we determine its *probability distribution* $p(E_a)\,dE_a$. Let us classify the micro-states (1.20), of which there are altogether $W'(U, B_a, B_b, N_a, N_b)$, according to the energies E_a and E_b of the two subsystems. For one of the subsystems, a or b, we counted in § 1.2.3 the number of micro-states between two neighbouring energy values and found expression (1.16). The number of micro-states of the system a+b such that the energy of a lies between E_a and $E_a + dE_a$ while the energy of b lies between E_b and $E_b + dE_b$ is thus, as function of the expressions defined by (1.17) and (1.17′),

$$A_a\,dE_a\,e^{S_a/k}\,A_b\,dE_b\,e^{S_b/k}. \tag{1.21}$$

We assume that the systems are sufficiently large so that the spectra can be treated as continua; if we want greater rigour we replace the variables E_a and E_b by E_a/N_a and E_b/N_b which behave as continuous variables in the limit of large systems. The total number W' is the integral of (1.21) over the domain $U < E_a + E_b < U + \Delta U$. The number of micro-states of the class (1.20) such that the energy of a lies between E_a and $E_a + dE_a$ is found simply by replacing dE_b by ΔU in (1.21), since ΔU is rather small while dE_a is infinitesimal. Using the fact that the accessible micro-states (1.20) are equiprobable we find the required probability density:

$$p(E_a) = \frac{A_a A_b\,e^{(S_a+S_b)/k}}{\int dE_a\,A_a A_b\,e^{(S_a+S_b)/k}}. \tag{1.22}$$

In both the denominator and the numerator of (1.22) E_a and E_b are related through $E_a + E_b = U$. Moreover, their values cannot go beyond the ranges $|E_a| < N_a \mu_B B_a$ and $|E_b| < N_b \mu_B B_b$. We were able to get rid of the integration over E_b in the denominator of (1.22) thanks to our assumptions about the size of the systems and the magnitude of ΔU.

In order to analyze the behaviour of (1.22) we shall distinguish between *extensive* quantities, such as E_a and E_b, which are proportional to N_a or N_b, and *intensive* quantities, such as B_a, B_b, ϱ_a, or ϱ_b, which are finite in the limit as $N_a \to \infty$, $N_b \to \infty$. Expression (1.17) shows that S is an extensive quantity so that the probability (1.22) is completely *dominated by its exponent*. In terms of the intensive variable $x \equiv E_a/N_a$, and for given values of B_a, B_b, N_b/N_a, and U/N_a, we have

$$\ln p \sim N_a\,f(x),$$

where f is an intensive quantity. In the domain $|\varrho_a| \leq 1$, $|\varrho_b| \leq 1$ where the function f, which is a sum of two expressions like (1.17), is defined, it is concave and its derivative changes continuously from $+\infty$ to $-\infty$. It thus attains a maximum at some value x_0 which determines the energy $U_a = N_a x_0$. Expanding f around x_0, where $df/dx_0 = 0$ and where $d^2f/dx_0^2 \equiv -\sigma^{-2} < 0$, we find for the dominant behaviour of p, apart from a multiplicative constant,

$$p(E_a) \propto e^{-N_a(x-x_0)^2/2\sigma^2} = e^{-(E_a-U_a)^2/2N_a\sigma^2}. \tag{1.23}$$

In the limit considered the statistical distribution of the energy E_a is thus *very strongly peaked* around the value U_a with a fluctuation $\sigma\sqrt{N_a}$, the relative value of which is of the order of $1/\sqrt{N_a}$ and thus negligible, and it is approximately a *Gaussian.* This property is a consequence of the enormous disparity between the number of configurations with energies E_a near U_a and the number of other configurations with total energy U. The former are so numerous that the system remains in them practically all the time during an evolution where all states with energy U have the same chance of being reached. The maximum of p is so pronounced that the probability for a noticeable deviation is infinitesimal: when two macroscopic samples are in thermal contact, everything happens as if their energies have *well defined values* U_a and $U_b = U - U_a$; fluctuations, which are associated with the existing freedom to exchange energy, only give rise to a *negligible relative error.* This is another example of the law of large numbers.

In order to determine U_a more explicitly we use (1.17) and (1.15) to rewrite the dominant contribution to (1.22) as follows:

$$k \ln p \sim S(E_a, B_a, N_a) + S(U - E_a, B_b, N_b). \tag{1.24}$$

The value U_a at thermal equilibrium is obtained by looking for the *maximum* of (1.24) with respect to E_a. This remark will help us to identify – in § 1.3.3 – the function S with the *entropy* of the thermodynamicists. We introduce the quantity

$$\beta \equiv \frac{1}{k} \frac{\partial S(U,B,N)}{\partial U} = \frac{\partial}{\partial U} \ln W(U,B,N), \tag{1.25}$$

which is associated with each of the subsystems a or b at thermal equilibrium. To express the fact that (1.24) reaches its maximum when $E_a = U_a$ we can then write down the condition that the energies U_a and $U_b = U - U_a$ take on values such that the *parameters* β_a *and* β_b *become equal* when a and b are in thermal contact. This will enable us, in § 1.3.1, to identify β with a relative temperature scale.

Summarizing, the microscopic solution of the problem of energy partition has shown us two important facts:

- Because the two systems in contact are large, each one takes a well defined fraction U_a and U_b of the total available energy, while the probabilistic nature of the partition is invisible at the macroscopic scale: $1/\sqrt{N}$ is of the order 10^{-10}.

- The values of U_{a} and U_{b} are such that (1.24) is a maximum or that the state variables β_{a} and β_{b}, defined by (1.25), are equal in thermal equilibrium.

As an exercise, this study can be completed by calculating explicitly the statistical fluctuations in the energy E_{a} for the microcanonical equilibrium of the a+b system, and expressing it in terms of the specific heats of the two parts of the system, given in § 1.4.4.

1.2.5 Absence of Correlations Between Magnetic Moments at Thermal Equilibrium

As a further exercise we shall evaluate the probability law $p_{12}(\sigma_1, \sigma_2)$ for two magnetic moments at thermal equilibrium for the microcanonical ensemble defined in § 1.2.3. Using a method which can be generalized to apply to p_{12}, we shall first rederive the probability law $p_1(\sigma_1)$ for the magnetic moment 1. As all admissible micro-states are equally probable, $p_1(+1)$ is equal to the ratio W_+/W of the number W_+ of micro-states satisfying (1.11) and such that $\sigma_1 = +1$, to the total number W calculated in § 1.2.3 as function of U and N. To calculate W_+ we note that when $\sigma_1 = +1$, inequality (1.11) becomes

$$U - \mu_{\mathrm{B}}B \;\leq\; \mu_{\mathrm{B}}B \sum_{i=2}^{N} \sigma_i \;<\; U - \mu_{\mathrm{B}}B + \Delta U.$$

As a result W_+ follows from the expression for W by replacing U by $U - \mu_{\mathrm{B}}B$ and N by $N-1$. In the limit where N and U are large, and introducing S through (1.16) and (1.17), we have thus

$$\begin{aligned} k \ln p_1(+1) \;&\sim\; S(U-\mu_{\mathrm{B}}B, B, N-1) - S(U, B, N) \\ &\sim\; -\mu_{\mathrm{B}}B\, \frac{\partial S}{\partial U} - \frac{\partial S}{\partial N}. \end{aligned}$$

A simple calculation then leads again to (1.18).

The same method enables us now to evaluate $p_{12}(+1,+1)$. The number of micro-states satisfying (1.11) and such that $\sigma_1 = \sigma_2 = +1$ is the number of configurations of $\sigma_3, \ldots, \sigma_N$ for which

$$U - 2\mu_{\mathrm{B}}B \;\leq\; \mu_{\mathrm{B}}B \sum_{i=3}^{N} \sigma_i \;<\; U - 2\mu_{\mathrm{B}}B + \Delta U.$$

We must replace U by $U - 2\mu_{\mathrm{B}}B$ and N by $N-2$ in the expression for W and we now find

$$k \ln p_{12}(+1,+1) \;\sim\; -2\mu_{\mathrm{B}}B\, \frac{\partial S}{\partial U} - 2\, \frac{\partial S}{\partial N}.$$

In the limit where N and U are large we have thus

$$\ln p_{12}(+1,+1) \;=\; 2 \ln p_1(+1) \;=\; \ln p_1(+1) + \ln p_2(+1),$$

or

$$p_{12}(+1,+1) = p_1(+1)\,p_2(+1).$$

Similarly one can show that

$$p_{12}(-1,-1) = p_1(-1)p_2(-1),$$

while

$$p_{12}(+1,-1) = p_{12}(-1,+1)$$

follows from the normalization. Altogether we have therefore

$$p_{12}(\sigma_1,\sigma_2) = p_1(\sigma_1)\,p_2(\sigma_2), \tag{1.26}$$

so that the magnetic moments on different sites are statistically independent.

In the canonical ensemble (Exerc.4c) the absence of correlations between magnetic moments at thermal equilibrium is exact and will correspond to maximum disorder. The microcanonical equilibrium ensemble considered in the present chapter is characterized by the constraint (1.11) which correlates the N ions; this constraint produces correlations between two ions, which are finite when N is finite, but which tend to zero for large N so that in the infinite-N limit *the two equilibrium ensembles are equivalent.*

1.2.6 Spin Thermostat

In § 1.2.4 we studied the thermal equilibrium between two paramagnetic samples which were both macroscopic. Let us assume now that the system b is much larger than a so that the energy which it can exchange with the latter represents only a negligible part of its energy U_b which remains close to U; hence β_b which is defined by (1.25) hardly changes when b is put in contact with a. As far as magnetic energies are concerned, the system b thus plays the rôle of a thermostat and the energy of a adjusts itself in such a way that β_a takes the given value β_b.

In case the system a remains microscopic, one cannot avoid the use of a probabilistic description for it. Let us find in that case the probability p_m for a micro-state m, with energy ε_m, of the system a. This probability is proportional to the number of micro-states of the system b such that

$$U \le \varepsilon_m + E_\text{b} < U + \Delta U,$$

that is, to the number W_b evaluated for the energy $U - \varepsilon_m$. Taking into account the large size of the system b we can express W_b by using (1.16) and (1.17) and expanding in powers of ε_m, which is much smaller than U; this leads, apart from an additive constant, to

$$k \ln p_m \sim -\frac{\partial S_\text{b}}{\partial U}\,\varepsilon_m.$$

Using (1.25) we thus find that

$$p_m = \frac{1}{Z_\mathrm{a}} \, \mathrm{e}^{-\beta_\mathrm{b}\varepsilon_m}, \tag{1.27}$$

where Z_a is a normalization constant.

The exponential form (1.27) which we found for the probability for the micro-state m in the thermostat b is a first example of a Boltzmann-Gibbs distribution. Equation (1.29) below is a particular case of (1.27), where a is the spin 1 placed in the same field $\boldsymbol{B}$ as the $N-1$ other spins, which form the system b. Similarly, the property (1.26) follows directly from (1.27) and the fact that the energies of the spins 1 and 2 are additive.

When the size of the system a becomes macroscopic – though remaining small as compared to that of b – we find again the probability distribution (1.23) for the energy E_a, starting from (1.27). We must bear in mind that $p(E_\mathrm{a})\,dE_\mathrm{a}$ is the product of p_m and the number (1.16) of micro-states in the range $E_\mathrm{a}, E_\mathrm{a}+dE_\mathrm{a}$; since N_a is large we can use the same approximations as in § 1.2.4.

1.3 Identification of the Thermodynamic Quantities

1.3.1 Relative Temperature

Empirically the concept of a relative temperature was defined on the macroscopic scale by the observation of two systems in thermal contact, evolving until thermal equilibrium is reached. One associates with each system a variable depending on its state, the temperature, such that the temperatures of the two systems become equal when thermal equilibrium is established. On the microscopic scale thermal contact can be interpreted as a weak interaction which allows energy exchanges, and the establishment of thermal equilibrium as an evolution towards maximum disorder. This is just the problem studied in § 1.2.4, where we postulated that the most disordered macro-state possible is the one where all accessible micro-states are equally probable. The microscopic and statistical analysis provided us with the equilibrium condition that *the parameters* β, defined as function of U, B, and N, *must be equal* in the two parts.

On the microscopic scale the quantity β is connected with the statistical nature of the macro-state of the system in thermal equilibrium; however, on the macroscopic scale it can thus be interpreted as a relative temperature. If we use (1.25), (1.17), and (1.15) we can easily express it as a function of the field B applied to the sample and of its energy U, in the form

$$\boxed{\beta \equiv \frac{1}{k}\frac{\partial S}{\partial U}}$$

$$= \frac{1}{2\mu_\mathrm{B}B} \ln \frac{1-\varrho}{1+\varrho} = \frac{1}{2\mu_\mathrm{B}B} \ln \frac{N\mu_\mathrm{B}B - U}{N\mu_\mathrm{B}B + U}. \tag{1.28}$$

This expression shows that β is an *intensive* variable, like B and U/N.

1.3.2 The Boltzmann-Gibbs Distribution

Consider the probability distribution for one of the magnetic moments, which at thermal equilibrium is given by (1.18), and replace the mean energy U/N in (1.28) by its expression in terms of the probabilities $p_1(\sigma_1)$. We find

$$\beta = \frac{1}{2\mu_B B} \ln \frac{p_1(-1)}{p_1(+1)},$$

that is,

$$\frac{p_1(+1)}{e^{-\beta\mu_B B}} = \frac{p_1(-1)}{e^{\beta\mu_B B}}.$$

Since the energy of the magnetic ion 1 is $\varepsilon_1 = \mu_B B \sigma_1$, we see that

$$\boxed{p_1(\sigma_1) = \frac{1}{Z_1} e^{-\beta\varepsilon_1}}, \tag{1.29}$$

where Z_1 is a normalization constant.

Therefore, for the system consisting of *one magnetic moment in equilibrium with the rest of the sample*, which plays the rôle of a thermostat, the probability for each micro-state is proportional to an *exponential of the energy* of this state. In Chap.4 we shall find this exponential form for the probability as a quite general result: it is the *Boltzmann-Gibbs distribution.* The quantity β that on the macroscopic scale can be interpreted as a relative temperature thus appears on the microscopic scale as a *parameter which is the conjugate of the energy* and which characterizes the probability distribution of one of the magnetic moments.

We saw in § 1.2.6 and we shall find as a general result in § 5.7.2 that, if the sample considered is in thermal equilibrium with a much larger thermostat, the probability distribution $p(\sigma_1, \ldots, \sigma_N)$ has the so-called "*canonical*" form

$$p(\sigma_1, \ldots, \sigma_N) = \frac{1}{Z} e^{-\beta E(\sigma_1, \ldots, \sigma_N)},$$

whereas here we worked in the so-called "*microcanonical*" statistical ensemble (1.12). Notwithstanding the difference of the form of these two probability laws, they are practically equivalent for the evaluation of macroscopic quantities. In particular, both lead to the same expression (1.29) for the reduced probability $p(\sigma_i)$ for the i-th ion and to the same expression (1.26) for the reduced probability for two ions. Technically, it is more convenient to use the canonical ensemble, but in the present chapter we have preferred to use the microcanonical ensemble which needs fewer new concepts (Exerc.4c).

Note finally that the normalization constant Z_1, which is equal to

$$Z_1 = \sum_{\varepsilon_1} e^{-\beta\varepsilon_1} = e^{\beta\mu_B B} + e^{-\beta\mu_B B}, \tag{1.30}$$

is nothing but the partition function for the site 1, as we can see from the definition (1.13).

1.3.3 Entropy and Absolute Temperature

The macroscopic state of our system is characterized in thermal equilibrium by three variables: the internal energy U, the applied magnetic field B, and the number N of paramagnetic sites. In thermodynamics one shows that in an infinitesimal transformation, where the field B changes by dB, while N remains constant, the *magnetic work done on the sample* equals

$$\delta \mathcal{W} = -M\,dB, \tag{1.31}$$

where M is the total magnetic moment. On the microscopic scale the latter can be identified with the expectation value $\langle M \rangle$, while the work (1.31) can be interpreted as the expectation value of the infinitesimal change in the Hamiltonian (1.2) when B changes.

Expression (1.31) for the magnetic work done is not an obvious one because the electromagnetic forces have a long range and the definition of the system studied itself is, as a result, ambiguous. Accordingly, the thermodynamic definition of the internal energy of a dielectric or magnetic substance and of the work done on it is not unique and depends on how one wants to take into account the field in which the sample is placed (§ 6.6.5). Here we use the method which is best suited for connecting microscopic and macroscopic physics: the field B is the one created by the same system of external currents as when there is no paramagnetic salt present, and the work (1.31) does not contain the electric work necessary to maintain these given currents (Eq.(6.107)).

One should also compare (1.31) with the work $-\mathcal{P}\,d\Omega$ done on a fluid when its volume Ω changes by $d\Omega$. The field B, as well as Ω, are parameters which are controlled from the outside and which occur in the Hamiltonian – the volume Ω occurs through an external potential, confining the particles to that volume. The work done is proportional to the changes in these parameters.

During any infinitesimal transformation, the balance of the internal energy (1.7) gives for the *heat received*

$$\delta \mathcal{Q} = dU - \delta \mathcal{W} = dU + M\,dB = -B\,dM. \tag{1.32}$$

We shall be able to identify on the microscopic scale the entropy and the absolute temperature of the thermodynamicists, if we are able to express the change $\delta \mathcal{Q}$ between equilibrium states in the form $T\,dS$, where T is a function of the relative temperature β and where dS is the total differential of some function S of the variables U, B, N which characterize the macrostate of the system. In thermodynamics one shows that if one can make this identification, it uniquely defines T and S, up to a multiplicative factor for T and S and an additive constant for S.

Anticipating the result which we want to prove, we differentiate expression (1.17), keeping N constant:

$$\begin{aligned} dS &= \frac{\partial S}{\partial U}\,dU + \frac{\partial S}{\partial B}\,dB = \frac{dS}{d\varrho}\left[\frac{\partial \varrho}{\partial U}\,dU + \frac{\partial \varrho}{\partial B}\,dB\right] \\ &= \frac{1}{2}kN\,\ln\,\frac{1-\varrho}{1+\varrho}\,d\varrho = -\,k\beta B\,dM. \end{aligned}$$

We have simplified the calculation by noting that S is a function of $M = -\mu_B N\varrho = -U/B$ only, and we have used the definition (1.28) of β and Eq.(1.15). Comparison with (1.32) gives us

$$\delta Q = \frac{1}{k\beta}\,dS. \tag{1.33}$$

Through a suitable choice for the constant k depending on the system of units in which the entropy is measured, we can thus identify the quantity S, defined by (1.17) and (1.19), with the *entropy.* We can also identify the *absolute temperature* T in terms of the parameter β, which so far had been recognized as being a relative temperature, by writing

$$\boxed{\beta = \frac{1}{kT}}\ . \tag{1.34}$$

If the temperature is measured in kelvin – the unit defined by stating that the temperature of the triple point of water is 273.16 K – and the entropy in joules per kelvin, k can be found from experiments which combine microscopic and macroscopic physics. The value of k, in SI units, is

$$k = 1.38 \times 10^{-23}\ \mathrm{J\,K^{-1}}; \tag{1.35}$$

it is called the *Boltzmann constant.* A system of units, better suited for microscopic physics, consists in putting $k = 1$; the entropy will then be a dimensionless quantity and the temperature is measured in energy units. The fact that the value (1.35) of k, in SI units, is small shows that in the natural units the entropies will always be very large – of the order of the Avogadro number, as can be seen from (1.17) – and the temperatures very low: 1 K, in fact, corresponds to 1.38×10^{-23} J, or to 8.5×10^{-5} eV. It is useful to remember that *room temperature* of 300 K in energy units corresponds to $\frac{1}{40}$ eV.

Implicitly we have put the additive constant, which appears when we identify S with the thermodynamic entropy, equal to zero. Doing this we satisfy the Third Law automatically, since (1.17) vanishes as $\varrho \to -1$ or $\beta \to \infty$, that is, in the limit of low temperatures $T = 1/k\beta \to 0$.

Finally, note that the equilibrium state of the composite system considered in § 1.2.4 is obtained by looking for the *maximum* of the function $S_a + S_b$ which characterizes the *total disorder* of a and b, taking into account the constraint $U_a + U_b = U$ imposed by energy conservation. Let us compare this idea with the formulation of the Laws of Thermodynamics as given in Chap.6: there we postulate the existence of a function of the variables characterizing the macroscopic state of a homogeneous system, namely the entropy, which possesses the following property. To determine the equilibrium state of a composite system we only need look for the maximum of its total entropy with respect to the parameters which can vary. Here, we have *proved* this property for two paramagnetic samples which can exchange energy, by identifying our function S with the entropy. Moreover, we have established the *explicit* form (1.17) of the entropy; this cannot be done by thermodynamics.

When the energy U changes from $-N\mu_B B$ to $+N\mu_B B$, the parameter β changes from $+\infty$ to $-\infty$, and T changes from $+0$ to $+\infty$ and then from $-\infty$ to -0: if we put energy into a system of spins, the temperature first increases, but after that it may become negative (Exerc.1a). These negative temperatures should be considered higher than temperatures for which $T > 0$. The inverse temperature scale β is thus more natural than the ordinary temperature scale.

1.4 Experimental Checks

1.4.1 Equations of State

We have seen that, if we restrict ourselves to magnetic phenomena, the macroscopic state of our paramagnetic solid is characterized by three thermodynamic variables, for instance, U, B, and N. To make comparisons with experiments easier, it is more convenient to choose the absolute temperature T as state variable rather than the internal magnetic energy.

We can then express the various macroscopic quantities as functions of T, B, and N. First of all, we invert relation (1.28) in order to evaluate the magnetic energy at a given temperature:

$$U = -N\mu_B B \, \frac{e^{2\mu_B B/kT} - 1}{e^{2\mu_B B/kT} + 1}. \tag{1.36}$$

It varies from $-N\mu_B B$ at $T = 0$ to 0 when $T \gg \mu_B B/k$. For 1 kg of $\underline{Cu}SO_4 . K_2SO_4 . 6H_2O$ in a field of 1 T this change equals 12.7 J, which is small compared to other forms of energy in solid state theory.

We can use (1.7) to find from U the total magnetic moment $M \equiv \langle M \rangle$,

$$M = N\mu_B \tanh \frac{\mu_B B}{kT}, \tag{1.37}$$

a relation which we can also check directly, using (1.6), (1.29), and (1.30). Finally, expression (1.29) easily gives us the entropy as function of the temperature, if we use (1.29), (1.30), and (1.34):

$$S = Nk \ln\left(2 \cosh \frac{\mu_B B}{kT}\right) - \frac{N\mu_B B}{T} \tanh \frac{\mu_B B}{kT}. \tag{1.38}$$

1.4.2 Curie's Law

The magnetization of a body is equal to its magnetic moment per unit volume so that the magnetic susceptibility in our model follows from (1.1) and (1.37):

$$\chi = \frac{N}{\Omega}\frac{\mu_B^2}{kT}. \tag{1.39}$$

The magnetic susceptibility is *inversely proportional to the absolute temperature.* We have thus established a theoretical justification of Curie's law of which we showed in Fig.1.1 an experimental example. The constant of proportionality, or *Curie's constant*, can also be evaluated using (1.39) and be compared with experiments. Remember that in SI units

$$\mu_B = \frac{e\hbar}{2m} = 0.927 \times 10^{-23}\ \mathrm{J\,T^{-1}}. \tag{1.40}$$

In the example of $\underline{Cu}SO_4.K_2SO_4.6H_2O$ which has a molecular mass of 442, we thus find for the Curie constant per unit mass, in SI units,

$$\frac{6 \times 10^{23}}{442 \times 10^{-3}} \times \frac{1}{1.38 \times 10^{-23}} \left(0.927 \times 10^{-23}\right)^2 = 8.5,$$

whereas the experimental curve led to $\chi_m T = 10.3$. Our model gives a good explanation of Curie's law, but the agreement with experiment is only qualitative, as far as the value of Curie's constant is concerned.

1.4.3 Saturation

Whereas when the field is weak, the magnetization is proportional to the applied field, Eq.(1.37) shows that it must increase less and less rapidly as the field increases and that finally it must tend to a *maximum value.* The latter corresponds to a *complete alignment* of all the $\widehat{\boldsymbol{\mu}}_i$ moments in the same direction as $\boldsymbol{B}$. The change in behaviour occurs for a field of the order of kT/μ_B.

Measurements have been made for several paramagnetic salts. Figure 1.4 shows the values in Bohr magnetons obtained for the mean magnetic moment per ion. Here again, our model gives a qualitative understanding of the saturation effect. In particular, the observed magnetization only depends, as expected from (1.37), on the ratio B/T of the field to the absolute temperature. However, the quantitative agreement is not good, as the $\tanh(\mu_B B/kT)$ behaviour predicted by (1.37) and shown by the dashed curve in Fig.1.4 tends to 1 for strong fields, and not to 3, 5, or 7, and altogether does not have the same shape as the experimental curves, even if we forget about the different normalization (Exerc.1b and 1c).

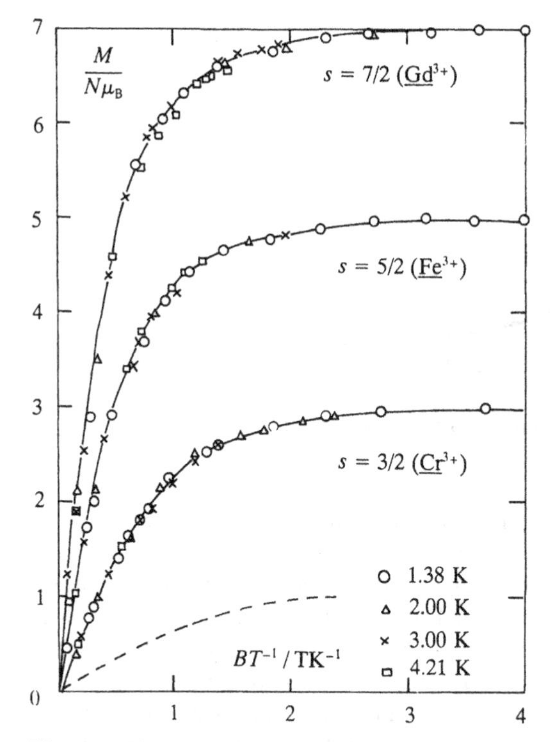

Fig. 1.4. The magnetization as function of the field and of the temperature for various paramagnetic salts: Brillouin curves

1.4.4 Thermal Effects; Characteristic Temperature

When taking the derivative of (1.36) or (1.38) with respect to T we get the *specific heat* – at constant magnetic field:

$$C = \frac{\partial U}{\partial T} = T\,\frac{\partial S}{\partial T} = \frac{N\mu_B^2 B^2}{kT^2 \cosh^2(\mu_B B/kT)}. \tag{1.41}$$

It is shown in Fig.1.5. It has a maximum for $T \simeq 0.8\Theta$, where Θ is a temperature which depends on the applied field and which is defined by

$$\Theta = \frac{\mu_B B}{k}; \tag{1.42}$$

Θ is small, even for a strong applied field: when $B = 1$ T, we have $\Theta = 0.67$ K.

When we cool the system down to a temperature of the order of magnitude of Θ, *order sets in.* One calls Θ the *characteristic temperature* associated with the paramagnetism of the system. When $T \gg \Theta$, the spins are disordered; the corresponding specific heat is small, as a change in temperature hardly changes the state of the spins. On the other hand, when $T \ll \Theta$, the spins are nearly completely oriented parallel to the field; one says that they are

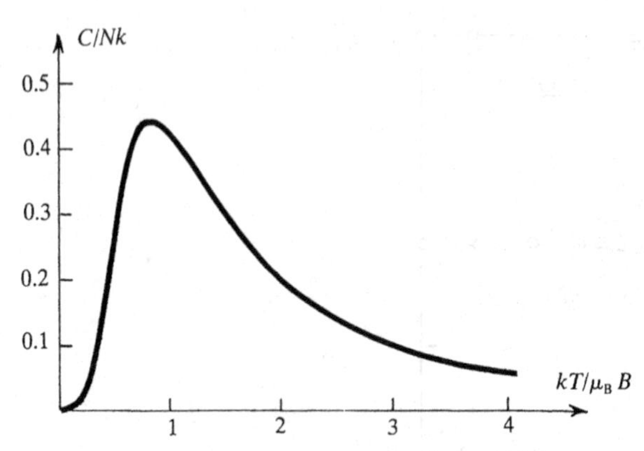

Fig. 1.5. The paramagnetic contribution to the molar specific heat

"*frozen in*". The specific heat again is small, as the system hardly reacts to changes in the temperature.

Experimental checks are difficult, as the total specific heat contains other contributions than the peak shown in Fig.1.5, which come from the other degrees of freedom of the crystal, such as the vibrations of the ions around their equilibrium positions, and which dominate at room temperatures.

This thermal coupling between the moments of the paramagnetic ions and the other degrees of freedom leads, nevertheless, to an interesting technical application, often used in laboratories: it produces, in fact, one of the best methods for *obtaining very low temperatures*, the *adiabatic demagnetization* of a paramagnetic salt. One starts from a situation where the paramagnetic salt is magnetized, thanks to a rather strong initial magnetic induction B_i, say of 1 T, at an initial temperature T_i which is already rather low, say 1 K, so that $\mu_B B_i/kT_i \gtrsim 1$. Then the salt being in contact with the sample which we want to cool down, while the total system is isolated from the outside world, one reduces the magnetic field to a final value B_f which is as small as possible (Fig.1.6). If the transformation is sufficiently slow, the total entropy remains constant. In order to understand how the temperature changes, let us first argue as if the magnetic moments did not interact with the rest of the system. As the entropy (1.17) then only depends on the magnetization, the latter remains constant while B decreases from B_i to B_f – the term "demagnetization" traditionally used to denote this technique is an improper one, as the magnetization remains fixed when the field decreases, at least as long as we ignore the other degrees of freedom. As the magnetization depends only on the ratio B/T, the temperature of the spins *decreases by a factor* B_f/B_i. We could also have reached this conclusion from Eq.(1.38) for the entropy. Let us now include the thermal contact between the spins and the rest of the system and let us write down the condition that the total entropy remains almost constant. If we denote the sample which has to be cooled as well as the non-magnetic degrees of freedom of the salt by a and the

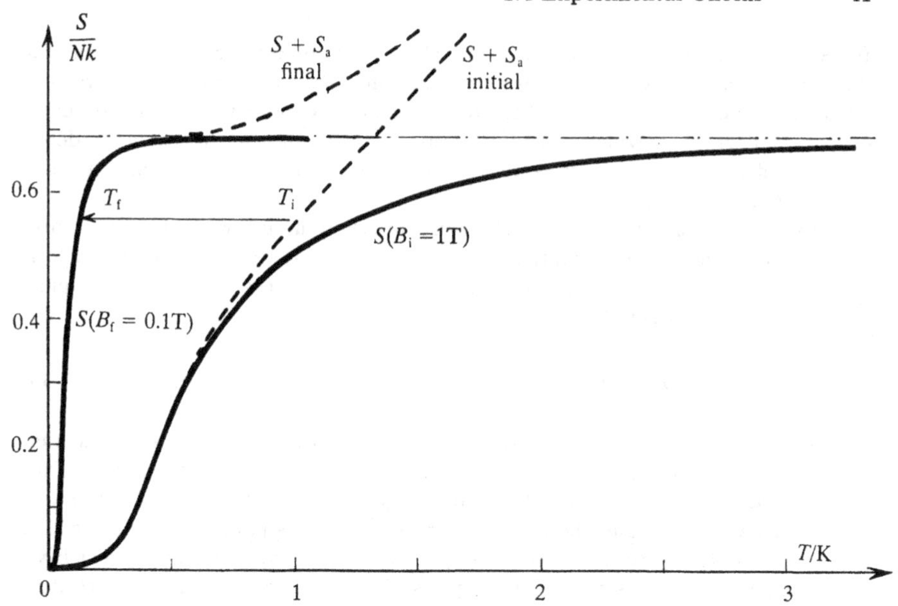

Fig. 1.6. Cooling by adiabatic demagnetization

corresponding entropy by S_a, we find that the final temperature T_f is given by the equation

$$S(T_i, B_i, N) + S_a(T_i) \quad = \quad S(T_f, B_f, N) + S_a(T_f). \tag{1.43}$$

We know that the entropy is an increasing function of the temperature, which tends to 0 for $T = 0$ and we shall see later on why this is the case. If the sample to be cooled down is sufficiently small and if the initial temperature is sufficiently low, S_a is rather small and the total entropy $S + S_a$ has the shape shown in Fig.1.6. The horizontal arrow shows the cooling from T_i to T_f, a temperature slightly higher than $T_i B_f / B_i$. Altogether, the very large initial magnetic *order is transferred* to the other degrees of freedom and their ordering is reflected in the lowering of the temperature. The process is quite efficient: it enables one to cool down to temperatures of the order of millikelvin substances which were first cooled down to liquid helium temperatures of about 4.2 K (Debye 1926, Giauque 1927).

1.4.5 Magnetic Resonance

So far we have only studied the thermodynamic behaviour of a paramagnetic sample at equilibrium. We restricted ourselves to handwaving arguments when describing how equilibrium is established. Nonetheless, it is important to elaborate the theory for the dynamics of the spins as many experiments are based upon the detection of oscillating magnetic moments – measuring the field they produce. In fact, magnetic resonance under the influence of an oscillating field applied to a sample constitutes an extremely powerful tool for exploring the structure and the properties of substances, whether we are dealing with the paramagnetic resonance of electrons or the nuclear magnetic resonance of the nuclear magnetic moments – which are much smaller, but which provide vital information. As examples we mention the determination of molecular structures and scanning of the brain with recent studies of the metabolism of some of its parts.

The study of the dynamics of the spins makes it necessary to take the three components of the operator $\widehat{\boldsymbol{\sigma}}$ into account. In a – possibly time-dependent – field $\boldsymbol{B}$ the equation of motion of their mean values, which is governed by the Hamiltonian $\widehat{H} = \mu_{\mathrm{B}}(\boldsymbol{B}\cdot\widehat{\boldsymbol{\sigma}})$, is (Exerc.2a)

$$\frac{d\langle\boldsymbol{\sigma}\rangle}{dt} = \frac{2\mu_{\mathrm{B}}}{\hbar}\Big[\boldsymbol{B}\times\langle\boldsymbol{\sigma}\rangle\Big]. \tag{1.44}$$

If $\boldsymbol{B}$ is a steady field along the z-axis, the magnetic moment precesses around that axis at the *Larmor frequency* $\omega_{\mathrm{L}} = 2\mu_{\mathrm{B}}B/\hbar$. This frequency also corresponds to the difference $\hbar\omega_{\mathrm{L}} = 2\mu_{\mathrm{B}}B$ between the two energy eigenlevels of the spin in the field $\boldsymbol{B}$. In thermal equilibrium, the precession ceases, as $\langle\boldsymbol{\sigma}\rangle$ is parallel to $\boldsymbol{B}$. Note that we cannot use (1.44) to describe the approach towards equilibrium for which we must include extra terms representing spin-spin and spin-other degrees of freedom interactions.

A magnetic resonance experiment can thus be described as follows. Starting from a paramagnetic system at equilibrium in a field $\boldsymbol{B}$ along the z-axis, we perturb it by applying a second oscillating field $\boldsymbol{b}\sin\omega t$ along the x-axis. This induces transitions between the levels $\sigma_i = +1$ and $\sigma_i = -1$ of each spin. The transition rate has a very pronounced maximum when the frequency of the perturbation agrees with the energy difference between the levels, that is, when $\omega = \omega_{\mathrm{L}}$. Bearing in mind the form (1.44) of the equation of motion, we can understand this behaviour by seeing it as a resonance effect of a periodic perturbation of a system which has a natural oscillation frequency. When $B = 1$ T, the Larmor frequency is equal to

$$\frac{2\mu_{\mathrm{B}}B}{2\pi\hbar} = \frac{2\times 0.93\times 10^{-23}}{6.6\times 10^{-34}} = 2.8\times 10^{10}\ \mathrm{Hz},$$

and the frequency of the perturbing field lies in the radio-frequency band with wave-lengths of the order of cm. By varying the frequency of the perturbing field we thus observe a sharp resonance and its position gives us information about the energy difference between the magnetic levels, while its shape gives us a great deal of further information.

This kind of experiment provides, moreover, a direct measurement of the probability for the occupation of each level and hence an *experimental check of the Boltzmann-Gibbs distribution* (1.29). Let us describe the spin dynamics by a balance method, details of which will be given in Chap.15. Let λ be the transition rate per unit time, which is the same for the $\sigma_i = -1 \to \sigma_i = +1$ transition as for the inverse $\sigma_i = +1 \to \sigma_i = -1$ transition. We study the evolution during a time dt of the probabilities $p_i(\sigma_i)$, by examining the change in $p_i(+1)$. This probability decreases by $\lambda\, dt\, p_i(+1)$ due to the $\sigma_i = +1 \to \sigma_i = -1$ transitions, but it increases also by $\lambda\, dt\, p_i(-1)$ due to the $\sigma_i = -1 \to \sigma_i = +1$ transitions, so that

$$\frac{dp_i(+1)}{dt} = -\lambda\Big[p_i(+1) - p_i(-1)\Big]. \tag{1.45}$$

If we now use Eqs.(1.6) and (1.7) for the energy, we see that the solid absorbs an amount of power equal to

$$\frac{dU}{dt} = 2\lambda N \mu_{\rm B} B \Big[p_i(-1) - p_i(+1) \Big]. \tag{1.46}$$

One measures the strength of the resonance lines by observing the attenuation of the radio-frequency field after it has passed through the paramagnetic sample. When the latter is kept at thermal equilibrium, one can determine the absorbed power (1.46) and hence the probabilities $p_i(\pm 1)$ and one can check that their variation with temperature is given by (1.29) and (1.34).

1.4.6 Discussion of the Model

Notwithstanding the gross and somewhat arbitrary simplifications which we have made in § 1.1.2 our model produces results much as we had expected from it. Being rather similar to the paramagnetic substances which we want to study it has enabled us to understand their properties, to make reasonable predictions about the order of magnitude of various measurable quantities and about functional dependences. Its study has thus been justified *a posteriori* by its explanatory value and by its qualitative agreement with experiments.

However, one can be more ambitious and try to construct a theory which *quantitatively* accounts for the experimental facts. In that case our model would appear as the first stage in the working out of such a theory.

To do this we must first analyze more seriously *the origin of the magnetic moments* $\boldsymbol{\mu}_i$. In the case of an *isolated* ion whose ground state has an angular momentum $\boldsymbol{J} = \boldsymbol{L} + \boldsymbol{S}$, with l, s, j, and j_z being good quantum numbers, the magnetic moment associated with the orbital and spin angular momenta of the electrons is the operator (see end of § 1.1.2)

$$\widehat{\boldsymbol{\mu}} = -\frac{\mu_{\rm B}(\widehat{\boldsymbol{L}} + 2\widehat{\boldsymbol{S}})}{\hbar}.$$

One can prove that the matrix elements of $\widehat{\boldsymbol{\mu}}$ between the lowest, degenerate, energy states are proportional to those of $\widehat{\boldsymbol{J}}$:

$$\widehat{\boldsymbol{\mu}} = -\frac{g\mu_{\rm B}\widehat{\boldsymbol{J}}}{\hbar}, \quad \text{with} \tag{1.47}$$

$$g = \frac{3}{2} + \frac{s(s+1) - l(l+1)}{2j(j+1)}.$$

An approximation which is more realistic than our model thus consists in assuming that each of the paramagnetic ions of the solid in the field $\boldsymbol{B}$ has the same energy levels,

$$-(\boldsymbol{\mu} \cdot \boldsymbol{B}) = g\mu_{\rm B} B j_z, \qquad j_z = -j, \ldots, +j,$$

as if it were isolated. Our earlier results can now easily be generalized, by replacing the two levels with $\sigma_i = \pm 1$ for each ion by $2j+1$ equally spaced levels (Exerc.1b). One can then successfully derive the Curie constants of many substances from the values for l, s, and j of their paramagnetic ions. One can also explain quantitatively and with good accuracy the saturation curves of Fig.1.4. For instance, in the case of Gd^{+3} we have $l = 0, j = s = \frac{7}{2}$, $g = 2$; the saturation curve has the asymptote $\langle M \rangle / N\mu_B = 7$, as predicted. Figure 1.4, moreover, shows the full-drawn theoretical curves – *Brillouin curves* – associated with the various values of s; their agreement with the experimental data is remarkable.

A question now arises about the Cu^{2+} salt for which we successfully compared in § 1.4.2 the Curie constant with experiment. The ground state of the Cu^{2+} ion is a $^2D_{5/2}$ state, that is, $l = 2$, $s = \frac{1}{2}$, $j = \frac{5}{2}$, $g = 1.2$; the Curie constant calculated from (1.47) starting from these values would be 36 SI units, which is in much worse agreement with experiment than the result of our simplistic theory where the magnetic moment was that of a spin-$\frac{1}{2}$. Everything happens as if the orbital motion of the electrons does not contribute to the magnetic moment. To explain this, we must take into account the fact that the Cu^{2+} ions are not isolated, but in a crystal. The neighbouring ions produce an anisotropic potential which perturbs the state of the Cu^{2+} ion. In particular, the rotational invariance is broken, and l and j are no longer good quantum numbers; the ground state is no longer $2j+1 = 6$ times degenerate, but only $2s + 1 = 2$ times, corresponding to the electron spin. Only two levels with $s_z = \pm\frac{1}{2}$ are involved for each ion and we find again the results of the simplified model. One says that the orbital degrees of freedom of the electrons, which do not contribute, are *quenched*, or frozen in. Moreover, one can show that the $(\boldsymbol{L} \cdot \boldsymbol{S})$ coupling which modifies the magnetic moment of the two-fold degenerate ground state explains the deviation of the experimental value from the Curie constant predicted by the $s = \frac{1}{2}$ theory, a deviation which we noted in § 1.4.2.

Many other experimental facts can still change our theoretical ideas and those will, in turn, suggest the study of new effects. For instance, the anisotropy of the crystal potential seen by the paramagnetic ions, which we mentioned a moment ago, leads to an *anisotropy* of the magnetic susceptibility. More exactly, the latter is a 3×3 tensor which connects the components of $\boldsymbol{B}$ with those of $\boldsymbol{M}$. In order to measure it one must study a single crystal rather than a powder and vary the orientation of $\boldsymbol{B}$. The results are depicted in Fig.1.1 where the solid curve shows the susceptibility of the powder and the dashed curves the three eigenvalues of the susceptibilities, which are measured by applying the field $\boldsymbol{B}$ successively in the three principal directions. We see thus that the susceptibility of a $\underline{Cu}SO_4.K_2SO_4.6H_2O$ single crystal is anisotropic, in agreement with our theoretical predictions based upon the anisotropy of the crystal potential.

A study of the deviations from the Curie law, illustrated in Fig.1.7 which gives both experimental values and theoretical curves, gives us a means for investigating the structure of the substance under consideration. They can be explained either by the effect of excited levels of each paramagnetic ion, or by the interactions between these ions and the other atoms in the crystal, or by their interactions with one another.

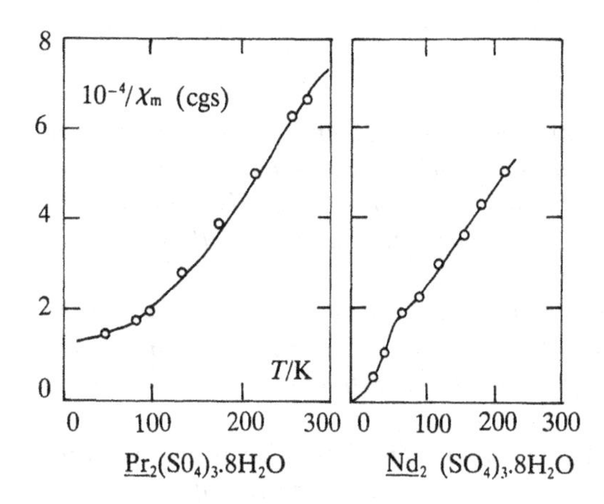

Fig. 1.7. Deviations from the Curie law

Note finally that at very low temperatures these interactions between ions start to play an important rôle, especially in a weak or zero magnetic field. In fact, we see from (1.26) and (1.29) that the magnetic moments will, for $B = 0$, remain perfectly disordered at all temperatures if there is no interaction between paramagnetic ions. Interactions of the form $-(\widehat{\boldsymbol{\mu}}_i \cdot \widehat{\boldsymbol{\mu}}_j)$ tend to align the magnetic moments parallel to one another. At very low temperatures the energy must diminish, which implies that the magnetic moments *become spontaneously correlated with one another*: all magnetic moments align in the same direction and *ferromagnetic order* appears. The applications of ferromagnetism to electrical engineering are very crucial – alternators, transformers, motors, electromagnets. As far as physics is concerned, its theoretical explanation will pass through stages which are similar to the ones of the present chapter: solve a model with interactions like $-(\boldsymbol{\mu}_i \cdot \boldsymbol{\mu}_j)$ between the magnetic moments in order to understand qualitatively the way the moments align spontaneously (Exercs.9a, 9b, and 11f); then discuss the origin of the localized magnetic moments and of their interactions to explain, in particular, why these interactions have the necessary sign to produce ferromagnetism in a given substance.

Summarizing, our model has, on the one hand, helped us to introduce the basic ideas of statistical mechanics which we shall again meet with in the remainder of the present book and, on the other hand, has enabled us to understand qualitatively some phenomena connected with the paramagnetism of salts. Through improvements and modifications it, finally, has enabled us to predict or to explain quantitatively the experimental data. We have here a characteristic example of scientific reasoning: our knowledge progresses thanks to a give-and-take between theory and experiment and the present chapter has shown the first stages of such a process.

Summary

After having shown some experimental facts about paramagnetism and defined a schematic model of a paramagnetic solid on the microscopic scale, we have characterized a macroscopic state of this solid by a probability law for the various possible microscopic states. The expectation values of all observables can be derived from this probability law.

The evolution of a macroscopic system to thermal equilibrium can on the microscopic probabilistic scale be interpreted as an evolution in time of the macro-state to the maximum disorder compatible with energy conservation. If the energy is assumed to be well defined initially, the macro-state reached at thermal equilibrium is characterized by equal probabilities for all micro-states with that energy: the "microcanonical ensemble". Its degree of disorder is measured by (1.17) or (1.19); it can be identified with the macroscopic thermodynamic entropy. Comparing the microscopic probabilistic description of the thermal contact between two spin systems with the Laws of Thermodynamics, we can interpret the absolute temperature as a parameter characterizing the probability distribution (1.29), (1.34) of one of the magnetic ions – or the Boltzmann-Gibbs distribution – and we can explain why the statistical nature of the underlying microscopic theory does not show up at the macroscopic level.

This theory enables us to explain experimental facts: Curie's law (1.39), saturation of the magnetic moment in strong fields, cooling by adiabatic demagnetization. In passing we introduced the technique of using the partition function (1.13) and the concept of a characteristic temperature (§ 1.4.4) in the vicinity of which order sets in.

Exercises

1a Negative Temperatures

How do the parameter β, the entropy S, and the temperature T vary when the internal energy (1.36) of a system of paramagnetic ions, which are assumed not to interact with the other degrees of freedom of the solid, changes continuously from its minimum value $-N\mu_{\rm B}B$ to its maximum value $+N\mu_{\rm B}B$? Note that the temperatures $T = \pm\infty$ should be regarded as being identical, but that $T = +0$ corresponds to the ground state, while $T = -0$ corresponds to the highest state.

Using the fact that the entropy is a concave function of the energy, show that one can distinguish two kinds of systems, the first, and most common one, such that the energy spectrum is bounded only from below and for which β varies from $+\infty$ to $+0$, and the second such that the spectrum has both an upper and a lower bound, while β varies from $+\infty$ to $-\infty$. Show that thermal contact between two systems, of either type, which initially are at different temperatures β_a and β_b, leads to a final equilibrium state with an intermediate temperature β, such that $\beta_a < \beta < \beta_b$, if $\beta_a < \beta_b$. Discuss this result in terms of the temperatures T for each of the three possible combinations of temperature signs.

Hint. Consider the curves $S_a(U_a)$ and $S_b(U_b)$; start from the final equilibrium state, where $\beta_a = \beta_b = \beta$, and study the change in the temperatures when energy is transferred from one system to the other.

Comments. Equilibrium states with negative spin temperatures can be observed in paramagnetic substances in the common cases where the coupling between the spins and the "lattice" – that is, the other degrees of freedom, such as the vibrations of the atoms in the crystal which have a spectrum without an upper bound – is much weaker than the spin-spin coupling: for a fairly long time a quasi-equilibrium of the spin system occurs which can have any temperature, positive or negative, depending on the value of the magnetic energy. There coexist therefore in the same substance two different temperatures, as would be the case for two weakly coupled thermostats. Of course, ultimately the combination of all the degrees of freedom reaches equilibrium at a positive temperature.

If $\beta < 0$, there is a population inversion: it follows from (1.29) that there are more spins with positive energies than with negative energies. In contrast, a Boltzmann-Gibbs distribution with positive temperature shows a preference for micro-states with lower energy; this explains microscopically why β must be positive if the spectrum has no upper bound.

1b Brillouin Curves

Study the paramagnetism of a salt containing paramagnetic ions with a magnetic moment (1.47) corresponding to arbitrary spin j. Start from the Boltzmann-Gibbs distribution (1.29) which describes each of the magnetic moments at thermal equilibrium at a temperature T. Calculate the normalization constant Z_1 and hence derive – by differentiating with respect to B – the average magnetic moment of an ion as function of the field and of the temperature and the susceptibility as function of the temperature. Show that Curie's law is satisfied and that the saturation curves have the Brillouin shape shown in Fig.1.4. Check the results by comparing them with those in the text for $l = 0$, $j = s = \frac{1}{2}$.

Results:

$$Z_1 = \frac{\sinh\left[\left(j+\frac{1}{2}\right)\beta g\mu_{\mathrm{B}}B\right]}{\sinh\left(\frac{1}{2}\beta g\mu_{\mathrm{B}}B\right)},$$

$$\begin{aligned}\langle\mu_z\rangle &= \sum_{j_z=-j}^{+j} p_1(j_z)\left[-g\mu_{\mathrm{B}}j_z\right] = \frac{1}{\beta}\frac{\partial}{\partial B}\ln Z_1\\ &= g\mu_{\mathrm{B}}\left\{\left(j+\tfrac{1}{2}\right)\coth\left[\left(j+\tfrac{1}{2}\right)\beta g\mu_{\mathrm{B}}B\right] - \tfrac{1}{2}\coth\left(\tfrac{1}{2}\beta g\mu_{\mathrm{B}}B\right)\right\}.\end{aligned}$$

$$\chi = \frac{N}{\Omega}\frac{\mu_{\mathrm{B}}^2}{kT}\frac{g^2 j(j+1)}{3}.$$

1c Langevin Paramagnetism

Before the birth of quantum mechanics, Langevin explained paramagnetism by assuming that each paramagnetic ion possessed a permanent magnetic moment, a classical vector $\boldsymbol{\mu}$ which could freely align itself in any direction. Its energy $E = -(\boldsymbol{\mu}\cdot\boldsymbol{B})$ depends on the direction $\boldsymbol{n} = \boldsymbol{\mu}/\mu$ of $\boldsymbol{\mu}$. The probability law for $\boldsymbol{n}$ in thermal equilibrium is the Boltzmann-Gibbs distribution

$$p(\boldsymbol{n})\,d^2\boldsymbol{n} = \frac{1}{Z}\,\mathrm{e}^{-\beta E(\boldsymbol{n})}\,d^2\boldsymbol{n}.$$

Calculate Z and hence find $\langle\boldsymbol{\mu}\rangle$ and χ. Compare these results with quantum theory: show that the results of Exerc.1b reduce to classical paramagnetism in the limit as $j\to\infty$, $\mu = \mu_{\mathrm{B}}g(j+\frac{1}{2})$. The Langevin saturation curve is quantitatively incorrect.

Results:

$$Z = \int_0^{2\pi} d\varphi \int_{-1}^{+1} d\cos\theta\;\mathrm{e}^{\beta\mu B\cos\theta} = \frac{4\pi}{\beta\mu B}\sinh(\beta\mu B).$$

$$\langle\mu_z\rangle = \int d^2\boldsymbol{n}\;p(\boldsymbol{n})\,\mu_z = \frac{1}{\beta}\frac{\partial}{\partial B}\ln Z = \mu\left(\coth(\beta\mu B) - \frac{1}{\beta\mu B}\right).$$

$$\chi = \frac{N}{\Omega}\frac{\mu^2}{3kT}.$$

In the classical limit $j\to\infty$ and $\mu = \mu_{\mathrm{B}}g(j+\frac{1}{2})$ we find that the Brillouin expressions for $\langle\mu_z\rangle$ and χ tend to the Langevin expressions. However, there is in Z an extra factor $4\pi/(2j+1)$ which comes from the normalizations which are different: $\sum_{j_z} 1 = 2j+1$ in the Brillouin case, and $\int d^2\boldsymbol{n} = 4\pi$ in the Langevin case.

2. Probabilistic Description of Systems

"Une cause très petite qui nous échappe détermine un effet considérable que nous ne pouvons pas ne pas voir, et alors nous disons que cet effet est dû au hasard."

Henri Poincaré, Calcul des Probabilités, 1912

"Probabilitatem esse deducendam."

Letter from Einstein to Pauli, 1932

"We meet here in a new light the old truth that in our description of nature the purpose is not to disclose the real essence of the phenomena but only to track down, so far as it is possible, relations between the manifold aspects of our experience."

Niels Bohr

In the present chapter we start the study of the general formalism of statistical physics by showing how one can mathematically represent a system which is not well known on the microscopic scale. We deliberately use *quantum* mechanics for several reasons. On the one hand, microscopic physics is basically quantal; classically there would exist neither atoms nor molecules with discrete bound states. On the other hand, we shall see that, notwithstanding a few conceptual difficulties to begin with, quantization brings about simplifications by replacing integrals by discrete sums. Last and not least, many important phenomena, such as the very existence of solids or magnetic substances, the properties of black-body radiation, or even the extensivity of matter, can only be explained by a quantum-mechanical approach. Even in the case of gases or liquids, classical statistical mechanics is insufficient; it does not enable one to elucidate the Gibbs paradox or to understand the values of the specific heats.

We assume that the reader is familiar with the elements of quantum mechanics such as one can find in introductory textbooks. In order to fix the notation and to remind ourselves of a few results for future reference, we start with a survey of the mathematical tools; we review the basic postulates in the Dirac formalism, only considering, as is usually done, completely prepared states, the so-called micro-states; these are characterized by their ket (§ 2.1).

In quantum statistical mechanics, however, the preparations are always incomplete and a system is not in a state characterized by a well defined ket. The state of an incompletely prepared system, a so-called macro-state, is only partially known and corresponds to assigning a probability law for all possible kets. The synthesis of this probabilistic nature with the probabilities arising from the quantum mechanics of kets will be made using a mathematical entity, the *density operator*, which generalizes the probability concept to quantum statistical mechanics. In fact, the density operator formalism, introduced independently in 1927 by Lev Davidovich Landau and Johann (John) von Neumann (Budapest 1903–Washington 1957), enables us to reformulate the laws of quantum mechanics more generally than with the formalism using kets or wavefunctions, and to get rid of these concepts. Even in microphysics, it will turn out that density operators are more basic and easier to understand than state vectors. We develop the various aspects of density operator theory in § 2.2, and collect for convenience in § 2.2.7 the main results, the only ones to be used in what follows.

In classical mechanics a micro-state of a system of particles is characterized once we know their positions and momenta; a macro-state – which is imperfectly known at the microscopic level – is characterized by a probability measure in the phase space representing these positions and momenta (J.Willard Gibbs). All the same, we must define this measure with some care in order that classical statistical mechanics appears, as it should, as a limiting case of quantum statistical mechanics. We shall use the classical formalism only in Chaps.7 to 9 and in Chap.15; one can postpone the study of § 2.3 until we start applying statistical physics to gases or liquids.

2.1 Elements of the Quantum Formalism

2.1.1 Hilbert Spaces

One can associate with any *physical system*, described in a given frame of reference, a Hilbert space $\mathcal{E}_{\rm H}$ with elements, or *kets*, denoted by $|\psi\rangle$. The physical significance of this space and of the entities connected with it will be made clearer starting from § 2.1.3, but for the moment we shall only be concerned with its mathematical properties. A Hilbert space possesses the structure of a vector space, involving addition, $|\psi_1\rangle+|\psi_2\rangle$, and multiplication by a complex number c, $c|\psi\rangle$. Moreover, one associates with each ket $|\psi\rangle$ a *bra* $\langle\psi|$ which is an element of the dual space of $\mathcal{E}_{\rm H}$, $\mathcal{E}_{\rm H}^*$; the bra which is the conjugate of $c|\psi\rangle$ is $\langle\psi|c^*$ and we define a scalar product, $\langle\varphi|\psi\rangle$, of bras and kets, which satisfies the relation $\langle\varphi|\psi\rangle = \langle\psi|\varphi\rangle^*$.

For some systems the space $\mathcal{E}_{\rm H}$ has a finite number of dimensions. For instance, the dimensionality of $\mathcal{E}_{\rm H}$ is 2 for the spin-$\frac{1}{2}$ of an electron, if we forget about all other degrees of freedom. Usually, its dimensionality is infinite. For instance, for a particle in one dimension $\mathcal{E}_{\rm H}$ is the space of all square integrable complex functions of x, which may possibly be confined to a box, $0 < x < L$. To avoid mathematical complications without any physical relevance, we disregard in this book all convergence problems which are connected with the infinite dimensionality of $\mathcal{E}_{\mathrm{H}}$.

In order to define coordinates in $\mathcal{E}_{\mathrm{H}}$ we choose in it a *complete orthonormal base* $\{|k\rangle\}$ of vectors $|k\rangle$, which are characterized by an index, or by a set of indices, k and which satisfy the *orthonormality* relations,

$$\langle k|k'\rangle = \delta_{k,k'}, \tag{2.1}$$

and the *closure* relation,

$$\sum_k |k\rangle\langle k| = \widehat{I}; \tag{2.2}$$

here $\widehat{I}$ denotes the identity operator of the space defined by $\widehat{I}|\psi\rangle = |\psi\rangle$. Every ket $|\psi\rangle$ can thus be represented by its coordinates $\langle k|\psi\rangle$:

$$|\psi\rangle = \sum_k |k\rangle\langle k|\psi\rangle, \tag{2.3}$$

and we find for the scalar product $\langle\varphi|\psi\rangle$

$$\langle\varphi|\psi\rangle = \sum_k \langle\varphi|k\rangle\langle k|\psi\rangle. \tag{2.4}$$

For instance, for a spin-$\frac{1}{2}$ (Exerc.2a) the two base vectors $|\pm\rangle$ are characterized by the quantum numbers $\sigma_z = \pm 1$. For a one-dimensional particle, we can choose for $\{|k\rangle\}$ the enumerable base of eigenfunctions of the harmonic oscillator, with $k = 0, 1, 2, \ldots$. An alternative base $\{|x\rangle\}$ of kets localized at the point x is physically convenient, but encounters some mathematical difficulties as $|x\rangle$ is not normalizable; it is characterized by a continuous index x, in which case the Kronecker δ in (2.1) should be replaced by a Dirac δ-function and the sum in (2.3) by an integral.

When a *composite physical system* a+b consists of two parts a and b, with which the spaces $\mathcal{E}_{\mathrm{H}}^{\mathrm{a}}$ and $\mathcal{E}_{\mathrm{H}}^{\mathrm{b}}$ are associated, its Hilbert space $\mathcal{E}_{\mathrm{H}}$ is the *tensor product* $\mathcal{E}_{\mathrm{H}} = \mathcal{E}_{\mathrm{H}}^{\mathrm{a}} \otimes \mathcal{E}_{\mathrm{H}}^{\mathrm{b}}$. A base $\{|k_{\mathrm{a}}\rangle\}$ of the space $\mathcal{E}_{\mathrm{H}}^{\mathrm{a}}$ and a base $\{|l_{\mathrm{b}}\rangle\}$ of $\mathcal{E}_{\mathrm{H}}^{\mathrm{b}}$ generate the base $\{|k_{\mathrm{a}} l_{\mathrm{b}}\rangle\}$ of $\mathcal{E}_{\mathrm{H}}$ labelled by the combination of the quantum numbers k_{a} of system a with the quantum numbers l_{b} of system b. The same procedure can be extended to the case where a and b denote independent *degrees of freedom.* For instance, the Hilbert space of a particle moving in the x, y-plane can be constructed as the direct product of the Hilbert spaces associated, respectively, with the x- and the y-dimensions. For an electron, $\mathcal{E}_{\mathrm{H}}$ is the direct product of the space of the wavefunctions $\psi(\boldsymbol{r}) \equiv \langle\boldsymbol{r}|\psi\rangle$ and the two-dimensional space associated with the spin. More generally, the systems we have to deal with in quantum statistical mechanics will usually have a very large number of degrees of freedom; it will be convenient to choose as a base spanning their Hilbert space the tensor product of the bases associated with each of the degrees of freedom. For instance, in Chap.1 we used the base $|\sigma_1, \ldots, \sigma_N\rangle$ associated with N spin-$\frac{1}{2}$ particles.

The *direct sum* $\mathcal{E}_\mathrm{H} = \mathcal{E}_\mathrm{H}^1 \oplus \mathcal{E}_\mathrm{H}^2$ of two spaces $\mathcal{E}_\mathrm{H}^1$ and $\mathcal{E}_\mathrm{H}^2$ is the set of linear combinations of the kets of $\mathcal{E}_\mathrm{H}^1$ and $\mathcal{E}_\mathrm{H}^2$. The union of the base vectors of $\mathcal{E}_\mathrm{H}^1$ and $\mathcal{E}_\mathrm{H}^2$ forms the base of $\mathcal{E}_\mathrm{H}$. For instance, the Hilbert space of a particle moving in three dimensions is the direct sum of its subspaces, chacterized by the values $l = 0$, $l = 1$, $l = 2$, ... of the angular momentum. In Chaps.4 and 10 we shall define the Fock space associated with a system consisting of an arbitrary number of indistinguishable particles as a direct sum of spaces representing N-particle systems ($N = 0, 1, 2, \ldots$).

2.1.2 Operators

Still discussing only purely mathematical questions let us review the properties of operators. An operator $\widehat{X}$ of the Hilbert space $\mathcal{E}_\mathrm{H}$ defines a *linear transformation* $\widehat{X}|\psi\rangle$ of the kets. Its action on the bras follows from the associativity of the scalar product: $(\langle\varphi|\widehat{X})|\psi\rangle = \langle\varphi|(\widehat{X}|\psi\rangle) \equiv \langle\varphi|\widehat{X}|\psi\rangle$. In the base $\{|k\rangle\}$ an operator is represented by a *matrix* $\langle k|\widehat{X}|k'\rangle$; we can write the operator in terms of this matrix as the linear combination

$$\widehat{X} = \sum_{kk'} |k\rangle \langle k|\widehat{X}|k'\rangle \langle k'|, \tag{2.5}$$

and the action of $\widehat{X}$ on a ket $|\psi\rangle$ is represented by the matrix product

$$\langle k|\widehat{X}|\psi\rangle = \sum_{k'} \langle k|\widehat{X}|k'\rangle \langle k'|\psi\rangle. \tag{2.6}$$

The operators form an *algebra*. This algebra has a specific structure which is characteristic for the physical system we are studying. It also has general properties, involving (i) *linear operations*: a sum $\widehat{X}_1 + \widehat{X}_2$ and a product $c\widehat{X}$ with a complex number; (ii) a *non-commutative* product $\widehat{X}_1\widehat{X}_2$ which produces commutators $[\widehat{X}_1, \widehat{X}_2]$; and (iii) the operation of *Hermitean conjugation* $\widehat{X} \Longleftrightarrow \widehat{X}^\dagger$ such that the bra $\langle\psi|\widehat{X}^\dagger$ is the Hermitean conjugate of $\widehat{X}|\psi\rangle$; this implies $(\widehat{X}_1\widehat{X}_2)^\dagger = \widehat{X}_2^\dagger\widehat{X}_1^\dagger$. For a given base $\{|k\rangle\}$ the product is represented by matrix multiplication, and Hermitean conjugation by $\langle k|\widehat{X}^\dagger|k'\rangle = \left(\langle k|\widehat{X}|k'\rangle\right)^*$. The expansion (2.5) shows that the linear structure of the set of operators $\widehat{X}$ is isomorphic with that of the vector space $\mathcal{E}_\mathrm{H} \otimes \mathcal{E}_\mathrm{H}^*$. Indeed, the *dyadics* $|k\rangle\langle k'|$ are operators which form a base for the operator space and the coordinates of $\widehat{X}$ in that base are, according to (2.5), the matrix elements $\langle k|\widehat{X}|k'\rangle$.

One can choose other bases in the $\mathcal{E}_\mathrm{H} \otimes \mathcal{E}_\mathrm{H}^*$ space such that any operator $\widehat{X}$ can be expressed as a linear combination of the operators of the base considered.

This defines the *Liouville representations* which are more general than the representations in the Hilbert space and which are often useful.[1] For instance, for a spin-$\frac{1}{2}$, instead of representing the operators as 2×2 matrices, it can be of interest to represent them as linear combinations of the Pauli matrices and the identity (Exerc.2a).

Similarly, for a one-dimensional particle one can write any operator $\widehat{X}$ in the form

$$\widehat{X} = \int dx\, dp\, X_{\mathrm{W}}(x,p)\, \widehat{\Omega}(x,p), \tag{2.7}$$

as a linear combination of the operators $\widehat{\Omega}(x,p)$ defined by

$$\widehat{\Omega}(x,p) \equiv \frac{1}{4\pi^2} \int d\alpha\, d\beta\, \mathrm{e}^{\mathrm{i}\alpha(\widehat{x}-x)+\mathrm{i}\beta(\widehat{p}-p)}. \tag{2.8}$$

It is important to make a clear distinction between the position and momentum operators $\widehat{x}$, $\widehat{p}$ and the variables x, p which are a pair of continuous indices characterizing each operator (2.8). The set of operators $\widehat{\Omega}(x,p)$ form a base spanning operator space. Their use rests upon the identity

$$\mathrm{e}^{\mathrm{i}\alpha\widehat{x}+\mathrm{i}\beta\widehat{p}} = \mathrm{e}^{\mathrm{i}\alpha\widehat{x}}\, \mathrm{e}^{\mathrm{i}\beta\widehat{p}}\, \mathrm{e}^{\mathrm{i}\alpha\beta\hbar/2} = \mathrm{e}^{\mathrm{i}\beta\widehat{p}}\, \mathrm{e}^{\mathrm{i}\alpha\widehat{x}}\, \mathrm{e}^{-\mathrm{i}\alpha\beta\hbar/2},$$

which follows from the commutation relation $[\widehat{x},\widehat{p}] = \mathrm{i}\hbar$. For instance, this can be used to prove that (2.7) can be inverted to give

$$X_{\mathrm{W}} = 2\pi\hbar\, \mathrm{Tr}\, \widehat{\Omega}(x,p)\widehat{X}, \tag{2.9}$$

a relation enabling us to express any operator $\widehat{X}$ as (2.7) in the base $\widehat{\Omega}(x,p)$. The equations (2.7), (2.9) define the *Wigner representation*, a special case of a Liouville representation. Each operator $\widehat{X}$ is then represented not by a matrix as in Hilbert space, but by a *function* $X_{\mathrm{W}}(x,p)$ *of commuting variables* which play the rôle of coordinates in a classical phase space. The Wigner representation is particularly simple for operators of the form $f(\widehat{x}) + g(\widehat{p})$ or $\exp(\mathrm{i}\alpha\widehat{x} + \mathrm{i}\beta\widehat{p})$, the transforms (2.9) of which are, respectively, $f(x) + g(p)$ and $\exp(\mathrm{i}\alpha x + \mathrm{i}\beta p)$. The Wigner transform of the operator $\widehat{\Omega}(x',p')$ itself is, according to (2.7), just $\delta(x - x')\delta(p - p')$. This enables us to interpret $\widehat{\Omega}(x',p')$ as an observable which probes, within the limits allowed by the Heisenberg relations, whether the particle is at the point x', p' of phase space. The occurrence of ordinary variables x and p, even when $\hbar \neq 0$, thus introduces a classical structure in the bosom of quantum mechanics. Hence, the Wigner representation is a convenient way to study the *classical limit* of quantum mechanics (§2.3.4), in which case $[\widehat{x},\widehat{p}] \to 0$ and $\widehat{\Omega}(\widehat{x},\widehat{p}) \to \delta(\widehat{x} - x)\delta(\widehat{p} - p)$, and to calculate the lowest-order *quantum corrections*, for instance, for gases at low temperatures. To do this we note that the algebraic relations in the Wigner representation are summarized by the identity

[1] N.L. Balasz and B.K. Jennings, Phys.Repts **104**, 347 (1984); M. Hillery, R.F. O'Connell, M.O. Scully, and E.P. Wigner, Phys.Repts **106**, 121 (1984); R. Balian, Y. Alhassid, and H. Reinhardt, Phys.Repts **131**, 1 (1986).

$$\widehat{\Omega}(x,p)\,\widehat{\Omega}(x',p') = \frac{1}{4\pi^2}\int d\alpha\, d\beta\, \widehat{\Omega}\left(\frac{x+x'+\beta\hbar}{2}, \frac{p+p'-\alpha\hbar}{2}\right) \times \mathrm{e}^{\mathrm{i}\alpha(x-x')+\mathrm{i}\beta(p-p')} \tag{2.10}$$

for the operators of the $\widehat{\Omega}$ base. This relation implies (see Eq.(2.80) below) that the Wigner representations of a product $\widehat{X}\widehat{Y}$ or of a function $f(\widehat{X})$ are different from $X_{\mathrm{W}}(x,p)Y_{\mathrm{W}}(x,p)$ or from $f(X_{\mathrm{W}})$ when $\hbar \neq 0$, but not when $\hbar = 0$. In the classical limit, the Wigner transforms X_{W} of quantum observables thus become identical with the classical observables which are simply functions of x and p.

For a composite system the *tensor product of two operators*, $\widehat{X}_{\mathrm{a}}$ operating in the space $\mathcal{E}_{\mathrm{H}}^{\mathrm{a}}$ associated with the system a, and $\widehat{Y}_{\mathrm{b}}$ acting in $\mathcal{E}_{\mathrm{H}}^{\mathrm{b}}$, is an operator $\widehat{X}_{\mathrm{a}} \otimes \widehat{Y}_{\mathrm{b}}$ acting in $\mathcal{E}_{\mathrm{H}}$, represented by the matrix

$$\langle k_{\mathrm{a}} l_{\mathrm{b}} | \widehat{X}_{\mathrm{a}} \otimes \widehat{Y}_{\mathrm{b}} | k'_{\mathrm{a}} l'_{\mathrm{b}} \rangle = \langle k_{\mathrm{a}} | \widehat{X}_{\mathrm{a}} | k'_{\mathrm{a}} \rangle \, \langle l_{\mathrm{b}} | \widehat{Y}_{\mathrm{b}} | l'_{\mathrm{b}} \rangle . \tag{2.11}$$

In particular, the operator $\widehat{X}_{\mathrm{a}}$ in $\mathcal{E}_{\mathrm{H}}^{\mathrm{a}}$ acting on the system a can also be regarded as an operator acting on the system a+b; this extension is effected by identifying the operator $\widehat{X}_{\mathrm{a}}$ of $\mathcal{E}_{\mathrm{H}}$ with the tensor product $\widehat{X}_{\mathrm{a}} \otimes \widehat{I}_{\mathrm{b}}$ where $\widehat{I}_{\mathrm{b}}$ is the identity operator in $\mathcal{E}_{\mathrm{H}}^{\mathrm{b}}$, that is, by writing

$$\langle k_{\mathrm{a}} l_{\mathrm{b}} | \widehat{X}_{\mathrm{a}} | k'_{\mathrm{a}} l'_{\mathrm{b}} \rangle = \langle k_{\mathrm{a}} | \widehat{X}_{\mathrm{a}} | k'_{\mathrm{a}} \rangle \, \delta_{l_{\mathrm{b}}, l'_{\mathrm{b}}} . \tag{2.12}$$

Unitary operators $\widehat{U}$ are those which conserve the scalar products, $\langle\varphi|\psi\rangle = \big(\langle\varphi|\widehat{U}^\dagger\big)\big(\widehat{U}|\psi\rangle\big)$. They thus satisfy the relations $\widehat{U}\widehat{U}^\dagger = \widehat{U}^\dagger\widehat{U} = \widehat{I}$. By a unitary transformation an orthonormal base $\{|k\rangle\}$ becomes another orthonormal base $\{\widehat{U}|k\rangle\}$. The algebra of the $\widehat{X}$ operators also remains invariant under a unitary transformation, $\widehat{X} \Longrightarrow \widehat{U}^\dagger\widehat{X}\widehat{U}$. For a given $\widehat{X}$ such a transformation leaves the determinant of the matrix $\widehat{X} - \lambda\widehat{I}$ unchanged, that is, it leaves the eigenvalues of $\widehat{X}$ invariant.

Hermitean or *self-adjoint operators* $\widehat{A}$ are those satisfying the relation $\widehat{A} = \widehat{A}^\dagger$. The matrix representing a Hermitean operator $\widehat{A}$ in a given base $\{|k\rangle\}$ can be *diagonalized* by a unitary transformation; in other words, there exists a base $\{\widehat{U}|k\rangle\}$ such that the matrix representing $\widehat{A}$ is diagonal. This base constitutes a set of eigenvectors of $\widehat{A}$, defined by the equation $\widehat{A}|\psi_\alpha\rangle = a_\alpha|\psi_\alpha\rangle$; the associated eigenvalues a_α are real. Hermitean operators which *commute* with one another can be *diagonalized by a single unitary transformation.*

A *positive operator* is a Hermitean operator, all eigenvalues of which are positive, or, equivalently, an operator the diagonal elements of which, $\langle\varphi|\widehat{X}|\varphi\rangle$, are real and positive in any base. For a *non-negative operator*, such as $\widehat{A}^2$, the eigenvalues and the diagonal elements are positive or zero.

If $f(x)$ is an entire function, the *operator function* $f(\widehat{X})$ *of the operator* $\widehat{X}$ is defined by its series expansion. For instance, the operator $\exp(\widehat{X})$ or $\mathrm{e}^{\widehat{X}}$ is positive when $\widehat{X}$ is Hermitean and unitary when $\mathrm{i}\widehat{X}$ is Hermitean.

Alternatively, one can define the operator $f(\widehat{A})$ through diagonalization when $\widehat{A}$ is Hermitean: $f(\widehat{A})$ and $\widehat{A}$ have the same eigenvectors $|\psi_\alpha\rangle$ and the eigenvalue of $f(\widehat{A})$ corresponding to the eigenvalue a_α of $\widehat{A}$ is equal to $f(a_\alpha)$. This definition can be extended to arbitrary functions $f(x)$ of a real variable. It enables us to define the logarithm, $\ln \widehat{A}$, of a positive operator $\widehat{A}$, or its inverse, $\widehat{A}^{-1}$. Similarly, the *projection* $\widehat{P}_\alpha$ *onto the eigenvalue* a_α (or onto several eigenvalues) of $\widehat{A}$ is of the form $f(\widehat{A})$, where $f(x)$ equals 1 for $x = a_\alpha$ (or for the selected set of eigenvalues) and vanishes for the other eigenvalues; this definition holds even when the eigenvalue a_α is degenerate, in which case the dimension of $\widehat{P}_\alpha$ is the multiplicity of a_α. One can decompose the operator $\widehat{A}$ in its eigenprojections,

$$\widehat{A} = \sum_\alpha a_\alpha \widehat{P}_\alpha. \tag{2.13}$$

Finally, the *trace of an operator* is defined by

$$\mathrm{Tr}\, \widehat{X} = \sum_k \langle k|\widehat{X}|k\rangle, \tag{2.14}$$

which is independent of the base $\{|k\rangle\}$. In what follows we shall need some of its properties. The trace of a dyadic is equal to

$$\mathrm{Tr}\, |\psi\rangle\, \langle\varphi| = \langle\varphi|\psi\rangle. \tag{2.15}$$

The trace of a projection is equal to the dimensionality of the space it spans. A trace is invariant under a cyclic permutation:

$$\mathrm{Tr}\, \widehat{X}\widehat{Y} = \mathrm{Tr}\, \widehat{Y}\widehat{X} = \sum_{kk'} \langle k|\widehat{X}|k'\rangle\, \langle k'|\widehat{Y}|k\rangle. \tag{2.16}$$

The complex conjugate of a trace satisfies the relation

$$\left(\mathrm{Tr}\, \widehat{X}\right)^* = \mathrm{Tr}\, \widehat{X}^\dagger. \tag{2.17}$$

In a space $\mathcal{E}_\mathrm{H} = \mathcal{E}_\mathrm{H}^\mathrm{a} \otimes \mathcal{E}_\mathrm{H}^\mathrm{b}$ the trace can be factorized as

$$\mathrm{Tr} = \mathrm{Tr_a}\, \mathrm{Tr_b},$$

where $\mathrm{Tr_b}\widehat{X}$ is an operator in $\mathcal{E}_\mathrm{H}^\mathrm{a}$ defined by

$$\langle k_\mathrm{a}|\mathrm{Tr_b}\, \widehat{X}|k'_\mathrm{a}\rangle = \sum_{l_\mathrm{b}} \langle k_\mathrm{a} l_\mathrm{b}|\widehat{X}|k'_\mathrm{a} l_\mathrm{b}\rangle \tag{2.18}$$

in a tensor product base. If we change the operator $\widehat{X}$ infinitesimally by $\delta\widehat{X}$, the trace of an operator function $f(\widehat{X})$ can be differentiated as if $\widehat{X}$ and $\delta\widehat{X}$ commuted:

$$\delta\, \mathrm{Tr}\, f(\widehat{X}) = \mathrm{Tr}\, \delta\widehat{X}\, f'(\widehat{X}). \tag{2.19}$$

The equation

$$\operatorname{Tr} \widehat{X}\, \delta\widehat{A} = 0, \qquad \forall \quad \delta\widehat{A} = \delta\widehat{A}^{\dagger}, \tag{2.20}$$

implies that $\widehat{X} = 0$.

2.1.3 Observables

Having surveyed the mathematical formalism we shall now recapitulate the general principles of quantum mechanics.

One can associate with any physical quantity A an observable $\widehat{A}$ which is a Hermitean operator in the Hilbert space $\mathcal{E}_{\mathrm{H}}$ of the system considered. When this system interacts with an apparatus which can either prepare it or carry out measurements on it, *the only possible values* for the quantity A, controlled by this apparatus, *are the eigenvalues a_α of the observable $\widehat{A}$.* This postulate is the *principle of quantization.*

One associates with the physical quantity $f(A)$ the observable $f(\widehat{A})$, which is a function of the operator $\widehat{A}$. Consider now a quantity P, taking the value 1, if a particular proposition is true, and the value 0, if it is false. It satisfies the identity $P^2 = P$, and is thus represented by an observable $\widehat{P}$ satisfying the relation $\widehat{P}^2 = \widehat{P}$ which characterizes projections. A *proposition* or a *"true or false" experiment* is thus associated with a *projection operator.* In particular, the proposition $A < \lambda$ is represented by the projection $\theta(\lambda - \widehat{A})$ onto the eigenvalues $a_\alpha < \lambda$. We shall denote by θ the Heaviside step function, $\theta(x) = 1$ when $x > 0$, $\theta(x) = 0$ when $x < 0$.

If two observables commute, they are *compatible* in the sense that they can be *controlled simultaneously* in a preparation or measurement process. A *complete set of commuting observables* is characterized by the following property. Any set of eigenvalues of these observables defines a *single* common eigenket – apart from a multiplying phase factor.

Not every Hermitean operator in Hilbert space necessarily represents a physical observable. On the practical level it is certainly not true that every quantity can, in fact, be controlled. This is especially true for the large systems studied in statistical mechanics where the number of accessible quantities is small. Moreover, even as a matter of principle, there are – microscopic – systems for which some Hermitean operators do not make sense as observables. For instance, in the case of a system of *two indistinguishable particles* 1 and 2, any physical observable is necessarily symmetric under an exchange of 1 and 2, even if it is mathematically legitimate to deal with other operators such as the position of particle 1. In the case of a particle of charge q in an *electromagnetic field* the position $\widehat{\boldsymbol{r}}$ is a physical observable, but the momentum $\widehat{\boldsymbol{p}}$ is not an observable. In fact, the choice of the *gauge* in which the system is described is arbitrary and formal differences occurring in the theoretical description when we carry out a gauge transformation are not observable. Especially, the momentum $\widehat{\boldsymbol{p}}$ and the vector potential $\boldsymbol{A}(\widehat{\boldsymbol{r}})$ change under such a transformation and are thus just mathematical tools; only the invariant combination $\widehat{\boldsymbol{p}} - q\boldsymbol{A}(\widehat{\boldsymbol{r}})$ is an observable. A final example is given by the *superselection rules* associated, for instance, with the charge, the baryon number, or the parity of the angular momentum. The Hilbert space is then a direct sum of spaces $\mathcal{E}_{\mathrm{H}}^{1} \oplus \mathcal{E}_{\mathrm{H}}^{2} \oplus \cdots$ such that the operators which have matrix elements across two of them are not observables. In particular, all physical observables commute with the charge, a quantum number distinguishing the subspaces $\mathcal{E}_{\mathrm{H}}^{1}$, $\mathcal{E}_{\mathrm{H}}^{2}$, $\ldots$. Under those conditions the superposition principle does not hold since one cannot prepare any state vector which would be a linear combination of kets associated with different charge values.

2.1.4 Pure States, Preparations and Measurements

Most elementary courses assume that the state of a system is *at all times represented by a ket* $|\psi\rangle$, normalized according to $\langle\psi|\psi\rangle = 1$. In fact, kets represent only a special class of states which we shall denote by the names *pure states* or *micro-states*, in order to distinguish them from the more general states introduced in § 2.2.1. Identifying the pure states with the vectors in a Hilbert space constitutes the *superposition principle*, according to which each normalized linear combination of pure states is again a pure state.

A *preparation* of a system is a process that produces it in a certain state thanks to the control of an observable $\widehat{A}$, or of several commuting observables, associated with the preparation apparatus. This control is generally carried out through *filtering* of an eigenvalue a_α of $\widehat{A}$, with all other eigenstates eliminated. The pure states are those which result from *complete preparations* where one controls a complete set of commuting observables. In that case the ket $|\psi\rangle$, produced by the preparation, is the common eigenvector of those observables, which is defined by the chosen set of eigenvalues.

A *measurement* aims to determine the value of the quantity A in a state, prepared beforehand. One is always dealing with a *statistical* process, the probabilistic nature of which is implied by quantum mechanics. Even if the preparation is complete and one knows that the system is in the micro-state $|\psi\rangle$ just before the measurement, the outcome of the latter is not certain. In fact, a ket does not describe the properties of *one* single system, but *of a statistical ensemble of systems* prepared under the same conditions. Knowing $|\psi\rangle$ then determines not an exact value of the quantity A, but its *expectation value* given by

$$\langle A\rangle = \langle\psi|\widehat{A}|\psi\rangle. \tag{2.21}$$

This is a basic expression in the formalism of the kets, since together with the equation of motion it makes it possible to make predictions about any physical quantity A for a system in the state $|\psi\rangle$.

The general expression (2.21) implies the existence of *statistical fluctuations* of a quantum origin. They are, in fact, given by the variance of A,

$$\langle A^2\rangle - \langle A\rangle^2 = \langle\psi|\big(\widehat{A} - \langle A\rangle\big)^2|\psi\rangle,$$

which vanishes only when $|\psi\rangle$ is an eigenket of $\widehat{A}$, with the eigenvalue $\langle A \rangle$. In particular, if two observables $\widehat{A}$ and $\widehat{B}$ do not commute, and $[\widehat{A}, \widehat{B}] = \mathrm{i}\widehat{C}$, we have *Heisenberg's uncertainty inequality*

$$\langle A^2 \rangle \langle B^2 \rangle \geq \frac{1}{4} \left[\langle C \rangle^2 + \langle AB + BA \rangle^2 \right].$$

Moreover, expression (2.21) *implies the quantization laws*. Indeed, the expectation value of any function $f(\widehat{A})$ equals

$$\langle f(A) \rangle = \langle \psi | f(\widehat{A}) | \psi \rangle,$$

and knowing it for any f is equivalent to knowing the probability distribution of A. In particular, if we choose for $f(x)$ a function which vanishes for all eigenvalues $x = a_\alpha$, the expectation value of $f(\widehat{A})$ vanishes. This implies that the probability distribution of A vanishes outside the points a_α of the spectrum; hence only the values a_α can be obtained in a measurement.

Finally, the probability $\mathcal{P}(a_\alpha)$ that the measurement provides one or other of the possible values a_α also appears as a consequence of (2.21). In fact, the *projection* $\widehat{P}_\alpha$ *onto the eigenvalue* a_α is an observable which takes the value 1, if one finds the value a_α, and the value 0 otherwise. Its expectation value,

$$\mathcal{P}(a_\alpha) = \langle \psi | \widehat{P}_\alpha | \psi \rangle, \tag{2.22}$$

can thus be identified with the probability for finding a_α. This expresses the *spectral decomposition law*. Conversely, (2.21) follows from (2.13) and (2.22).

Immediately following a measurement the system is no longer in the same state as before. Some measurements, such as the detection of a particle, may even destroy the system; at the other extreme, *ideal measurements* perturb it as little as possible (§ 2.2.5). However, since finding a_α is a probabilistic phenomenon, even an ideal measurement process produces, in general, a change in the state of the system which cannot be foreseen *a priori*. The *wavepacket reduction law* states that, if one has observed the eigenvalue a_α in an ideal measurement of A and if the system was in a pure state $|\psi\rangle$ before its interaction with the measuring apparatus, it will afterwards be in the pure state

$$\widehat{P}_\alpha |\psi\rangle \left[\langle \psi | \widehat{P}_\alpha | \psi \rangle \right]^{-1/2}. \tag{2.23}$$

This property enables one to use an ideal measuring process with filtering as a preparation.

2.1.5 Evolution, Transformations, Invariances

When the *evolution* of a system is known on the microscopic scale, it is generated by the *Hamiltonian* $\widehat{H}$, which is a Hermitean operator. The change with time of a micro-state $|\psi\rangle$ of this system is governed by the *Schrödinger equation*

$$\mathrm{i}\hbar \frac{d}{dt} |\psi(t)\rangle = \widehat{H} |\psi(t)\rangle. \tag{2.24}$$

Formally, we can integrate this equation to obtain

$$|\psi(t)\rangle \;=\; \widehat{U}(t)\,|\psi(0)\rangle, \tag{2.25}$$

where $|\psi(0)\rangle$ is the micro-state prepared at the initial time $t = 0$, and where $\widehat{U}(t)$ is the unitary *evolution operator*, defined by the equations

$$i\hbar\,\frac{d\widehat{U}}{dt} \;=\; \widehat{H}\,\widehat{U}, \qquad \widehat{U}(0) \;=\; \widehat{I}. \tag{2.26}$$

If $\widehat{H}$ does not explicitly depend on the time, $\widehat{U}(t)$ is given by

$$\widehat{U}(t) \;=\; \mathrm{e}^{-i\widehat{H}t/\hbar}. \tag{2.27}$$

The Hamiltonian of an *isolated* system is, in general, independent of the time. Nevertheless, we shall have to consider systems in interaction with sources exerting *known external forces* which may depend on the time. For instance, this is the case when we have a fluid in a cylinder, compressed by a moving piston, or a substance in an oscillating electromagnetic field. The evolution of the system itself remains governed by (2.24), but with a *Hamiltonian which depends on the time* through parameters controlled from the outside.

The evolution (2.24) of a micro-state implies that the expectation value (2.21) of an observable $\widehat{A}$, in general, changes with time. In order to find this evolution we need the conjugate of Eq.(2.24), that is,

$$-i\hbar\,\frac{d}{dt}\,\langle\psi(t)| \;=\; \langle\psi(t)|\,\widehat{H}, \tag{2.28}$$

and we find

$$i\hbar\,\frac{d}{dt}\,\langle\psi(t)|\widehat{A}|\psi(t)\rangle \;=\; \langle\psi(t)|(\widehat{A}\widehat{H} - \widehat{H}\widehat{A})|\psi(t)\rangle + \left\langle\psi(t)\left|\frac{\partial\widehat{A}}{\partial t}\right|\psi(t)\right\rangle.$$

The last term, which is quite often absent, takes the possible explicit time-dependence of the observable $\widehat{A}$ into account. From this equation we find the *Ehrenfest theorem*

$$\boxed{\frac{d}{dt}\,\langle A\rangle \;=\; \left\langle\frac{1}{i\hbar}\,[\widehat{A},\widehat{H}]\right\rangle + \left\langle\frac{\partial A}{\partial t}\right\rangle}\;, \tag{2.29}$$

which relates the change in time of the expectation value of a physical quantity to the expectation value of its commutator with the Hamiltonian.

To describe the evolution of a system we have used so far the Schrödinger picture. As any measurable quantity involves both a state and an observable in the combination (2.21), it amounts to the same whether we change $|\psi\rangle$ into $\widehat{U}|\psi\rangle$, keeping $\widehat{A}$ unchanged, or change, with time, the observables $\widehat{A}$ into $\widehat{U}^\dagger\widehat{A}\widehat{U}$, leaving $|\psi\rangle$ unchanged. In the Heisenberg picture, defined in this way, $|\psi\rangle$ represents the state at all times, the evolution is reflected through the Heisenberg equation $i\hbar\, d\widehat{A}/dt = [\widehat{A}, \widehat{H}]$ for the change in the observables, and (2.29) follows directly.

The evolution is a special case of a transformation of the state vector of a system; it corresponds to displacements with time. More generally, the laws of quantum mechanics must specify how the observables and the states of a system transform under physical operations such as translations, rotations, dilatations, parity changes, and so on. Actually, we have defined a ket of $\mathcal{E}_{\mathrm{H}}$ as representing a state, not in an absolute sense, but in some given frame of reference. A so-called *passive transformation* is thus induced by changing the reference frame – or the units of measurement, the sign convention for the charges, and so on. The spaces $\mathcal{E}_{\mathrm{H}}$ associated with the two frames of reference are isomorphic; they can be identified, and the transformation is described as a mapping from the ket $|\psi\rangle$ in the first frame to the ket $|\psi'\rangle$ representing the *same* state of the system in the second frame. Usually, this mapping is *linear* and *unitary*, $|\psi'\rangle = \widehat{U}|\psi\rangle$. The observables attached to the reference frame, such as the position with respect to the origin, remain unchanged, whereas the absolute observables transform according to $\widehat{A}' = \widehat{U}\widehat{A}\widehat{U}^\dagger$. The unitarity of $\widehat{U}$ implies that any *algebraic relation* connecting observables *remains unchanged.* For instance, a *parity* transformation or reflection in space changes the signs of $\widehat{\boldsymbol{r}}$ and $\widehat{\boldsymbol{p}}$, but the relation $[\widehat{\boldsymbol{r}}, \widehat{\boldsymbol{p}}] = i\hbar$ does not change. Transformations can alternatively be considered as being *active*, that is, applied to the system itself, while the reference frame remains unchanged.

Systematically taking advantage of *symmetries* or *invariances* under transformations is, together with the use of statistical methods, an efficient way to study complex systems. We shall see in Chap.14 how non-equilibrium thermodynamics uses it to reach non-trivial results. Let us here already mention *conservation laws*, which follow from a symmetry or invariance principle and on which the idea of thermodynamic equilibrium relies (Chaps.4 and 5). For instance, thermodynamics stresses *conservation of energy.* This is associated with the invariance of the theory under translations in time; indeed, the evolution (2.24) is in this case produced by a time-independent Hamiltonian $\widehat{H}$, and Ehrenfest's equation (2.29) then implies that $d\langle\widehat{H}\rangle/dt = 0$. *We find similarly a conservation law corresponding to each invariance.* In the simplest cases, invariance under a time-independent transformation $\widehat{U}$ is reflected by the invariance $\widehat{H} = \widehat{U}\widehat{H}\widehat{U}^\dagger$ of the Hamiltonian. It follows that $[\widehat{U}, \widehat{H}] = 0$ and hence, if we use (2.29), that the expectation value of any function $f(\widehat{U})$ remains constant with time. For instance, *conservation of the number of particles* $\widehat{N}$ is associated with the invariance of the theory under the transformations $\widehat{U} = e^{i\lambda\widehat{N}}$ which change the relative phases of the kets describing systems with different numbers of particles. In the case of Galilean invariance, the argument is changed, as $\widehat{U} = \exp[i(\boldsymbol{v} \cdot \{\widehat{\boldsymbol{P}}t - \widehat{\boldsymbol{R}}M\})/\hbar]$ depends on the time; $\widehat{\boldsymbol{P}}$ is the total momentum, M the total mass, and $\widehat{\boldsymbol{R}}$ the position of the centre of mass. Here we have $[\widehat{U}, \widehat{H}] + i\hbar d\widehat{U}/dt = 0$, and the conserved quantity is $\langle \widehat{\boldsymbol{P}}t - \widehat{\boldsymbol{R}}M \rangle$.

When the transformations $\widehat{U}$ form a continuum which includes the identity $\widehat{I}$, they can be produced as a succession of infinitesimal transformations represented by a Hermitean *generator*. The *Hamiltonian* appears in this way as the *generator of the evolution operator* (2.26). Similarly, an *infinitesimal spatial translation* displacing the reference frame by $\boldsymbol{\delta r}$ transforms in the $\boldsymbol{r}$-representation the wavefunction $\psi(\boldsymbol{r})$ into $\widehat{U}\psi(\boldsymbol{r}) = \psi(\boldsymbol{r} + \boldsymbol{\delta r}) = \psi(\boldsymbol{r}) + \big(\boldsymbol{\delta r} \cdot \nabla\psi(\boldsymbol{r})\big)$. It is thus represented by the translation operator $\widehat{U} = \widehat{I} + \frac{i}{\hbar}(\widehat{\boldsymbol{P}} \cdot \boldsymbol{\delta r})$ so that *the total momentum* $\widehat{\boldsymbol{P}}$ *is the generator of translations.* The invariance under translation, $[\widehat{\boldsymbol{P}}, \widehat{H}] = 0$, is then equivalent to the *conservation of the momentum* $\langle \boldsymbol{P} \rangle$. The *rotation operators* similarly are generated by the *angular momentum operators* and invariance under rotation is equivalent to conservation of angular momentum.

When a system of *identical particles* is described by labeling its particles, a permutation of their indices is a transformation with which a symmetry law is associated. In fact, not only $\widehat{H}$, but all physical observables must commute with this permutation; this just expresses the *indistinguishability* of the particles in all experiments (§ 10.1.1). The *Pauli principle* (§ 10.1.2) postulates that, depending on the nature of the particles, the Hilbert space $\mathcal{E}_\mathrm{H}$ contains only kets which are either *symmetric* or *antisymmetric* under the exchange of the indices of two indistinguishable particles, that is, they remain unchanged or change their sign. In Chaps.10 to 13 we shall review many important consequences of the Pauli principle in statistical mechanics. *Bose-Einstein statistics* must be used for indistinguishable particles with integer spin, *bosons*, for which the kets are *symmetric*, and *Fermi statistics* for indistinguishable particles with half-odd-integral spin, *fermions*, for which the kets are *antisymmetric*.

For completeness let us mention here another category of transformations which are represented by *antilinear* and *unitary* operators. The prototype is *time reversal*; from a passive point of view it describes a change in the convention of the sign of t with respect to the initial time. This transformation amounts to changing the signs of the momentum and spin observables without changing the position observables; these operations must be accompanied by changing i into $-$i as well as inverting the magnetic fields. Such an anti-unitary transformation associates $c_1^*|\psi_1'\rangle + c_2^*|\psi_2'\rangle$ with $c_1|\psi_1\rangle + c_2|\psi_2\rangle$, if $|\psi'\rangle$ is associated with $|\psi\rangle$; it preserves the commutation relations and the algebra of the observables. One can show that, conversely, the only transformations of observables which leave their algebra and their traces unchanged are the unitary and anti-unitary transformations.

2.1.6 Shortcomings of the State Vector Concept

For various reasons the representation of states by kets is not completely satisfactory. This representation is not unique, since a state vector is only defined within an *arbitrary phase vector*, which drops out of (2.21). This phase has no more physical meaning than the choice of gauge in electromagnetism. Note, however, that the relative phase of two kets is important when we superpose them.

The ket formalism is inadequate when we want to represent the state of a *subsystem*. Consider an ideal preparation of a composite system a+b, which leaves it in a pure state,

$$|\psi\rangle \equiv \sum_{k,l} |k_a l_b\rangle \langle k_a l_b|\psi\rangle, \tag{2.30}$$

of the $\mathcal{E}_H = \mathcal{E}_H^a \otimes \mathcal{E}_H^b$ space. Using (2.12) we can construct the set of expectation values $\langle\psi|\widehat{A}_a \otimes \widehat{I}_b|\psi\rangle$ of the observables $\widehat{A}_a$ pertaining to the subsystem a. If there existed, for the subsystem a, some state vector $|\psi_a\rangle$ of $\mathcal{E}_H^a$ describing the same physics as $|\psi\rangle$ for the measurements carried out on a, it would have to satisfy the relations

$$\langle\psi_a|\widehat{A}_a|\psi_a\rangle = \langle\psi|\widehat{A}_a \otimes \widehat{I}_b|\psi\rangle, \qquad \forall\ \widehat{A}_a.$$

However, there is, in general, no ket $|\psi_a\rangle$ satisfying these conditions, as Exercs.2a and 3b show. It is thus *impossible to describe the state of a part of a quantum system in the formalism of kets.* Two systems which have interacted in the past and thus have become correlated cannot separately be represented by kets: the state vector language does not allow us to isolate a system from the rest of the Universe.

In particular, the *theory of measurement* in quantum mechanics implies that one treats the object of the experiment together with the measuring apparatus during their interaction as a single system; afterwards one separates them in order to talk about the state of the object. This separation is precluded by the representation of states by kets.

Moreover, the ket formalism is not suited to retrieve *classical mechanics* in the $\hbar \to 0$ limit, starting from quantum mechanics. In general, wavefunctions do not have a classical limit: a finite momentum gives rise to oscillations with an infinite wavenumber.

Finally, it is necessary to change the formalism to treat circumstances where an ordinary random feature occurs which must be added to the intrinsic random nature of quantum mechanics. This occurs for *incompletely prepared systems* like unpolarized or partly polarized spins. Such a situation is the rule in statistical physics, since for macroscopic systems one controls only a few observables, far fewer than those of a complete set of commuting observables. A *random evolution* where $\widehat{H}$ is badly known can also not be described in the ket language. Section 2.2 will deal with all these requirements, both for microscopic systems and for the macroscopic systems which we want to study.

It was the impossibility to describe a subsystem by a state vector – one of the difficulties of the state vector formalism – which led Landau in 1927 to introduce the density operator concept. On the other hand, von Neumann was led to this both in order to apply quantum mechanics to macroscopic systems and in order to construct a theory of measurement. He also noted the ambiguity of the representation (2.33).

2.2 Quantum Systems: Density Operators

2.2.1 Pure States and Statistical Mixtures

We shall in the present section construct a new formalism of quantum mechanics by enlarging the concept of a state. We have just seen that the kets represent states which are known as well as is allowed by quantum mechanics. Depending on the context we call them *pure states* when we want to stress that their preparation was ideal, or *micro-states* when we use them as the building blocks for constructing a macro-state.

In thermodynamics or in statistical physics we want to describe statistical ensembles of systems which are defined solely by the specification of macroscopic variables. To simplify the language, we shall usually talk about "*a system*" to denote *an element of a statistical ensemble of systems*, all prepared under the same conditions. Such a system is on the microscopic scale represented by a *macro-state*, that is, by a state which is not completely known, or ideally prepared. The representative ket is thus not completely defined; all we can give is the probability that the system is described by one state vector or another. The multiplicity of the possible ket is reflected by the terminology "*statistical mixture*", an alternative name for a macro-state.

A macro-state thus appears as the set of possible micro-states $|\psi_\lambda\rangle$ each with its own probability q_λ for its occurrence. The probabilities q_λ are positive and normalized,

$$\sum_\lambda q_\lambda \;=\; 1. \tag{2.31}$$

The various micro-states $|\psi_\lambda\rangle$ are arbitrary kets of $\mathcal{E}_\mathrm{H}$ which are normalized, $\langle\psi_\lambda|\psi_\lambda\rangle = 1$, but not necessarily orthogonal to one another. The values of λ can be discrete or continuous; in the latter case, the notation $\sum_\lambda \ldots$ must be understood to mean $\int d\lambda \ldots$ and q_λ must be interpreted as a general normalized measure in the space $\mathcal{E}_\mathrm{H}$.

Because of the quantum nature of the kets $|\psi_\lambda\rangle$, the numbers q_λ can not truly be interpreted as ordinary probabilities. In classical probability theory the various possible events to which one can assign probabilities must be mutually exclusive. Here, the various possible micro-states of $\mathcal{E}_{\rm H}$ are *distinguishable*, but *not exclusive*: the overlap $|\langle\varphi|\psi\rangle|^2$ represents the probability that the system be observed in $|\varphi\rangle$, if it had been prepared in $|\psi\rangle$. The set of micro-states, obtained in a given ideal preparation, forms an orthonormal base of exclusive events, but there exist other micro-states which are linear superpositions of these and which can be produced by another preparation apparatus. If then the $|\psi_\lambda\rangle$ are not orthonormal, we must understand by the weights q_λ the relative *frequencies* in a population which mixes these various micro-states. Expression (2.34) for the mean value of an observable thus combines quantum expectation values with a weighted average. Anyway, we shall see that in a statistical mixture the micro-states $|\psi_\lambda\rangle$ hardly have a meaning. The concept of a density operator will enable us to get rid of these difficulties by introducing a *new probability theory* imposed by quantum mechanics.

2.2.2 The Expectation Value of an Observable

Let us consider a macro-state described by a set of micro-states $|\psi_\lambda\rangle$ with probabilities q_λ. This set is represented by a statistical ensemble of systems, all prepared under similar conditions. In this ensemble a fraction q_λ is in the ket $|\psi_\lambda\rangle$. For them the expectation value $\langle A\rangle$ of a physical quantity A equals $\langle\psi_\lambda|\widehat{A}|\psi_\lambda\rangle$ by virtue of (2.21). For the whole population represented by our macro-state we have thus for the expectation value of A:

$$\langle A\rangle = \sum_\lambda q_\lambda \, \langle\psi_\lambda|\widehat{A}|\psi_\lambda\rangle. \tag{2.32}$$

In order to evaluate (2.32) we note that, taking (2.15) into account, (2.32) is the trace of the operator

$$\sum_\lambda |\psi_\lambda\rangle \, q_\lambda \, \langle\psi_\lambda| \, \widehat{A}.$$

Let us now associate with the macro-state considered, which so far was characterized by the possible micro-states $|\psi_\lambda\rangle$ and their probabilities q_λ, the *density operator*

$$\widehat{D} = \sum_\lambda |\psi_\lambda\rangle \, q_\lambda \, \langle\psi_\lambda|. \tag{2.33}$$

We can then rewrite Eq.(2.32) in the form

$$\boxed{\langle A\rangle = \mathrm{Tr}\, \widehat{D}\widehat{A}} \; . \tag{2.34}$$

We are thus led to represent the macro-state by the density operator $\widehat{D}$ or, in a given base $\{|k\rangle\}$, by the *density matrix* $\langle k|\widehat{D}|k'\rangle$. The results (2.34) of any possible measurements at a given instant will only involve $\widehat{D}$. *All information about the macro-state of the system is contained in the density operator* $\widehat{D}$. This density operator, which represents the macro-state, *combines* in a compact form its probabilistic nature due to *quantum mechanics*, and connected with the $|\psi_\lambda\rangle$, with that coming from our *incomplete knowledge* of the micro-state, and included through the q_λ.

Several seemingly different descriptions or preparations of statistical mixtures can lead to the same density operator. Consider, for instance, a spin-$\frac{1}{2}$ with a probability $\frac{1}{2}$ of being oriented in the $+z$-direction and the same probability of being oriented in the $-z$-direction. Its density operator is (Exerc.2a)

$$\widehat{D} = \tfrac{1}{2}\left[\,|+\rangle\langle+| + |-\rangle\langle-|\,\right] = \tfrac{1}{2}\widehat{I}. \tag{2.35}$$

One finds the same operator if the spin is oriented along the $+x$- or the $-x$-directions with probabilities $\frac{1}{2}$, since

$$\frac{1}{2}\left[\frac{1}{\sqrt{2}}\big(|+\rangle+|-\rangle\big)\big(\langle+|+\langle-|\big)\frac{1}{\sqrt{2}}\right.$$
$$\left. +\frac{1}{\sqrt{2}}\big(|+\rangle-|-\rangle\big)\big(\langle+|-\langle-|\big)\frac{1}{\sqrt{2}}\right] = \frac{1}{2}\widehat{I}. \tag{2.35$'$}$$

One finds also the same operator if the probability that the spin is prepared to lie within a solid angle $d^2\boldsymbol{\omega}$ around the direction $\boldsymbol{\omega}$ is equal to $d^2\boldsymbol{\omega}/4\pi$. In fact, let us denote by $|\boldsymbol{\omega}\rangle$ the ket representing a spin oriented in the direction $\boldsymbol{\omega}$; apart from a phase, it can be obtained from $|+\rangle$ through a rotation which brings the z-axis along $\boldsymbol{\omega}$. (We are dealing with the case where $\lambda = \boldsymbol{\omega}$ is a continuously changing variable and where the $|\psi_\lambda\rangle = |\boldsymbol{\omega}\rangle$ are not orthogonal onto one another.) One can by a direct calculation check that

$$\frac{d^2\boldsymbol{\omega}}{4\pi}\,|\boldsymbol{\omega}\rangle\langle\boldsymbol{\omega}| = \frac{1}{2}\widehat{I}; \tag{2.35$''$}$$

more simply, one can note that the operator (2.35″) is invariant under a rotation and has unit trace and that, apart from a multiplicative factor, $\widehat{I}$ is the only rotationally invariant operator. In a Stern-Gerlach type experiment, one may imagine that before any preparation the spin, oriented completely at random, is described by the probability (2.35″), which is clearly isotropic. However, if we combine the two emerging beams without filtering them, it looks as if the state of the spin is described either by (2.35) or by (2.35′), depending on the orientation of the apparatus. In actual fact, however, in contrast to what one expects intuitively, all these various situations are completely indistinguishable. The *equivalent descriptions* (2.35), (2.35′), and (2.35″) all represent the *same macro-state*, that of an *unpolarized spin.*

Let us, more generally, consider two equivalent macro-states, $\{|\psi_\lambda\rangle, q_\lambda\}$ and $\{|\psi'_\mu\rangle, q'_\mu\}$, such that

$$\sum_\lambda |\psi_\lambda\rangle q_\lambda\langle\psi_\lambda| = \sum_\mu |\psi'_\mu\rangle q'_\mu\langle\psi'_\mu| = \widehat{D}.$$

According to (2.34), *no experiment can enable us to distinguish between them.* The two descriptions $\{|\psi_\lambda\rangle, q_\lambda\}$ and $\{|\psi'_\mu\rangle, q'_\mu\}$ thus correspond to the *same physical reality* just as two kets which differ by a phase factor describe one and the same pure state. The existence of apparently different descriptions of the same statistical ensemble of systems indicates that the state vector formalism becomes inadequate, as soon as the ket of the ensemble is not uniquely defined. Such a description cannot be given a physical interpretation, and the only *meaningful representation of macro-states is given by density operators.*

The density operator formalism includes pure states as special cases. A pure state, represented in the formalism of § 2.1.4 by the ket vector $|\psi\rangle$ in the Hilbert space $\mathcal{E}_{\rm H}$, is now represented by an operator in that space, namely, the density operator

$$\widehat{D} = |\psi\rangle\langle\psi|, \tag{2.36}$$

which, moreover, reduces to the projection onto $|\psi\rangle$. Note that the description of a micro-state by the density operator $|\psi\rangle\langle\psi|$ rather than by the ket $|\psi\rangle$ has the advantage of eliminating the arbitrary phase which is inherent to the ket formalism.

Finally, we see that the example of Chap.1 falls into the new formalism. The density operator representing the macro-state defined in § 1.1.3 can, in fact, be written as

$$\widehat{D} = \sum |\sigma_1, \ldots, \sigma_N\rangle\, p(\sigma_1, \ldots, \sigma_N)\, \langle\sigma_1, \ldots, \sigma_N|, \tag{2.37}$$

where the sum is over $\sigma_1 = \pm 1$, ..., $\sigma_N = \pm 1$. In Chap.1 we restricted ourselves (see the remark at the end of § 1.1.3) to special density operators and observables that were diagonal in the $|\sigma_1, \ldots, \sigma_N\rangle$ representation, in which case (2.34) reduces to (1.3). In fact, this restriction was not crucial for the study of thermal equilibrium: as we shall see, the thermal equilibrium macro-state is always represented by a density operator which is diagonal in a representation where the Hamiltonian is diagonal.

2.2.3 Characteristic Properties of Density Operators

In the foregoing we have introduced the density operator $\widehat{D}$ representing a macro-state by means of (2.33). This expression has the following properties:

(a) $\widehat{D}$ *is Hermitean*: $\widehat{D} = \widehat{D}^\dagger$. This follows from the fact that the q_λ are real. By using (2.34), (2.17), and (2.16), we see that this property means that the *expectation value of any observable is real.*

(b) $\widehat{D}$ *has unit trace*: $\mathrm{Tr}\,\widehat{D} = 1$. This follows from the definition (2.33), the normalization of the micro-states $|\psi_\lambda\rangle$, and the normalization (2.31) of the q_λ. This property means that the *expectation value of the unit operator* $\widehat{I}$ *is equal to* 1.

(c) $\widehat{D}$ *is non-negative:* $\langle\varphi|\widehat{D}|\varphi\rangle \geq 0, \quad \forall\ |\varphi\rangle \in \mathcal{E}_{\text{H}}$. This follows from the fact that the q_λ are positive, as it follows from (2.33) that we have $\langle\varphi|\widehat{D}|\varphi\rangle = \sum_\lambda q_\lambda |\langle\varphi|\psi_\lambda\rangle|^2$. This property means that the expectation value (2.34) of any positive operator is positive, or, if we bear in mind that the variance $\langle A^2\rangle - \langle A\rangle^2$ of A is the expectation value of $(\widehat{A} - \langle A\rangle)^2$, that *every variance is positive.*

Conversely, consider an operator $\widehat{D}$ which satisfies these properties. The Hermiticity of $\widehat{D}$ means that we can write it as a function of its orthonormalized eigenvectors $|m\rangle$ and its real eigenvalues:

$$\boxed{\widehat{D} = \sum_m |m\rangle\, p_m\, \langle m|} \ . \tag{2.38}$$

From (b) and (c) it follows that the numbers p_m are positive or zero, and that their sum equals 1. We can thus regard them as probabilities for the micro-states $|m\rangle$ so that (2.38) is a realization of (2.33) and $\widehat{D}$ is, indeed, a density operator. Note that in the diagonal representation (2.38) of $\widehat{D}$ the micro-states $|m\rangle$ are orthonormal, whereas in (2.33) the $|\psi_\lambda\rangle$ were not necessarily orthogonal onto one another.

The representation (2.38) of the density operator in terms of an orthonormal base $\{|m\rangle\}$ plays a special rôle amongst the various possible equivalent representations (2.33). Its arbitrariness is smaller, being reduced to unitary transformations in each of the subspaces for which the eigenvalues p_m are equal, as in the case of (2.35) and (2.35′). It is more convenient for practical purposes since the density matrix is diagonal if one chooses $\{|m\rangle\}$ for the representation base, as we did implicitly in Chap.1. *We shall use it whenever we need the explicit form of* $\widehat{D}$.

Finally, the physical interpretation of (2.38) is simpler than that of (2.33). In fact, imagine an ideal preparation process in which the eigenkets of the controlled observable are the $|m\rangle$. We can then regard $\widehat{D}$ as the density operator of an ensemble of systems which have interacted with the apparatus without filtering of the results; the fraction of systems for which we have found the result m is p_m. We can then interpret the p_m as probabilities for *exclusive events*, and the $|m\rangle$ as pure states which can be obtained by a preparation of the system. In the general case (2.33) the non-orthogonal kets $|\psi_\lambda\rangle$ are associated with distinguishable, but not exclusive, events – such as the polarization of a spin in arbitrary directions; they cannot be prepared by a single apparatus, but the q_λ can again be interpreted as the fraction of systems prepared to be in the micro-state $|\psi_\lambda\rangle$.

Even though the density operators and the observables have in common the property of being Hermitean operators in the Hilbert space $\mathcal{E}_{\text{H}}$ – forgetting about difficulties connected with the infinite dimensionality of this Hilbert space – one should note that they play a completely different rôle, not only in their physical meaning, but also in a mathematical sense:

$\widehat{D}$ defines a *quantum macro-state as a mapping* (2.34) *of the observables onto their expectation values* (end of § 2.2.7). Quantum statistical mechanics thus introduces a new kind of probability calculus where *random variables are replaced by non-commuting objects, the observables*, and where the *probability distribution is replaced by the density operator.*

Let us finally note that amongst the various possible density operators, satisfying the properties (a), (b), and (c), those which represent a *pure state* are characterized by the fact that *one of their eigenvalues equals* 1, while all others vanish. This special feature is reflected, for instance, by the relation $\widehat{D}^2 = \widehat{D}$; in this case $\widehat{D}$ is a projection.

2.2.4 The Density Operator of a Subsystem

Statistical mechanics often uses the concept of a subsystem, by taking out of a composite system one of its parts which is described independently of the remainder. For instance, one might be interested in the study of systems *in thermal contact with a heat bath* or in *volume elements* of an extensive substance. Under such circumstances and in other cases which we shall mention at the end of the present subsection, it is necessary, starting from the state of the composite system a+b, to eliminate the subsystem b in order to focus on the state of the system a in which we are interested. Let us therefore assume that the Hilbert space $\mathcal{E}_{\mathrm{H}}$ is the tensor product (§§ 2.1.1 and 2.1.2) of two spaces $\mathcal{E}_{\mathrm{H}}^{\mathrm{a}}$ and $\mathcal{E}_{\mathrm{H}}^{\mathrm{b}}$ spanned by the bases $\{|k_{\mathrm{a}}\rangle\}$ and $\{|l_{\mathrm{b}}\rangle\}$. A macro-state of the system is characterized by its density matrix $\langle k_{\mathrm{a}} l_{\mathrm{b}}|\widehat{D}|k'_{\mathrm{a}} l'_{\mathrm{b}}\rangle$ in $\mathcal{E}_{\mathrm{H}}$. Amongst the various observables $\widehat{A}$, those $\widehat{A}_{\mathrm{a}}$ that describe *measurements relating to the subsystem* a are described in $\mathcal{E}_{\mathrm{H}}$ by matrices such as (2.12) which represent $\widehat{A}_{\mathrm{a}} \otimes \widehat{I}_{\mathrm{b}}$. Their expectation values,

$$\begin{aligned} \langle A_{\mathrm{a}}\rangle &= \mathrm{Tr}\, \widehat{D}\big(\widehat{A}_{\mathrm{a}} \otimes \widehat{I}_{\mathrm{b}}\big) = \sum_{k_{\mathrm{a}},k'_{\mathrm{a}},l_{\mathrm{b}},l'_{\mathrm{b}}} \langle k_{\mathrm{a}} l_{\mathrm{b}}|\widehat{D}|k'_{\mathrm{a}} l'_{\mathrm{b}}\rangle\, \langle k'_{\mathrm{a}} l'_{\mathrm{b}}|\widehat{A}_{\mathrm{a}} \otimes \widehat{I}_{\mathrm{b}}|k_{\mathrm{a}} l_{\mathrm{b}}\rangle \\ &= \sum_{k_{\mathrm{a}},k'_{\mathrm{a}},l_{\mathrm{b}}} \langle k_{\mathrm{a}} l_{\mathrm{b}}|\widehat{D}|k'_{\mathrm{a}} l_{\mathrm{b}}\rangle\, \langle k'_{\mathrm{a}}|\widehat{A}_{\mathrm{a}}|k_{\mathrm{a}}\rangle, \end{aligned}$$

involve only the matrix in $\mathcal{E}_{\mathrm{H}}^{\mathrm{a}}$

$$\langle k_{\mathrm{a}}|\widehat{D}_{\mathrm{a}}|k'_{\mathrm{a}}\rangle = \sum_{l_{\mathrm{b}}} \langle k_{\mathrm{a}} l_{\mathrm{b}}|\widehat{D}|k'_{\mathrm{a}} l_{\mathrm{b}}\rangle,$$

or, in a form independent of the base, only the operator

$$\boxed{\widehat{D}_{\mathrm{a}} = \mathrm{Tr}_{\mathrm{b}}\, \widehat{D}}\,, \tag{2.39}$$

defined through (2.18) in the partial Hilbert space $\mathcal{E}_{\mathrm{H}}^{\mathrm{a}}$. This $\widehat{D}_{\mathrm{a}}$ operator is Hermitean, has unit trace, and is non-negative. One can thus define the *density operator* $\widehat{D}_{\mathrm{a}}$ *of the subsystem* a, starting from the density operator $\widehat{D}$

of the global system a+b. The expectation value of any operator $\widehat{A}_{\rm a}$ relating to this subsystem is, in fact, found from the equation

$$\langle A_{\rm a}\rangle = {\rm Tr}\,\widehat{D}\,\widehat{A}_{\rm a} = {\rm Tr}_{\rm a}\,\widehat{D}_{\rm a}\,\widehat{A}_{\rm a}, \tag{2.40}$$

without going beyond the subspace $\mathcal{E}_{\rm H}^{\rm a}$. For the set of measurements that only involve the subsystem a, one does not need to know the whole $\widehat{D}$: the partial density operator $\widehat{D}_{\rm a}$ suffices to characterize all the properties of this subsystem.

As an example consider two spins-$\frac{1}{2}$, a and b, coupled into a singlet state with zero angular momentum:

$$|\psi\rangle = \frac{1}{\sqrt{2}}\left(\,|+-\rangle - |-+\rangle\,\right).$$

The density operator $\widehat{D}$ of a+b is given by (2.36) so that the density operator of the spin a following from (2.39) is

$$\widehat{D}_{\rm a} = \tfrac{1}{2}\left(\,|+\rangle\langle+| + |-\rangle\langle-|\,\right). \tag{2.41}$$

It describes an unpolarized spin. Similarly, the spin b is in an unpolarized state. Note that the operator $\widehat{D}$ is not equal to the tensor product $\widehat{D}_{\rm a}\otimes\widehat{D}_{\rm b}$; it represents not only the individual states of the two spins a and b, described by $\widehat{D}_{\rm a}$ and $\widehat{D}_{\rm b}$, but also their correlations. It is important to note that (2.41) does not have the form (2.36) so that the state of the system a cannot be represented by a ket, even though a+b are represented in that way.

More generally, in agreement with the remarks in § 2.1.6, it is *impossible to describe the state of part of a quantum system, remaining in the framework of pure states.* Considering subsystems *forces* us to introduce the density operator concept (Exercs.2a and 3b): even if a system is in a pure state, its subsystems are, in general, statistical mixtures. A complete preparation for a composite quantum system is not complete for its subsystems, in contrast to what happens in classical physics.

Let us finally note that with our definitions, the subsystems of a given system are *not necessarily separated in space.* One may be dealing with *different degrees of freedom* chosen to be studied separately for the sake of simplicity. For instance, in Chap.1 we studied the spins of the paramagnetic ions of a solid. The density operator used, (2.37), represented the state of a subsystem of the solid consisting only of the spin degrees of freedom. The density operator of the solid would act in a much larger Hilbert space and would involve all the degrees of freedom of this solid, such as the vibrations of the atoms around their equilibrium positions. The partial density operator (2.37) should be regarded as the result of using the trace (2.39) to eliminate the irrelevant degrees of freedom.

This is a very general observation and we shall often have to consider density operators of subsystems. For instance, for a gas, we shall introduce the density operator relating to the degrees of freedom of the centres of mass of its molecules, and forget about the internal motions of the molecules; on the other hand, we shall also introduce the internal density operator for one molecule, which describes the behaviour of its elementary constituents relative to the centre of mass. In this way we can study separately one or other part of a system. In order that this separation be adequate, the coupling between these parts should be weak. Under such conditions exchanges of energy and angular momentum between them are similar to heat exchanges. Each part behaves like a system put in a heat bath, but in this case the bath is superimposed in space upon the system studied.

2.2.5 Measurements in Quantum Mechanics

The concept of a density operator and its use for describing the state of a subsystem can usefully clarify the theory of measurements in quantum mechanics.[2] The analysis of a measurement process assumes that one separates the system considered, that is, the apparatus plus the object, into two subsystems. Neither the apparatus, not the object will, in general, be in a pure state after the measurement, and one is led to describe them in terms of density operators.

In order better to understand the postulates for quantum measurements (§ 2.1.4) it is useful to make a mental decomposition of the experiment into two stages, one during which the apparatus and the object interact and then separate, and the other corresponding to the observation of the result. Before the first stage, the object and the apparatus are not correlated; the density operator of the system is the tensor product of those of the two parts. The interaction changes the density operator of the system, creating correlations between object and apparatus; this modification is specific of the observable that the apparatus is meant to measure. However, a measurement has another characteristic: after the interaction the apparatus and the object are *separated* and the only experiments that one eventually carries out refer to observables relating to either one or the other. One should thus replace the global density operator by the only relevant density operators, those of the *object and apparatus subsystems.*

Let us remember the postulate (2.22), (2.23) of the wavepacket reduction, which we shall justify at the end of the present subsection. If the object is initially in the pure state $|\psi_\lambda\rangle$, an ideal measurement of the observable $\widehat{A}$ forces it to change into one or other of the states $\widehat{P}_\alpha|\psi_\lambda\rangle\left[\langle\psi_\lambda|\widehat{P}_\alpha|\psi_\lambda\rangle\right]^{-1/2}$ with the respective probabilities $\mathcal{P}_\lambda(a_\alpha) = \langle\psi_\lambda|\widehat{P}_\alpha|\psi_\lambda\rangle$. This means that, forgetting about the apparatus, the first stage of the measurement before the result is observed replaces the initial pure state (2.36)

[2] J.A.Wheeler and W.H.Zurek (Eds.), *Quantum Theory and Measurement*, Princeton University Press, 1983; M.Cini and J.-M.Lévy-Leblond (Eds.), *Quantum Theory without Reduction*, Hilger – IOP Publishing, 1990.

$$|\psi_\lambda\rangle\langle\psi_\lambda| \tag{2.42}$$

of the object by the statistical mixture

$$\sum_\alpha \hat{P}_\alpha|\psi_\lambda\rangle\langle\psi_\lambda|\hat{P}_\alpha; \tag{2.43}$$

the normalization factor of each possible final micro-state has been compensated by its probability $\mathcal{P}_\lambda(a_\alpha)$. More generally, if before measurement the system is prepared in the state (2.42) with the probability q_λ, it will have the same probability q_λ for being in the final state (2.43). Altogether, *the interaction of the object with the apparatus measuring A and their subsequent separation* transform the initial density operator $\hat{D}$ of the object into

$$\hat{D}_A = \sum_\alpha \hat{P}_\alpha \hat{D} \hat{P}_\alpha. \tag{2.44}$$

In a base where $\hat{A}$ is diagonal this change amounts to *truncating* the density matrix, $\hat{D}$, by replacing all the off-diagonal elements which connect two different eigenvalues a_α and a_β by zero, while leaving all diagonal blocks relating to a given eigenvalue unchanged. If $\hat{D}$ commutes with $\hat{A}$, it is not changed by the operation (2.44).

The second stage, the *observation* of a particular result a_α on the measuring apparatus and selection of the state corresponding to α in (2.43) or (2.44), is of the same nature as the observation of any random process. The specifically quantum stage is not the *observation*, but rather the *separation* of the apparatus and the object, after the interaction, which changes the density operator of the object into (2.44), even if the result of the measurement has not been observed. In a base which diagonalizes $\hat{A}$, the final state selected by the observation of a_α is represented by the diagonal block, associated with a_α, in the truncated density matrix (2.44); taking the normalization into account we find that this final state is represented by the density operator

$$\hat{D}_\alpha = \frac{\hat{P}_\alpha \hat{D} \hat{P}_\alpha}{\mathrm{Tr}\, \hat{D} \hat{P}_\alpha}. \tag{2.45}$$

In terms of density operators we can finally express the postulate of the wavepacket reduction as follows. *If an ideal measurement of $\hat{A}$ carried out on a system in the macro-state $\hat{D}$ has given the result a_α, it leaves the system in the state* (2.45) which has thus been prepared. If there is no observation, the macro-state of the system after it has been separated from the measurement apparatus is (2.44). Of course, when we talk about *the* system, we are always dealing with a statistical ensemble of systems prepared under similar conditions.

The concept of a state as a mapping (2.34) of observables on their expectation values can, moreover, help us to understand the quantum phenomenon of wavepacket reduction, expressed by (2.44) and (2.45), and thus reduce the number of quantum-mechanical postulates to a minimum. Let us, first of all, note that the reasoning of § 2.1.4, which *showed the quantization* of the measurement results, can without any difficulties be extended to statistical mixtures by replacing (2.21) by (2.34) so that we have now

$$\langle f(A) \rangle = \mathrm{Tr}\, \widehat{D}\, f(\widehat{A}).$$

Similarly, the *probability* $\mathcal{P}(a_\alpha)$ that *the measurement gives the result* a_α is the expectation value (2.34) of the projection $\widehat{P}_\alpha$, which equals

$$\mathcal{P}(a_\alpha) = \mathrm{Tr}\, \widehat{D}\, \widehat{P}_\alpha, \tag{2.46}$$

replacing (2.22).

Let us next consider the statistical ensemble of systems initially described by $\widehat{D}$. We want to determine the final state $\widehat{D}_\alpha$ that, after we have measured A and selected the result a_α, we should attribute to our resulting subensemble of systems. We assume the measurement to be ideal, which implies that *repeating it*, after having observed a_α, *does not modify the state.* Any subsequent measurement of A will thus necessarily again give a_α. This can be expressed by the condition

$$\mathrm{Tr}\, \widehat{D}_\alpha \left(\widehat{I} - \widehat{P}_\alpha\right) = 0, \tag{2.47}$$

since the expectation value of the projection $\widehat{I} - \widehat{P}_\alpha$ is the probability that we observe something different from a_α. From the identity $\widehat{I} - \widehat{P}_\alpha = (\widehat{I} - \widehat{P}_\alpha)^2$, (2.47), and the cyclic invariance of the trace, it follows that $(\widehat{I} - \widehat{P}_\alpha)\widehat{D}_\alpha(\widehat{I} - \widehat{P}_\alpha)$ has zero trace. This operator must therefore vanish, since it is non-negative. The fact that $\widehat{D}_\alpha$ itself is non-negative then implies that its off-diagonal matrix elements connecting the two subspaces spanned by $\widehat{P}_\alpha$ and $\widehat{I} - \widehat{P}_\alpha$ are zero. (To prove this we only need note that any 2×2 diagonal submatrix taken from the matrix representing $\widehat{D}_\alpha$ in a representation where $\widehat{A}$ is diagonal is non-negative.) As a result, $\widehat{D}_\alpha$ contains only the diagonal block associated with a_α:

$$\widehat{D}_\alpha = \widehat{P}_\alpha\, \widehat{D}_\alpha\, \widehat{P}_\alpha.$$

The next stage consists in relating the $\widehat{D}_\alpha$ to $\widehat{D}_A$. The population which originally was described by $\widehat{D}$ has been split by the measurement into sub-populations, each associated with a particular result a_α; each sub-population contains a fraction of systems $\mathcal{P}(a_\alpha)$, given by (2.46). As the $\widehat{D}_\alpha$ realize the mapping $A \Rightarrow \langle A \rangle$ in the form (2.34) for the various sub-ensembles, whereas $\widehat{D}_A$ is associated with the whole ensemble, we get by weighting

$$\widehat{D}_A = \sum_\alpha \mathcal{P}(a_\alpha)\, \widehat{D}_\alpha. \tag{2.48}$$

There remains finally the task to express $\widehat{D}_\alpha$ and $\widehat{D}_A$ in terms of $\widehat{D}$. To do this we use the fact that commuting observables represent compatible physical quantities. Slightly generalizing the principle given in § 2.1.3, we postulate that in an ideal measurement the expectation value of any observable which commutes with $\widehat{A}$ is the same before and after measuring A. This identity applies to the whole population, originally described by $\widehat{D}$ and afterwards by $\widehat{D}_A$. Hence, if $\widehat{B}$ is an arbitrary observable, measuring A has no effect on the expectation value of $\widehat{P}_\alpha \widehat{B} \widehat{P}_\alpha$, which commutes with $\widehat{A}$. Using (2.48), we thus find

$$\mathrm{Tr}\,\widehat{D}\widehat{P}_\alpha\widehat{B}\widehat{P}_\alpha \;=\; \mathrm{Tr}\,\widehat{D}_A\widehat{P}_\alpha\widehat{B}\widehat{P}_\alpha \;=\; \mathcal{P}(a_\alpha)\,\mathrm{Tr}\,\widehat{D}_\alpha\widehat{P}_\alpha\widehat{B}\widehat{P}_\alpha,$$

and hence

$$\mathrm{Tr}\left[\widehat{P}_\alpha\widehat{D}\widehat{P}_\alpha - \mathcal{P}(a_\alpha)\widehat{D}_\alpha\right]\widehat{B} \;=\; 0.$$

As this identity holds for *any* Hermitean operator $\widehat{B}$, it gives us the required expressions (2.45) for $\widehat{D}_\alpha$ and (2.44) for $\widehat{D}_A$.

Altogether, *the wavepacket reduction is simply a consequence of the compatibility of ideal measurements on commuting observables*, of their *repeatability*, of the expression (2.34) for the expectation values, which can be regarded as the definition of $\widehat{D}$, and of the property that *the density operators are non-negative*, which expresses the fact that any variance is non-negative.

2.2.6 Evolution with Time

The evolution of the micro-states of an isolated quantum system (or a quantum system subject to forces which vary in a well defined manner) is governed by the Schrödinger equation (2.24) where the Hamiltonian $\widehat{H}$ (which may possibly depend on the time) is completely known. Each component $|\psi_\lambda\rangle$ of a statistical mixture evolves according to the same equation. The probabilities q_λ are independent of the time, as they simply describe the composition of the population of micro-states of which the macro-state is made up. Combining (2.33), (2.24), and (2.28) we find that the evolution with time of a density operator is governed by

$$\boxed{\frac{\mathrm{d}\widehat{D}}{dt} \;=\; \frac{1}{\mathrm{i}\hbar}\,[\widehat{H},\widehat{D}]}\;, \tag{2.49}$$

called the *Liouville-von Neumann equation.*

One can formally solve this equation, using the *evolution operator* $\widehat{U}(t)$ given by (2.26) or (2.27), in terms of the density operator $\widehat{D}(0)$, assumed to be known at time $t = 0$. In fact, one can easily check that

$$\widehat{D}(t) \;=\; \widehat{U}(t)\,\widehat{D}(0)\,\widehat{U}(t)^\dagger \tag{2.50}$$

is the solution of (2.49).

The *evolution of the expectation value of an observable* follows from (2.34) and (2.49) which yield

$$\frac{d}{dt}\,\mathrm{Tr}\,\widehat{D}\widehat{A} \;=\; \frac{1}{\mathrm{i}\hbar}\,\mathrm{Tr}\,[\widehat{H},\widehat{D}]\widehat{A} \;=\; \frac{1}{\mathrm{i}\hbar}\,\mathrm{Tr}\,\widehat{D}[\widehat{A},\widehat{H}].$$

We thus find again the *Ehrenfest equation* (2.29). In particular, the observables $\widehat{A}$ which commute with $\widehat{H}$ and which do not contain the time explicitly are *constants of the motion*: their expectation values $\langle A\rangle$ remain constant during the evolution of the system.

It is important to note that the Liouville-von Neumann equation (2.49) is written in the *Schrödinger picture*, where the states evolve and the observables remain fixed, provided they do not depend explicitly on the time. On the other hand, as we saw in § 2.1.5, in the *Heisenberg picture* these observables evolve according to

$$\frac{d\widehat{A}}{dt} = \frac{1}{i\hbar}[\widehat{A},\widehat{H}],$$

which is an equation with the opposite sign of (2.49), while the density operator of the system remains unchanged in time during the Hamiltonian evolution.

Considering statistical mixtures enables us to describe situations where the *evolution is not well known*, that is, situations which cannot be described in the conventional framework of quantum mechanics. This was the case in Chap.1 where the interactions between the magnetic ions were treated as small random perturbations producing transitions between the micro-eigenstates of the Hamiltonian of independent spins. More generally, this is the case for the macroscopic systems considered by statistical physics which are so complicated that the Hamiltonian always contains small random parts. Since the system is assumed to be *isolated* it evolves according to a law such as (2.49), but this evolution is not well known, that is, there exists a class of possible Hamiltonians $\widehat{H}_j$ each with a probability μ_j (j may be a continuous index). If we assume that the Hamiltonian is $\widehat{H}_j$, the evolution operator will be $\widehat{U}_j(t)$ and, according to (2.50), the density operator at time t will be equal to

$$\widehat{D}_j(t) = \widehat{U}_j(t)\,\widehat{D}(0)\,\widehat{U}_j(t)^\dagger,$$

where $\widehat{D}(0)$ is its initial value at time $t=0$. The expectation value of A at time t will then be $\mathrm{Tr}\,\widehat{D}_j(t)\widehat{A}$. This situation occurs with a probability μ_j so that the expectation value of A at time t for all possible evolutions, starting from time $t=0$, is

$$\langle A\rangle_t = \sum_j \mu_j\,\mathrm{Tr}\,\widehat{D}_j(t)\,\widehat{A} = \mathrm{Tr}\left[\sum_j \mu_j\widehat{D}_j(t)\right]\widehat{A}.$$

This equation defines the density operator at time t, which is thus given by

$$\widehat{D}(t) = \sum_j \mu_j\widehat{D}_j(t) = \sum_j \mu_j\,\widehat{U}_j(t)\,\widehat{D}(0)\,\widehat{U}_j(t)^\dagger. \tag{2.51}$$

Therefore, we can simply find the density operator resulting from a *random evolution* by taking the *average over the various possible evolutions.* Here again, the ket formalism would be inadequate, as a state which is pure at time $t=0$ will not remain pure for ever afterwards.

2.2.7 Summary: Reformulation of Quantum Mechanics

Starting from the idea that the state vector of a system is not well known when the preparation is not ideal, we have noted that the random nature of the predictions has two origins, on the one hand, the quantum nature of each of the possible micro-states $|\psi_\lambda\rangle$ and, on the other hand, poor knowledge of those micro-states which is reflected in the weights q_λ. These two kinds of probabilities merge together into a single object, the density operator $\widehat{D}$. Once we have made this synthesis, however, the distinction between the two different kinds of randomness can no longer be disentangled unambiguously. Not only is it clear, that the use of the $|\psi_\lambda\rangle$ kets and their probabilities q_λ was uselessly complicated, but also that these quantities $|\psi_\lambda\rangle$ and q_λ themselves hardly had a meaning because the decomposition of $\widehat{D}$ into (2.33) is not unique. It is therefore natural to forget our starting point and to *reformulate the laws of quantum mechanics in terms of the density operator* and no longer in terms of the kets. The main changes concern § 2.1.4 where expression (2.21) for the expectation value $\langle A\rangle$ of A is replaced by (2.34) and the wavepacket reduction (2.23) by (2.45). The transformations $\widehat{U}$ which acted in § 2.1.5 on the kets as $\widehat{U}|\psi\rangle$ act as $\widehat{U}\widehat{D}\widehat{U}^\dagger$ upon the density operators. For the sake of convenience we summarize below the principles upon which the remainder of this book will be based. Recall that quantum mechanics has a statistical nature: when we talk about *one* system, we are, in fact, dealing with an *ensemble* of systems prepared under similar conditions.

> With each system is associated a Hilbert space $\mathcal{E}_\text{H}$. The physical quantities A relating to this system have a probabilistic character and are represented by observables $\widehat{A}$ which are Hermitean operators in $\mathcal{E}_\text{H}$. The states of the system are represented by density operators $\widehat{D}$ which are Hermitean, non-negative, and have unit trace, while the expectation value of a quantity A in the state $\widehat{D}$ equals
>
> $$\langle A\rangle = \text{Tr}\,\widehat{D}\,\widehat{A}. \tag{2.34}$$
>
> The states of an isolated system with a known Hamiltonian $\widehat{H}$ evolve according to
>
> $$\frac{d\widehat{D}}{dt} = \frac{1}{i\hbar}\,[\widehat{H},\widehat{D}]. \tag{2.49}$$

Amongst the consequences of these principles we shall use the concept of *density operator of the subsystem* a *of a composite system* a+b, described by $\widehat{D}$,

$$\widehat{D}_\text{a} = \text{Tr}_\text{b}\,\widehat{D}, \tag{2.39}$$

the *Ehrenfest theorem* (2.29) which is especially useful to express the *conservation laws*, the form (2.45) of a state *prepared by an ideal measurement* and expression (2.51) for a *random evolution.*

Finally, in practice, for most of the applications we shall be dealing with, we shall know the *base* $\{|m\rangle\}$ *which diagonalizes* $\widehat{D}$ and the eigenvalues p_m of $\widehat{D}$. In that base (2.34) becomes

$$\boxed{\langle A \rangle = \sum_m p_m \, \langle m|\widehat{A}|m\rangle} \; . \tag{2.52}$$

The *expectation value* of A can then be calculated as if the micro-states $|m\rangle$ were *ordinary probabilistic events* with probabilities p_m and as if the quantity A were an ordinary *random variable* $\langle m|\widehat{A}|m\rangle$. The non-commutability will only play an important rôle when we are dealing with the evolution of a system or for problems involving strong interactions when it is impossible to diagonalize $\widehat{D}$. The reason why we have in the present section discussed a few quantum subtleties is not so much because we are going to use them, but in order to understand better the significance of the concepts introduced.

Note that for the large systems, in which we are interested, the micro-states $|m\rangle$ are characterized, as in the example (2.37) of Chap.1, by a large number of quantum numbers over which we must sum in order to calculate traces such as (2.34) or (2.52). Note also that, depending on which problem we are considering, the number N of particles may either be fixed, or occur amongst the quantum numbers m characterizing the $|m\rangle$ micro-states. In the first case the trace means a sum over *micro-states with fixed* N, whereas in the second case *the trace includes a summation over* N (§ 2.3.6).

The formulation of the principles of quantum mechanics in terms of density operators is useful as much for the physics of microscopic systems as for the statistical physics of macroscopic samples. We have seen that it enables us to describe states which are incompletely prepared or which are not well known, to consider subsystems, and to treat random evolutions; it also helps us better to understand the theory of quantum measurements and preparations. We have also noted that a density operator is a *faithful* representation of a state in the sense that it contains exactly the information needed to characterize the set of all measurements which one may perform on that state. Finally, the formalism is closer to experiments than the wavefunction or ket formalism: indeed, stress is laid here on the density operator as a *tool to predict the expectation values for all observables* in the form (2.34). Giving $\widehat{D}$ is equivalent to giving these expectation values, since its matrix elements $\langle k|\widehat{D}|k'\rangle$ can, according to (2.15) and (2.34), be identified with the expectation values of the dyadics $|k'\rangle\langle k|$.

The *algebraic formulation of quantum statistics* which was started by Jordan, von Neumann, and Wigner, is based upon this idea.[3] Even more general than the

[3] W.Thirring, *A Course in Mathematical Physics*, Springer, New York, Vol.3, 1981; G.G. Emch, *Mathematical and Conceptual Foundations of 20th Century Physics*, North-Holland, Amsterdam, 1984.

density operator formalism, it disregards Hilbert space. Its starting point is the algebra of the observables representing the physical quantities which are associated with the system (§ 2.1.3). *A state is then defined directly as the collection of expectation values* $\langle A \rangle$ *of all the observables*, that is, as a mapping from the algebra of observables onto the set of real numbers. This mapping must be linear, it must associate the number 1 with the unit observable, and a non-negative number to any observable of the form $\widehat{A}^2$. Such a definition has a conceptual advantage in that it exhibits the similarities and the differences between classical and quantum statistics: the classical *random variables* are here replaced by non-commuting *operators*, while the concept of a state as a mapping of the observables $\widehat{A}$ onto their expectation values $\langle A \rangle$ remains the same. Moreover, this definition of a state *involves directly the measurable quantities.*

If we disregard mathematical difficulties arising when the algebra of observables has an infinite dimension, we can represent the operators by matrices and in this way construct the Hilbert space. Taking as base in the vector space of the observables the dyadics $|k\rangle\langle k'|$, according to (2.5), and denoting the expectation value of $|k\rangle\langle k'|$ by $\langle k'|\widehat{D}|k\rangle$, we then realize the mapping $\widehat{A} \Rightarrow \langle A \rangle$ in the form (2.34) and we can associate a density operator with this mapping. However, the general definition of a state as a set of average values $\langle A \rangle$ does not need the introduction of a matrix representation for the algebra of observables. In fact, it is sufficient for the characterization of a state to specify the expectation values of a set of operators, forming an arbitrary base for this algebra; the expectation values are the *Liouville representation of the state* for the base considered (§ 2.1.2). Exercise 2b illustrates this point, showing that if we express for a spin-$\frac{1}{2}$ the observables in the base made up from the Pauli matrices and the unit operator, a state is simply represented by the polarization vector. Similarly, in the Wigner representation, the expectation values of the observables (2.8) constitute a representation of states which is well suited for the *classical limit* of quantum mechanics (§ 2.3.4).

In the algebraic formulation, the states appear as elements of a vector space which is the *dual* of that of the observables; as a matter of fact, the mapping $\widehat{A} \Rightarrow \langle A \rangle$ is linear, and the space which is the dual of the space of observables is, by its very definition, the set of linear forms. This remark fixes the difference in mathematical structure between the observables and the density operators mentioned at the end of § 2.2.3. For the finite systems in which we are interested in the present book, we shall be satisfied with representing these two kinds of objects by matrices. Nonetheless, if one wants to pass to the infinite volume limit, the Hilbert space becomes mathematically pathological whereas the algebra of localized observables remains manageable. The density operator concept loses its meaning and *must* be replaced by that of state as a *mapping*. The algebraic formulation also enables one to short-circuit the difficulties mentioned at the end of § 2.1.3. In what follows we avoid these difficulties by always studying large systems and letting their volumes tend to infinity at the end, and by working in the Hilbert space.

Note finally that the algebraic formulation of quantum mechanics not only makes the concept of state, as we have just seen, or the analysis of measuring processes (§ 2.2.5), but also the evolution equation and the Pauli principle, easier to understand. In the Heisenberg picture, the equation of motion for $\widehat{A}$ simply expresses the fact that the *algebraic relations* between the observables remain *unchanged*,since this invariance implies the existence of a unitary operator $\widehat{U}(t)$ which transforms $\widehat{A}(0)$ into $\widehat{A}(t) = \widehat{U}(t)^\dagger \widehat{A}(0)\widehat{U}(t)$. As to the Pauli principle (Chap.10), it follows from the hypothesis that *all* Hermitean matrices of the Hilbert space of a system of indistinguishable particles can represent physical observables.

2.3 Classical Systems: Densities in Phase

In the general chapters 3 to 6 we shall mainly use the quantum formalism which is simpler and more general. Nevertheless, for the sake of giving a coherent picture we right now introduce the formalism of classical statistical mechanics that we shall need in Chaps.7, 8, 9, and 15 for applications to the physics of gases and liquids. In fact, we must even in such cases take into account the quantum nature of the underlying microscopic physics.

2.3.1 Phase Space and Micro-states

Whereas the motion of electrons in atoms, the relative motions of atoms in molecules, or the properties of solids can only be described by quantum mechanics, the *translational degrees of freedom of the molecules in a gas or in a liquid* are described with great accuracy by the laws of classical mechanics. In such a classical system of N molecules the physical quantities which play the rôle of *classical observables* are the positions $\boldsymbol{r}_i$ $(i = 1, 2, \ldots, N)$ of their centres of mass, their momenta $\boldsymbol{p}_i$ and, more generally, *functions* $f(\boldsymbol{r}_1, \boldsymbol{p}_1, \cdots, \boldsymbol{r}_N, \boldsymbol{p}_N)$ *of these* $6N$ *variables.* (In fact, when the particles are indistinguishable, the physical quantities only consist of functions which are symmetric under permutations of the particles.)

A *micro-state*, which is a state of the system about which everything is known, is defined by giving the values of all the observables. It is thus characterized by *giving the* $6N$ *coordinates and momenta* of the particles. We introduce the *phase space* $\mathcal{E}_\mathrm{P}$, *a* $6N$*-dimensional space* with as coordinates $\boldsymbol{r}_1, \boldsymbol{p}_1, \ldots, \boldsymbol{r}_N, \boldsymbol{p}_N$. A micro-state is then represented by a point in phase space.

More generally, we shall show that when the temperature is sufficiently high some internal degrees of freedom of the gas molecules, such as the orientation of these molecules while they are rotating, can be treated classically. Under such conditions the classical observables are functions not only of the coordinates $\boldsymbol{r}_i$ and the momenta $\boldsymbol{p}_i$, but also of these internal variables and their conjugate momenta. In general, we shall denote the *configurational variables*, the positions of the molecules and their internal coordinates, by q_k and their *conjugate momenta* by p_k. The phase space $\mathcal{E}_\mathrm{P}$ is spanned by the q_k, p_k coordinates , the observables are functions defined in $\mathcal{E}_\mathrm{P}$, and the micro-states correspond to points in $\mathcal{E}_\mathrm{P}$.

2.3.2 Classical Macro-states

A *classical macro-state*, which is not well known or completely prepared, is characterized by giving a probability law for the possible micro-states, that is, for the set of possible points in phase space. A classical macro-state is thus represented by a probability measure in phase space:

$$D_N(\boldsymbol{r}_1,\boldsymbol{p}_1,\ldots,\boldsymbol{r}_N,\boldsymbol{p}_N)\; d\tau_N, \tag{2.53}$$

where D_N is called *density in phase* or *phase density.* The quantity (2.53) is the probability that the representative point of the N particles of the system is within the volume element $d\tau_N$ of phase space. The phase density is real, normalized, and positive, properties of which we have listed the quantum counterpart in § 2.2.3.

A classical observable $A(\boldsymbol{r}_1,\boldsymbol{p}_1,\ldots,\boldsymbol{r}_N,\boldsymbol{p}_N)$ then becomes a *random variable* governed by the probability law (2.53) when the system is in the macro-state represented by the density in phase D_N. Its *mean* or *expectation value* is equal to

$$\langle A\rangle \;=\; \int d\tau_N\; D_N(\boldsymbol{r}_1,\boldsymbol{p}_1,\ldots,\boldsymbol{r}_N,\boldsymbol{p}_N)\; A(\boldsymbol{r}_1,\boldsymbol{p}_1,\ldots,\boldsymbol{r}_N,\boldsymbol{p}_N), \tag{2.54}$$

which is the classical equivalent of (2.34). The trace is replaced by integration over the phase space $\mathcal{E}_\mathrm{P}$ and the density operator by an ordinary probability measure.

Strictly staying within the framework of classical mechanics, one could choose for $d\tau_N$ any normalization whatever, as the mean values (2.54) remain unchanged when one multiplies $d\tau_N$ and divides D_N by the same factor. Nevertheless, one wants classical statistical mechanics to appear as a limiting case of quantum mechanics. When constructing the density in phase as a classical limit of the density operator, we shall see in § 2.3.4 that we need *normalize the phase space volume element* $d\tau_N$ as follows:

$$\boxed{d\tau_N \;=\; \frac{1}{N!}\prod_{i=1}^{N}\left(\frac{d^3\boldsymbol{r}_i d^3\boldsymbol{p}_i}{h^3}\right)}\;, \tag{2.55}$$

in the case of N indistinguishable particles, where $h = 2\pi\hbar$ is Planck's constant.

The exact form of (2.55) will be crucial, even in the study of classical gases (see, for instance, § 8.1.5 or § 8.2.1). It can only be justified, if we start from quantum mechanics, which retains its imprint even in the classical limit (§ 2.3.4). We can, however, use rules of thumb to get an intuitive feeling for (2.55).

On the one hand, the factor h^{-3N} appears as a natural means to define the volume element $d\tau_N$, and hence the density in phase D_N, as a *dimensionless* quantity, since Planck's constant has the same dimension as the product $dx\,dp$.

On the other hand, the factor $1/N!$ is associated with the *indistinguishability* of the particles. This idea is ambiguous in a purely classical context. It would, for example, be permitted to imagine that the particles, though they are strictly identical, are labelled initially and after that can be followed as they move along their orbits, even during collisions, so that we can always distinguish between them. Nevertheless, quantum mechanics suggests, in contrast to this, that two points in phase space which derive from one another by a permutation of the indices i of the particles represent the *same physical micro-state*. Moreover, this is consistent with the fact that even in classical mechanics all physical observables are symmetric under particle permutations. From this point of view phase space is not a faithful representation of the space of micro-states: each of them appears, in general, $N!$ times when one sweeps through the whole of phase space. The factor $1/N!$ in (2.55) compensates for this redundancy in the counting. Of course, the density in phase itself must be *symmetric* under any interchange of indistinguishable particles.

The density in phase can be a function; more generally, it is a measure, that is, a positive distribution. Let us, for instance, consider N particles with *perfectly well known* positions and momenta, $\boldsymbol{r}_1^0, \boldsymbol{p}_1^0, \ldots, \boldsymbol{r}_N^0, \boldsymbol{p}_N^0$. Using the *normalization of the probability*

$$\int d\tau_N \; D_N(\boldsymbol{r}_1, \boldsymbol{p}_1, \ldots, \boldsymbol{r}_N, \boldsymbol{p}_N) \;=\; 1, \tag{2.56}$$

and the indistinguishability of the particles, this state is represented by the density in phase

$$\begin{aligned} D_N(\boldsymbol{r}_1, \boldsymbol{p}_1, \ldots, \boldsymbol{r}_N, \boldsymbol{p}_N) \;&=\; h^{3N} \sum \delta^3(\boldsymbol{r}_1 - \boldsymbol{r}_{i_1}^0)\, \delta^3(\boldsymbol{p}_1 - \boldsymbol{p}_{i_1}^0) \\ &\ldots\; \delta^3(\boldsymbol{r}_N - \boldsymbol{r}_{i_N}^0)\, \delta^3(\boldsymbol{p}_N - \boldsymbol{p}_{i_N}^0), \end{aligned} \tag{2.57}$$

where the sum is over the $N!$ permutations of the indices $\{i_1, i_2, \ldots, i_N\} = \{1, 2, \ldots, N\}$. This sum is necessary in the present framework: nothing in our classical formalism should enable us to distinguish the particles and, indeed, (2.57) preserves the symmetry under their permutations.

Another example is the density in phase of a *perfect gas* in thermal equilibrium. On the microscopic scale, a perfect gas is defined as a classical macro-state in which the N particles are uncorrelated. The probability measure thus reduces to a *product* of factors relating to the separate particles,

$$D_N(\boldsymbol{r}_1, \boldsymbol{p}_1, \ldots, \boldsymbol{r}_N, \boldsymbol{p}_N) \;=\; h^{3N}\, N! \prod_{i=1}^{N} \varrho(\boldsymbol{r}_i, \boldsymbol{p}_i), \tag{2.58}$$

where all factors $\varrho(\boldsymbol{r}, \boldsymbol{p})$ are the same because of the indistinguishability of the particles.

The extension to arbitrary configuration variables q_k and their conjugate momenta p_k is straightforward. The phase densities $D\{q,p\}$ and the observables $A\{q,p\}$ are functions of the point q_k, p_k in phase space. The volume element $d\tau$ occurring in the expression $\int d\tau\, DA$ for the mean value of A equals

$$d\tau = \frac{1}{\mathcal{S}} \prod_k \left(\frac{dq_k\, dp_k}{h} \right). \tag{2.59}$$

It contains a factor $h^{-1} \equiv (2\pi\hbar)^{-1}$ for each pair of conjugated variables; also it contains a factor $1/\mathcal{S}$, where $\mathcal{S}$ is a *symmetry number*, equal to the number of configurations in phase space which describe the same situation, if we take into account the indistinguishability of the particles. For instance, if the system is a mixture of fluids consisting of N_1 molecules of the first kind and N_2 molecules of the second kind, $\mathcal{S}$ equals $N_1!N_2!$. Another example concerns the *rotation of polyatomic molecules*; we shall it study in Chap.8 in order to calculate the specific heats of gases and the chemical equilibrium constants. If we can model a molecule as a rigid classical rotator, the integration variables q, p are the Euler angles and their conjugate momenta; the symmetry number $\mathcal{S}$ for each molecule is the order of the group which leaves this molecule invariant, for instance, $\mathcal{S} = 2$ for O_2 or H_2O, $\mathcal{S} = 3$ for NH_3, and $\mathcal{S} = 12$ for CH_4 or C_6H_6.

2.3.3 Evolution with Time

On the microscopic scale the dynamics are described by analytical mechanics and we shall briefly recall its laws.[4] In the *Lagrangian formalism* the instantaneous micro-state of the system is characterized by the configuration variables q_k – which in the simplest case reduce to the $3N$ particle coordinates $\boldsymbol{r}_1, \ldots, \boldsymbol{r}_N$ – and their velocities $\dot{q}_k \equiv dq_k/dt$. We look for the physical trajectory $q(t)$ in configuration space which starts from a given point $q(t_0)$ at time t_0 and ends at a given point $q(t_1)$ at time t_1. To do that we introduce arbitrary virtual trajectories $q(t)$ connecting these two points. The equations of motion can be deduced from a *Lagrangian* $L\{q, \dot{q}\}$, which is a function of the positions q_k and of the velocities $\dot{q}_k$, and possibly of the time. The Lagrangian generates the *action*

$$S(\{q(t)\}) = \int_{t_0}^{t_1} dt\, L\{q, \dot{q}\}, \tag{2.60}$$

which is a functional of the virtual trajectories $q(t)$. (Other types of action are defined in mechanics, but in the present book we shall use only (2.60).) According to *Hamilton's variational principle*, the action (2.60) must be stationary along the trajectory actually followed by the system. More precisely, we

[4] H.Goldstein, *Classical Mechanics*, Addison-Wesley, 1950; D.ter Haar, *Elements of Hamiltonian Mechanics*, Pergamon, 1971; R.Abraham and J.E.Marsden, *Foundations of Mechanics*, Benjamin/Cummings, 1978.

must have $\delta S = 0$ for arbitrary variations $\delta q(t)$ such that $\delta q(t_0) = \delta q(t_1) = 0$, that is,

$$\delta S = \int_{t_0}^{t_1} dt \sum_k \left[\frac{\partial L}{\partial q_k} \delta q_k(t) + \frac{\partial L}{\partial \dot{q}_k} \delta \dot{q}_k(t) \right] = 0.$$

Integrating by parts we then get the equations of motion in their *Euler-Lagrange* form,

$$\frac{\partial L}{\partial q_k} - \frac{d}{dt} \frac{\partial L}{\partial \dot{q}_k} = 0. \tag{2.61}$$

For instance, for a particle of mass m and charge e in an electromagnetic field, the Lagrangian is equal to

$$L(\boldsymbol{r}, \dot{\boldsymbol{r}}) \equiv \tfrac{1}{2} m \dot{\boldsymbol{r}}^2 + e\big(\boldsymbol{A}(\boldsymbol{r}) \cdot \dot{\boldsymbol{r}}\big) - e\Phi(\boldsymbol{r}), \tag{2.62}$$

where Φ is the scalar potential and $\boldsymbol{A}$ the vector potential; the Lorentz force can be derived from this. For interacting particles, L is the difference between their kinetic energy and their interaction potential.

The Lagrangian formalism is well suited to changes of variables; they are carried out by writing down that the action remains invariant. We shall use this formalism in non-equilibrium statistical mechanics to establish the local conservation laws (§ 14.3.1). However, in the case of equilibrium statistical mechanics we wish to retain a parallelism between classical and quantum systems and also to stress the conservation of the total energy. To do this we shall base ourselves mainly on the *Hamiltonian formalism*. The motion is then described not in configuration space but in the phase space $\mathcal{E}_{\text{P}}$ which has twice as many dimensions. It is generated by the *Hamiltonian* $H\{q,p\}$, a function of the positions q_k and their conjugate momenta p_k which are defined by eliminating the $\dot{q}_k$, considered to be independent variables, between

$$H\{q,p\} = \sum_k p_k \dot{q}_k - L\{q, \dot{q}\} \tag{2.63}$$

and

$$\frac{\partial L}{\partial \dot{q}_k} = p_k. \tag{2.63'}$$

This relation between the Lagrangian and the Hamiltonian is a Legendre transformation (§ 6.3.1). The equations (2.61) are then equivalent to *Hamilton's equations*

$$\frac{dq_k}{dt} = \frac{\partial H}{\partial p_k}, \qquad \frac{dp_k}{dt} = -\frac{\partial H}{\partial q_k}, \tag{2.64}$$

which describe the motion of a micro-state in $\mathcal{E}_{\text{P}}$. The Hamiltonian for a system of N non-relativistic particles of masses m_i and charges e_i has the form

$$
\begin{aligned}
H_N &= \sum_{i=1}^{N} \frac{\left[\boldsymbol{p}_i - e_i \boldsymbol{A}(\boldsymbol{r}_i)\right]^2}{2m_i} + \sum_{i=1}^{N} V(\boldsymbol{r}_i) \\
&+ \sum_{i,j=1;i<j}^{N} W\left(|\boldsymbol{r}_i - \boldsymbol{r}_j|\right),
\end{aligned} \tag{2.65}
$$

where W is the potential for the interaction between the particles, $V(\boldsymbol{r})$ the potential accounting for the box in which the particles are contained, and, possibly, also for an external, gravitational or electric, field, while $\boldsymbol{A}(\boldsymbol{r})$ is a possible electromagnetic vector potential.

Let us now find the equations of motion for classical macro-states. In quantum mechanics we derived the equation of motion for the density operator from that for the micro-states, noting that the probability for each of the latter was conserved during the motion. Here again, the probability (2.53) is conserved along the trajectories, which are defined by the Hamiltonian equations (2.64). However, the evolution transforms a volume element $d\tau_N$ (around a point q_k, p_k of phase space) at time t into another volume element $d\tau'_N$ (around the point q'_k, p'_k) at time $t + dt$. In view of (2.53) we must, if we want to describe the evolution of the density in phase D_N, know not only how the representative point of a micro-state in phase space changes, but also how a volume element around that point changes.

For this we shall use the important *Liouville theorem*: *the volume of a region in phase space remains constant* when one follows this region during the temporal evolution.

For a proof of this theorem it is sufficient to show that the Jacobian J of the infinitesimal transformation in $\mathcal{E}_{\mathrm{P}}$

$$
q_k \Longrightarrow q'_k = q_k + \frac{dq_k}{dt}\,dt, \qquad p_k \Longrightarrow p'_k = p_k + \frac{dp_k}{dt}\,dt, \tag{2.66}
$$

which changes a point at time t to its position at time $t+dt$, equals 1. The Jacobian J is the determinant of the $6N \times 6N$ matrix of the derivatives of the q'_k, p'_k with respect to the q_k, p_k. This matrix is of the form $1 + K\,dt$ and, to first order in dt,

$$
\begin{aligned}
J &= \det\,(1 + K\,dt) = 1 + \mathrm{Tr}\,K\,dt \\
&= 1 + \sum_k \left(\frac{\partial}{\partial q_k}\frac{dq_k}{dt} + \frac{\partial}{\partial p_k}\frac{dp_k}{dt} \right) dt,
\end{aligned}
$$

or, if we use the equations of motion (2.64),

$$
\frac{dJ}{dt} = \sum_k \left(\frac{\partial}{\partial q_k}\frac{\partial H_N}{\partial p_k} - \frac{\partial}{\partial p_k}\frac{\partial H_N}{\partial q_k} \right) = 0. \qquad QED
$$

According to (2.54), (2.56), or (2.69), *a volume in phase space* plays in classical mechanics the same rôle as the *number of states* in quantum mechanics. Liouville's theorem is thus the classical analogue of the following property: a set of $\mathcal{N}$ orthonormal states in $\mathcal{E}_{\mathrm{H}}$ is with time transformed into $\mathcal{N}$ other orthonormal states, because the unitary transformation (2.27) conserves the scalar product.

The Liouville theorem is the reason why we chose to work in phase space. In fact, it expresses the homogeneity of $\mathcal{E}_{\mathrm{P}}$: one can always imagine an evolution in time leading from some point in $\mathcal{E}_{\mathrm{P}}$ to any other one; whatever pair of points we are dealing with, the surrounding corresponding volume elements are the same. This property would not be valid, if we had chosen to characterize the system by an arbitrary set of $6N$ coordinates, for instance, velocities, rather than momenta, and positions, except for some special forms of the Hamiltonian.

The equation of motion of the density in phase can then be derived from the *conservation of the probability* (2.53) along the trajectories in phase space. By virtue of the Liouville theorem we have $d\tau_N = d\tau'_N$, and hence

$$\begin{aligned} D_N(\{q_k,p_k\},t) &= D_N(\{q'_k,p'_k\},t+dt) \\ &= D_N\left(\left\{q_k + \frac{dq_k}{dt}\,dt, p_k + \frac{dp_k}{dt}\,dt\right\}, t+dt\right). \end{aligned}$$

If we use the Hamiltonian equations (2.64), we find to first order in dt

$$\begin{aligned} D_N(\{q_k,p_k\},t) &= D_N(\{q_k,p_k\},t+dt) \\ &\quad + \sum_k \left(\frac{\partial D_N}{\partial q_k}\frac{\partial H_N}{\partial p_k} - \frac{\partial D_N}{\partial p_k}\frac{\partial H_N}{\partial q_k} \right) dt. \end{aligned}$$

The *Poisson bracket* of two functions of a point in phase space is defined by

$$\{A,B\} \equiv \sum_k \left(\frac{\partial A}{\partial q_k}\frac{\partial B}{\partial p_k} - \frac{\partial A}{\partial p_k}\frac{\partial B}{\partial q_k} \right), \tag{2.67}$$

so that we can write the equation for the evolution of the density in phase, the so-called *Liouville equation*, in the form

$$\frac{\partial D_N}{\partial t} = \{H_N, D_N\}. \tag{2.68}$$

2.3.4 The Classical Limit of Quantum Statistics

It is important to know how the formalism of classical statistical mechanics that we have just given can be derived starting from quantum mechanics which is the theory best suited for the microscopic scale. This is essential for a discussion of the validity of the classical approach, for an evaluation of possible quantum corrections, and for a justification of expressions (2.55) and (2.59) for the volume element. The classical approximation of quantum mechanics is valid when Planck's constant h *is negligible as compared to quantities with the same dimensions* which appear in the problem. These quantities typically are the product of a characteristic length and momentum, or the product of a characteristic energy and time. A complete discussion of this limit would go beyond the confines of the present textbook, but we shall show here how the contents of §§ 2.3.1 to 2.3.3 can be derived from the principles laid down in § 2.2.7.

Note, first of all, that the algebra of observables is generated by the position operators $\widehat{q}_k$ and the conjugate momentum operators $\widehat{p}_k$, the commutators of which are equal to $i\hbar$. Any observable can be written as a function of those operators. In the classical limit the algebra becomes *commutative*: the elements $\widehat{q}_k$ and $\widehat{p}_k$ can be regarded as ordinary variables and the observables as classical *functions* of these variables. Similarly, the density operators $\widehat{D}$, constructed from the generators $\widehat{q}_k$, $\widehat{p}_k$ of this algebra, become functions of a point in phase space.

In order to justify, starting from (2.34), the classical expression (2.54) for the expectation value of an observable, we must calculate the *classical limit of the trace* of an operator $\widehat{F}$ which in the present case is $\widehat{D}\widehat{A}$. We shall in what follows show that for a system of N indistinguishable particles this limit is equal to

$$\boxed{\operatorname{Tr} \widehat{F} \equiv \sum_m \langle m|\widehat{F}|m\rangle \to \int d\tau_N \; F(\boldsymbol{r}_1,\boldsymbol{p}_1,\ldots,\boldsymbol{r}_N,\boldsymbol{p}_N)} \, , \tag{2.69}$$

where $F(\boldsymbol{r}_1,\boldsymbol{p}_1,\ldots,\boldsymbol{r}_N,\boldsymbol{p}_N)$ is the function which is derived from $\widehat{F}$ by neglecting all commutators, and where $d\tau_N$ is defined by (2.55). More generally, for systems of the kind considered at the end of § 2.3.2, the trace tends to an integral with (2.59) as the volume element. Expression (2.69) of the classical limit of a trace can be interpreted by saying that the *measure* $\int_\omega d\tau_N$ *of a domain* ω in phase space is asymptotically, for sufficiently large ω, equal to the *number of orthogonal quantum micro-states* located in the domain ω, within the margins of the uncertainty relations. We can now understand the factor $[N!]^{-1}$ in $d\tau_N$ as reflecting the *Pauli principle.* In fact, if the N particles were distinguishable, the Hilbert space would be the direct product of the spaces associated with each particle and the volume element would be the product of the volume elements $dq_k\, dp_k/h$ for each degree of freedom. However, the indistinguishability obliges us to restrict the Hilbert space and to retain only either symmetric or antisymmetric wavefunctions. Those represent only one in $N!$ of the functions produced by the permutations of the particles, if we start from a generic wavefunction without any particular symmetry properties; this gives us an intuitive idea of the quantum origin of the factor $[N!]^{-1}$ in (2.55) or of $\mathcal{S}^{-1}$ in (2.59).

Note that one can take the classical limit only for *some of the degrees of freedom*, for which the characteristic lengths and momenta are the largest, whereas other degrees of freedom, for which the product of a characteric length and a characteristic momentum is of order $\hbar$, must still be treated quantum-mechanically. For instance, for a diatomic gas, we can for the molecular translations use the classical approximation, whereas their vibrations must be treated by quantum statistical mechanics (Chap.8). Similarly, if the N elementary constituents have spin s, there is no classical equivalent of the latter. When there is no coupling with the other degrees of freedom, the trace over the spins gives in the classical limit an *extra multiplying degeneracy factor* $d = (2s+1)^N$ for $d\tau_N$. These remarks play an important rôle in chemical thermodynamics (§ 8.2.2).

In order to prove (2.69) we proceed in stages. Let us, to begin with, consider a system consisting of a one-dimensional particle for which the operator algebra is generated by the two operators $\widehat{x}$ and $\widehat{p}$. Let us examine first the special operators of the form

$$\widehat{F} = f(\widehat{p})g(\widehat{x}). \tag{2.70}$$

Classically, the function $F(x,p) = f(p)g(x)$ corresponds to $\widehat{F}$. Let us calculate the trace of (2.70) in the base $|x\rangle$, each element of which is an eigenstate of $\widehat{x}$ representing a particle localized at the point x:

$$\mathrm{Tr}\,\widehat{F} = \int dx\,\langle x|f(\widehat{p})g(\widehat{x})|x\rangle = \int dx\,\langle x|f(\widehat{p})|x\rangle\, g(x).$$

We use the closure relation (2.2) for the base of plane waves which are normalized as follows:

$$\langle x|p\rangle = \frac{1}{\sqrt{2\pi\hbar}}\, \mathrm{e}^{ipx/\hbar};$$

this leads to

$$\begin{aligned}\mathrm{Tr}\,\widehat{F} &= \int dx\,\langle x|f(\widehat{p})|p\rangle\, dp\langle p|x\rangle\, g(x) = \int dx\, dp\, |\langle x|p\rangle|^2\, f(p)\, g(x)\\ &= \int \frac{dx\, dp}{h}\, F(x,p).\end{aligned}$$

Equation (2.69) is thus exact for the special operators of the form (2.70)

Let us now consider an arbitrary operator $\widehat{F}$ of the algebra. As the commutators are small in the classical limit, we are allowed to change inside $\widehat{F}$ the order of the operators $\widehat{p}$ and $\widehat{x}$, and neglect terms which come from commuting them. One can thus, through successive commutations, bring all $\widehat{x}$ operators to the right so that in the classical limit every operator is equivalent to a sum of terms of the form (2.70). Its trace can then be expressed as an integral over phase space of the corresponding classical function $F(x,p)$ with a weight h^{-1} associated with the two degrees of freedom (x,p). The extension to an arbitrary number of degrees of freedom gives us the factor h^{-3N} of (2.55), (2.69).

This calculation applies to Cartesian coordinates and their conjugate momenta. If we want to describe the rotations and vibrations of molecules in the classical limit (Chap.8) we must extend the result we have obtained in order to justify (2.59). We shall restrict ourselves here to the case of a rotation in a plane, where $\widehat{q}$ is an angle varying from 0 to 2π and $\widehat{p}$ the associated angular momentum, taking on values which are integral multiples of $\hbar$. Using the above reasoning the integration over

p is replaced by a sum, and $\langle q|p\rangle$ is an eigenfunction of the angular momentum, normalized as $\mathrm{e}^{ipq/\hbar}/\sqrt{2\pi}$. The replacement of the summation over p by an integral gives again the above-mentioned measure $dq\,dp/2\pi\hbar$.

A more rigorous approach to the classical approximation, which enables us to evaluate the quantum corrections when they are small, is provided by the *Wigner representation* of quantum mechanics already outlined in § 2.1.2. We saw for a one-dimensional system that the observables $\widehat{A}$ can be represented, even when $\hbar$ is finite, by the functions $A_{\mathrm{W}}(x,p)$ defined by (2.9). Let us similarly define the Wigner representation of the density operator $\widehat{D}$ as the function

$$D_{\mathrm{W}}(x,p) \;\equiv\; 2\pi\hbar\ \mathrm{Tr}\ \widehat{D}\,\widehat{\Omega}(x,p). \tag{2.71}$$

We can evaluate the trace in the Wigner representation using the relations

$$\mathrm{Tr}\ \widehat{\Omega}(x,p) \;=\; \frac{1}{h}, \qquad \mathrm{Tr}\ \widehat{\Omega}(x,p)\,\widehat{\Omega}(x',p') \;=\; \frac{1}{h}\,\delta(x-x')\,\delta(p-p'), \tag{2.72}$$

which follow from the definition (2.8) of the $\widehat{\Omega}(x,p)$ observables and expression (2.10) for their product. From this we get for the expectation value of an observable $\widehat{A}$

$$\langle A\rangle \;=\; \mathrm{Tr}\ \widehat{D}\,\widehat{A} \;=\; \int \frac{dx\,dp}{2\pi\hbar}\ D_{\mathrm{W}}(x,p)\,A_{\mathrm{W}}(x,p). \tag{2.73}$$

The transform $D_{\mathrm{W}}(x,p)$ of $\widehat{D}$ can thus be interpreted as a *quantum density in phase*, with Eq.(2.73) formally having the same structure as the relation (2.54) for classical probabilities with the weight (2.55). Nevertheless, the quantum nature of the theory, when $\hbar$ is finite, subsists in a hidden manner in (2.73); indeed, the fact that the operator $\widehat{D}$ is positive does not imply that $D_{\mathrm{W}}(x,p)$, defined by (2.71), is a positive probability measure. However, we indicated in § 2.1.2 that, in the limit when $\hbar$ is negligible as compared to any other quantity of the same dimensions, the Wigner transform of a product of operators becomes the product of their transforms. Then A_{W} can, indeed, be identified as a function of the classical random variables x and p and D_{W} as a classical phase density. For instance, the Wigner transform of the density operator $\widehat{D} \propto \mathrm{e}^{-\beta\widehat{H}}$, which describes a quantum equilibrium macro-state, tends in this limit to $\mathrm{e}^{-\beta H_{\mathrm{W}}}$ which describes the corresponding classical equilibrium macro-state, and (2.73) can be identified with a classical average.

Going over to three dimensions is straightforward. For N distinguishable particles, the integral of $D_{\mathrm{W}}A_{\mathrm{W}}$ is then a trace over a Hilbert space which is the direct product of N one-particle spaces. However, when the particles are *indistinguishable* the trace which we must calculate is only over either symmetric or antisymmetric wavefunctions. We can take this restriction into account by writing

$$\langle A\rangle \;=\; \mathrm{Tr}\ \widehat{\Pi}\,\widehat{D}\,\widehat{A}, \tag{2.74}$$

where the trace is still over all wavefunctions, but where the *projection* $\widehat{\Pi}$ *over the symmetric or antisymetric states* eliminates those which are undesirable. For two particles, we can write this projection in the form

$$\widehat{\Pi} \;=\; \tfrac{1}{2}\,(\widehat{I} \pm \widehat{E}), \tag{2.75}$$

where $\widehat{E}$ is the operator which exchanges the labels of the two particles and where the + and − signs correspond, respectively, to bosons and fermions. A simple calculation gives us the Wigner representation of this exchange operator; in three dimensions it is equal to

$$E_{\mathrm{W}}(\boldsymbol{r}_1, \boldsymbol{r}_2, \boldsymbol{p}_1, \boldsymbol{p}_2) = h^3 \delta^3(\boldsymbol{r}_1 - \boldsymbol{r}_2) \delta^3(\boldsymbol{p}_1 - \boldsymbol{p}_2). \tag{2.76}$$

In the classical limit the exchange term in (2.75) is thus negligible and there remains only the first term which contains a factor $\frac{1}{2}$. More generally, in the case of N particles the projection $\widehat{\Pi}$ is equal to

$$\widehat{\Pi} = \frac{1}{N!} \sum_{\alpha} (\pm)^{n_\alpha} \widehat{P}_\alpha, \tag{2.77}$$

where the sum is over the $N!$ permutations $\widehat{P}_\alpha$ of the N particles and where n_α denotes the parity of the permutation $\widehat{P}_\alpha$. Since these permutations are generated as products of two-particle exchange operators, the classical limit of the projection $\widehat{\Pi}$ only involves the identity permutation, and reduces to $\widehat{I}/N!$. The indistinguishability of the N particles thus introduces in (2.74) the factor $(N!)^{-1}$ that we anticipated in (2.55).

Note finally that the volume element $d\tau$ is invariant under canonical changes in variables, that is, those which conserve the conjugation of the coordinates q with the momenta p. In fact, an infinitesimal canonical transformation is produced by a Poisson bracket, like the evolution (2.66), and the proof of the Liouville theorem can immediately be adapted. This obvious invariance appears as the classical limit of the invariance of the trace under unitary changes of the base.

We still must prove that the Liouville equation (2.68) is the classical limit of the Liouville-von Neumann equation (2.49) which governs the evolution of the density operator. This rests upon the following property that we shall prove in a moment. When one replaces in the limit as $\hbar \to 0$ the observables and the density operators by functions of the q_k, p_k variables, the *commutators* are to dominant order equivalent to the *Poisson brackets* (2.67), as follows:

$$\boxed{\frac{1}{\mathrm{i}\hbar} [\widehat{X}, \widehat{Y}] \to \{X, Y\}} \,. \tag{2.78}$$

This property is also useful to find the classical limit of the transformations $\widehat{U}\widehat{A}\widehat{U}^\dagger$ or $\widehat{U}\widehat{D}\widehat{U}^\dagger$ of observables or density operators, when $\widehat{U}$ is produced from a generating operator $\widehat{G}$ (§2.1.5). In fact, the infinitesimal transformation $\widehat{U} = \widehat{I} + \mathrm{i}\widehat{G}\varepsilon/\hbar$ changes $\widehat{A}$ to $\widehat{A} + \mathrm{i}[\widehat{G}, \widehat{A}]\varepsilon/\hbar$, whose classical limit is $A - \{G, A\}\varepsilon$. Similarly, the classical limit of the Ehrenfest theorem (2.29) is

$$\frac{d}{dt} \langle A \rangle = \langle \{A, H\} \rangle + \left\langle \frac{\partial A}{\partial t} \right\rangle, \tag{2.79}$$

and the *constants of the motion* are the mean values of the observables which have vanishing Poisson brackets with H.

In order to prove (2.78) we return to the Wigner representation. Consider a one-dimensional system – the generalization to an arbitrary dimensionality does not present any problems. If we use (2.7) and (2.10) we can write the product of two operators in the form

$$\begin{aligned}\widehat{X}\widehat{Y} &= \frac{1}{(2\pi)^2}\int dx\, dp\, dx'\, dp'\, d\alpha\, d\beta\; X_\mathrm{W}(x,p)\, Y_\mathrm{W}(x',p') \\ &\quad\times \mathrm{e}^{\mathrm{i}\alpha(x-x')+\mathrm{i}\beta(p-p')}\, \widehat{\Omega}\left(\frac{x+x'+\beta\hbar}{2}, \frac{p+p'-\alpha\hbar}{2}\right).\end{aligned}$$

The *Wigner representation of the product* $\widehat{X}\widehat{Y}$ can be derived from this expression and, after changing variables, can be written in the form

$$\begin{aligned}(\widehat{X}\widehat{Y})_\mathrm{W} &= \frac{1}{(2\pi)^2}\int d\lambda\, d\mu\, d\xi\, d\eta\; X_\mathrm{W}\left(x+\frac{1}{2}\lambda\hbar, p+\frac{1}{2}\mu\hbar\right) \\ &\quad\times Y_\mathrm{W}(x+\xi, p+\eta)\, \mathrm{e}^{\mathrm{i}\lambda\eta-\mathrm{i}\mu\xi}. \end{aligned} \tag{2.80}$$

We must expand this expression to first order in $\hbar$, which leads to

$$\begin{aligned}(\widehat{X}\widehat{Y})_\mathrm{W} &= X_\mathrm{W}(x,p)\, Y_\mathrm{W}(x,p) + \frac{\mathrm{i}\hbar}{2}\left[\frac{\partial X_\mathrm{W}(x,p)}{\partial x}\frac{\partial Y_\mathrm{W}(x,p)}{\partial p}\right. \\ &\quad\left. - \frac{\partial X_\mathrm{W}(x,p)}{\partial p}\frac{\partial Y_\mathrm{W}(x,p)}{\partial x}\right] + \mathcal{O}(\hbar^2),\end{aligned}$$

and the expected result (2.78) follows. This proof enables us to find the domain of validity of the classical approximation. In fact, each factor $\hbar$ is accompanied by derivatives of X_W and Y_W with respect to x or p. The classical limit is thus valid if $\delta x\, \delta p \gg \hbar$, where δx and δp denote characteristic distances and momenta over which the physical quantities change significantly.

2.3.5 Reduced Densities

Many physical quantities involve *one-particle observables* of the form $\sum_i a(\boldsymbol{r}_i, \boldsymbol{p}_i)$. For instance, the kinetic energy is found for $a(\boldsymbol{r},\boldsymbol{p}) = \boldsymbol{p}^2/2m$ and the particle density at the point $\boldsymbol{r}_0$ for $a(\boldsymbol{r},\boldsymbol{p}) = \delta^3(\boldsymbol{r}-\boldsymbol{r}_0)$. We shall also encounter two-particle observables, such as the interaction energy $\sum_{i<j} W(|\boldsymbol{r}_i - \boldsymbol{r}_j|)$ in (2.65), but rarely does one have to calculate averages of more complicated observables. We do not need therefore the complete density in phase D_N which includes, in addition to the information of interest, also information about correlations between 3, 4, ..., N particles. Just as in Chap.1 we defined reduced probabilities (Eqs.(1.5) and (1.8)) and as in quantum mechanics we introduced density operators of subsystems, we can here define *one- and two-particle reduced densities*

$$\begin{aligned}f(\boldsymbol{r},\boldsymbol{p}) &\equiv \frac{1}{h^3}\int d\tau_{N-1}(2,\ldots,N) \\ &\quad\times D_N(\boldsymbol{r},\boldsymbol{p},\boldsymbol{r}_2,\boldsymbol{p}_2,\ldots,\boldsymbol{r}_N,\boldsymbol{p}_N),\end{aligned} \tag{2.81}$$

$$f_2(\boldsymbol{r},\boldsymbol{p},\boldsymbol{r}',\boldsymbol{p}') \equiv \frac{1}{h^6}\int d\tau_{N-2}(3,\ldots,N) \times D_N(\boldsymbol{r},\boldsymbol{p},\boldsymbol{r}',\boldsymbol{p}',\boldsymbol{r}_3,\boldsymbol{p}_3,\ldots,\boldsymbol{r}_N,\boldsymbol{p}_N). \tag{2.82}$$

To evaluate the expectation value (2.54) of a one-, or two-, particle observable, we take into account the symmetry under permutation of the labels of the particles; the integration over the variables of the remaining $N-1$, or $N-2$, particles can then be done beforehand, introducing f, or f_2, directly in the calculation of the averages. For instance, the *internal energy* U, which is the average of the Hamiltonian (2.65), can be written in the form

$$U = \int d^3\boldsymbol{r}\, d^3\boldsymbol{p}\, f(\boldsymbol{r},\boldsymbol{p}) \left[\frac{\left(\boldsymbol{p}-e\boldsymbol{A}(\boldsymbol{r})\right)^2}{2m} + V(\boldsymbol{r})\right] + \frac{1}{2}\int d^3\boldsymbol{r}\, d^3\boldsymbol{p}\, d^3\boldsymbol{r}'\, d^3\boldsymbol{p}'\, f_2(\boldsymbol{r},,\boldsymbol{p},\boldsymbol{r}',\boldsymbol{p}')\, W(|\boldsymbol{r}-\boldsymbol{r}'|). \tag{2.83}$$

The one-particle reduced density $f(\boldsymbol{r},\boldsymbol{p})$ can be interpreted as the *particle density in the one-body phase space*, since the average number of particles in a volume element $d^3\boldsymbol{r}\, d^3\boldsymbol{p}$ around the point $\boldsymbol{r},\boldsymbol{p}$ of this phase space is equal to the expectation value

$$\left\langle \sum_i \delta^3(\boldsymbol{r}_i-\boldsymbol{r})\,\delta^3(\boldsymbol{p}_i-\boldsymbol{p})\, d^3\boldsymbol{r}\, d^3\boldsymbol{p} \right\rangle = f(\boldsymbol{r},\boldsymbol{p})\, d^3\boldsymbol{r}\, d^3\boldsymbol{p}.$$

Similarly, the particle density in ordinary space equals $\int d^3\boldsymbol{p}\, f(\boldsymbol{r},\boldsymbol{p})$. The integral of f over all its coordinates gives the total number of particles. One can check that for a perfect gas in thermal equilibrium with the phase density (2.58) $f(\boldsymbol{r},\boldsymbol{p})$ reduces to $N\varrho(\boldsymbol{r},\boldsymbol{p})$; we shall use this fact in Chap.7 when we study gases at equilibrium.

We shall also calculate in Chap.15 the non-equilibrium properties by studying the *evolution equation of the one-particle reduced density.* It can easily be written down when the Hamiltonian (2.65) reduces to a sum of one-particle terms,

$$H_N = \sum_{i=1}^{N} h(\boldsymbol{r}_i,\boldsymbol{p}_i), \qquad h(\boldsymbol{r},\boldsymbol{p}) = \frac{\left(\boldsymbol{p}-e\boldsymbol{A}(\boldsymbol{r})\right)^2}{2m} + V(\boldsymbol{r}). \tag{2.84}$$

In fact, in this case the Hamiltonian equations (2.64) become

$$\frac{d\boldsymbol{r}}{dt} = \frac{\boldsymbol{p}-e\boldsymbol{A}(\boldsymbol{r})}{m} \equiv \boldsymbol{v}, \qquad \frac{d\boldsymbol{p}}{dt} = -\nabla_{\boldsymbol{r}}\, h \equiv \boldsymbol{\varphi}, \tag{2.85}$$

where $(\boldsymbol{r},\boldsymbol{p})$ must successively be replaced by $(\boldsymbol{r}_1,\boldsymbol{p}_1),\ldots,(\boldsymbol{r}_N,\boldsymbol{p}_N)$, and where $\boldsymbol{\varphi}$ represents the force produced by the scalar and vector potentials. The problem has been simplified considerably since Eqs.(2.85) no longer couple the motions of different particles and since all particles now have the same equations of motion. This enables us to work in the six-dimensional *single-particle phase space*, rather than in the $6N$-dimensional full phase space. The reduced density $f(\boldsymbol{r},\boldsymbol{p})$ represents the particle density in that space. The *number* of particles $f\,d^3\boldsymbol{r}\,d^3\boldsymbol{p}$ within a volume element is *conserved* during the motion (2.85) so that we can obtain the evolution equation for f in the single-particle space by the same arguments as that for D_N in the N-particle space. The Liouville theorem is valid for the Hamiltonian evolution (2.85) in the single-particle phase space, so that the density f is conserved along the trajectories in that space:

$$\frac{d}{dt}\,f\big(\boldsymbol{r}(t),\boldsymbol{p}(t);t\big) \;=\; 0. \tag{2.86}$$

Hence we find the evolution equation for f:

$$\frac{\partial f}{\partial t}+\left(\boldsymbol{v}\cdot\nabla_{\boldsymbol{r}}\right)f+\left(\boldsymbol{\varphi}\cdot\nabla_{\boldsymbol{p}}\right)f \;=\; 0. \tag{2.87}$$

In the general case where the particles interact, the presence of the two-particle potential W in the Hamiltonian (2.65) gives rise to collisions which change the momenta of the particles. The number of particles within a volume element in the single-particle phase space is no longer conserved, as a *collision can make a particle appear in or disappear from the volume element* $d^3\boldsymbol{r}\,d^3\boldsymbol{p}$ around the point $(\boldsymbol{r},\boldsymbol{p})$. The total derivative (2.86) of f is therefore no longer equal to zero, and Eq.(2.87) is changed to

$$\frac{\partial f}{\partial t}+\left(\boldsymbol{v}\cdot\nabla_{\boldsymbol{r}}\right)f+\left(\boldsymbol{\varphi}\cdot\nabla_{\boldsymbol{p}}\right)f \;=\; \left(\frac{\partial f}{\partial t}\right)_{\text{coll}}. \tag{2.88}$$

The term on the right-hand side represents the balance from collisions. It can be evaluated by using various approximations, depending on the physical situation. In Chap.15 we shall establish the form of that term in the case of a rarefied gas with short-range forces; the resulting equation is called the *Boltzmann equation.*

By integrating the Liouville equation (2.68) we can obtain an *exact* expression for the change in time of the reduced density (2.81); it is, however, useless without extra approximations. A preliminary remark will help us to simplify the integration of the right-hand side: consider a term in the Poisson bracket relating to the pair of conjugate variables q_k and p_k and assume that we integrate over those variables; noting that the density in phase D is, in general, zero at infinity – otherwise (2.56) would not converge – integration by parts leads to

$$\begin{aligned}&\int dq_k\,dp_k\,\left(\frac{\partial H}{\partial q_k}\frac{\partial D}{\partial p_k}-\frac{\partial D}{\partial q_k}\frac{\partial H}{\partial p_k}\right)\\&\quad=\int dq_k\,dp_k\,\left(-\frac{\partial^2 H}{\partial p_k\,\partial q_k}\,D+\frac{\partial^2 H}{\partial q_k\,\partial p_k}\,D\right)\;=\;0.\end{aligned}$$

The only terms in the Poisson bracket that remain after integration are thus those corresponding to variables over which we do not integrate. In particular, integrating the Liouville equation over all variables, we see that the normalization (2.56) of a density in phase is conserved in time. If we integrate over the variables which are associated with particles 2, ..., N, we find

$$h^3 \frac{\partial f(\boldsymbol{r},\boldsymbol{p},t)}{\partial t} = \int d\tau_{N-1}(2,\ldots,N) \left[\left(\nabla_r H \cdot \nabla_p D\right) - \left(\nabla_p H \cdot \nabla_r D\right)\right].$$

On the right-hand side, the one-particle terms in H do not involve the variables $\boldsymbol{r}_2, \boldsymbol{p}_2, \ldots, \boldsymbol{r}_N, \boldsymbol{p}_N$ so that we can integrate over those variables, using (2.81); this produces the left-hand side of (2.88). Amongst the two-particle terms in H, only those survive which come from $W(|\boldsymbol{r}-\boldsymbol{r}_k|)$ with $k=2,\ldots,N$; because of symmetry their contributions, of which there are $N-1$, are all equal to

$$\int \frac{d^3r_2\, d^3p_2}{(N-1)h^3}\, d\tau_{N-2}(3,\ldots,N) \left(\nabla_r W(|\boldsymbol{r}-\boldsymbol{r}_2|) \cdot \nabla_p D\right),$$

which, if we use (2.82), can be expressed as a function of the two-particle reduced density. Altogether we find

$$\begin{aligned} &\frac{\partial f}{\partial t} + \left(\boldsymbol{v} \cdot \nabla_r\right) f + \left(\boldsymbol{\varphi} \cdot \nabla_p\right) f \\ &\quad = \int d^3r'\, d^3p' \left(\nabla_r W(|\boldsymbol{r}-\boldsymbol{r}'|) \cdot \nabla_p f_2(\boldsymbol{r},\boldsymbol{p},\boldsymbol{r}',\boldsymbol{p}')\right). \end{aligned} \tag{2.89}$$

The exact form of the collision term thus introduces a new unknown function, the two-particle reduced density. Similarly, if we write down the evolution equation for f_2 by integrating the Liouville equation as above over the variables relating to particles 3, ..., N, we find a coupling with the three-particle reduced density, and so on. The chain of equations that we obtain in this way is called the *BBGKY* (Bogolyubov, Born, Green, Kirkwood, Yvon) *hierarchy*. It replaces the Liouville equation which involves a very large, $6N$, number of variables by a chain of equations with a finite number, 6, 12, ..., of variables. On the other hand, solving the chain implies *approximations* which allow us to reduce the number of unknown functions to the number of equations.

For instance, the approximation which consists in neglecting the correlations between particles in phase space amounts to assuming that

$$f_2(\boldsymbol{r},\boldsymbol{p},\boldsymbol{r}',\boldsymbol{p}') = f(\boldsymbol{r},\boldsymbol{p})\, f(\boldsymbol{r}',\boldsymbol{p}').$$

One can show that it is often justified for a gas of charged particles because of the long-range nature of the Coulomb potential W. In this approximation (2.89) leads to a closed equation, called the *Vlasov equation*, for the one-particle density. This equation has the same form Eq.(2.87) as for a single particle in an external one-particle potential $V(\boldsymbol{r})$, but we must include here in $V(\boldsymbol{r})$ the *effective potential*

$$\int d^3r'\, W(|\boldsymbol{r}-\boldsymbol{r}'|) \int d^3p'\, f(\boldsymbol{r}',\boldsymbol{p}'), \tag{2.90}$$

produced on the average by the other particles; the last factor is the particle density at the point $\boldsymbol{r}'$ in ordinary space.

2.3.6 Uncertain Particle Number

In many applications we shall need to assume that not only quantities such as the particle density or the energy, but also the number of particles N itself is a random variable. One can easily generalize the above formalism to such situations.

In the *quantum* case, the Hilbert space $\mathcal{E}_{\mathrm{H}}$ is then the direct sum (§ 2.1.1) of the Hilbert spaces $\mathcal{E}_{\mathrm{H}}^{N}$ each associated with a fixed number of particles N. This space $\mathcal{E}_{\mathrm{H}}$ is called the *Fock space* (§ 10.2). The number N occurs amongst the quantum numbers characterizing the base states and the trace (2.34) includes a summation over N.

In the *classical* case, phase space is the union of the N-particle phase spaces. The density in phase D is a set of functions D_N of the kind considered earlier which are normalized in such a way that

$$\int d\tau_N \; D_N(\boldsymbol{r}_1, \boldsymbol{p}_1, \ldots, \boldsymbol{r}_N, \boldsymbol{p}_N) \tag{2.91}$$

is the *probability that the system contains N particles.*

A physical quantity A is also represented by a set of symmetric functions A_N – for instance, the energy is represented by the set of functions H_N given by (2.65) – and the expectation value of A is equal to

$$\langle A \rangle \;=\; \sum_N \int d\tau_N \; D_N \, A_N; \tag{2.92}$$

this formula is analogous to the general quantum expression (2.34) if we make the substitution

$$\boxed{\mathrm{Tr} \;\to\; \sum_N \int d\tau_N} \; . \tag{2.93}$$

In particular, the one- and two-particle reduced densities are now obtained by summing (2.81) and (2.82) over N.

Summary

All predictions – of a statistical nature – that one can make at a given time about a physical system can be found once we know its density operator. The density operator formalism, which we review in § 2.2.7, enables us to extend quantum mechanics to the description of statistical mixtures, or quantum macro-states, representing systems which are not well known or subsystems (Eq.(2.39)). The eigenvalues of a density operator $\widehat{D}$ can be interpreted as the probabilities for its eigenkets; the average values (2.34), (2.52) can be *evaluated in the base which diagonalizes $\widehat{D}$ as expectation values in probability theory.*

The classical limit is obtained by replacing the observables by functions defined in phase space, the density operator by the density in phase, and the trace by the integration (2.69) with the measure (2.55). The evaluation of most observed quantities involves, in general, only the one-particle reduced density. The particle number can possibly be a random variable.

Exercises

2a Density Operators for a Spin-$\frac{1}{2}$

Recall that the observables $\widehat{S}_x, \widehat{S}_y, \widehat{S}_z$ which represent the three components of the angular momentum of a spin-$\frac{1}{2}$ particle can be written in the form $\widehat{\boldsymbol{S}} = \frac{1}{2}\hbar\widehat{\boldsymbol{\sigma}}$, where the $\widehat{\sigma}_x$, $\widehat{\sigma}_y$, $\widehat{\sigma}_z$ operators are represented by the Pauli matrices

$$\widehat{\sigma}_x = \begin{pmatrix} 0 & 1 \\ 1 & 0 \end{pmatrix}, \qquad \widehat{\sigma}_y = \begin{pmatrix} 0 & -\mathrm{i} \\ \mathrm{i} & 0 \end{pmatrix}, \qquad \widehat{\sigma}_z = \begin{pmatrix} 1 & 0 \\ 0 & -1 \end{pmatrix},$$

in the base, $|+\rangle, |-\rangle$, which diagonalizes $\widehat{S}_z$.

1. Evaluate the components of the polarization vector $\boldsymbol{\varrho} = \langle\widehat{\boldsymbol{\sigma}}\rangle$ which is proportional to the expectation value of the spin in an arbitrary pure state $|\psi\rangle = \alpha|+\rangle + \beta|-\rangle$. On how many independent real parameters does this state depend? What is the locus of the vector $\boldsymbol{\varrho}$ in three-dimensional space? Compare $\langle \boldsymbol{S}\rangle^2$ with $\langle \boldsymbol{S}^2\rangle$. Is it possible to characterize, inversely, a ket by giving $\boldsymbol{\varrho}$? What kets are represented by two vectors $\boldsymbol{\varrho}$ in opposite directions? Write down the density matrix

$$D \equiv \begin{pmatrix} a & b \\ c & d \end{pmatrix},$$

which represents the pure state $|\psi\rangle$, first as function of α, β and then as function of $\boldsymbol{\varrho}$.

2. Answer the same questions for an arbitrary statistical mixture. Use the representation of a density operator $\widehat{D}$ in terms of a vector $\boldsymbol{\varrho}$ to show that the decomposition $\widehat{D} = \sum |\psi_\lambda\rangle q_\lambda \langle\psi_\lambda|$ is, in general, not unique. What is the statistical mixture described by a ket $\alpha e^{\mathrm{i}\varphi}|+\rangle + \beta e^{\mathrm{i}\psi}|-\rangle$, where the phases φ and ψ are completely random? Compare this mixture with the pure state $\alpha|+\rangle + \beta|-\rangle$.

3. Starting from the Pauli matrices algebra, find directly the expression for the density operator $\widehat{D}$ as function of the polarization $\boldsymbol{\varrho}$. What is the geometric meaning of the eigenvectors and the eigenvalues of $\widehat{D}$? Express the entropy $S = -\mathrm{Tr}\widehat{D}\ln\widehat{D}$ as function of $\boldsymbol{\varrho}$. How does it change with ϱ?

4. Consider a two-spin system in the pure state $|\psi\rangle = \alpha|+-\rangle + \beta|-+\rangle$. What is the density operator for the first spin? Can its state be described by a ket? Answer the same question for $|\psi\rangle = \alpha|+-\rangle + \beta|++\rangle$ and for $|\psi\rangle = \alpha|+-\rangle + \beta|--\rangle$.

5. Larmor precession: The Hamiltonian of the spin of an electron in a magnetic field $\boldsymbol{B}$ is $H = -(\boldsymbol{B}\cdot\widehat{\boldsymbol{\mu}})$, where the magnetic moment is given by $\widehat{\boldsymbol{\mu}} = -(e\hbar/2m)\widehat{\boldsymbol{\sigma}}$. Write down the evolution equation for $\widehat{D}$ and use it to find the evolution equation for $\boldsymbol{\varrho}$. How does the entropy vary?

6. What happens for a spin placed in a magnetic field which either has a random strength, or a random direction? More precisely, consider a spin population where each spin sees a different field. In that case there is a probability law for the field seen by an arbitrarily chosen spin; this field is fixed in time, but not well known (§ 15.4.5).

7. What happens for a spin placed in a field which evolves in a well known manner? or in a random manner? Study only the case where the field lies in the z-direction.

8. Consider a two-level system where the Hilbert space is spanned by $|1\rangle$ and $|2\rangle$ and where the Hamiltonian $v\big[|1\rangle\langle 2| + |2\rangle\langle 1|\big]$ couples the levels, inducing transitions between $|1\rangle$ and $|2\rangle$. Show that the system spends as much time in $|1\rangle$ as in $|2\rangle$.

Answers:

1. $$\varrho_x = \alpha^*\beta + \beta^*\alpha, \qquad \varrho_y = -\mathrm{i}\alpha^*\beta + \mathrm{i}\beta^*\alpha, \qquad \varrho_z = |\alpha|^2 - |\beta|^2.$$

Two parameters, as the phase of $|\psi\rangle$ is irrelevant, and as $|\alpha|^2 + |\beta|^2 = 1$.
The surface of the sphere $\boldsymbol{\varrho}^2 = 1$.

$$\langle \boldsymbol{S}\rangle^2 = \frac{1}{4}\hbar^2, \qquad \langle \boldsymbol{S}^2\rangle = \frac{3}{4}\hbar^2.$$

If $\boldsymbol{\varrho}$ is characterized by the Euler angles θ, φ with $\varrho = 1$, $|\psi\rangle$ is – apart from its phase – determined by $|\alpha| = \cos\frac{1}{2}\theta$, $|\beta| = \sin\frac{1}{2}\theta$, $\arg(\beta/\alpha) = \varphi$.

$\boldsymbol{\varrho}_1 = -\boldsymbol{\varrho}_2$ correspond to two orthogonal kets.

$$a = |\alpha|^2 = \tfrac{1}{2}\big(1+\varrho_z\big) = 1-d, \qquad b = c^* = \alpha\beta^* = \tfrac{1}{2}\big(\varrho_x - \mathrm{i}\varrho_y\big).$$

2. $$\varrho_x = b + b^*, \qquad \varrho_y = \mathrm{i}\big(b - b^*\big), \qquad \varrho_z = a - d.$$

Three real parameters: $b = c^*$, $a + d = 1$, $ad \geq |b|^2$.
The interior of the sphere $\boldsymbol{\varrho}^2 = 1$.

$$\langle \boldsymbol{S}\rangle^2 = \tfrac{1}{4}\hbar^2\varrho^2.$$

$$a = \tfrac{1}{2}\big(1+\varrho_z\big), \qquad b = \tfrac{1}{2}\big(\varrho_x - \mathrm{i}\varrho_y\big).$$

$\widehat{D} = \sum_\lambda |\psi_\lambda\rangle q_\lambda \langle\psi_\lambda|$ is represented by the barycentre, with weights q_λ, of the points representing $|\psi_\lambda\rangle$ on the surface of the sphere. For this reason ϱ lies inside this sphere. Conversely, any point inside this sphere can be considered in an infinite number of ways as the barycentre of points on the surface of the sphere; this leads to an infinity of representations $|\psi_\lambda\rangle$, q_λ. These various, discrete or continuous, representations are, however, of no interest whatever, as they cannot be distinguished physically.

$\widehat{D} = |+\rangle|\alpha|^2\langle+| + |-\rangle|\beta|^2\langle-|$. The off-diagonal elements of $\widehat{D}$ are associated with a *phase coherence* between the components of the ket in the base considered. The two states give the same polarization ϱ_z along the z-axis; however, in the one case $\varrho_x = \varrho_y = 0$, and in the other $\varrho_x^2 + \varrho_y^2 = 1 - \varrho_z^2$ takes the maximum value which characterizes a pure state.

3. The Pauli matrix algebra is characterized by the relations $\widehat{\sigma}_i\widehat{\sigma}_j = \delta_{ij} + \mathrm{i}\sum \varepsilon_{ijk}\widehat{\sigma}_k$, $\mathrm{Tr}\widehat{\sigma}_i = 0$. Any observable can be written in the form $\widehat{A} = u\widehat{I} + (\boldsymbol{v}\cdot\widehat{\boldsymbol{\sigma}})$, where u and $\boldsymbol{v}$ are real; the average of A is $\langle A\rangle = u + (\boldsymbol{v}\cdot\boldsymbol{\varrho})$. The fact that $\langle A^2\rangle$ is positive implies that $\varrho \le 1$ and $\langle A\rangle = \mathrm{Tr}\widehat{D}\widehat{A}$ leads to $\widehat{D} = \frac{1}{2}\big(\widehat{I} + (\boldsymbol{\varrho}\cdot\widehat{\boldsymbol{\sigma}})\big)$.

The eigenvectors of $\widehat{D}$ are represented by the endpoints of the diameter through $\boldsymbol{\varrho}$ and its eigenvalues are equal to $\frac{1}{2}(1 \pm \varrho)$. The entropy

$$S = k\left(\frac{1+\varrho}{2}\ln\frac{2}{1+\varrho} + \frac{1-\varrho}{2}\ln\frac{2}{1-\varrho}\right)$$

decreases from $k\ln 2$ to 0 as ϱ increases from 0 to 1.

4. $\quad \varrho_x = \varrho_y = 0, \qquad \varrho_z = |\alpha|^2 - |\beta|^2.$

There is no ket, except when $\alpha\beta = 0$.

For the last two examples the states are pure: $|+\rangle$ and $\alpha|+\rangle + \beta|-\rangle$.

5. It follows from $\widehat{D} = \frac{1}{2}\big(\widehat{I} + (\boldsymbol{\varrho}\cdot\widehat{\boldsymbol{\sigma}})\big)$ that $d\boldsymbol{\varrho}/dt = [\boldsymbol{\omega}\times\boldsymbol{\varrho}]$, with $\boldsymbol{\omega} \equiv e\boldsymbol{B}/m$. The end of the vector $\boldsymbol{\varrho}$ precesses in a plane perpendicular to $\boldsymbol{B}$ with the Larmor frequency $eB/2\pi m$. The entropy remains constant.

6. When we state that the field is random, we say, in fact, that various evolutions are possible, each with its own probability. Each evolution leads to a point $\boldsymbol{\varrho}(t)$ on the sphere with radius $\varrho(t=0)$. However, the average spin at time t which is the barycentre of the points $\boldsymbol{\varrho}(t)$ lies inside the sphere: due to the random evolution the spin gets depolarized and the entropy increases. If, for instance, we consider the various, non-interacting, spins in a paramagnetic solid as constituting a statistical ensemble, the local field to which each of them is subject is proportional to the applied field, but varies from one site to another; if all spins were initially oriented in the same way, the statistical average – that is, the total magnetization of the solid – decreases in length: we end up with a disordered state. We refer, however, to the § 15.4.5 for a discussion of spin echo experiments.

When the field $\boldsymbol{B}$ along the z-axis has a random strength, the Larmor rotation around the z-axis equals $\varphi = eBt/m$ after a time t; this is also a random variable with a fluctuation $\Delta\varphi = e\Delta Bt/m$. When t is sufficiently large, $\Delta\varphi$ is large as compared to 2π so that the points corresponding to different evolutions are distributed evenly over the circle $\varrho_z(t) = \varrho_z(0)$, $\varrho(t) = \varrho(0)$ and that ϱ_x and ϱ_y tend to zero. When the direction of $\boldsymbol{B}$ is also random, $\boldsymbol{\varrho}$ tends to zero in the final state.

This discussion also applies to the total magnetic moment of a set of spins which interact neither with one another nor with the surroundings and which are placed in an inhomogeneous magnetic field $\boldsymbol{B}$ (§§ 1.4.5 and 15.4.5).

7. The precession equation from 5. remains valid and the polarization $\boldsymbol{\varrho}$ evolves on a sphere while the entropy remains constant. If $\boldsymbol{B}(t)$ varies randomly, we must

again take an average over the various possible evolutions; this makes ϱ smaller. Let $\omega(t)$ be the instantaneous Larmor frequency; if $\boldsymbol{B}$ is along the z-axis, ϱ_z remains constant; the other components $\varrho_x + \mathrm{i}\varrho_y \equiv \zeta$ vary as

$$\zeta(t) = \zeta(0) \exp\left[\mathrm{i} \int_0^t dt\, \omega(t)\right].$$

Let us assume that $\omega(t)$ is a Gaussian random function; its probability law is then described by the characteristic function

$$\left\langle \exp\left[\mathrm{i} \int_{-\infty}^{+\infty} dt'\, \lambda(t')\omega(t')\right]\right\rangle$$
$$= \exp\left[\mathrm{i}\overline{\omega} \int dt'\, \lambda(t') - \frac{1}{2} \int dt'\, dt''\, \lambda(t')\lambda(t'')g(t'-t'')\right],$$

which is a functional of $\lambda(t)$ and which is defined by an average taken over the various evolutions; $\overline{\omega}$ is the average of ω; the autocorrelation function $g(t'-t'')$ of the signal $\omega(t)$ is non-vanishing only for very small values of $t'-t''$, as $\omega(t')$ and $\omega(t'')$ are no longer correlated when the time interval $t'-t''$ becomes large. We find thus the average of $\zeta(t)$ over the random evolution by taking $\lambda(t') = \theta(t')\theta(t-t')$. This yields

$$\langle\zeta(t)\rangle = \zeta(0) \exp\left[\mathrm{i}\overline{\omega}t - \frac{1}{2}\int_0^t \int_0^t dt'\, dt''\, g(t'-t'')\right]$$
$$\simeq \zeta(0) \exp\left[\mathrm{i}\overline{\omega}t - \Gamma t\right],$$

$$\Gamma = \frac{1}{2}\int_{-\infty}^{+\infty} dt\, g(t).$$

An exponential *relaxation* of $\varrho_x^2 + \varrho_y^2$ with a characteristic time $1/\Gamma$ is superimposed upon the precession (§ 14.1.2). In other words, the off-diagonal elements of $\widehat{D}$ in the base which diagonalizes $\widehat{H}$ tend to disappear.

The part of $\boldsymbol{B}(t)$ which varies randomly with time can be considered as a model of the effect, on the spin considered, of other spins with which it interacts, as these spins themselves change in a complex manner (§ 1.2.2).

8. This problem is formally similar to the Larmor precession of a spin around the x-axis, if we identify $|1\rangle$ with $|+\rangle$ and $|2\rangle$ with $|-\rangle$. The rotation implies that $\langle 1|\psi\rangle$ and $\langle 2|\psi\rangle$ oscillate in a complementary way. If two levels with the same unperturbed energy are coupled by a small random potential v, their populations tend to become equal, the density operator being $\frac{1}{2}\widehat{I}$. More generally, for N levels with close-lying energies, this remark can justify the use of a microcanonical ensemble, as in Chap.1.

2b Density Fluctuations

Disregarding the velocities, the macro-state of a gas is characterized by the probability distribution of the positions of the molecules. In equilibrium the N particles contained in a vessel of volume Ω are statistically independent and uniformly distributed in Ω. The number n of particles contained in a small volume v is a random variable. Check that $\overline{n} = vN/\Omega$. Calculate the variance of n, defined by

$$\Delta n^2 = \left\langle (n - \overline{n})^2 \right\rangle = \langle n^2 \rangle - \langle n \rangle^2.$$

Determine for a gas under normal conditions the values of v such that the relative fluctuation $\Delta n/\overline{n}$ be smaller than 10^{-6}.

What is the probability distribution p_n of n? What is it in the limit as $N \to \infty$, $\Omega \to \infty$, v fixed, N/Ω fixed? What in the same limit, with moreover $\overline{n} \gg 1$, $|n - \overline{n}| \ll \overline{n}^{2/3}$? What in the limit as $N \to \infty$, $\overline{n} \to \infty$, $|n - \overline{n}| \ll \overline{n}^{2/3}$, $|n - \overline{n}| \ll (N - \overline{n})^{2/3}$?

What is the probability that $|n - \overline{n}|/\overline{n} > 10^{-9}$ in a volume v of 1 cm^3, under normal conditions?

Results:

$$\Delta n^2 = N \frac{v}{\Omega} \left(1 - \frac{v}{\Omega}\right) \simeq \overline{n}.$$

$$\overline{n} > 10^{12}, \qquad v > 3.7 \times 10^{-14} \text{ m}^3.$$

If one measures the densities within a margin of 10^{-6}, the gas is homogeneous only over distances larger than 0.03 mm.

Binomial law : $$p_n = \frac{N!}{n!(N-n)!} \left(\frac{v}{\Omega}\right)^n \left(1 - \frac{v}{\Omega}\right)^{N-n}.$$

Poisson law : $$p_n \to \frac{1}{n!} \overline{n}^n \mathrm{e}^{-\overline{n}}.$$

Gaussian law : $$p_n \to \frac{1}{\sqrt{2\pi\overline{n}}} \exp\left[-\frac{(n-\overline{n})^2}{2\overline{n}}\right].$$

Gaussian law : $$p_n \to \sqrt{\frac{N}{2\pi\overline{n}(N-\overline{n})}} \exp\left[-\frac{N(n-\overline{n})^2}{2\overline{n}(N-\overline{n})}\right].$$

The probability we are looking for is

$$\sqrt{\frac{2}{\pi}} \int_{10^{-9}\sqrt{\overline{n}}}^{\infty} dx\, \mathrm{e}^{-x^2/2} = 2.2 \times 10^{-7},$$

so that n is practically certain.

Hints. The above results can be derived by alternative methods.

(i) *Probability Law Method*: By counting arguments, one can first establish the probability law p_n, and from this one finds $\overline{n}$ and Δn, as well the asymptotic forms of p_n.

(ii) *Indicator Method*: For each particle i we introduce a random variable a_i which is equal to 1 if the particle is in v and 0 if it is not in v, and we use the fact that $n = a_1 + \ldots + a_N$, and that $\langle a_i \rangle = \langle a_i^2 \rangle = v/\Omega$, $\langle a_i a_j \rangle = \langle a_i \rangle \langle a_j \rangle$ for $i \neq j$. This defines the probability law for the variables a_i, from which $\overline{n}$, Δn, and p_n are deduced.

(iii) *Characteristic Function Method* (cf Eq.(1.13)): Start from

$$Z(\alpha) = \left\langle \mathrm{e}^{\alpha n} \right\rangle = \prod_i \left(\frac{v}{\Omega} \mathrm{e}^{\alpha} + 1 - \frac{v}{\Omega} \right) = \left(\frac{v}{\Omega} \mathrm{e}^{\alpha} + 1 - \frac{v}{\Omega} \right)^N ,$$

and take twice the derivative of $\ln Z$ for $\alpha = 0$; this gives us $\overline{n}$ and Δn^2. The probability p_n is found by expanding Z in powers of e^{α}.

The first two methods are common in probability theory, and the third one is the prototype of the partition function method, which we shall use systematically in the present book for calculating most of the physical quantities in equilibrium statistical mechanics. However, the function $Z(\alpha)$ introduced here is not the partition function of the gas, which takes into account the velocity distribution and which will be introduced in Chap.7.

3. Information Theory and Statistical Entropy

"La science est un produit de l'esprit humain, produit conforme aux lois de notre pensée et adapté au monde extérieur. Elle offre donc deux aspects, l'un subjectif, l'autre objectif, tous deux également nécessaires, car il nous est aussi impossible de changer quoi que ce soit aux lois de notre esprit qu'à celles du Monde."

Bouty, La Vérité Scientifique, 1908

"– Vous savez, naturellement, ce que c'est que l'entropie?
– Oui, dit-elle, essayant, sans succès, de s'en remémorer la formule.
– Eh bien, l'entropie, c'est-à-dire, en gros, l'usure, la décadence de l'énergie, guette l'érotisme comme l'univers tout entier."

E.Arsan, Emmanuelle, 1973

In Chap.2 we made a synthesis of quantum mechanics, which governs on the microscopic scale the systems that we are studying, and statistics, which is necessary because systems are macroscopic and thus incompletely known. Here we shall complete the formalism by introducing information, a mathematical concept associated with probability theory, which in the following chapters will turn out to be essential in statistical mechanics as it will enable us to understand the microscopic significance of the entropy.

The density operator in quantum mechanics or the phase density in classical mechanics represent our knowledge about the system. This knowledge is more or less complete: clearly our information is a maximum when we can make predictions with full certainty, and it is larger when the system is in a pure state than when it is in a statistical mixture. Moreover, this system is better known when the number of possible micro-states is small or when the probability for one of them is close to unity than when there are a large number of possible micro-states with all approximately the same probability. The object of the present chapter is to formulate mathematically these intuitive ideas: we shall show how we can *quantify the amount of information that we are missing* due to the fact that the only knowledge we have about the system is a probability law. Using a less subjective language we shall identify this missing information with a quantitive measure for the *degree of disorder* existing in a system the preparation of which has a random nature. We shall first of all introduce the concept of information in the context of probability calculus and of communication theory (§ 3.1). We shall then consider its application to quantum or classical statistical mechanics. This will enable us to associate with each density operator, or density in phase, a number which measures our uncertainty arising from the probabilistic nature of our description (§§ 3.2 and 3.3). This number, called the *statistical entropy*, has various mathematical properties which confirm its interpretation as a measure of disorder (§§ 3.1.2 and 3.2.2). In Chap.5 we shall identify it with the macroscopic entropy introduced by the Second Law of thermodynamics.

We shall end this chapter with a historical survey of the entropy concept from various angles and of the rôle it has played in the evolution of science since the middle of the nineteenth century (§ 3.4).

3.1 Information and Probability Theory

3.1.1 Statistical Entropy: The Measure of Missing Information

We consider a set of M events with respective probabilities $p_1, \ldots, p_m, \ldots, p_M$, which are non-negative and add up to 1. The degree of predictivity of such a probability law is an intuitive concept which we want to make more quantitative by *assigning a number to the amount of information which is missing* because the expected event is random. When all probabilities, except $p_m = 1$, vanish, we are certain that the event m will occur; there is no information missing. On the other hand, when all possible events have the same probability, our perplexity increases with the number M of such events. The *lack of information* which will characterize the random nature of the probabilities $p_1, \ldots, p_M$ is a certain function $S(p_1, \ldots, p_M)$ of those probabilities, also called the *statistical entropy*, or the *uncertainty*, or the *dispersion*, associated with the probability law considered.

We shall in § 3.1.2 construct an expression for S, starting from a small number of postulated intuitive properties. We give this result here. Apart from a multiplicative factor k, the *expression for the statistical entropy* is the following:

$$S(p_1, \ldots, p_M) = -k \sum_{k=1}^{M} p_m \log p_m. \tag{3.1}$$

An important special case is the one where the M events are *equiprobable*. The statistical entropy then reduces to

$$\sigma_M = S\left(\frac{1}{M}, \ldots, \frac{1}{M}\right) = k \log M. \tag{3.2}$$

These expressions were, in fact, first introduced in statistical mechanics. They also appear in the context of *communication theory* where one tries to quantify the amount of information contained in a message. One regards each *message* m which can be transmitted from an emitter to a receptor as an *event* with a probability p_m. The set of probabilities p_m is a characteristic of the *language*, that is, of the collection of all conceivable messages. The quantity of information, I_m, acquired in the reception of the message m must satisfy some simple properties. In particular, the information brought by the reception of two statistically independent messages must be *additive*. One can then show (§ 3.1.3) that the quantity of information carried by the message m should be defined by

$$I_m = -k \log p_m. \tag{3.3}$$

This quantity may also be considered as measuring the *surprisal* produced by the occurrence of the event m: the less probable a message, the smaller p_m, the larger I_m, and the more surprised we are to receive it.

Expression (3.1) can in this framework be interpreted as the information which is *gained on average* by receiving one of the possible messages; the least expected messages with a small probability p_m are those which carry most information, but their contribution to the mean information (3.1) is small, precisely because there is little chance of receiving them: as $p_m \to 0$, $-\log p_m \to \infty$, but $-p_m \log p_m \to 0$. The maximum of each term in (3.1) is reached when $p_m = 1/\mathrm{e}$.

Even though the same concept, called statistical entropy, lack of information, uncertainty, or disorder, appears in different guises in probability calculus, in communication theory, or, as we shall see, in statistical mechanics, the points of view differ, depending on whether one is looking at things before or after the realization of the event. In communication theory, expression (3.3) refers to the situation *after* the reception of the message m. Accordingly, expression (3.1) can be interpreted as the average *gain* in information associated with the transmission of a message. However, one can also look at (3.1) *before* the transmission, while the recipient is waiting for it: from that standpoint the number S appears as the entropy of the language, that is, as an *uncertainty* about the set of all messages which may be received. We shall always adopt that point of view in statistical mechanics as one is, in practice, never getting to know the "message" of a macroscopic physical system completely, that is, never carrying out a complete measurement. The statistical entropy will characterize, for a macro-state prepared according to a given probability law, our uncertainty about the set of all microscopic experiments which are conceivable, but which will not be performed.

The information is defined apart from a multiplying factor k which determines the *unit of information*. In communication or information theory, one often chooses $k = 1$ with logarithms to the base 2. The information unit, called the "bit" or "binary digit", is then the one acquired in a head or tail draw, since $\sigma_2 = 1$ for a simple alternative with equal probabilities. With that choice the quantity of information associated with a message written in binary characters, 0 or 1, and containing n uncorrelated characters is equal to n bits, since there are 2^n equiprobable possible messages and $\sigma_{2^n} = n$.

In statistical mechanics, the systems are so complex and the possibilities so numerous that it will be more convenient to choose k very small (Exerc.3a). The discussion in § 1.3.3 suggests already that, if we take for k Boltzmann's constant, $k = 1.38 \times 10^{-23}$ J K^{-1}, and natural logarithms, we can identify S with the entropy of thermodynamics measured in J K^{-1} (§ 5.3.2). In fact, the choice of the kelvin as the unit of temperature and the joule as the unit of energy leads to thermodynamic entropies of macroscopic bodies which are numerically neither too large nor too small. For instance, the statistical entropy of one mole of spins with 2^{N_A} equiprobable configurations (§ 1.2.1) equals N_A bits, where $N_\text{A} = 6 \times 10^{23}$ mol^{-1} is Avogadro's number, but reduces to $kN_\text{A} \ln 2 = 5.76$ J K^{-1} mol^{-1} in thermodynamic units. Had we chosen to measure entropies in dimensionless units such as bits, their typical values would have been large like N_A for macroscopic objects. It is thus no accident that the value of Boltzmann's constant in J K^{-1} is of the order of the inverse of Avogadro's number.

3.1.2 Properties of the Statistical Entropy

In order to make expression (3.1) for the statistical entropy more translucent we list below some of its properties which support the idea that we are dealing with a *measure of uncertainty* or of lack of information. We shall prove these properties in § 3.2.2 where we shall return to the form that they take in the statistical mechanics framework.

(a) *Maximum.* For a fixed number M of events, S reaches a maximum, equal to $\sigma_M = k \log M$, when these events are *equiprobable.* Indeed, we know least about the system in this case. More generally, S increases whenever two probabilities p_m and p_n get closer to one another.

(b) *Minimum.* The quantity S reaches a minimum, equal to *zero*, when one of the events is a *certainty*:

$$S(1, 0, \ldots, 0) = 0. \tag{3.4}$$

One gains no information, when one knows in advance which message one is about to receive.

(c) *Increase.* In the case of M *equiprobable* events the statistical entropy σ_M increases with M. If all possible events are equally probable, the more their number increases, the more information we lack. The logarithmic behaviour of the increase in σ_M is connected with the fact that information acquired when reading a message is proportional to its length, and that the number of possible equiprobable messages increases exponentially with the number of characters in the message, provided the latter are equally probable and not correlated.

(d) *Impossibility of Events.* If some events have zero probability, they can just as well be left out of the reckoning when we evaluate the uncertainty:

$$S(p_1,\ldots,p_M,0,\ldots,0) \;=\; S(p_1,\ldots,p_M). \tag{3.5}$$

(e) *Symmetry.* As the M events are treated on the same footing, $S(p_1,\ldots,p_M)$ is a symmetric function of its M arguments.

(f) *Additivity.* If we consider the occurrence of any one among the set of events 1, ..., m as a single event A, with probability $q_A = p_1 + \ldots + p_m$, and similarly one of the events $m+1$, ..., M as a single event B, it follows from (3.1) that

$$\begin{aligned} &S(p_1,\ldots,p_m,p_{m+1},\ldots,p_M) \\ &\quad \equiv \; S(q_A,q_B) + q_A S\left(\frac{p_1}{q_A},\ldots,\frac{p_m}{q_A}\right) + q_B S\left(\frac{p_{m+1}}{q_B},\ldots,\frac{p_M}{q_B}\right). \end{aligned} \tag{3.6}$$

This equation expresses that the *average information acquired in two stages must be added*: the first term corresponds to the choice between A and B, and the next two terms to the choices inside the groups A and B. We shall comment in detail on the form of these terms and their interpretation in § 3.1.3.

Another form of the additivity of information refers to the particular case of MN *equiprobable* events each characterized by a pair of indices, $1 \le m \le M$ and $1 \le n \le N$. When one of the events (m,n) is realized, we acquire the same quantity of information by finding m and n *simultaneously*, as by finding them *successively*; this can be expressed by

$$\sigma_{MN} \;=\; \sigma_M + \sigma_N. \tag{3.7}$$

(g) *Sub-additivity.* Consider a *composite event* corresponding to the combination of two events belonging one to a set $a = \{\ldots,m,\ldots,M\}$ and the other to a set $b = \{\ldots,n,\ldots,N\}$. If the composite event m,n has a probability p_{mn}, the probability of the event m is

$$p_m^a \;=\; \sum_n p_{mn},$$

and that of n

$$p_n^b \;=\; \sum_m p_{mn}.$$

One can show (§3.2.2e) that the statistical entropies associated with these probability laws satisfy the so-called sub-additivity inequality:

$$\begin{aligned} &S(p_1^a,\ldots,p_m^a,\ldots,p_M^a) + S(p_1^b,\ldots,p_n^b,\ldots,p_N^b) \\ &\qquad\qquad \ge \; S(p_{11},\ldots,p_{mn},\ldots,p_{MN}). \end{aligned} \tag{3.8}$$

The equal sign holds when the sets a and b are *statistically independent*, that is, when there is no correlation between the events m and n, in which case the factorization of the probabilities,

$$p_{mn} = \overset{a}{p}_m \overset{b}{p}_n,$$

entails the additivity of the statistical entropies. Inequality (3.8) thus expresses that we *lack less information about a composite system when we know the correlations* between the events m and n than when we only know the separate probabilities of these events.

(h) *Concavity.* Consider two probability laws $p_1, \ldots, p_M$ and $p'_1, \ldots, p'_M$, relating to the *same events.* Let λ be a number such that $0 < \lambda < 1$; from p and p' we can form a new probability law $p''_1 = \lambda p_1 + (1-\lambda)p'_1, \ldots, p''_M = \lambda p_M + (1-\lambda)p'_M$. The statistical entropy is then a *concave function of the probabilities*:

$$S(p''_1, \ldots, p''_M) > \lambda\, S(p_1, \ldots, p_M) + (1-\lambda)\, S(p'_1, \ldots, p'_M). \tag{3.9}$$

The proof of this inequality follows from the fact that each term of (3.1) is concave. Its interpretation is simple: it means that if we *combine two statistical ensembles* relating to the same events into a single new ensemble, the uncertainty is larger than the average of the initial uncertainties. Let us, for instance, consider two bags containing balls of M different colours: the probabilities $p_1, \ldots, p_M$ and $p'_1, \ldots, p'_M$ to draw a ball of a given colour from each bag are proportional to the number of balls of each colour in the two bags; if λ and $1-\lambda$ are proportional to the total number of balls in the two bags, the probabilities $p''_1, \ldots, p''_M$ correspond to the mixture of their contents. The concavity of the statistical entropy, considered as a measure of the disorder, means simply that *this mixture has increased the disorder*, and it gives a quantitative measure for this increase.

(i) *Shannon's Theorems.* In communication theory, one is mainly interested in expression (3.1) for the statistical entropy because of Shannon's theorems (1948). We shall here only give the main ideas.[1] One can distinguish the following elements in a transmission system, for texts, numbers, images, sound, and so on. A *source* produces messages, in a language characterized by a probability law with which we can associate a statistical entropy. This entropy (3.1) increases with the number of possible messages; it decreases if the elementary constituents of the message have unequal frequencies or are correlated with one another. The messages are transformed by a *transmitter* into signals which are going to be emitted. This operation, which involves *coding*, that is, a translation into a new language, is necessary to make the message suitable for being sent over long distances by a physical process, such as electromagnetic waves, electric signals, optical signals, acoustic signals, ... Moreover, different codings can be implemented using the same physical process, such as amplitude or frequency modulation in radio waves. This flexibility is essential for optimizing the communication system, as the elementary signals which are to be transmitted have a probability distribution depending both on that of

[1] C.E.Shannon and W.Weaver, *The Mathematical Theory of Communication*, University of Illinois Press, 1949.

the original messages and also on that of the chosen code. The *channel* through which the signals are then sent is characterized by the flow rate of the elementary signals and by its trustworthiness. In fact, the transmission can be accompanied by errors where some of the signals are randomly lost or altered; such *noise* thus destroys signals with a certain probability law, and the *capacity* of the channel is defined in a way which is derived from that of the statistical entropy (3.1), taking into account both the flow rate of the signals and the probability law for the noise. The chain is completed by a *receiver* which decodes the received signals for use by the addressee.

Shannon's theorems give a limit to the efficiency of such a communication system in the form of an upper bound for the amount of information received, on average, per unit time. In particular, there exists an *optimum coding* which depends on the initial language and the characteristics of the channel; this makes it possible to transmit messages with a mean rate of errors which can be arbitrarily small notwithstanding the presence of noise, provided the flow of information received is smaller than the capacity of the channel. In order to reduce the transmission time, the signals encoding the most frequent elements of the message must be the shortest. In Morse telegraphy, the letter E which has the largest frequency – 0.10 in English and 0.14 in French – has a single point as its code, whereas the rare letters have four elements: for instance, the letter Z which has a frequency of 5×10^{-4} in English and 6×10^{-4} in French is – – · ·, and we have – – · – for Q which has a frequency of 8×10^{-4} in English and 10^{-2} in French. Similarly, in stenography, words or groups of letters which occur frequently are coded by single signs. In order to overcome the noise, the coding must be *redundant*; repetition of the signals and their correlations make it possible to compensate for their destruction, as the probability for multiple faults is small. In a written text the redundancy comes from the existence of groups of correlated letters or words; this enables us to understand the meaning of a message without consciously seeing all its letters, rapid reading playing the rôle of transmission with noise. The genetic code contains redundancies which make it possible to overcome errors when the DNA is duplicated or the proteins are produced. Similarly, the control letter added to the number code of cheques reduces banking errors. In all cases, the optimum coding realizes a compromise between a compression of the message and redundancies.

3.1.3 The Statistical Entropy Derived from the Additivity Postulate

We have given earlier (3.1) as the definition of the statistical entropy and we have derived from it several properties to support its interpretation as a measure for the lack of information. Conversely, we can construct (3.1) from a few intuitive requirements which must be satisfied by the information.

One possible approach consists in first constructing I_m, the quantity of information acquired in the reception of a particular message m. To justify its expression (3.3), we postulate the following properties. This information is a *function of the probability* of the event considered; it depends neither on the language used, nor on the probabilities of the other messages which could have been received. The function $I_m = I(p_m)$ is a *decreasing* function on the interval [0,1]: the more stereotyped a message is, that is, the larger its probability for being received, the less information it imparts. Finally, the information acquired from two independent messages is *additive*: let us assume that we receive two messages n and m, where one is part of a set $1, \ldots, N$ with probabilities $q_1, \ldots, q_N$, and the other part of a set $1, \ldots, M$ with probabilities $p_1, \ldots, p_M$. The information associated with these two messages is $I(q_n)$ and $I(p_m)$. Let us consider the composite message (n, m) as a single message and let us assume that the two sets are statistically independent, so that the probability for the composite message (n, m) is $q_n p_m$. It is natural to postulate that the information $I(q_n p_m)$ is the sum of the informations $I(q_n)$ and $I(p_m)$ acquired separately. One could equally well have postulated another kind of additivity, grouping the messages as in § 3.1.1f and considering, for instance, as the first message, with probability q_A, the specification of the group A, and as the second message, with probability p_m/q_A, the specification of m inside the group A, the composite message being the message m itself; the resulting equation for I would have been the same.

One is thus led to construct a function $I(p)$ which decreases in the interval (0,1) and which satisfies the identity

$$I(pq) = I(p) + I(q). \tag{3.10}$$

One can easily prove that the only solution of this problem is given by (3.3). In fact, if one restricts oneself first of all to integer $1/p = M$ and puts $I(1/M) \equiv \sigma_M$, (3.10) implies (3.7). This is an equation which produces the σ_M for integer M, starting from the σ_M for prime M, with $\sigma_1 = 0$. In order to determine σ_M for prime M we take advantage of the increase of σ as follows: let M and N be given and let m be a large integer given beforehand; there then exists an integer n such that

$$\frac{n}{m} \leq \frac{\ln N}{\ln M} < \frac{n}{m} + \frac{1}{m},$$

that is, such that

$$M^n \leq N^m < M^{n+1}.$$

From the increasing nature of σ we find that

$$\sigma_{M^n} \leq \sigma_{N^m} < \sigma_{M^{n+1}},$$

or, if we use (3.7) and divide by m,

$$\frac{n}{m} \leq \frac{\sigma_N}{\sigma_M} < \frac{n}{m} + \frac{1}{m}.$$

From the definition of n it then follows that

$$\left| \frac{\sigma_N}{\sigma_M} - \frac{\ln N}{\ln M} \right| < \frac{1}{m}, \qquad \forall\, m,$$

which proves (3.2). The constant k is positive, since σ_M increases with M, but it remains otherwise arbitrary. As we thus know $I(1/M) = k \ln M$ for integer M, (3.10) with $q = 1/M$ and $pq = 1/N$ gives us $I(p) = -k \ln(M/N)$ when $p = M/N$ is rational. Finally, the decreasing nature of $I(p)$ extends this function to irrational values of p, which justifies (3.3). The lack of information (3.1) follows by taking the average over the various messages m.

The proof is simpler, if we require that $I(p)$ be a function with a derivative. In fact, one can deduce from (3.10) that $pI'(pq) = I'(q)$, that is, if we put $x = pq$, that $I'(x) = -k/x$, where $k \equiv qI'(q)$ is a constant which does not depend on x. By integration we then find again (3.3); the integration constant follows from $I(1) = 0$, itself a consequence of (3.10).

Alternatively, we can try to construct expression (3.1) for the statistical entropy directly. In that case the additivity of the entropy for composite events is not sufficient for its complete determination. One can, for instance, check that the equality (3.8) for uncorrelated events is satisfied not only by (3.1), but also by the so-called α-entropies

$$S_\alpha = \frac{k}{1-\alpha} \ln \sum_{m=1}^{M} (p_m)^\alpha,$$

where α is positive and arbitrary. We find the ordinary statistical entropy as $\alpha \to 1$. We shall, nevertheless, prove that, in order to justify (3.1), it is sufficient to postulate some of the other properties listed in § 3.1.2. To do this we shall, for illustrative purposes, reproduce the reasoning which enabled Shannon to introduce this expression.

The most important postulate is the *additivity* of the statistical entropy in the form (3.6), the interpretation of which we have briefly indicated. We shall add a few comments in order to make this property obvious and to explain the rôle played by the various terms in (3.6). The first term, $S(q_A, q_B)$, expresses that one does not know whether the random event is part of the group A or of the group B, and it quantifies that uncertainty. We must then estimate the extra uncertainty that, if the event is in the group A, we do not know whether it is event 1, or ..., or event m, and, if it is in the group B, whether it is $m+1$, or ..., or M. We have thus two alternatives. Let us assume that the event pertains to group A. The respective probabilities of the events $1, \ldots, m$ are in that case $p_1/q_A, \ldots, p_m/q_A$; in fact, we are dealing with conditional probabilities for 1, ..., M, for the case when A is realized. Under that alternative we obtain $S(p_1/q_A, \ldots, p_m/q_A)$ for the *conditional* lack of information, for the case when A is realized. Similarly, if we knew that the event was part of group B, we would for the associated conditional lack of information get $S(p_{m+1}/q_B, \ldots, p_M/q_B)$. On the other hand, there is a probability q_A for the group A and a probability q_B for the group B. For this reason the second term on the right-hand side of (3.6) has the weight q_A and the third term the weight q_B: the ratio of the conditional lack of information to the contribution to the global lack of information is thus equal to the ratio of the conditional probabilities to the global probabilities p_m. If, in fact, the group A is practically impossible, so that $q_A \simeq 0$, one does not care much whether one knows what are the relative chances for the events 1, ..., m to be realized, as one will hardly ever encounter that case. Conversely, if the group A is almost a certainty, so that $q_A \simeq 1$, the lack of information $S(p_1/q_A, \ldots, p_m/q_A)$ about the events in that group must be almost the total lack of information $S(p_1, \ldots, p_m, p_{m+1}, \ldots, p_M)$. Similarly, the factor q_B of the last term on the right-hand side of (3.6) can be dropped if the events $m+1$, ..., M are impossible; it must play the leading rôle if they are the only possible ones. Altogether one can thus interpret (3.6) by stating that one gains the same quantity of information on average whether the observation is made in one or in two stages.

One can, moreover, postulate the *continuity* of $S(p_1, \ldots, p_M)$ with respect to its M arguments and the *increase* of σ_M with M. The proof of (3.1) is then carried out in two stages. First, we determine the statistical entropy (3.2) for equiprobable events and then we derive the general expression (3.1).

By iterating over the number α of groups of events, we first derive from (3.6) a more general form of the additivity,

$$S(p_1, \ldots, p_{m_1}, p_{m_1+1}, \ldots, p_{m_2}, \ldots, p_{m_\alpha}) = S(q_1, q_2, \ldots, q_\alpha)$$
$$+ q_1 S\left(\frac{p_1}{q_1}, \ldots, \frac{p_{m_1}}{q_1}\right) + \ldots + q_\alpha S\left(\frac{p_{m_{\alpha-1}+1}}{q_\alpha}, \ldots, \frac{p_{m_\alpha}}{q_\alpha}\right), \tag{3.11}$$

where $q_1 = p_1 + \ldots + p_{m_1}, \ldots, q_\alpha = p_{m_{\alpha-1}+1} + \ldots + p_{m_\alpha}$. A consequence of (3.11) when all events have the same probability is then that

$$\sigma_{MN} = \sigma_M + \sigma_N. \tag{3.7}$$

The construction of σ_M which satisfies (3.7) has been carried out at the beginning of this subsection and led to (3.2).

One then uses (3.11) to determine $S(q_1, \ldots, q_\alpha)$ when the probabilities $q_1, \ldots, q_\alpha$ are rational. In fact, let us write $q_1 = m_1/M, \ldots, q_\alpha = m_\alpha/M$. The left-hand side of (3.11) corresponds to all probabilities being equal to $1/M$ and is thus equal to σ_M. The first term on the right-hand side is the unknown. The other terms all correspond to equal probabilities and thus also follow from (3.2), which leads us to

$$S\left(\frac{m_1}{M}, \ldots, \frac{m_\alpha}{M}\right) = \sigma_M - \frac{m_1}{M}\sigma_{m_1} - \ldots - \frac{m_\alpha}{M}\sigma_{m_\alpha}$$
$$= -k\left(\frac{m_1}{M}\log\frac{m_1}{M} + \ldots + \frac{m_\alpha}{M}\log\frac{m_\alpha}{M}\right). \tag{3.12}$$

Finally, continuity enables us to use (3.12) to find the value of $S(p_1, \ldots, p_M)$ for irrational probabilities p_m and this determines S in the form (3.1).

3.1.4 Continuous Probabilities

So far we have only considered the case of a finite number of possible events. The definition (3.1) can without difficulties be generalized as an infinite series to the case of countable events. To extend (3.1) to continuous events, such as the random position of a point x along the interval $[a, b]$, we reduce the problem to the case of discrete events by cutting this interval up into sections Δx_m, centred around a finite number of points x_m. For a continuous probability density $p(x)$, the probability p_m that x lies within Δx_m is $p(x_m)\Delta x_m$, if Δx_m is sufficiently small. One then lets the Δx_m tend to zero, by taking for them the form $\Delta x_m = \varepsilon\chi(x_m)$, where $\varepsilon \to 0$ and $\chi(x)$ is a continuous function. We can then use (3.1) to write for the statistical entropy

$$S = -k \sum_m p(x_m)\,\Delta x_m \log[p(x_m)\,\Delta x_m]$$
$$= -k \sum_m \Delta x_m\, p(x_m) \log[p(x_m)\chi(x_m)] - k \log \varepsilon. \tag{3.13}$$

The last term tends to infinity. This can be understood: one acquires incomparably more information by observing the *exact* position of a random point along a segment than by making measurements with a finite accuracy. We can therefore no longer define S absolutely.

If we are only interested in the *relative changes* in S from one probability density to another, we can drop the constant term $-k \log \varepsilon$ and limit ourselves to the first term on the right-hand side of (3.13), which tends to

$$S' = -k \int dx\, p(x) \log[p(x)\chi(x)]. \tag{3.14}$$

There is still an arbitrary function $\chi(x)$ which is, moreover, modified when we change the variable x.

To choose this function $\chi(x)$ in order to define the lack of information associated with a continuous probability density in a natural manner, *invariance* arguments are indispensable for determining the *a priori weight* $\chi(x)$. If, for instance, the precision of the measurements is the same all along the segment, we must, of course, choose the Δx_m to be equal, that is, put $\chi(x) = 1$. The same choice is also compelling in the case of a random signal for which x is the time, because in that case the phenomena are invariant when the time origin changes. Apart from an additive constant the lack of information is then

$$S' = -k \int dx\, p(x) \log p(x). \tag{3.15}$$

3.2 The Statistical Entropy of a Quantum State

3.2.1 Expression as Function of the Density Operator

Let there be a macro-state represented by the density operator $\widehat{D}$ which can be written as

$$\widehat{D} = \sum_m |m\rangle\, p_m\, \langle m|, \tag{3.16}$$

in terms of its eigenstates and eigenvalues, and which plays for the system a rôle, similar to that of a probability law. We want to associate with $\widehat{D}$ a number S which quantifies the missing quantity of information, that is, the quantity of information that one might acquire by knowing the system better through microscopic measurements. However, quantum measurements *differ from the events considered in probability theory* as, in general, they perturb the state of the system (§§ 2.1.4 and 2.2.5). The results of these measurements do not directly give us information about the initial state $\widehat{D}$, but about the state which is reduced through interaction with the measuring apparatus, $\sum_\alpha \widehat{P}_\alpha \widehat{D} \widehat{P}_\alpha$. In order to use the results of § 3.1.1 we must thus restrict ourselves to considering those particular measurements which do not reduce $\widehat{D}$, that is, the ideal measurements represented by the observables,

$$\widehat{A} = \sum_m |m\rangle\, a_m\, \langle m|,$$

which commute with $\widehat{D}$. In this case an event is an observation of a_m and the probability that one observes a_m is p_m, in the general case where the a_m are discrete. We shall examine the case of other measurements in § 3.2.4.

One is thus led to an ordinary probabilistic problem and it is natural to define, as above, the *quantity of missing information*, or the *dispersion*, or the *statistical entropy associated with the density operator* $\widehat{D}$ by (von Neumann, 1927)

$$\boxed{S(\widehat{D}) \;=\; -k \sum_m p_m \ln p_m \;=\; -k\ \mathrm{Tr}\,(\widehat{D} \ln \widehat{D})}\ . \tag{3.17}$$

One obtains the last expression by noting that the matrix $\widehat{D}$ is diagonal in the base $\{|m\rangle\}$ and that its eigenvalues are the probabilities p_m. If the latter are all equal and if there are W of them, (3.17) reduces to $k \ln W$. The constant defining the units, k, is from now on chosen to be *Boltzmann's constant*, $k = 1.38 \times 10^{-23}$ J K^{-1} as discussed in § 3.1.1.

The statistical entropy (3.17) is invariant when one changes the density operator $\widehat{D}$ through a *unitary transformation* as it depends only on the eigenvalues p_m of $\widehat{D}$. This property which reflects the invariance of the structure of Hilbert space under a unitary transformation is the quantum generalization of the invariance of (3.1) under a permutation (§ 3.1.2e).

One can also interpret the statistical entropy (3.17) as a *measure of the disorder* which exists in the state represented by the density operator $\widehat{D}$ because of the random nature of that state. It may seem shocking to identify the lack of information, a quantity of a subjective nature measuring our uncertainty about an incompletely known system, with the degree of disorder, a quantity seemingly objective and to be connected with the system independent of whether we observe it (see § 3.4.3). A comparison will help us to understand this identification. Let us consider a game of 52 cards which are initially arranged in the natural order: ace of spades, king of spades, queen of spades, ... After a thorough shuffling the pack of cards looks to us as being perfectly disordered and this transformation seems irreversible. However, if the shuffling was done by a nimble conjurer who has arranged the cards in an order known only to him, the final configuration is just as ordered *for him* as the initial configuration was for us. Moreover, he can reconstruct the former without looking at the cards. The nature of the order is not the same in the original and the final situations, but the quantity of disorder is the same: it is zero, as knowledge is perfect. The shuffled pack is disordered *for us* because we consider it to be just one possible realization of the 52! possible permutations; we assume these to have the same probability since we do not know how the particular configuration was brought about. We shall meet with a similar situation in physics when we analyze spin echo experiments (§ 15.4.5).

The following table shows the correspondence between the various aspects of diverse concepts from information theory, depending on the context:

	Probabilities	*Communications*	*Statistical Physics*
m	event	message	micro-state
p_m	probability law	language	macro-state
S	uncertainty, dispersion	missing information	entropy, disorder

3.2.2 Inequalities Satisfied by the Statistical Entropy

We shall review some properties of the statistical entropy in statistical mechanics, like we did in § 3.1.2 in the framework of information theory. This will turn out to support the interpretation of (3.17) as a measure of the incomplete nature of the probabilistic knowledge or as a measure of disorder. We shall before that prove an inequality which will be useful several times in the present book.

Lemma. For any pair of non-negative operators $\widehat{X}$ and $\widehat{Y}$ we have

$$\boxed{\mathrm{Tr}\,\widehat{X}\ln\widehat{Y} - \mathrm{Tr}\,\widehat{X}\ln\widehat{X} \;\le\; \mathrm{Tr}\,\widehat{Y} - \mathrm{Tr}\,\widehat{X}}\,. \tag{3.18}$$

The equal sign holds only when $\widehat{X} = \widehat{Y}$. The right-hand side vanishes when $\widehat{X}$ and $\widehat{Y}$ are normalized density operators.

Let $|m\rangle$ and X_m be the eigenvectors and eigenvalues of $\widehat{X}$, and $|q\rangle$ and Y_q those of $\widehat{Y}$. Let us first assume that the X_m and the Y_q are positive. By writing $\widehat{X}$ and $\widehat{Y}$ in their respective eigenbases we can write the left-hand side of (3.18) in the form

$$\sum_{m,q} X_m\langle m|q\rangle\,\ln Y_q\,\langle q|m\rangle - \sum_m X_m\ln X_m.$$

Using the closure property of the base $|q\rangle$, that is, the fact that

$$\sum_q \langle m|q\rangle\langle q|m\rangle \;=\; 1,$$

we can also write this expression in the form

$$\sum_{m,q} |\langle m|q\rangle|^2\,X_m\ln\frac{Y_q}{X_m}.$$

We use the inequality $\ln x \le x - 1$, where the equal sign holds only when $x = 1$. This leads to

$$\begin{aligned}\mathrm{Tr}\,\widehat{X}\ln\widehat{Y} - \mathrm{Tr}\,\widehat{X}\ln\widehat{X} &\le \sum_{m,q} |\langle m|q\rangle|^2\,\left(Y_q - X_m\right),\\ &= \sum_q Y_q - \sum_m X_m = \mathrm{Tr}\,\widehat{Y} - \mathrm{Tr}\,\widehat{X},\end{aligned}$$

where we have again used the closure property of the bases $|m\rangle$ and $|q\rangle$. Using continuity we can extend the result to non-negative operators $\widehat{X}$ and $\widehat{Y}$; if some of the eigenvalues Y_q are zero, the left-hand side of (3.18) may be $-\infty$. The equal sign holds only when $Y_q = X_m$ for all pairs m, q such that $|\langle m|q\rangle|^2 \neq 0$. This implies that $\langle m|q\rangle\big(Y_q - X_m\big) = 0$, $\forall\,(m,q)$, whence

$$\sum_{m,q} |m\rangle\langle m|q\rangle\,\big(Y_q - X_m\big)\,\langle q| = 0,$$

which gives $\widehat{Y} - \widehat{X} = 0$. *QED*

(a) *Maximum.* If the possible kets are those of a finite W-dimensional subspace $\mathcal{E}_{\mathrm{H}}^{W}$ of $\mathcal{E}_{\mathrm{H}}$, the *statistical entropy is a maximum and equal to*

$$\boxed{S = k\log W} \tag{3.19}$$

if the probabilities of all the kets are equal to one another.

To prove this property it is sufficient to apply lemma (3.18), taking for $\widehat{X}$ an arbitrary density operator and for $\widehat{Y}$ the operator $\widehat{I}_W/W$, where $\widehat{I}_W$ denotes the unit matrix in the subspace $\mathcal{E}_{\mathrm{H}}^{W}$ considered. The density operator corresponding to the state of *maximum disorder* in the space $\mathcal{E}_{\mathrm{H}}^{W}$ is thus $\widehat{I}_W/W$, which means that all the kets of an orthonormal base of $\mathcal{E}_{\mathrm{H}}^{W}$ have the same probabilities, equal to $1/W$. For instance, for a spin $\frac{1}{2}$ the most disordered state is the unpolarized state (2.35). In Chap.1 we stated intuitively for the example considered that the microcanonical thermal equilibrium state, that is, the most disordered macro-state with an energy between U and $U + \Delta U$, was the one corresponding to equal probabilities p_m. We see now that this choice corresponded to looking for the maximum degree of disorder as measured by the statistical entropy S, and we regain expression (1.19) or (3.19) for the maximum of S.

(b) *Minimum.* The statistical entropy S is a *minimum and equal to zero in a pure state.*

Inversely, the vanishing of S can be used to characterize a pure state, as (3.17) is zero only when all p_m except one vanish. The pure states, the ones where our knowledge is best, even though they involve quantum uncertainties, are those for which the degree of disorder is zero.

(c) *Additivity.* Let us consider a physical system consisting of *two statistically independent subsystems* a *and* b with density operators $\widehat{D}_{\rm a}$ in the Hilbert space $\mathcal{E}_{\rm H}^{\rm a}$ and $\widehat{D}_{\rm b}$ in the Hilbert space $\mathcal{E}_{\rm H}^{\rm b}$, respectively. *The statistical entropy of the composite system is the sum,*

$$S(\widehat{D}) \;=\; S(\widehat{D}_{\rm a}) + S(\widehat{D}_{\rm b}), \tag{3.20}$$

of the entropies of its parts.

We can prove this equality by writing the density operator $\widehat{D} = \widehat{D}_{\rm a}\otimes\widehat{D}_{\rm b}$ of the composite system in the space $\mathcal{E}_{\rm H} = \mathcal{E}_{\rm H}^{\rm a}\otimes\mathcal{E}_{\rm H}^{\rm b}$ in the factorized base which diagonalizes $\widehat{D}_{\rm a}$ and $\widehat{D}_{\rm b}$. It expresses the fact that the amount of disorder contained in a system consisting of two statistically independent parts is the *sum of the amounts of disorder in each of the parts.*

(d) *Correlations.* If a composite system is described by the density operator $\widehat{D}$ in the space $\mathcal{E}_{\rm H} = \mathcal{E}_{\rm H}^{\rm a}\otimes\mathcal{E}_{\rm H}^{\rm b}$, the statistical entropies of its parts satisfy the *sub-additivity inequality*

$$S(\widehat{D}_{\rm a}) + S(\widehat{D}_{\rm b}) \;\geq\; S(\widehat{D}), \tag{3.21}$$

where the equal sign holds only when $\widehat{D} = \widehat{D}_{\rm a}\otimes\widehat{D}_{\rm b}$.

We saw in § 2.2.4 that the density operator of the subsystem a, given by

$$\widehat{D}_{\rm a} \;=\; {\rm Tr}_{\rm b}\,\widehat{D},$$

is sufficient to characterize the state of the subsystem a as it enables us to evaluate the mean values of all observables $\widehat{A}_{\rm a}$ relating to the subsystem a. Similarly,

$$\widehat{D}_{\rm b} \;=\; {\rm Tr}_{\rm a}\,\widehat{D}$$

characterizes the state of the subsystem b. We introduce a new density operator

$$\widehat{D}' \;\equiv\; \widehat{D}_{\rm a}\otimes\widehat{D}_{\rm b}. \tag{3.22}$$

The state $\widehat{D}'$ is equivalent to $\widehat{D}$ for all measurements carried out on a and b separately and represented by operators which have the factorized form $\widehat{A}_{\rm a}\otimes\widehat{A}_{\rm b}$; but it does not contain the information, included in $\widehat{D}$, about the correlations between a and b. To prove (3.21) we express $S(\widehat{D}_{\rm a})$ as a trace in the space $\mathcal{E}_{\rm H}$:

$$S(\widehat{D}_{\rm a}) \;=\; -k\;{\rm Tr}_{\rm a}\,\widehat{D}_{\rm a}\ln\widehat{D}_{\rm a} \;=\; -k\;{\rm Tr}\,\widehat{D}\ln\left(\widehat{D}_{\rm a}\otimes\widehat{I}_{\rm b}\right),$$

and we note that

$$\ln\left(\widehat{D}_{\rm a}\otimes\widehat{I}_{\rm b}\right) + \ln\left(\widehat{I}_{\rm a}\otimes\widehat{D}_{\rm b}\right) \;=\; \ln\left(\widehat{D}_{\rm a}\otimes\widehat{D}_{\rm b}\right),$$

an obvious identity in the factorized base of $\mathcal{E}_{\rm H}$ which diagonalizes $\widehat{D}_{\rm a}\otimes\widehat{D}_{\rm b}$. Using (3.20) and (3.22) we have thus

$$S(\widehat{D}') \;=\; S(\widehat{D}_{\rm a}) + S(\widehat{D}_{\rm b}) \;=\; -k\;{\rm Tr}\,\widehat{D}\ln\widehat{D}', \tag{3.23}$$

and applying lemma (3.18) with $\widehat{X} = \widehat{D}$ and $\widehat{Y} = \widehat{D}'$ leads immediately to the stated result.

Reciprocally, one can prove that the statistical entropy (3.17) is *the only* function of the density operator which is invariant under a unitary transformation and which satisfies the sub-additivity condition (3.21) and the additivity condition (3.20).

The interpretation of (3.21) is obvious: $\widehat{D}_{\rm a}$ *and* $\widehat{D}_{\rm b}$ *together contain less information than* $\widehat{D}$, which describes as well the *correlations* between the subsystems a and b. In other words, a system is *more disordered when its parts are statistically independent than when they are correlated.* For instance, in the case of two spin-$\frac{1}{2}$ particles which are coupled in the singlet state (see Eq.(2.41)), we have $S(\widehat{D}_{\rm a}) = S(\widehat{D}_{\rm b}) = k \ln 2$ and $S(\widehat{D}) = 0$: each of the spins a and b is unpolarized, that is, completely disordered, whereas the correlation due to their coupling makes the global system completely ordered.

In Chap.1 we had another example of an approximate additivity of the type (3.20) and of a sub-additivity of the type (3.21). In thermal equilibrium, each paramagnetic ion is in a state characterized by the probabilities $p_i(\pm 1)$ given by (1.18). In the density operator formalism, the density operator of this ion is

$$\widehat{D}_i = \sum_{\sigma_i = \pm 1} |\sigma_i\rangle \, p_i(\sigma_i) \, \langle \sigma_i|,$$

and its statistical entropy equals

$$S_i = -k \sum_{\sigma_i = \pm 1} p_i(\sigma_i) \ln p_i(\sigma_i).$$

Equation (1.19) then simply expresses the fact that the global statistical entropy behaves, in the limit as $N \to \infty$, as the sum of the statistical entropies of the paramagnetic ions

$$S = k \ln W \sim \sum_i S_i. \tag{3.24}$$

This was to be expected as in that limit the magnetic moments tend to be statistically independent in thermal equilibrium (§ 1.2.5).

Moreover, (1.19) contains corrections for finite N. As there are small, but for finite N non-negligible, correlations between the magnetic moments, we expect, by virtue of (3.21), that $S < \sum_i S_i$. Indeed, for large N, the quantity $S = k \ln W$, given by (1.14), (1.17′), differs from (1.17) by

$$\sum_i S_i - k \ln W \sim k \ln A \Delta U \sim \tfrac{1}{2} k \ln N. \tag{3.25}$$

This expression, which is clearly positive, represents the order associated with the weak correlations between the spins resulting from the fact that their sum is fixed, with a finite dispersion when $N \to \infty$.

The additivity (3.20) and the sub-additivity (3.21) are generalizations of (3.8), but quantum mechanics brings in new, sometimes unexpected, features. For instance, one might think that a partial statistical entropy $S(\widehat{D}_{\rm a})$ is smaller than the global statistical entropy $S(\widehat{D})$, and that there is less disorder in part of the system than in the whole system. This is not always so in quantum mechanics. Let us, for instance, consider two spin-$\frac{1}{2}$ particles coupled in the completely ordered singlet state; the global entropy S is zero. Nonetheless the spin a, which is unpolarized, is completely disordered ($S_{\rm a} = k \ln 2$). If, on the other hand, one considers a composite system, the density matrix $\widehat{D}$ of which is diagonal in a factorized base of $\mathcal{E}_{\rm H}^{\rm a} \otimes \mathcal{E}_{\rm H}^{\rm b}$, one can easily prove that $S \geq S_{\rm a}$. Because of correlations of a quantum nature between the subsystems a and b one cannot state in general that the whole contains more disorder than one of its two parts. Inequality (3.21) shows, on the other hand, that the whole always contains less disorder than the sum of its two parts.

(e) *Concavity.* The statistical entropy is a *concave* function on the set of density operators in a given Hilbert space $\mathcal{E}_{\rm H}$: for any pair $\widehat{D}_1$ and $\widehat{D}_2$ we have, if $0 < \lambda < 1$,

$$S\big(\lambda\widehat{D}_1 + (1-\lambda)\widehat{D}_2\big) \;\geq\; \lambda S(\widehat{D}_1) + (1-\lambda)S(\widehat{D}_2), \tag{3.26}$$

where the equal sign holds only when $\widehat{D}_1 = \widehat{D}_2$.

The proof is less direct than in § 3.1.2h because $\widehat{D}_1$ and $\widehat{D}_2$ may not commute. If we write

$$\widehat{D} \;=\; \lambda\widehat{D}_1 + (1-\lambda)\widehat{D}_2,$$

and apply lemma (3.18) successively to the two terms on the right-hand side of (3.26) first with $\widehat{X} = \widehat{D}_1$, $\widehat{Y} = \widehat{D}$, and then with $\widehat{X} = \widehat{D}_2$, $\widehat{Y} = \widehat{D}$, we find that

$$\lambda S(\widehat{D}_1) + (1-\lambda)S(\widehat{D}_2) \;\leq\; -k\lambda \,\mathrm{Tr}\, \widehat{D}_1 \ln \widehat{D} - k(1-\lambda)\,\mathrm{Tr}\, \widehat{D}_2 \ln \widehat{D},$$

whence (3.26) follows. The equal sign holds only when $\widehat{D} = \widehat{D}_1$ and $\widehat{D} = \widehat{D}_2$.

The concavity inequality, on the interpretation of which we have already commented using an example in § 3.1.2h, means that *combining two states of the same system in a single statistical mixture increases the disorder* as measured by the statistical entropy. For instance, if we include in the same statistical ensemble spin particles which are differently polarized, we lose information.

Inequality (3.26) can immediately be generalized, by iteration, to any number of states $\widehat{D}_j$ relating to the same system. If one regroups within the same statistical ensemble several populations described by density operators $\widehat{D}_j$ with respective weights μ_j, where $\mu_j > 0$ and $\sum_j \mu_j = 1$, the resulting statistical mixture is on average more disordered than the initial statistical mixtures:

$$S\left(\sum_j \mu_j \widehat{D}_j\right) \geq \sum_j \mu_j \, S(\widehat{D}_j). \tag{3.27}$$

3.2.3 Changes in the Statistical Entropy

The change in an operator function $f(\widehat{D})$, such as $\widehat{D} \ln \widehat{D}$, which results from an infinitesimal change $\delta\widehat{D}$ in the density operator, is not necessarily equal to $f'(\widehat{D})\delta\widehat{D}$, because $\widehat{D}$ and $\delta\widehat{D}$ may not commute. However, when one considers the trace, taking the derivative is justified (Eq.(2.19)). In fact, we have

$$\delta \operatorname{Tr} \widehat{X}^n = \sum_{m=0}^{n-1} \operatorname{Tr} \widehat{X}^m \, \delta\widehat{X} \, \widehat{X}^{n-m-1} = n \operatorname{Tr} \delta\widehat{X} \, \widehat{X}^{n-1},$$

whatever the value of n, and we can imagine $f(\widehat{D})$ to be expanded in powers of $\widehat{X} = \widehat{I} - \widehat{D}$. We get thus

$$\delta S(\widehat{D}) = -k \operatorname{Tr} \delta\widehat{D} \, (\ln \widehat{D} + 1) = -k \operatorname{Tr}(\delta\widehat{D} \ln \widehat{D}), \tag{3.28}$$

where the last term in the second part of this equation vanishes since $\operatorname{Tr} \widehat{D} = 1$.

In particular, if the system evolves freely according to (2.49), with a *Hamiltonian* $\widehat{H}$ *which is completely known*, we have, if we use the cyclic invariance (2.16) of the trace,

$$\frac{i\hbar}{k} \frac{dS}{dt} = -\operatorname{Tr}\,[\widehat{H}, \widehat{D}] \ln \widehat{D} = -\operatorname{Tr} \widehat{H}\,[\widehat{D}, \ln \widehat{D}] = 0, \tag{3.29}$$

and its *statistical entropy remains constant in time*: when an isolated system evolves according to a perfectly known equation of motion, its density operator may change, but the missing amount of information associated with it remains unchanged. We know, in fact, that during such an evolution $\widehat{D}$ changes according to a unitary transformation (2.50) and we have indicated in § 3.2.1 that the statistical entropy is invariant under a unitary transformation.

Let us now consider an *isolated system, with an evolution which we do not know completely.* The density operator changes according to (2.51). Each of the possible evolutions j would conserve the information:

$$S\big(\widehat{D}_j(t)\big) = S\big(\widehat{D}(0)\big). \tag{3.30}$$

However, the concavity inequality (3.27) together with (3.30) shows immediately that

$$\boxed{S\big(\widehat{D}(t)\big) \geq S\big(\widehat{D}(0)\big)} \, . \tag{3.31}$$

Even though the Schrödinger equation is invariant under time reversal, statistics has thus introduced a preferred time direction. One *loses information* during the evolution of an isolated system, if its Hamiltonian is not completely known. *This evolution is irreversible*: the statistical entropy, that is the *disorder, increases.* Of course, the time necessary for this loss of memory, or for this increase in disorder, may be very long if the Hamiltonian is nearly certain.

Let us finally note that the statistical entropy of a system which is *not isolated* may either increase or decrease: such a system can, in fact, gain order, when it interacts with an external system which is well ordered, by yielding to it part of its statistical entropy.

3.2.4 Information and Quantum Measurements

A measurement always involves a system which is not isolated, since the object considered and the apparatus must interact at some time and afterwards be separated. Let us analyze the changes in the statistical entropy of the object during this process, distinguishing as in § 2.2.5 two stages. The first stage, which is specifically quantum-mechanical, consists in letting the object interact with the apparatus and then separating them in such a way that one no longer will be interested in the observables which would couple them. We have seen (Eq.(2.44)) that in an ideal measurement this stage brings the object to a state $\widehat{D}_A = \sum_\alpha \widehat{P}_\alpha \widehat{D} \widehat{P}_\alpha$ through an irreversible transformation. Let us prove that the *statistical entropy has increased through this truncation of the density operator*:

$$S(\widehat{D}_A) - S(\widehat{D}) \geq 0. \tag{3.32}$$

If we use the cyclic invariance of the trace we can write $S(\widehat{D}_A)$ in the form

$$-k \operatorname{Tr} \left(\sum_\alpha \widehat{P}_\alpha \widehat{D} \widehat{P}_\alpha \ln \widehat{D}_A \right) = -k \operatorname{Tr} \left\{ \widehat{D} \sum_\alpha \widehat{P}_\alpha \left(\ln \widehat{D}_A \right) \widehat{P}_\alpha \right\}.$$

In a base which diagonalizes $\widehat{A}$, the matrices $\widehat{D}_A$ and thus $\ln \widehat{D}_A$ consist of diagonal blocks each relating to a projector $\widehat{P}_\alpha$ so that $\ln \widehat{D}_A$ remains unchanged under a new truncation. We have thus

$$\begin{aligned} S(\widehat{D}_A) &= -k \operatorname{Tr} \left\{ \widehat{D} \sum_\alpha \widehat{P}_\alpha \left(\ln \widehat{D}_A \right) \widehat{P}_\alpha \right\} \\ &= -k \operatorname{Tr} \left(\widehat{D} \ln \widehat{D}_A \right). \end{aligned}$$

The use of lemma (3.18) with $\widehat{X} = \widehat{D}$ and $\widehat{Y} = \widehat{D}_A$ then proves inequality (3.32) which reduces to an equality only when $\widehat{D}$ commutes with $\widehat{A}$. The reduction of the density operator thus increases the disorder *irreversibly*; using communication theory language, it makes us *lose information.* This is natural, as by separating the apparatus from the object we are forced to neglect certain correlations which exist between them and which appeared during their interaction. These correlations, of a quantum nature, are related to the presence in $\widehat{D}$ of off-diagonal elements which prevent it from commuting with $\widehat{A}$.

After this stage there remain between the apparatus and the object other correlations, of a classical nature, which mean simply that if one observes a_α in the apparatus, the object is in the state

$$\widehat{D}_\alpha = \frac{\widehat{P}_\alpha \widehat{D} \widehat{P}_\alpha}{\mathrm{Tr}\, \widehat{D} \widehat{P}_\alpha}.$$

These correlations refer to each individual experiment performed on the statistical ensemble described by $\widehat{D}$; they are essential when one uses an ideal measurement as a preparation of a state $\widehat{D}_A$, since they allow one to sort out the systems of this ensemble according to the outcome a_α of the measurement. The statistical entropy of the final state $\widehat{D}_\alpha$ characterizes the degree of disorder reached by the systems on which a_α has been observed. In the case of a complete ideal observation, where the space α has one dimension, the final state is pure with zero entropy and the quantity of information gained by the observation is $S(\widehat{D}_A)$. In the case of an incomplete observation, the uncertainty $S(\widehat{D}_\alpha)$ about the final state is not necessarily smaller than $S(\widehat{D})$, nor even than $S(\widehat{D}_A)$. This fact may seem surprising, but it also occurs in communication or probability theory where an incomplete observation may also lead to a loss of information. For instance, if one draws a card from a pack containing 1000 aces of hearts, 1 ace of spades, and 1 ace of clubs, the initial uncertainty, 0.02 bit, is small; however, in the case where a partial observation indicates that the card drawn is black, the remaining uncertainty would equal 1 bit. Nevertheless, *on average, observation will increase our information* as is natural. In fact, the mean information gained by reading the results a_α of measuring A when the system was initially in the state $\widehat{D}$ equals, according to (3.1),

$$S_A = -k \sum_\alpha \mathcal{P}(a_\alpha) \ln \mathcal{P}(a_\alpha), \tag{3.33}$$

where $\mathcal{P}(a_\alpha) = \mathrm{Tr}\, \widehat{D} \widehat{P}_\alpha$ is given by (2.46). The identity

$$S_A = S(\widehat{D}_A) - \sum_\alpha \mathcal{P}(a_\alpha)\, S(\widehat{D}_\alpha) \tag{3.34}$$

expresses that this information gained by observing the results a_α is equal to the average decrease in the uncertainty between the final state $\widehat{D}_A$ of the whole statistical ensemble and the states $\widehat{D}_\alpha$ after the sorting out. Identifying $S(\widehat{D})$ with the thermodynamic entropy, (3.34) means therefore that an observer may *make the entropy of a statistical ensemble of systems decrease, at the expense of some information* which he possesses.

The statistical entropy satisfies several other relations related to the transfer of information in a quantum measurement.[2] We shall here restrict ourselves to summarizing them with a short interpretation, and without any proofs. The inequality

$$S(\widehat{D}) \le S_A + \sum_\alpha \mathcal{P}(a_\alpha)\, S(\widehat{D}_\alpha) \tag{3.35}$$

[2] R.Balian, Europ. J. Phys. **10**, 208 (1989).

shows, on comparison with (3.11), that quantum mechanics spoils the additivity of the information. Acquiring information about A must in a quantum measurement be paid for by the destruction of some information (3.32) when $\widehat{D}$ does not commute with $\widehat{A}$.

On the other hand, one can show that

$$S(\widehat{D}) \geq \sum_{\alpha} \mathcal{P}(a_\alpha)\, S(D_\alpha), \tag{3.36}$$

where the equal sign holds only when $S(\widehat{D}_\alpha) = S(\widehat{D})$ for all α. According to this inequality an ideal quantum measurement makes the statistical entropy of the systems that we observe decrease on average. In other words, the gain of information S_A through observation is larger than the loss $S(\widehat{D}_A) - S(\widehat{D})$ due to the reduction of the wavepacket. In the special case where the initial state $\widehat{D}$ is pure, (3.36) implies that the reduced states $\widehat{D}_\alpha$ themselves also are pure; one can check this from (2.45).

As we noted towards the end of § 2.2.5, the definition of an ideal measurement implies that the perturbation produced in the system by the measuring apparatus is minimal and that it involves only the observables which do not commute with $\widehat{A}$. Let us introduce the class of density operators $\widehat{D}'$ which are equivalent to $\widehat{D}$ as far as all observables $\widehat{C}$ which are compatible with $\widehat{A}$ are concerned:

$$\text{Tr}\, \widehat{D}'\widehat{C} = \text{Tr}\, \widehat{D}\widehat{C}, \qquad \forall\, \widehat{C}, \text{ such that } [\widehat{C}, \widehat{A}] = 0. \tag{3.37}$$

The state $\widehat{D}_A$ belongs to this class and one sees easily that

$$S(\widehat{D}') \leq S(\widehat{D}_A), \tag{3.38}$$

where the equal sign holds only when $\widehat{D}' = \widehat{D}_A$. As a result $S(\widehat{D}_A)$ is the upper bound of the entropies $S(\widehat{D}')$. The reduction of $\widehat{D}$ to $\widehat{D}_A$ not only leads to a loss of information, but that *loss* (3.32) is the *largest possible* one, taking into account the conserved information (3.37).

In classical physics we have $\widehat{D}_A = \widehat{D}$ so that according to (3.34) the information gained by the observation cannot be larger than the initial uncertainty $S(\widehat{D})$. This is not necessarily so in quantum mechanics. For instance, a complete measurement will always leave the object in a pure state and S_A then reduces to $S(\widehat{D}_A)$: by a complete observation one gains on average the information $S(\widehat{D}_A)$ which is larger than $S(\widehat{D})$. This apparent paradox, peculiar to quantum mechanics, is related to the fact that an observation does not give us information directly about the initial state, but about the reduced state $\widehat{D}_A$ which itself retains a certain memory of the former, but which is more disordered.

One should note finally that the quantity S characterizes the lack of information about the set of measurements of *all* the observables associated with the object. One may be inclined to drop certain measurements which cannot be realized, for instance, those which imply correlations between a large number of particles; one thus discards information which, in principle, is contained in the density operator but which in practice is out of reach. An example of such an omission was provided by the first stage of the measurement of $\widehat{A}$ with which the loss of information (3.32) was associated; the density operator $\widehat{D}_A$ retains the same information as $\widehat{D}$ about the observables that commute with $\widehat{A}$ but some information associated with other observables has been dropped. The *apparent disorder* of a system relative to the set of measurements considered is thus larger than the *absolute disorder* defined by (3.17). In particular, the latter remains constant in time (Eq.(3.29)) whereas a relative disorder $S(\widehat{D}_A)$ can grow if some quantity of information, connected with practicable measurements, flows towards those which are not practicable. These remarks enable us to understand one possible origin of *irreversibility* (§ 3.4.3, Exerc.3c, and § 15.4).

3.3 The Statistical Entropy of a Classical State

3.3.1 Classical Statistical Entropy

Taking the limit (2.69), (2.93) which enables us to change from quantum mechanics to classical mechanics, we find that the statistical entropy can be expressed in terms of the phase density function as follows:

$$\boxed{S(D) = -k \sum_N \int d\tau_N \, D_N \log D_N} \, , \tag{3.39}$$

which is the classical limit of (3.17). If the number of particles N is well defined, (3.39) reduces to a single integral over the $6N$ variables of phase space.

The density in phase is a continuous probability measure so that the direct justification of (3.39) without starting from quantum mechanics runs into the difficulties underlined in § 3.1.4. In order to lift the arbitrariness of formula (3.14) we should choose a division of phase space into volume elements considered to be equivalent *a priori*. The equivalence of equal volumes $d\tau_N$ of phase space has, in fact, been justified in § 2.3.3 by Liouville's theorem which expresses the *invariance* of the volume $d\tau_N$ during any temporal evolution. As to a comparison between different values of N, we have given in § 2.3.2 an interpretation of the factor $1/N!$ in $d\tau_N$: since the particles are indistinguishable, the information that would consist in stipulating which of the particles is at a given point of phase space cannot be made available. Quantum mechanics provides us with a simpler justification, as it identifies the volume $d\tau_N$ with a number of orthonormal quantum states, which all have the same *a priori* weight.

It may appear surprising that the *continuous* structure of $\mathcal{E}_{\rm H}$ has not led to any similar difficulties in quantum mechanics, and that the statistical entropy (3.17) has the same form as the lack of information (3.1) for *discrete* probabilities. In fact, one has used implicitly the *invariance under unitary transformations* which means that $S(\widehat{D})$ is independent of the orthonormal base $|m\rangle$ of the eigenstates of $\widehat{D}$.

3.3.2 Properties

In contrast to the quantum statistical entropy which is non-negative and vanishes for a pure state, the classical statistical entropy (3.39) has *no lower bound.* As a result, if we get a negative $S(D)$, it means that the density in phase D cannot be considered to be the classical limit of a density matrix $\widehat{D}$ and that the classical limit therefore is not valid. For instance, $S \to -\infty$ for the density in phase (2.57) which represents a completely known classical state; this kind of state, for which $\Delta x \Delta p = 0$, clearly violates the principles of quantum mechanics and cannot be allowed.

The other properties of the classical statistical entropy are like those of the quantum statistical entropy provided we replace traces by integrations over phase space. In particular, we note that for a system, the representative point of which is constrained to lie inside a domain ω of the N-particle phase space, the entropy is a *maximum when the density* D_N *is constant inside* ω, and vanishes outside it. We denote by

$$W = \int_{\omega} d\tau_N \tag{3.40}$$

the *volume of the domain* ω *in phase space*, measured in units which depend on Planck's constant through the definition (2.55) of $d\tau_N$. If W is large, it can be interpreted as the number of orthogonal quantum micro-states in ω. The density in phase D_N, which is constant in ω, is normalized by (2.56) as $D_N = 1/W$; its corresponding statistical entropy is thus

$$S(D_N) = k \log W, \tag{3.41}$$

which is the classical equivalent of (3.19).

3.4 Historical Notes About the Entropy Concept

It has often been said that expression (3.41), which is written on Boltzmann's grave (Fig.3.1), symbolizes one of the major advances in science. Indeed, it was an essential beacon for the understanding of the important concept of entropy. This concept plays a central part in the present book, where we are following a logical path,in contrast to the historical order in which the various ideas were introduced. In order better to understand the ideas in question we shall sketch how, during the last century and a half, they developed thanks to a fruitful interplay of many different disciplines, namely, thermodynamics, analytical mechanics, statistical physics, and communication theory. As so often in science, progress has been made through gradually enlarging our horizons; however, in the present case the road has been a particularly difficult one.[3]

[3] S.G.Brush, *The Kind of Motion we call Heat, Studies in Statistical Mechanics*, Vol.VI, North-Holland, 1976; C.C.Gillespie (Ed.), *Dictionary of Scientific*

Fig. 3.1. The Boltzmann memorial in Vienna. Photograpn by courtesy of the Picture Archive of the Austrian National Library, Vienna

3.4.1 Entropy in Thermodynamics

The creation of thermodynamics can be attributed to Sadi Carnot (Paris 1796–1832) who in his "Réflexions sur la puissance motrice de feu" (1824) introduced the *Second Law* in the following form: "La puissance motrice de la chaleur est indépendante des agens mis en œuvre pour la réaliser; sa quantité est fixée uniquement par les températures des corps entre lesquels se fait en dernier résultat le transport du calorique". This statement contains the germs of a possible construction of the entropy, but for this to be done it was necessary that the equivalence between work and heat was recognized.

At that time scientists were wondering about the *nature of heat.* Was it a substance made of particles or was it an imponderable fluid, called "calorique" by Guyton de Morveau, Lavoisier, Berthollet, and Fourcroy in 1787, which could be conserved during exchanges? Was it a kind of radiation, perhaps similar to light – that Young and Fresnel had just shown to have a wave nature – as Herschel's experiments on infra-red rays suggested? Was it connected with the motion of molecules or was it even a "mode of motion" as was proposed by Rumford and

Biography, 16 Volumes, Scribner, New York, 1981; the complete scientific works of most authors quoted below have been gathered in book form, and extensive references are given in this dictionary.

Davy, who relied on their experiments on the boring of cannon (1798) or the melting of ice by friction?

Carnot himself understood between 1825 and 1832 that "la chaleur n'est autre que la puissance motrice ou plutôt que le mouvement qui a changé de forme, c'est un mouvement"; he stressed that the quantity which was conserved was not the heat, as he had still thought in 1824, but the energy: "Partout où il y a destruction de puissance motrice dans les particules des corps, il y a en même temps production de chaleur en quantité précisément proportionnelle à la quantité de puissance détruite .... On peut donc poser en thèse générale que la puissance motrice est en quantité invariable dans la nature, qu'elle n'est jamais à proprement parler ni produite ni détruite. A la vérité elle change de forme". Unfortunately, Carnot's posthumous notes which stated the position so clearly were not published until after 1878. It was therefore only after 1847, the year when *conservation of energy* was definitely established (§ 5.2.1), that the Second Law could be expressed in a mathematical form.

Between 1848 and 1854, William Thomson, the later Lord Kelvin (Belfast 1824–Netherhall, Scotland 1907) and Rudolf Clausius (Cöslin, Prussia 1822–Bonn 1888) analyzed Carnot's principle of the maximum efficiency of heat engines in the light of energy conservation. They only knew Carnot's work through a publication of Clapeyron who had in the meantime introduced the concept of reversibility. They reformulated and numbered the First and the Second Laws in a logical order. In 1848, Thomson defined the *absolute temperature*, which made it possible to give a simple expression for the efficiency of reversible engines. Clausius, in 1854, made a decisive step forward by introducing under the term "Aequivalenzwerth" (equivalence value) or "Verwandlung" (conversion) what was to become the entropy. He defined the *equivalence value* in the conversion of work into heat $\mathcal{Q}$ at a temperature T as $\mathcal{Q}/T$, in such a way that for a reversible engine one has $\mathcal{Q}_1/T_1 + \mathcal{Q}_2/T_2 = 0$. More generally, he showed that $\oint \delta\mathcal{Q}/T$ vanishes for any reversible cyclic process and is positive otherwise.

The quantity $\delta\mathcal{Q}/T$ which is conserved in reversible processes was by some people identified with the old "caloric". In 1865, the terminology was a source of confusion and Clausius, after having proved that the equation $dS = \delta\mathcal{Q}/T$ defines a *function S of the state* of a system which *increases* in a spontaneous adiabatic process, gave it the name entropy; this name was based on the Greek τροπή (transformation) by analogy with the name energy. Entropy thus appeared as an indicator of evolution, giving a mathematical expression for the "arrow of time". Clausius's book ends with the famous words: "Die Energie der Welt ist konstant. Die Entropie der Welt strebt einem Maximum zu".

3.4.2 Entropy in Kinetic Theory

Having been introduced in this way in thermodynamics the entropy remained a rather mysterious quantity. There was nothing which enabled one to understand the significance of *irreversibility*, a concept directly coupled with the entropy, whereas energy conservation swept all before it. The situation gradually became clearer, at least as far as the physics of gases was concerned, thanks to the introduction of statistical methods in kinetic theory.

Starting in 1860, James Clerk Maxwell (Edinburgh 1831–Cambridge 1879) developed the theory of gases, that he described as assemblies of uncorrelated molecules with a statistical position and velocity distribution characterized by the single-particle density f (§ 2.3.5); slowly there began to appear connections between thermodynamics, dealing with macroscopic scales, and probability theory, dealing with microscopic scales. The irreversibility accompanying the conversion of work into heat was interpreted as an increase in the *disordered motion* of the molecules: while their total energy is fixed, the molecules can have, for instance, a distribution f peaked around the value $\boldsymbol{p} = m\boldsymbol{v}$ for a cold gas which is moving as a whole, or they can have an isotropic, but broader distribution f for a heated gas which does not move in bulk. The disorder is larger in the second case.

It fell to Ludwig Boltzmann (Vienna 1844–Duino near Trieste 1906) to make this argument a quantitative one. Being convinced that analytical mechanics should govern all natural phenomena, he set himself the task, right from the beginning of his career, to find a mechanical interpretation for the entropy so as to understand why it always increases; this long quest he pursued during the whole of his life. His first attempt, in 1866, resulted in associating an entropy with a molecular system which followed a periodic trajectory. Soon, however, Boltzmann recognized under the influence of Maxwell that in less exceptional situations, it is necessary to introduce *statistics*. This enabled him, between 1868 and 1871, to find an equation which is the forerunner of Eq.(4.30) that we shall find in the next chapter; this was the first time that the *entropy of a classical system in equilibrium* was expressed as a function of the Hamiltonian describing the molecular dynamics on the microscopic scale.

The problem of the increase in the entropy remained unsolved. In 1872, Boltzmann wrote down the equation that bears his name to describe the evolution of the single-particle distribution for a *rarefied gas* (Chap.15). He associated with f a quantity,

$$\int d^3\boldsymbol{r}\, d^3\boldsymbol{p}\, f(\boldsymbol{r},\boldsymbol{p},t)\ \ln f(\boldsymbol{r},\boldsymbol{p},t), \tag{3.42}$$

which he called H. He proved that H *decreases* with time, remaining constant only if f is the thermal equilibrium distribution introduced by Maxwell: the celebrated H*-theorem*. He noted that for a perfect gas at equilibrium, apart from the sign and additive and multiplying constants, H is identical with Clausius's entropy. This enabled him to conclude that he had extended this concept to non-equilibrium situations and that he had found – at least for gases – a microscopic explanation of the Second Law.

In 1877, Boltzmann returns to the problem of an *arbitrary system*, changing from kinetic theory to a more general statistical mechanics. Identifying the probability concept with that of the *number of microscopic configurations*, he poses that the macroscopic thermal equilibrium state is the "most probable" state; as we did in § 1.2.2 he suggests that this macro-state is on the microscopic scale the microcanonical equilibrium state where all W possible configurations, compatible with the macroscopic data, are equally probable. He thus identifies the entropy with a measure of *likelihood*: "so können wir diejenige Größe, welche man gewöhnlich als die Entropie zu bezeichnen pflegt, mit die Wahrscheinligkeit des betreffenden Zustandes identifizieren", which is represented by the famous formula (3.41). Irreversibility expresses the evolution from a less probable to a more probable state, where W and S are larger. Nonetheless, the significance of W and S was not yet completely clarified. Indeed, the probability concept itself had not been used in an absolutely clear manner. Moreover, it was difficult to define W in classical mechanics since the configurations form a continuum (§§ 3.1.4 and 3.3.1). We have seen that it is impossible to write down the correct expression (3.40) for W without considering classical mechanics as the limit of quantum mechanics. This problem does not arise for problems which are truly quantal (§ 1.2.2). In fact, Boltzmann himself replaced, as a mathematical artifice, the continuous energy spectra of classical mechanics by *discrete spectra*; this suggestive idea was going to help Planck in 1900 to introduce the quantization of radiation (§ 3.4.4).

In thermodynamics the entropy concept was from the beginning intimately connected with that of *heat*. Kinetic theory helped to decouple these two concepts and to demonstrate the more general nature of the entropy, as a measure of *disorder*. For instance, the mixing of two distinguishable gases at the same temperature and with the same density is not accompanied by any thermal effect, but nevertheless it leads to an increase in entropy. Maxwell had already in 1868 compared this mixture with that of black and white balls (§ 3.1.2h), stressing its irreversibility. The paradigm of a *mixture* has later on taken the place of that of heat transfer as a guide to an understanding of entropy. Moreover, between 1873 and 1878, Josiah Willard Gibbs (New Haven 1839–1903) extended the realm of entropy in thermodynamics to *open systems* which can exchange molecules: mixtures of several components, equilibria between phases, chemical equilibria, ... He came back to this question in 1902 in the framework of equilibrium statistical mechanics (§§ 2.3.6 and 4.3.2).

3.4.3 The Period of Controversies

The probabilistic and microscopic interpretation of the entropy was not accepted easily. In order to understand why the discussion of this topic during the last quarter of the nineteenth century was so animated, one must steep oneself in the ideas dominating that period. Continuum sciences, such as electromagnetism, thermodynamics, or acoustics, were all the vogue and atomism was therefore considered to be old-fashioned. Kinetic theory was subjected to serious scientific criticism: there was a lack of rigour, one used hypotheses which did not have a proper foundation, one adjusted the intermolecular forces empirically, depending on which effects one was studying, and there were serious failures, such as the impossibility to explain the specific heats of gases, other than monatomic ones. Moreover, the philosophical background had changed. Many scholars rejected the mechanistic vision of the world of which kinetic theory was the latest step, and were against introducing hypothetical objects such as the atoms were at that time. Energeticists like the physical chemist Wilhelm Ostwald considered energy rather than matter as the final reality; positivists like the physicist and physiologist Ernst Mach went even further by stating that science should restrict itself to establishing empirical relations between the perceptions of our senses. Under those circumstances, why should one worry and go beyond the framework of thermodynamics?

The opponents of kinetic theory were not the only ones to lodge objections. The advocates of the statistical interpretation of entropy, trying to understand it better, themselves encountered difficulties. They illustrated these by stating paradoxes which were not completely solved until much later – in the framework of information theory or of quantum theory.

The *Gibbs paradox* (1875) is based upon the following observation. When one mixes two gases, the entropy does not change if the molecules are identical, but it increases by a finite amount if they are distinguishable. Gibbs asked what would happen, if the molecules a and b of the two gases could be distinguished, but only through their interactions with a third kind of molecules, c. In so far as the existence of these last molecules is not known, the mixture of a and b behaves like that of two parts of the same gas; however, afterwards, one has an opportunity of separating the mixture a+b into its components through a reversible process involving c. From this Maxwell concluded that the entropy to be assigned to the mixture differs according to whether or not the c molecules have been discovered. The entropy is therefore *not a property only of the system, but depends also on our knowledge*: "confusion, like the correlative term order, is not a property of material things in themselves, but only in relation to the mind which perceives them". The Gibbs paradox was not unravelled, not even for perfect gases, without the help of quantum statistical mechanics (§ 8.2.1). As a matter of fact, it helped to sort out the problem of the indistinguishability of particles in quantum mechanics.

The *irreversibility paradox* (W.Thomson 1874, J.Loschmidt 1876) is based upon a qualitative difference between thermodynamics and analytical mechanics: how can one reconcile the macroscopic irreversibility which shows up in the increase of S with a microscopic dynamics which is invariant under time reversal? This contradiction led to passionate controversies, with kinetic theory as the stakes, as we have seen, and they went on long after kinetic theory was well established.

Even though they have not always been satisfactory, the solutions proposed for the irreversibility paradox have been essential for an understanding of statistical mechanics. Maxwell's idea was a subterfuge: he imagined that there were *stochastic perturbations* superimposed upon Hamiltonian dynamics; Eq.(3.31) shows in modern terminology that such perturbations increase the disorder as measured by the entropy. This, however, does not enable us to understand what happens to an isolated system governed by *deterministic and reversible microscopic dynamics.* W.Thomson (1874), and later Boltzmann (1877), using as an example a volume filled with air, noted that there is only an infinitesimal fraction of configurations such that the oxygen and nitrogen molecules are separated in space; taking the value of Avogadro's number which was used at that time, Thomson made a numerical estimate which came to $10^{-10^{12}}$. This demonstrates the improbability of a spontaneous separation of the mixture. The *law of large numbers* makes violations of the Second Law so rare that the increase in the entropy appears to be inescapable on the macroscopic scale.

Starting in 1877, Boltzmann considered another important idea which was later, in 1902, developed systematically by Gibbs. Statistical mechanics describes not a single object, but an *ensemble of systems*: the entropy is a *statistical concept* attached to this ensemble, rather than a function of the coordinates of the molecules of an individual system. Its increase is a probabilistic property, which is practically certain for macroscopic systems. However, the possibility that one might observe for some individual system an increase in order is not completely ruled out. For instance, it is not excluded that the molecules of a mixture at a later stage divide up, but only starting from *certain very special and extremely improbable initial conditions.*

The debate was sustained by Poincaré's *recurrence theorem* (1889) which states that almost all trajectories of a dynamic system return to the vicinity of the initial point after some shorter or longer period. The mathematician Ernst Zermelo concluded from this (1896) that, if thermodynamic systems obey the laws of mechanics on the microscopic scale, the entropy should behave periodically rather than monotonically. Boltzmann replied by coming back to the idea that the entropy was not simply a function of the dynamic variables, but that it had a probabilistic nature. He also noted that the Poincaré recurrence times are huge and unobservable for macroscopic systems; even for a few particles they already reach billions of years.

The irreversibility problem was also extended in this period to cosmology. Does the Universe show a recurrent behaviour like a dynamic system, or does it evolve towards a "*thermal death*" where the temperature and the density are uniform, as Clausius's statement requires? In that case, why has equilibrium not yet been reached? One should note that the current idea of the Universe at that time was one without a beginning and without boundaries. Boltzmann's reply (1897) is interesting. As certain specific initial conditions can lead to the creation of order – which eventually will disappear – there is nothing to prevent that that part of the Universe in which we find ourselves is the result of such a fluctuation; this is very improbable, but not prohibited for an infinite Universe.

These discussions of the irreversibility paradox remained, however, qualitative. In the only case studied with a certain amount of mathematical rigour – that of dilute gases – there remained a contradiction between the reversibility of microscopic mechanics and the irreversibility expressed by the H-theorem. This sowed doubts either about the Boltzmann equation from which the H-theorem followed directly, or about the microscopic and probabilistic interpretation of the entropy. This question was elucidated by Paul Ehrenfest (Vienna 1880–Amsterdam 1933) and his wife Tatyana. They proposed in 1906 a simple statistical model which helped to understand how a reversible and recurrent evolution like the ones in mechanics can be compatible with an irreversible progress to thermodynamic equilibrium (Exerc.4g).

They also discussed in 1911 the H-theorem and, more generally, the irreversibility of dynamic systems. In § 15.4 we shall come back to the modern explanation of this question, which is based upon their ideas; we give here the main outline of that discussion. As Boltzmann's equation is irreversible, it cannot be entirely exact. This, however, is not the difficulty, as the evolution of the reduced single-particle density f that it provides is an excellent approximation if the gas is dilute; the error does not become significant until after periods of the order of Poincaré's recurrence time. The essential point is that, away from equilibrium, one must consider simultaneously *several more or less detailed microscopic descriptions*. Each of those is associated with an entropy so that the *entropy is not a unique quantity*. If one were able to follow the evolution of *all* the observables, that is, of all the matrix elements of $\widehat{D}$ or of all details of the phase density D, the associated entropy $S(D)$ would remain constant, according to (3.29). However, if one only follows f, the evolution of which obeys Boltzmann's equation, one is led to introduce another statistical entropy, which measures only the *uncertainty relative to* f. It is *this* entropy, directly related to (3.42), of which the H-theorem proves the growth. It is the same as $S(D)$ when there are no correlations between the molecules, in which case D can be factorized (Eq.(2.58) and § 2.3.5). We shall also show that for a dilute gas it is identical with the entropy of non-equilibrium thermodynamics on not too short a time scale. Its increase can then be interpreted as a *leak of information* from the single-particle degrees of freedom, described by f, *towards the inaccessible degrees of freedom* associated with the correlations between a large number of molecules, produced during the evolution of the system. More generally, let us assume that the observables can be classified into two categories, the simpler ones being considered as the only relevant ones, while the other ones are too detailed to be significant. For example, a coarse-graining of space into little cells eliminates those observables which correspond to measurements with a greater precision than the size of the grains. One can then introduce a statistical entropy associated with the relevant observables (Exerc.3c). Its growth characterizes the fact that the evolution appears to be irreversible when one disregards the irrelevant variables.

The most famous paradox is the one of *Maxwell's demon* which Maxwell invented in 1867. He considered a vessel with two compartments A and B connected through a small hole. A tiny demon posted in front of the hole can open or shut it with a cover which is so light that one can neglect the work involved in manipulating it. Initially A and B contain gases at the same pressure and temperature. The demon lets only the fastest molecules pass from A to B and the slowest ones from B to A – or, alternatively, he only lets those molecules pass which go from A to B. The temperature, or the pressure, of B therefore increases at the expense of A so that the entropy of the A+B system decreases, violating the Second Law. It has taken more than a century to give a complete refutation of this paradox.

First, W. Thomson remarked (1870) that the demon, *being alive*, might perhaps not obey the laws of thermodynamics. This argument was not satisfactory, as the Second Law holds for biological processes; however, this validity was not established until much later through theoretical and experimental work by physicists, biochemists, and physiologists, and in the most diverse sectors of biology. In actual fact, in living organisms order increases, but the decrease of their entropy is more than compensated for by a larger increase in the entropy of their environment.

Avoiding this kind of problems, Marian Smoluchowski (Vienna 1872–Cracow 1917) returned in 1912 to the analysis of Maxwell's paradox, but with an *automatic demon*. Indeed, let us consider the situation where the demon and its cover are replaced by an automatic device, a valve which lets through only molecules travelling in one direction, from A to B, in the same way as diodes in electronics rectify a current, filtering the electrons according to their direction of propagation. Earlier, Smoluchowski had worked out the theory of fluctuations which always occur in small systems at non-zero temperatures (Exerc.2b and § 5.7). He noted that for all imaginable mechanisms of the Maxwell demon type the apparent violation of the Second Law occurs through *taking advantage of the fluctuations* of the system. For instance, the possibility to let the pressure in compartment B increase is based upon the existence of fluctuations of the molecular velocities around their – zero – average value; these fluctuations imply that at certain moments there are more molecules in the vicinity of the hole which move from A to B than from B to A, and this makes it possible to select them. If the demon is replaced by an automatic valve, it must be governed by the fluctuations themselves, opening or closing according to the dominant direction of the velocities. The mechanism must therefore have a very small inertia in order that the fluctuations in the gas can operate it. Smoluchowski noted, however, that the valve cannot be perfect, as *it must itself suffer thermal fluctuations*; these cause random openings and closings which are not governed by the gas fluctuations and which let molecules through with the wrong sense of direction. The required effect can thus only be produced provided the valve is sufficiently cold, in which case its fluctuations decrease and opening errors are rare. Smoluchowski showed that under those conditions the valve cannot operate if its temperature is equal to that of the system; if it is colder, it can provide a filtering, but the Second Law is not violated, as there are a hot and a cold source.

Similarly, in an electric circuit consisting of a resistance and a diode one might imagine that the diode acting as a rectifier would filter the thermal fluctuations of the current in the resistance (Exerc.5b), and thus give rise to a non-vanishing electromotive force by means of a single heat source; this is, however, prevented by the fact that the diode acts as an efficient filter only if it is colder than the resistance. Maxwell's demon can therefore not be made automatic. However, the paradox remains for an "*intelligent*" demon.

3.4.4 Entropy and Quantum Mechanics

It may seem that entropy is a concept alien to quantum mechanics; however, it has played a seminal rôle at the very birth of this science. The first time quantization appeared was at the end of an important series of papers devoted by Max Planck (Kiel 1858–Göttingen 1947) to black-body radiation (§ 13.2). When he started on this problem in 1895, Planck, who was a thermodynamicist and a pupil of Clausius, was still rejecting the statistical and mechanistic interpretation of entropy. He was involved in trying to find a new solution of the *irreversibility problem* to be based, not on kinetic theory, but on *electromagnetism* which he considered to be more fundamental. To do this, he studied a system of charged oscillators interacting with the electromagnetic field and enclosed in a vessel with reflecting walls. His main aim was to prove that this conservative system would evolve irreversibly to thermodynamic equilibrium; a by-product would be a microscopic derivation of the spectral distribution of the thermal radiation. His first attempt which left out statistics met in 1897 a rebuff through a remark by Boltzmann. The equations of electrodynamics, like those of analytical mechanics, are invariant under time reversal: one cannot get around the irreversibility paradox.

Planck then turned to an approach which combined macroscopic thermodynamics with the statistical ideas of Boltzmann. Writing the *entropy of an oscillator* as a function of its energy was one of the guidelines used in his subsequent work. He started by writing down an empirical expression for this entropy which should satisfy the Second Law, and from it he derived Wien's radiation distribution law (1899). He then modified this empirical law to take into account new experimental results; this led to the famous Planck law (1900). Finally, in his celebrated paper of December 1900 he tried to base his expression for the entropy of an oscillator upon a statistical foundation. To do this, he wrote down for the first time the form (3.19) for Boltzmann's entropy – he wrote W as $\mathfrak{R}_0$, "die Zahl der möglichen Complexionen". He then applied this formula to an assembly of N oscillators, all with the same frequency ν; in this way he justified the expression he had proposed earlier (Exerc.3e). In order to calculate W by combinatorial methods he had had to make the possible energy values discrete, as Boltzmann had done in 1877; however, he noted that one could not let the "energy elements" go to zero, but that one must take them to be of the form $h\nu$ in order to get agreement with the experimental results, and he thus introduced the constants h and k at the same time.

The entropy is also at the heart of Einstein's papers on radiation (1905) and on the specific heat of solids (1907), where he showed that photons and phonons behave like particles with energy $h\nu$. His theoretical results, confirmed by measurements by Walther Nernst (Briesen, Prussia 1864–Bad Muskau 1941), led the wave-particle dualism to be based upon the statistical form of the Second Law. Moreover, Nernst's principle (1906), the Third Law of thermodynamics, which made it possible to define an *absolute entropy*, provided a guidepost for quantum statistical mechanics.

Let us also note Ehrenfest's method for extending quantum mechanics to other systems than oscillators. Rather curiously, these papers by Ehrenfest (1911–6) are based as much upon thermodynamics as upon analytical mechanics. He identifies *very slow changes in the parameters* occurring in the Hamiltonian of a microscopic system with *adiabatic transformations.* As the entropy has to remain constant in those processes, the same must be true of the number W of quantal micro-states. That number must therefore be an adiabatic invariant in the sense of mechanics, to wit, a quantity which remains unchanged under the effect of slow perturbations. Ehrenfest conjectured more generally that the adiabatic invariants were the quantities which should be quantized. This *adiabatic principle*, used in 1914 by Einstein, clarified the quantization rules of N.Bohr (1913) and A.Sommerfeld (1916), and helped to establish the final form of quantum mechanics in the nineteen twenties.

This final form of quantum mechanics brought in an extension of the entropy concept. In order to understand the significance of the wavefunction in a measuring process it appeared essential to analyze the rôle of the observer, a rôle the importance of which was already clear from the Gibbs and the Maxwell paradoxes. When attempting to construct a consistent theory of *quantum-mechanical measurements* Johann von Neumann (Budapest 1903–Washington 1957) introduced in 1927 both the density operator concept and expression (3.17) for the *quantum-mechanical statistical entropy* which is, by the way, often called the von Neumann entropy. In this task he was guided by Boltzmann's expressions (3.41) and (3.42) and by Nernst's principle. We gave in §§ 2.2.5 and 3.2.4 his main ideas about the analysis of measuring processes (1932). In this context he established the distinction between Hamiltonian evolutions, which conserve S (Eq.(3.29)), and irreversible measuring processes, which lead to a growth of S (Eq.(3.32)). At the same time he laid the mathematical foundations of quantum statistical mechanics and showed that at equilibrium the statistical entropy (3.17) becomes identical with the thermodynamic entropy.[4]

3.4.5 Entropy and Information

Von Neumann foresaw that the statistical entropy which he had introduced could be used to evaluate the amount of information involved in a measurement. He based himself upon a paper by Leo Szilard (Budapest 1898–La Jolla 1964) who had in 1929 returned to the study of the Maxwell demon. Szilard analyzed in detail several thought experiments, where the entropy can be lowered without expenditure of work through the action of an *intelligent being.* In each case he showed up an essential aspect: if the intelligent being succeeds in violating the Second Law, this happens thanks to some knowledge he possesses about the microscopic state of the system, *knowledge which was acquired beforehand* and which he retains in his memory at the time when he manipulates the control mechanism. Szilard concluded from this quite generally that a decrease in entropy can only be accomplished through *exploiting information* which is memorized by the being who is acting upon the system. Smoluchowski's discussion suggested that acquiring this information

[4] This work by von Neumann is reprinted in the reference quoted in § 2.2.5.

had to be paid for through an increase in entropy elesewhere, but the information concept still remained vague and qualitative.

The progress which followed illustrates vividly the cross-fertilization of two disciplines: statistical physics and *communication science*. The latter, on the frontier of mathematical research and technology, contains, for instance, cybernetics which was founded in 1938 by Norbert Wiener and which was developed by von Neumann; it is not by accident that these two have also contributed to statistical mechanics. An essential step was the creation, in 1948, of information theory by Claude Shannon and Warren Weaver. Here also, it was the form of Boltzmann's expression (3.42) which guided them to introduce the *measure of a quantity of information*. We have summarized their theory in § 3.1.

The analogy between Shannon's statistical entropy (3.1) and the entropy of statistical mechanics remained formal and semantic until 1950 when Léon Brillouin (Sèvres 1889–New York 1969) once again resurrected Maxwell's demon. Szilard had shown that information could be transformed into negentropy, a word coined by Brillouin to denote the opposite of entropy. Brillouin, and many of his followers, examined the inverse problem of the *acquiring of information*: the demon was now replaced by a scientific observer. In order to be able to select the molecules the demon must measure their velocities with more or less precision, but the precision of any measurement is limited by the thermal fluctuations in the apparatus used for the observations – see the remarks at the end of Chap.5. Brillouin analyzed, for instance, how one can detect the molecules optically: in an isothermal enclosure the equilibrium radiation (§ 13.2) does not depend on the positions and velocities of the molecules, so that the demon *cannot see them*, in contrast to what Maxwell assumed. The demon must use radiation to illuminate the molecules with a frequency exceeding the range of frequencies of the equilibrium thermal radiation. This requires the use of a lamp, that is, a hot source, and the interaction of the photons emerging from this source with the gas molecules makes the total entropy grow. Using the very new information theory, which made it possible to quantify the information acquired, Brillouin analyzed the transformation of negentropy into information, which accompanies a measurement, and the inverse transformation of information into negentropy, previously discussed by Szilard. He showed not only that the Second Law is satisfied in the balance of the two processes combined, as Smoluchowski had already suggested, but also that each of the two stages can only be accompanied by loss, either of information or of negentropy. This enabled him to confirm that the two concepts are related, more or less like heat and work, as *information and negentropy* are measurable quantities which *can be interchanged one with the other*, possibly with losses. Equation (3.34) gives an example of such an equivalence.

Brillouin also considered the registration and the transport of messages through a physical system.[5] He showed that the material used cannot be in thermal equilibrium: its entropy is less than the maximum entropy by an amount, at least equal to the information which is recorded. This enabled him to *use entropy to define a quantity of information*, either as the minimum of entropy necessary to record it, or as the maximum of negentropy which can be created by exploiting this information. The establishing of equilibrium entails the loss of the message.Well known examples are the finite lifetimes of movie films or of magnetic tape recordings.

---

[5] L.Brillouin, *Science and Information Theory*, Academic Press, New York, 1956.

Conversely, the *entropy concept can be based upon the information concept*; this approach, introduced in 1957 by E.T.Jaynes, is the one followed by us in the first few chapters of the present book. It is based upon the *principle of maximum statistical entropy* used to assign a density operator to a statistical ensemble of systems. Actually, Boltzmann and Gibbs already used a maximum entropy principle as an inductive method, both in thermodynamics and in statistical mechanics, to make predictions about the state of a system in thermodynamic equilibrium. Information theory enables us to understand this procedure, thanks to the interpretation of the statistical entropy as a measure of the dispersion in probability laws. We shall see in Chaps.5 and 6 how the maximum of this quantity can be identified with the thermodynamic entropy.

3.4.6 Recent Trends

In the last few decades the use of the entropy concept has expanded in various forms and in new directions thanks to the application of statistical mechanics techniques to problems related to information theory: theory of memory, computer theory, studies in economics, traffic movement, operational research, control theory, ... In physics, the connections between information and entropy have been strengthened. For instance, the idea has been expressed that the irreversibility of a measuring process arises from the necessity of destroying some information to obtain other. The entropy of black holes in astrophysics appears to be related to their property of decisively trapping information. Even the old Maxwell paradox has been the subject of some recent developments: it has been shown that the final "exorcism" of the demon needs recourse to quantum mechanics.

Important developments have taken place in mathematical physics. The conditions for the validity of the thermodynamic limit (§ 5.5.2) have been established, showing under what circumstances the *entropy is an extensive quantity*. This enables us to understand the limitations that exist, for instance in astrophysics, on the stability of matter. We can also note the proof of some new properties of the entropy, for instance, an inequality which is stronger than the sub-additivity inequality (3.21).

However, the most striking modern progress related to the entropy concept has been made in another domain of mathematical physics, the theory of *dynamical systems*. Created by Henri Poincaré (Nancy 1854–Paris 1912), this branch of analytical mechanics aims at a qualitative study of the geometrical and topological structure of the set of trajectories produced by non-linear dynamics – in particular, by Hamiltonian dynamics, which in phase space satisfies the Liouville theorem of § 2.3.3. At the frontier of mathematics, mechanics, and physics, it has from its start made progress thanks to the work of people such as Alexandre M.Lyapunov or Jacques Hadamard, and later on of people such as J.von Neumann and George D.Birkhoff (ergodic theorem, 1931), but its great upsurge in development started in 1955. The major advances in the central question of the *stability* of trajectories when the initial conditions or the dynamics are slightly perturbed, have resulted from the proof of the KAM theorem (A.N.Kolmogorov, V.I.Arnold, and J.Moser, 1953–62) about the structural stability of small motions, and the theorem by D.Ruelle and F.Takens (1971) about the generic instability of trajectories, whenever the number of degrees of freedom of the system equals at least 3. Another point of view about the *chaotic* nature of the trajectories is provided by *ergodicity*. A dynamics is said to be ergodic, if each trajectory passes through practically the whole of phase space, in the sense that the time average of a function of a phase point over long periods, calculated along the trajectory, equals its average over phase space. Ya.Sinai showed in 1963 that kinetic theory is ergodic, on a constant energy subspace, if the molecules are described as hard spheres. At the present moment the extension of the ideas of chaotic dynamics to quantum mechanics is actively pursued.[6]

Such studies give, at least in classical mechanics, a new clue to the old *irreversibility problem*, which has already been clarified by other modern approaches mentioned in § 3.4.3. If the trajectories are chaotic, they are extremely sensitive to a change in the initial conditions, and two neighbouring points in the phase space $\mathcal{E}_P$ will almost always get much further apart from one another on a macroscopic time scale. Let us consider a phase density which initially is uniform in some compact volume W; according to Liouville's theorem, the measure of W, and hence the entropy (3.41), does not change with time. However, a volume element of W must become elongated in certain directions, while contracting in other directions, and then branch out, fold over, and produce many thinner and thinner sheets. Therefore, if the positions in $\mathcal{E}_P$ are not defined with infinite precision, but only within a more or less coarse graining, everything evolves as if the phase density in the final state did not have a finely multi-sheeted domain, of volume W, as support, but was spread out over a larger, more regular domain, showing a rather hazy occupation. Any coarse-graining, *even an infinitesimal one*, of $\mathcal{E}_P$ thus defines an entropy which increases over sufficiently long periods; information gets lost to details which *cannot be observed with the chosen accuracy*. In this context, the Second Law can be understood as a consequence of the combination of an *extremely complicated evolution* and an inevitable *coarse-graining*. Actually, for thermodynamic systems not all variables show chaotic motion; even on our time scales, the *conservative variables remain relevant* and their evolution will be the subject of our studies when we deal with non-equilibrium thermodynamics (Chap.14).

While statistical mechanics has benefited from the study of dynamical systems, in turn it has inspired the introduction of new entropy concepts. In particular, the *Kolmogorov entropy* (1949) appears as a useful measure for the chaotic nature of the *evolution* of a dynamical system, whereas the measure of the disorder existing at a certain moment is represented by the ordinary entropy.

Notwithstanding the many interrelations which have been established between the different kinds of entropy, the identification of the thermodynamic entropy and the statistical entropy has not yet been accepted universally. While the former can be measured more or less directly for systems in thermodynamic equilibrium and thus appears to be a property of the system itself, the latter refers to the knowledge of the system by an observer and does have a nature which is partially subjective, or at least anthropocentric and relative. It certainly may appear paradoxical that these two quantities would be equal to one another. However, an analysis of non-equilibrium processes (§ 14.3.3) and also the interpretation of such experiments as spin echo

[6] The proceedings of the sessions XXXVI, XLVI, and LII of the Les Houches Summer School (North-Holland, 1981, 1986, and 1989), and references therein, provide an introduction to these fields.

experiments, which are an actual realization of a Maxwell demon (§ 15.4.5), compel us to think nonetheless of the entropy in the framework of information theory.

Summary

Information theory enables us to associate a number with each state of a system represented by a density operator. This number, the lack of information or the statistical entropy (3.17), measures the degree of disorder, coming from the random nature of the state. The statistical entropy lies between 0, for a pure state, and $k \ln W$*, when all the kets in the* W*-dimensional Hilbert space are equiprobable. It increases when the disorder grows, in particular, when we discard correlations between subsystems, join up several statistical ensembles into a single one, or are dealing with an imperfectly known evolution. The entropies of statistically independent subsystems are additive. The classical limit is given by (3.39).*

Exercises

3a Information and Entropy: Orders of Magnitude

Assuming that the macro-state describing 1 mg of water at room temperature corresponds to a certain number of quantum micro-states which are all equally probable, evaluate that number, knowing that the entropy of water is 70 J K^{-1} mol^{-1}.

Estimate, in thermodynamic units, the order of magnitude of the quantity of information contained in the French National Library, which has 10 million volumes with an average of 350 pages, each containing on average 1500 characters per page. Compare this with the entropy of 1 mm^3 of a perfect gas, or with the increase in entropy when 1 mm^3 of ice melts; the latent melting heat for 1 g of water is 80 cal.

Compare it also with the order of magnitude of the genetic information of a living being – after development the DNA of the chromosomes of a bacterium measures 2 mm in total; its constituent nucleotides, of which there are 4, are at distances apart of 3.4 Å; the human genetic material is 1000 times larger. Evaluate the genetic information per unit mass for the chromosomes, where each nucleotide has a mass of about 300 u, and DNA is a double helix, and for the proteins: a protein is a long chain formed by 20 different amino-acids of average mass 100 u, and each acid is coded by three successive bases. Why is the information per unit mass smaller in DNA? What is the maximum number of different proteins which can exist in a bacterium?

The brain contains 10^{10} neurons, each connected with 1000 neighbours. If one assumes that the memory consists in activating or deactivating each bond, what is the maximum amount of information which may be stored in the brain? What is the quantity of information necessary to describe the way the neurons are connected? Show that this organization is too complex to be coded completely in the genetic material.

In order to realize how huge Avogadro's number is, or how small Boltzmann's constant is, evaluate the probability that air inhaled in one breath (1 l) contains one of the molecules from Julius Caesar's last breath, assuming that this last breath has been distributed uniformly over the whole of the atmosphere. Evaluate the length attained if the atoms of 1 mm^3 of iron were put in a straight line, rather than put on a lattice – the specific weight of ^{56}Fe is 7.9×10^{-3} kg m^{-3}; to estimate the distances between neighbouring atoms, assume that the lattice is cubic; compare the distance found with the distance between the Earth and the Moon. What is the mass of a monomolecular petrol film with an area of 1 km^2 on the surface of the sea? Assume that the molecules have a mass of 100 u and are at distances apart of 3 Å.

3b Subsystem of a Pure State

1. Consider a system, consisting of two subsystems a and b, with a density operator $\widehat{D}$ which represents a pure state: $S(\widehat{D}) = 0$. Show that the two subsystems have the same entropy: $S(\widehat{D}_a) = S(\widehat{D}_b)$. In particular, if ab and a are in pure states, b is also in a pure state.

2. Show that a system a with an arbitrary density operator $\widehat{D}_a$ can always be considered as a subsystem of a larger composite system ab which is in a pure state.

Hints. Write the pure state down in a factorized base, in the form $\sum_{kl} |k_a l_b\rangle C_{kl}$. Express $\widehat{D}_a$ and $\widehat{D}_b$ as functions of C, to be considered a rectangular matrix. Use the identity $C^\dagger f(CC^\dagger) = f(C^\dagger C)C^\dagger$ for any function f.

3c Relevant Entropy

The statistical entropy $S(\widehat{D})$ characterizes the lack of information associated with *all* conceivable observations on the macro-state, that is, with the whole algebra of the observables. We are solely interested in a set of relevant observables which span a vector subspace X of the space of the observables. These observables are, for instance, only those which can be measured in practice, or only those which represent the macroscopic quantities, or those observables which we define through a coarse-graining eliminating the finer details. Define a relevant statistical entropy $S_X(\widehat{D})$ *relative to the observables* X. This entropy needs to *take into account only the information associated with these observables.* To do that introduce the set of density operators $\widehat{\mathcal{D}}$ which lead to the same expectation values as $\widehat{D}$ for the observables $\widehat{A}$ belonging to X, but to arbitrary expectation values for other observables. Construct a density operator $\widehat{D}_X$ from among these density operators $\widehat{\mathcal{D}}$, which are equivalent to $\widehat{D}$ for the expectation values of the relevant observables, by looking for the maximum $S(\widehat{D}_X)$ of $S(\widehat{\mathcal{D}})$. This maximum defines the relevant entropy $S_X(\widehat{D})$ relative to the set X, since $\widehat{D}_X$ contains the least amount of information compatible with the data X.

1. Construct $\widehat{D}_X$ in the following cases: (a) X contains only the energy and the number of particles; (b) in classical statistical mechanics X is the set of single-particle observables; (c) given a set of orthogonal projectors $\widehat{P}_\alpha$, with unit sum, let X be the set of operators which commute with the $\widehat{P}_\alpha$; (d) X is the set of these $\widehat{P}_\alpha$.

2. Show that $S_X(\widehat{D})$ decreases when the space X gets larger and that $S_X(\widehat{D})$ is a concave function of $\widehat{D}$. Interpret this result.

In all these cases $S_X(\widehat{D})-S(\widehat{D})$, which is positive, defines the information which hypothetically could be gained if one were able to measure not only the relevant observables $\widehat{A}$ pertaining to X, but also all other observables, for the macro-state $\widehat{D}$ under consideration. The relevant entropy $S_X(\widehat{D})$ is an essential quantity in the discussion of irreversibility (§§ 3.2.4, 3.4.3, 4.1.5, 14.3.3, and 15.4).

Answers:

1. (a) If X contains solely $\widehat{H}$ and $\widehat{N}$, $\widehat{D}_X$ is the grand canonical equilibrium density operator (§ 4.3.2) which would be associated with the same values of $\langle H \rangle$ and $\langle N \rangle$ as $\widehat{D}$. At equilibrium, we have $S_X(\widehat{D}) = S(\widehat{D})$; the idea of a relative entropy is therefore only useful for non-equilibrium problems.

(b) If X represents the single-particle observables, D_X is a phase density without correlations, and $S_X(D)$ is the Boltzmann entropy S_B, the increase of which is expressed by the H-theorem (§§ 3.4.2 and 15.4.2).

(c) In this case, $\widehat{D}_X = \sum_\alpha \widehat{P}_\alpha \widehat{D}\, \widehat{P}_\alpha$ is obtained by cutting off the elements of $\widehat{D}$ which lie outside the diagonal blocs associated with the $\widehat{P}_\alpha$ in the base where these are diagonal. When the $\widehat{P}_\alpha$ are the projections onto the different eigenvalues of an observable $\widehat{A}$, $S(\widehat{D}_X)-S(\widehat{D})$ defines the information included in $\widehat{D}$ and associated with the set of observables which do not commute with $\widehat{A}$; this information is lost by measuring $\widehat{A}$ (§ 3.2.4). The suppression of the off-diagonal elements of $\widehat{D}$ thus increases the disorder, or induces a loss of information.

(d) The density operator $\widehat{D}_X$ is obtained by first of all truncating $\widehat{D}$ as above, and then replacing each diagonal bloc α by a matrix which is a multiple of the unit matrix, in such a way that the average value of the diagonal elements of the bloc α remains unchanged. This kind of reduction of the density operator $\widehat{D}$ to produce $\widehat{D}_X$ is called "coarse-graining"; the numbers $\langle P_\alpha \rangle$, which are the only information retained when changing from $\widehat{D}$ to $\widehat{D}_X$, are the probabilities that the observable $\widehat{A}$ has the value a_α in the macro-state considered. The evolution of these probabilities is given by an approximate equation, the so-called Pauli equation; its study is an approach to irreversibility when one restricts oneself to the observable $\widehat{A}$.

2. The introduction of a new constraint on $\widehat{\mathcal{D}}$ prevents $S(\widehat{\mathcal{D}})$ from attaining higher values. One gains information by giving an extra expectation value.

The definition of $S_X(\widehat{D}_1)$ and of $S_X(\widehat{D}_2)$ and the concavity (3.26) imply that

$$\begin{aligned}\lambda\, S_X(\widehat{D}_1) + (1-\lambda)\, S_X(\widehat{D}_2) &= \lambda\, S(\widehat{D}_{1X}) + (1-\lambda)\, S_X(\widehat{D}_{2X}) \\ &\leq S\big(\lambda \widehat{D}_{1X} + (1-\lambda)\widehat{D}_{2X}\big).\end{aligned}$$

The density operator $\lambda \widehat{D}_{1X} + (1-\lambda)\widehat{D}_{2X}$ is a member of the set of operators $\widehat{\mathcal{D}}$ which are equivalent to $\lambda \widehat{D}_1 + (1-\lambda)\widehat{D}_2$ as far as the expectation values of the relevant observables pertaining to X are concerned. Its entropy is thus at most equal to $S_X(\lambda \widehat{D}_1 + (1-\lambda)\widehat{D}_2)$. The mixture of the two populations in a single statistical ensemble makes us lose a certain amount of information relative to the relevant observables pertaining to X, which is a generalization of the ordinary concavity property (§ 3.2.2e) for which X was the set of all observables.

3d Inequalities Concerning the Entropy

1. Show that the statistical entropy of a mixture of ensembles, which has a lower bound given by (3.27), also has an upper bound as follows:

$$S\left(\sum_j \mu_j \widehat{D}_j\right) \leq \sum_j \mu_j\, S(\widehat{D}_j) + S(\mu),$$

$$S(\mu) \equiv -k \sum_j \mu_j \ln \mu_j,$$

where the equality holds only if $\widehat{D}_i \widehat{D}_j = 0, \forall\, i \neq j$. This expresses the fact that the increase in disorder through mixing is not larger than the dispersion of the relative weights μ_j of the ensembles j.

2. Consider the set of descriptions by wavefunctions $|\psi_\lambda\rangle$ and probabilities q_λ which are equivalent to a given density operator $\widehat{D}$ (§ 2.2.2). Associate with each the quantity $S_q = -k \sum q_\lambda \ln q_\lambda$. This would be the lack of information corresponding to the events λ if the latter were exclusive. Show that $S = -k\mathrm{Tr}\widehat{D} \ln \widehat{D}$ is the lower bound of the S_q. One can interpret this result by noting that the descriptions $|\psi_\lambda\rangle, q_\lambda$ contain, with respect to a particular description $|m\rangle, p_m$ on an orthonormal base, a disorder which has no physical meaning, as the events λ considered to be distinct in that description are, in fact, not exclusive.

Hints:

1. Show first of all that the operator $\ln(\widehat{X} + \widehat{Y}) - \ln \widehat{X}$ is positive, if $\widehat{X}$ and $\widehat{Y}$ are two positive operators. Hence, one finds that $\mathrm{Tr}\widehat{X} \ln(\widehat{X} + \widehat{Y}) > \mathrm{Tr}\widehat{X} \ln \widehat{X}$, an inequality which we can apply to $\widehat{X} = \mu_j \widehat{D}_j$, $\widehat{X} + \widehat{Y} = \sum_j \mu_j \widehat{D}_j$. Discuss the limiting cases where $\widehat{X}$ and $\widehat{Y}$ have zero eigenvalues.

2. Use the preceding result for $\widehat{D}_j = |\psi_\lambda\rangle\langle\psi_\lambda|$.

3e Entropy of Quantum Oscillators

Consider a set of N – distinguishable – oscillators, each of which has quantized energy levels $p\varepsilon$ ($p = 0, 1, 2, \ldots$). The total energy of this set equals $P\varepsilon$, where P is an integer. Assume that all possible micro-states with this energy, characterized by integers $p_1, p_2, \ldots, p_N$ such that $p_1 + p_2 + \ldots + p_n = P$, have the same probability. Evaluate the statistical entropy (3.19). Hence deduce, in the limit as $N \to \infty$, the average entropy per oscillator S/N as a function of its mean energy $E = P\varepsilon/N$. Write finally E in terms of the temperature T by identifying T with $N\,dE/dS$.

This calculation was the basis for Planck to establish his radiation law (§§ 3.4.4 and 13.2.2). Each mode of the electromagnetic field in a cavity is then a different oscillator. Planck introduced the hypothesis that its energy can only take the discrete values $p\varepsilon = ph\nu$, and he used the additivity of energy and entropy for all the modes.

Hint. Ehrenfest remarked that the number of ways in which we can choose N positive integers $p_1 + 1$, $p_2 + 1$, ..., $p_N + 1$ having a sum equal to $P + N$ is equal to the number of ways of distributing $N - 1$ points in $P + N - 1$ boxes.

Answer:

$$S = k \ln \frac{(P+N-1)!}{(N-1)!P!}$$

$$\sim Nk \left[\left(1+\frac{E}{\varepsilon}\right) \ln \left(1+\frac{E}{\varepsilon}\right) - \frac{E}{\varepsilon} \ln \frac{E}{\varepsilon}\right].$$

$$E = \frac{\varepsilon}{e^{\varepsilon/kT} - 1}.$$

4. The Boltzmann-Gibbs Distribution

"L'équilibre est la loi suprême et mystérieuse du grand Tout."

V. Hugo, Post-scriptum de ma Vie

"En remontant chez moi pour y passer la soirée à travailler de mon mieux, je me disais que le monde n'est pas construit pour l'équilibre. Le monde est désordre. L'équilibre n'est pas la règle, c'est l'exception."

G. Duhamel, Maîtres, 1937

The preceding two chapters helped us to set up the formalism of statistical mechanics. We introduced in Chap.2 the density operators $\widehat{D}$, and their classical limit, the densities in phase. They sum up our knowledge about the system and enable us to make predictions of a statistical nature about physical quantities, the expectation values of which we can calculate, starting from $\widehat{D}$. In Chap.3 we defined the statistical entropy $S(\widehat{D})$ which measures the random nature, or disorder, of a density operator. In those two chapters we assumed that the latter was given. However, in order actually to be able to calculate the properties of a system which has been prepared in some given way we must know how to assign to it a density operator representing the physical situation that we want to describe. This *problem of the choice of* $\widehat{D}$ will be solved in the present chapter for thermodynamic equilibrium states. In order to find the general form, the so-called Boltzmann-Gibbs distribution, of the density operators, or the densities in phase, describing these states, we shall use a postulate of a statistical nature which is similar to the criteria used in statistics to find the unbiased probability law for a set of random events. We introduce in this way a general prediction method (§ 4.1.3). This method leads us to represent a system in *thermodynamic equilibrium* by the *most disordered* macro-state compatible with the macroscopic data (§ 4.1).

We can then construct the density operator with the largest statistical entropy under the constraints accounting for the macroscopic equilibrium data (§ 4.2). Depending on the nature of these data, the equilibrium state may take on various specific forms (§ 4.3), all encompassed in the exponential Boltzmann-Gibbs expression (4.6). We introduce at the same time the important *partition function method* which we shall use systematically in the remainder of this book when studying the equilibrium properties of various physical systems. The main results of this chapter are gathered in § 4.2.6.

4.1 Principles for Choosing the Density Operators

4.1.1 Equal Probabilities

The problem of the choice of the density operator, or the density in phase, which in statistical mechanics describes a given physical situation, is similar to the problem of *statistical estimation* in probability calculus: how should one choose the probabilities for the various possible events that a random process can give rise to?

When one does not know anything the answer is simple. One is satisfied with *enumerating* the possible events and assigning *equal probabilities* to them: before throwing a die, there is no reason to suspect that it is loaded more heavily on one side than the other. Any probability law other than $p_1 = \ldots = p_6 = \frac{1}{6}$ would be *biased* and would introduce, without justification, preconceived ideas. Such a choice constitutes Laplace's *indifference principle*, that he called "*principe de raison insuffisante*". It rests upon the existence of a group, in this case the group of permutations of the faces of the die, which defines an equivalence between the elementary events expected at the start. This *invariance group* was already more or less implicitly involved in the definition of the entropy in §§ 3.1.2e, 3.1.4, and 3.3.1.

Similarly, when we studied in § 1.2.1 a perfectly random set of N spin-$\frac{1}{2}$ particles about which we knew nothing, we were led to assign equal probabilities to all possible micro-states. More simply, if we consider a single *unprepared* spin-$\frac{1}{2}$, there is no reason to assume that it is polarized in one direction rather than in another; we should assign to it the most disordered macro-state possible, where *all possible micro-states have the same probabilities*: the probability to find $s_z = \frac{1}{2}$ along some z-axis is equal to $\frac{1}{2}$.

Thus, if we do not know anything about a system it is natural to represent it by the statistical ensemble which is *completely disordered*, where the probabilities of the W possible micro-states are equal. We saw in §§ 3.2.2a and 3.3.2 that this state is characterized by the fact that its statistical entropy is a maximum and equal to $k \ln W$. It is also the only state which is invariant under a unitary transformation in quantum statistical mechanics, or under a canonical transformation in classical statistical mechanics; these invariances play the same rôle as the symmetry of the faces for the die. Our predictions remain unchanged under a unitary or canonical transformation of the observables.

4.1.2 Information About the System

The above argument cannot be applied when the state of the system is partially known. In general, we want to study the properties of a system as a function of *macroscopic parameters* such as the density of a given kind of particles or the total energy, and the density operator $\widehat{D}$, or the density in phase, must account for these. However, it is important to distinguish two different kinds of data at the microscopic level.

Some of the available information consists of *data given with certainty*, similar to the number of faces of a die: the spin and the nature of the particles, the shape and the volume of the box in which they are enclosed, ... Those we take into account through the *definition of the Hilbert space* or the phase space in which we are working. For instance, in §§ 1.2.2 and 1.2.4 we assumed that the energy of the system was lying between U and $U + \Delta U$. We had thus some certain information about the system, to wit, we knew that its representative ket was lying in a W-dimensional subspace $\mathcal{E}_{\mathrm{H}}^{W}$ of $\mathcal{E}_{\mathrm{H}}$, spanned by the micro-states $|\sigma_1, \ldots, \sigma_N\rangle$ satisfying (1.11). As in the preceding § 4.1.1, we were led to postulate that in thermal equilibrium the macro-state is the *most disordered one possible*, but in this case *in the allowed Hilbert space* $\mathcal{E}_{\mathrm{H}}^{W}$. Its density operator is the so-called "*microcanonical*" macro-state,

$$\widehat{D} = \frac{1}{W}\,\widehat{I}_W = \sum_{m=1}^{W} |m\rangle \frac{1}{W} \langle m|, \tag{4.1}$$

where $|m\rangle$ is a base in $\mathcal{E}_{\mathrm{H}}^{W}$; its statistical entropy, $k \ln W$, is a maximum within $\mathcal{E}_{\mathrm{H}}^{W}$.

There are, however, other situations where the information available about the system is in the form of *data of a statistical nature*, to wit, expectation values of some observables, averaged over the statistical ensemble of which the system considered is a member. For instance, one could ask in the model of Chap.1 what is the density operator of *a single* spin in thermal equilibrium, without first writing down the density operator of the system of N spins; we have one piece of information about the state of this spin, namely, the expectation value of its energy. More generally, when the system studied is put into contact with a large system with which it can exchange energy, its own energy is only fixed on average and it can fluctuate freely around that average. This is just what happens when we fix the temperature through an interaction with a heat bath, as we shall check in § 5.7.2. A statistical ensemble of this kind, where the energy at thermal equilibrium can fluctuate freely around the *given average value* $U = \mathrm{Tr}\,\widehat{D}\widehat{H}$ (§ 4.3.1), is called a "*canonical*" *ensemble*.

More generally, this kind of statistical data shows up when the state of the system on the macroscopic scale is characterized by our knowledge of the expectation values $\langle A_i\rangle$ of some observables $\widehat{A}_i$, the Hamiltonian operator, as a moment ago, or the particle number operator for a given kind of particles, or the momentum operator, ... Each of these data is reflected by the fact that the density operator is not arbitrary in the Hilbert space considered, but *must satisfy the constraint*

$$\mathrm{Tr}\,\widehat{D}\widehat{A}_i = \langle A_i\rangle. \tag{4.2}$$

Note that, depending on the circumstances, the value of the same physical quantity can be given *either with certainty, or as a probabilistic average*. We have just seen that the so-called micro-canonical and canonical ensembles are distinguished by the fact that the energy U is in the one case given with certainty – within a margin ΔU which is large compared to the distances between levels – and in the other case solely on average. Similarly, if the number of particles N in the system is exactly known, its value determines the Hilbert space in which the density operator acts, namely, that of N-particle states. If, on the other hand, only the average of this number is known, we must work in a larger Hilbert space, that of the states where the number of particles can take on all possible values, which is called the *Fock space*; the number of particles is now a random variable and we must impose on the density operator a constraint like (4.2) which expresses that the expectation value of the number of particles $\langle N \rangle$ is given. Similarly, in classical mechanics, giving N exactly leads to a description by a density in phase D_N in the $6N$-dimensional phase space, whereas giving the average $\langle N \rangle$ makes it necessary to use the formalism of § 2.3.6 with a constraint on $\langle N \rangle$. A statistical ensemble where both the energy and the number of particles are given on average is called a "*grand canonical*" *ensemble* (§ 4.3.2).

We shall see in § 5.5.3 that, in the limit of large systems, the two ways of giving the information which characterizes a state on the macroscopic scale lead to the same predictions for most physical properties. Nevertheless, it is essential to distinguish these two situations carefully, since the techniques for studying them are different, as we shall see in § 4.3.

4.1.3 Maximum of the Statistical Entropy

The information we have available or the conditions we impose on the state of the system are thus reflected by a certain number of restrictions and constraints on the density operator $\widehat{D}$, but they clearly are insufficient to determine it completely. In order to be able to make predictions about *other quantities* from the formula $\langle B \rangle = \mathrm{Tr}\, \widehat{D}\widehat{B}$, we must manage to determine $\widehat{D}$. To do this we must find a criterion which enables us to choose $\widehat{D}$ as reasonably as possible. This criterion must appear as an extension of the method of § 4.1.1 to the case where $\widehat{D}$ is subject to constraints of the kind (4.2).

The obvious concept of equal probability means, as we saw, that one chooses the most random probability law possible; any other choice is biased, introducing arbitrarily an order, for which there are no reasons of believing that it exists. On the other hand, the various properties of the statistical entropy that we proved in the preceding chapter suggest that we should consider that quantity as a *measure for the disorder*. It is thus natural to take it as an "estimator" in the sense of probability calculus and to use it to choose the density operator describing a given physical situation. We assume therefore that the best guess for $\widehat{D}$ is provided by the following prescription:[1]

[1] E.T.Jaynes, Phys.Rev. **106**, 620 (1957), **108**, 171 (1957); starting in 1979, the proceedings of an annual workshop on maximum entropy methods are being published (MIT Press, Reidel, Cambridge University Press, Kluwer).

Maximum Statistical Entropy Principle. Amongst all statistical distributions – density operators $\widehat{D}$ or phase densities D – compatible with the available data we must represent the system by that macro-state which has the largest value of the statistical entropy $S(\widehat{D})$.

This principle is commonly used not only in statistical physics but also for the most diverse applications of statistics, of which Exerc.4h gives an example; it generalizes the equal probabilities criterion which already led to the maximum of S. Here we must *maximize the statistical entropy taking the constraints into account*, so that the system be in the *most disordered macro-state possible compatible with the data.* In information theory terms, we would say that the various density operators reproducing the known information lack a more or less large amount of information. We must choose amongst them that one which contains *no more information than is strictly necessary to take the data into account*, that is, that one which corresponds to the largest amount of lack in information, S. Choosing any other density operator would mean that we assumed we possessed more information about the state of the system than the data actually supply us with, and thus would lead us to make biased predictions.

We have just tried to make the principle of maximum disorder intuitively understandable by using statistical arguments drawn from information theory, but it is possible to give a more convincing justification by relying solely on the equal probabilities principle (see § 5.7.2). We shall see in § 4.1.5 how an analysis of the dynamic processes which produce the disorder can also help us to understand this principle. Its best justification will come *a posteriori* from the remarkable agreement between the many predictions which it enables us to make and the experimental observations.

In fact, the maximum entropy principle is nothing but a method of statistical inference leading to predictions of a statistical nature and it is not precluded that it would not account properly of experimental facts. If that were the case, this disagreement would simply reveal that our premises were incomplete and that we had used an insufficient number of $\langle A_i \rangle$ parameters to describe the macroscopic state of the system. We shall progress in our understanding of the phenomena through introducing new parameters which impinge on the macroscopic physics, but we shall continue to rely on the principle of maximum statistical entropy. Examples of such "hidden variables" which we need to introduce in the theory in order to make our description adequate will be given in §§ 8.4.5, 9.3.3, 12.3.3, and 15.4.4.

4.1.4 Macroscopic Equilibrium and Conservation Laws

In § 4.2 we shall apply the maximum entropy principle and explicitly construct the density operator associated with a set of data. We discuss beforehand the choice of these data, a question, the importance of which we want to stress. They should characterize the state of the system on the macroscopic scale, whereas on the microscopic scale they give us constraints on the macro-state. For non-equilibrium systems we shall see in the last two chapters that their choice may be a subtle point. However, we shall mainly study in the present book equilibrium systems on the macroscopic scale. In that case the choice of state variables is guided, usually without any ambiguity, by thermodynamics. In fact, the data which on our scale characterize the equilibrium state of a system are, in general, the *values of the constants of the motion*, such as the energy, the momentum, the angular momentum, or the number of particles of a given kind. One must also add to these data *parameters* which one can control, such as the *volume* of the system or a *field* acting upon it.

We saw in § 4.1.2 that the constants of the motion can, depending on the statistical ensemble considered, be given either exactly, or statistically; in the latter case the quantity which remains constant from one sample to the next is the *expectation value of a conservative quantity.* Let us remind ourselves that the statistical ensemble is called canonical when the energy is determined on average, and it is called grand-canonical if both the energy and the number of particles are given on average. The control variables – the volume and the fields – will, in most examples that we shall be dealing with, be treated as exact data and they will occur as parameters in the Hamiltonian.

It is important to note that in an equilibrium situation the macroscopic quantities do not change. The corresponding macro-state, that is, the probability law describing in statistical mechanics an ensemble of systems, all prepared in the same way as regards the data considered, is stationary. However, the motion takes place on the microscopic scale for each of the samples in this statistical ensemble. If the system appears to us to be *stationary*, this is due to *statistical* reasons, with only the probability law and the macroscopic variables remaining constant. Microscopic observations (§ 5.7.3) reveal that a particular system, in fact, evolves rapidly while we have macroscopic equilibrium. One can consider its micro-states at successive times to be different instances of samples in the statistical ensemble. An example was given in Exerc.2a-8: the system evolves rapidly, spending as much time in each of the possible micro-states, so that successive sampling produces an ensemble of equally probable micro-states. More generally, consider a classical dynamic system with a representative point $\{q,p\}$ which moves through phase space. Given a function $F\{q,p\}$, let us associate with each trajectory $\{q(t),p(t)\}$, $0 < t < \tau$, of energy E its average value $\overline{F}$ over the period τ. The *ergodic theorem*, which is valid for a large class of systems, states that in the limit as

$\tau \to \infty$ this average $\overline{F}$ tends to the expectation value $\langle F \rangle$ in a microcanonical ensemble of energy E: provided the time τ is sufficiently long, the evolution realizes an almost complete covering of the energy surface E in phase space. Nevertheless, for a macroscopic system, the number of micro-states involved in an equilibrium macro-state is so huge (§ 1.2.3) that application of the ergodic theorem would need gigantic times τ, much longer than the age of the Universe. In actual fact, therefore, evolution in time enables us *only to explore an extremely small fraction* of the possible configurations.

A system at equilibrium can be isolated, but it can also be a *subsystem* of a larger system which itself is in equilibrium. For instance, an object in contact with a heat bath can be described as a macro-state of a canonical ensemble, characterized by giving the average value of its energy; exchanges with the thermostat allow the latter to fluctuate.

Whether a system be isolated or not, it can have macro-states which are *stationary, but non-equilibrium.* In the case of an isolated system the evolution is governed by (2.49) and any density operator commuting with the Hamiltonian – for instance, a projection on an eigenstate of the latter – will describe a stationary state, whereas the equilibrium states must correspond to a maximum of the disorder. This kind of stationary macro-state is not stable as its entropy increases (Eq.(3.31)) if there are small perturbations present. In the case of a non-isolated system the interaction with external systems which can evolve may give rise to non-equilibrium stationary solutions. For instance, a substance in contact at its two ends with two thermostats maintained at different temperatures will transfer heat from one to the other. As the thermostats are large, their temperatures only change imperceptibly over not too long a period. If one is interested only in the conductor itself, one ascertains that, after a short transitory period, it reaches a permanent non-equilibrium regime; such a macro-state is characterized not only by giving the constants of the motion, but also, for instance, the value of the energy *flux.* We refer to Chap.14 for a study of this kind of stationary non-equilibrium states.

Depending on the constants of motion involved, the concept of thermodynamic equilibrium covers a variety of physical situations. When the total energy is given, the state of maximum disorder is that of *thermal equilibrium.* When the system is, moreover, a fluid with a conserved number of particles, for instance a liquid and its saturated vapour in the field of gravity, we are dealing with *hydrostatic equilibrium.* In the case of particles in solution which can cross semipermeable membranes, we talk about *osmotic equilibrium.* When the system consists of several kinds of atoms which can combine to form molecules, the constants of motion are the numbers of atoms of each kind and we have *chemical equilibrium.* Finally, in the case of a system of charged particles, the state of maximum disorder compatible with conservation of charge is *electrostatic* or *magnetostatic equilibrium.*

The *identification of the quantities that are conserved* or nearly conserved (§ 4.1.6) is one of the important stages in the construction of the microscopic model suitable for the description of a system under the circumstances we are considering. Whereas the energy almost always occurs amongst these conserved quantities, the choice of the other variables sometimes raises problems, even in equilibrium situations. For instance, a mole of hydrogen under normal conditions is described by giving the number N of its molecules. A more fundamental description might involve the number of atoms, or even the numbers of protons and electrons as constants of the motion. This, however, would introduce useless complications, as the probability for observing a molecule which is dissociated into atoms or ionized is usually negligible. On the other hand, at very high temperatures, we must take the total number of atoms, whether isolated or bound into molecules, as a constant of the motion in order to characterize the state of chemical equilibrium $H_2 \leftrightarrows 2H$. A correct description of the same mole of hydrogen inside the sun similarly requires the consideration of its $2N$ protons and $2N$ electrons, and even of nuclear reactions. Depending on the circumstances, the conserved entities may thus be molecules, atoms, radicals, ions, or elementary particles. The choice of the constants of the motion is thus neither universal, nor always obvious; in some cases there even exist hidden constants of motion which only appear when we test the model experimentally, as we already indicated at the end of § 4.1.3.

4.1.5 Approach to Equilibrium

The maximum entropy principle which determines the equilibrium macro-state of a system relies upon purely statistical considerations, without worrying about how this macro-state has been reached in time. In thermodynamics, however, equilibrium is defined as the *final outcome of an evolution.* We are thus led to wonder about the problem of the approach to equilibrium. Let us assume that the system was initially prepared in a known non-equilibrium macro-state. How will it evolve starting from this state? The constants of motion will remain constant during that evolution. However, as far as the other variables are concerned, will the system reach after a more or less long period the equilibrium macro-state defined by the principle of a maximum statistical entropy?

Let us consider, for instance, two vessels, each filled with a different kind of gas at the same pressure and temperature, which we connect at the initial time. Experience indicates that after some time the N_a and N_b molecules of the two kinds will be perfectly mixed. The system evolves towards maximum disorder and macroscopic equilibrium is reached *irreversibly.* During this process the statistical entropy has increased by an amount ΔS. In § 8.2.1 we shall see that $\Delta S = k \ln \left[(N_a + N_b)!/N_a!N_b!\right]$; this expression can be interpreted as the information that would be acquired by learning which are the N_a atoms of type a amongst the $N_a + N_b$ molecules. However, the microscopic dynamic laws are *reversible* and they leave, according to (3.29), the *statistical entropy unchanged.* It is therefore not easy to understand the irreversibility of the observed evolution which is revealed by the fact that ΔS is positive. This contradiction is the so-called irreversibility paradox, the importance of which we discussed in § 3.4.3 and to which we shall return in Chaps.14 and 15.

Let us right now give a few indications about the approach to equilibrium. Two essential points are the *macroscopic size* of the system and the *statistical nature* of its description. Let us first of all consider a system the Hamiltonian of which contains small random parts. This happens, for instance, if we are interested in part of a substance which interacts with the remainder in a badly controlled way. We indicated in § 1.2.2 and we proved for the model of Exerc.2a that such random interactions tend after some time to make the probabilities for the micro-states with the same energy equal, and thus to *increase the disorder.* More generally, the irreversibility inequality (3.31) shows that if the Hamiltonian of a system is not known with certainty, one *loses information* during the evolution: as long as the system has not reached the most disordered state possible which is allowed by the evolution, its statistical entropy may grow under the effect of small random parts of the Hamiltonian. In this evolution the constants of motion conserve the value that they had been given during the preparation of the initial state, while the statistical entropy cannot decrease. We thus understand why the process can proceed *until the statistical entropy has reached its maximum, compatible with the constraints* on each of the constants of motion, and this provides us with a dynamic justification for the principle of maximum statistical entropy. However, it is assumed that we have taken into account *all* conservation laws which hold during the evolution; hence the only information about the initial state still existing in the final state is that about these constants of motion.

The explanation of the approach to equilibrium and of its irreversibility is more subtle when the Hamiltonian is completely determined. In fact, the system in that case keeps, in principle, a memory, not only of the constants of motion, but also of all other characteristics of the initial state. Nevertheless, since the system and its equations of motion are not very simple, this information is *transferred* with time to degrees of freedom that we cannot observe in practice. In the example of the mixing of two gases, the fact that the molecules a and b were initially in different containers is reflected after the mixing by extremely complicated correlations between the positions and velocities of the molecules. The evolution has transformed simple information into *information which in practice is inaccessible* since it is associated with correlations between a macroscopic number of particles. The irreversibility of the approach to equilibrium comes from the fact that we are justified in *discarding all information about the degrees of freedom which are too complicated to be observable* in any imaginable experiment. The increase in the relevant statistical entropy thus expresses this loss of accessible information (see Exerc.3c and § 15.4). The source of irreversibility for deterministic evolutions is therefore the complexity of the dynamics. In the framework of classical mechanics, we have already seen in § 3.4.6 how the *chaotic behaviour of trajectories* can produce a spreading of probability or a loss of information.

4.1.6 Metastable Equilibria

Many interesting physical systems are in *quasi-equilibrium* states. On a time-scale which can be long they hardly evolve at all. Such a kind of metastable equilibrium may be distinguished from true thermodynamic equilibria considered in § 4.1.4 by the fact that the macroscopic variables which characterize the state include, apart from the exact constants of motion, *approximate constants of motion.* The theoretical treatment, however, is based upon unchanged principles.

Let us as an example consider a system consisting of two parts a and b which are very weakly coupled, with a Hamiltonian $\widehat{H} = \widehat{H}_a + \widehat{H}_b + \widehat{V}$ where $\widehat{V}$ is extremely small. For definiteness we can imagine that a and b are two samples of a substance, separated by a wall which is practically impermeable to heat. If $\widehat{V} = 0$ the energies $\langle H_a \rangle$ and $\langle H_b \rangle$ of the two parts are two constants of motion which can be fixed independently and the macro-state of the system is obtained by maximizing the statistical entropy under these two constraints. If $\widehat{V}$ is very small, transfer of energy between a and b is very slow. If we wait sufficiently long we reach true equilibrium where $\langle H \rangle$ is the only constant of motion. However, if we restrict ourselves to a rather short period during which the energy has not had enough time to transfer between a and b, the situation is hardly different from the one where $\widehat{V} = 0$. The quantities $\widehat{H}_a$ and $\widehat{H}_b$ are no longer exactly conserved – this is true only for $\widehat{H}$ – but their expectation values vary slowly; at a given time we can treat them as if they were two independent data (Chap.14). The method for studying equilibrium systems, the principles of which we gave in §§ 4.1.3 and 4.1.4 and which we shall develop further in what follows, can thus be extended to a quasi-equilibrium state *during the period considered.*

The same situation occurs in a more subtle fashion in magnetic substances, especially in the case of *nuclear magnetism.* In this case the system a is represented by the degrees of freedom of the nuclear spins and the system b by the other degrees of freedom of the substance which are often just called the "lattice", since they mainly represent the vibrational modes of the crystal lattice. The mechanisms for establishing equilibrium can be classified into three categories, depending on whether they involve energy exchanges between the spins and the lattice, between one spin and another one, or between the various lattice degrees of freedom. These exchanges are, respectively, governed by $\widehat{V}$, by $\widehat{H}_a$, and by $\widehat{H}_b$. The characteristic time for the thermalization of the lattice itself, which is of the order of ps, is much shorter than the spin-lattice relaxation time τ_1 and the spin-spin relaxation time τ_2. These latter, which can be measured by magnetic resonance experiments, depend strongly on the temperature and the substance; they are the longer, the weaker their associated interactions. After times much longer than τ_1 and τ_2 the system is at equilibrium and the total energy is a constant of motion. However, spin-lattice interactions $\widehat{V}$ are often so weak that the time τ_1 is very long, typically between 0.1 s and several minutes, whereas τ_2 is of the order of 20 to 100 μs. In that case, *on intermediate time scales* between τ_1 and τ_2, $\widehat{V}$ does not play any rôle, and the two quantities $\langle H_a \rangle$ and $\langle H_b \rangle$, which are practically constant in time, can be controlled independently. Since the macroscopic state depends on the two energy variables $\langle H_a \rangle$ and $\langle H_b \rangle$, *two temperatures coexist* in the substance, the ordinary lattice temperature and the spin temperature, which, in fact, can also be negative (Exerc.1a).

Similar considerations apply to the number of constituent particles in systems which can *exchange matter* or which can *undergo chemical reactions.* Let us, for instance, consider a mixture of chlorine and hydrogen. At temperatures of a few hundreds of degrees, or even at room temperatures when the gas is illuminated, a stable chemical equilibrium is reached; the data which characterize it are, apart from the total energy and the volume, the *numbers of* H *and* Cl *atoms*, the only variables which are conserved in the reactions, and these data determine the number of Cl_2, H_2, and HCl molecules at equilibrium. However, at room temperature in the dark, the reaction rate is so low that a mixture of arbitrary amounts of Cl_2, H_2, and HCl will not react during several days. Thermal equilibrium and homogeneity are attained fast, but there is not enough time for chemical equilibrium to be reached and the *mixture is metastable.* During the period when the chemical reactions are inhibited, the *numbers of* Cl_2, H_2, *and* HCl *molecules* are practically conserved. These numbers are the data characterizing the metastable state and they can be fixed independently of one another – rather than the numbers of H and Cl atoms which in this case are insufficient to determine the state; of course, one also must give the energy and the volume as in the case of true equilibrium.

The characteristic times for *nuclear reactions* at room temperature are in most cases so large that as far as they are concerned, metastability is the rule. Under the usual conditions on Earth, a substance is adequately described if one uses as variables the numbers of molecules or of atoms, or, in the case of solids, in more detail, the numbers of electrons and nuclei of each species. If we want to take into account the possibility of nuclear reactions, we must descend to a still more microscopic scale and characterize the state by giving instead the numbers of electrons, of protons, and of neutrons. The true nuclear equilibrium is obtained by comparing the energies and entropies of the various elements which can be made up out of these constituents. We find then that the light elements are unstable with respect to fusion – for instance, hydrogen into helium with a change of nuclear energy into heat – and the heavy elements are unstable against fission, with the maximum stability range lying around the Fe nuclei. If we wanted to proceed with full rigour we should, for instance, consider hydrogen to be in metastable equilibrium as regards the formation of helium. However, the probability that such reactions would take place spontaneously is completely negligible at room temperatures: we must wait of the order of 10^{1000} years to obtain one reaction per mole of matter! As in the case of chemical reactions the evolution towards the most stable macroscopic equilibrium state is accelerated by increasing the temperature. However, it is necessary to reach temperatures of 10^8 K for nuclear reactions, such as the fusion of two deuterons into a helium nucleus, to have a chance to occur during a reasonable length of time. This happens inside the stars. This is also the reason why very high temperatures are needed to realize controlled fusion.

Another kind of metastability is associated with *phase changes.* For instance, depending on the temperature and the pressure, the equilibrium form of carbon can be graphite or diamond; even though we observe diamonds at room temperatures, only graphite is stable up to pressures of the order of 16 000 atm. Glycerine should crystallize towards 20° C, but it remains liquid in a *supercooled* state even in the worst winters. Similarly, glasses are metastable substances with a structure akin to that of liquids. They evolve spontaneously towards the stable, crystalline state: for instance, glasses produced in antiquity or even two or three centuries ago become opaque because of the formation of small crystals within them. The study of such metastable equilibria at the microscopic scale does not need, in general, the introduction of extra approximate constants of motion as in the case of the earlier examples. We can carry it out by looking for the maximum of the statistical entropy in a *restricted space containing only those configurations that can be reached within a limited time* when we take the dynamics of the system into account (§ 9.3.3). For instance, the theory of a supercooled liquid is hardly different from that of a stable liquid, provided we only consider configurations where the molecules are randomly distributed over space; in this way we exclude them from being arranged on a crystalline lattice which is the more stable equilibrium form.

4.2 Equilibrium Distributions

In order to apply the above principles we must first use those data which are certain so as to find the Hilbert space or the fraction of space in which we are operating. We then have to construct the density operator which produces maximum disorder while satisfying constraints such as (4.2) relating to the data given on average. In § 4.2.1 we shall carry out this construction by the Lagrangian multiplier method. We shall then directly check the results obtained employing another, variational, method, which is the basis of many often used approximation schemes (§ 4.2.2). The following subsections will introduce the general and powerful partition function technique.

4.2.1 Lagrangian Multipliers

Let us choose an arbitrary base $\{|k\rangle\}$ in *the Hilbert space defined by those data* about our system which are *certain.* The matrix elements $\langle k|\widehat{D}|k'\rangle$ of the density operator are the unknowns which we must determine by looking for the *maximum of* $S(\widehat{D})$, *taking into account the constraints* (4.2). We add to those the condition

$$\mathrm{Tr}\,\widehat{D} \;=\; 1, \tag{4.2'}$$

which has the same form as (4.2) with $\widehat{A}_0 \equiv \widehat{I}$ and $\langle A_0 \rangle \equiv 1$, and recall that $\widehat{D}$ must be Hermitean and non-negative.

We shall find a first solution of this problem by means of the *Lagrangian multiplier* method, which we shall describe beforehand, using a general formulation. Let $f(\{x_k\})$ be a function of K variables $x_1, \ldots, x_k, \ldots, x_K$. These variables are related to one another through J constraints $g_j(\{x_k\}) = a_j$ $(j = 1, 2, \ldots, J)$. We look for the values of $\{x_k\}$ for which f is stationary under the constraints satisfied by the $\{x_k\}$. To do this we associate with every constraint a new variable λ_j, called a Lagrangian multiplier, we introduce the expression

$$f(\{x_k\}) \;-\; \sum_j \lambda_j\, g_j(\{x_k\}),$$

and we look for the stationary values of this expression *for arbitrary changes* in the x_k, keeping the λ_j fixed. We must thus solve the K equations

$$\frac{\partial f}{\partial x_k} - \sum_j \lambda_j \frac{\partial g_j}{\partial x_k} \;=\; 0, \tag{4.3}$$

and in this way determine the $\{x_k\}$ as functions of the λ_j. The λ_j parameters can eventually be expressed as functions of the a_j by substituting the solution for the $\{x_k\}$ into the constraints $g_j(\{x_k\}) = a_j$.

The mathematical justification of the Lagrangian multipliers method is an exercise in linear algebra. Let us assume that we have found a solution $\{x_k\}$. If we consider its representative point $\boldsymbol{x}$, we have in its vicinity

$$df \;\equiv\; (\nabla f \cdot d\boldsymbol{x}) \;\equiv\; \sum_k \frac{\partial f}{\partial x_k}\, dx_k \;=\; 0 \tag{4.4a}$$

for all variations $\{dx_k\}$ which satisfy the constraints

$$dg_k \;\equiv\; (\nabla g_j \cdot d\boldsymbol{x}) \;\equiv\; \sum_k \frac{\partial g_j}{\partial x_k}\, dx_k \;=\; 0. \tag{4.4b}$$

The rank R of the $J \times K$ $(K > J)$ matrix $\partial g_j / \partial x_k$ is at most equal to J. To fix the ideas, let us assume that the determinant of the submatrix $\partial g_j / \partial x_k$ for $1 \le j \le R$ and $1 \le k \le R$ is non-vanishing. We can then imagine that the K variables dx_k are expressed as linear combinations of the R variables dg_j $(1 \le j \le R)$ and of $K - R$ variables $dh_k \equiv dx_k$ $(R + 1 \le k \le K)$. The differential df can be written in terms of these new infinitesimal independent variables in the form

$$df \;=\; \sum_{j=1}^{R} \lambda_j\, dg_j + \sum_{k=R+1}^{K} \mu_k\, dh_k;$$

if $R < J$, the differentials dg_j for $R < j \leq J$ are linear combinations of the dg_j for $1 \leq j \leq R$. We know that df must vanish identically for all changes in the dx_k, or equivalently of the dg_j and the dh_k, such that the dg_j vanish. This implies that $\mu_k \equiv 0$ $(R+1 \leq k \leq K)$; hence df is a linear combination of the dg_j for $j \leq R$. In terms of the original variables dx_k this condition is expressed by (4.3). If $R < J$, the existence of linear relations between the dg_j $(1 \leq j \leq J)$ implies that the λ_j are not unique. *QED*

A different approach consists in showing that the compatibility of (4.4a,b), considered as linear equations in the dx_k, implies that the rank of the $(J+1) \times K$ matrix $\partial f/\partial x_k$, $\partial g_j/\partial x_k$ is at most equal to J. Under those conditions, if we consider (4.3) as a set of linear equations for the λ_j, it will have at least one solution.

The Lagrangian multipliers method can also be justified by geometric considerations. The differentials $d\boldsymbol{x}$, with components $\{dx_k\}$, constitute a vector space $\mathcal{E}$ with a finite number K of dimensions. The forms (4.4a,b) define ∇f and ∇g_j as elements in the dual space $\mathcal{E}^*$. Let us consider the subspace $\mathcal{G}$ of $\mathcal{E}^*$ which is spanned by the vectors ∇g_j. The constraints (4.4b) express that the allowed variations $d\boldsymbol{x}$ are those which belong to the subspace $\mathcal{G}_0$ of $\mathcal{E}$ which is orthogonal to $\mathcal{G}$. Condition (4.4a) therefore means that ∇f is orthogonal to $\mathcal{G}_0$. However, one can show that the orthogonality of two subspaces from $\mathcal{E}$ and $\mathcal{E}^*$ is a symmetric property. Hence, ∇f belongs to $\mathcal{G}$: it must therefore be a linear combination of the ∇g_j.

In our problem, the x_k variables are the matrix elements of $\widehat{D}$ and the function f is the statistical entropy $S(\widehat{D})/k$, while the constraints $g_j = a_j$ are the relations (4.2) and (4.2'). We introduce Lagrangian multipliers λ_j associated with each of the constraints (4.2) and λ_0 associated with the normalization constraint (4.2'). We must thus ask for the stationarity of the quantity

$$-\operatorname{Tr}\widehat{D}\ln\widehat{D} - \sum_i \lambda_i \operatorname{Tr}\widehat{D}\widehat{A}_i - \lambda_0 \operatorname{Tr}\widehat{D}, \tag{4.5}$$

where we can now vary the matrix elements $\langle k|\widehat{D}|k'\rangle$ *freely*, except for the fact that $\widehat{D}$ is Hermitean. If we use (2.19) or (3.28), we obtain in this way

$$\begin{aligned} 0 &= -\operatorname{Tr}\left[\delta\widehat{D}\left(\ln\widehat{D} + 1 + \sum_i \lambda_i\widehat{A}_i + \lambda_0\right)\right] \\ &= \sum_{kk'} \langle k|\delta\widehat{D}|k'\rangle \left\langle k'\left|\left(\ln\widehat{D} + \sum_i \lambda_i\widehat{A}_i + \lambda_0 + 1\right)\right|k\right\rangle. \end{aligned}$$

In this expression we can choose as the real independent variations the quantities $\langle k|\delta\widehat{D}|k\rangle$, $\langle k|\delta\widehat{D}|k'\rangle + \langle k'|\delta\widehat{D}|k\rangle$, and $\mathrm{i}\langle k|\delta\widehat{D}|k'\rangle - \mathrm{i}\langle k'|\delta\widehat{D}|k\rangle$, the coefficients of which must vanish $(k < k')$. This means that

$$\left\langle k'\left|\left(\ln\widehat{D} + \sum_i \lambda_i\widehat{A}_i + \lambda_0 + 1\right)\right|k\right\rangle = 0,$$

for any k and k', or

$$\widehat{D} = e^{-\sum_i \lambda_i \widehat{A}_i - \lambda_0 - 1}.$$

Putting $Z \equiv e^{\lambda_0+1}$ we thus find

$$\boxed{\widehat{D} = \frac{1}{Z} e^{-\sum_i \lambda_i \widehat{A}_i}} . \tag{4.6}$$

This expression, which is called the *Boltzmann-Gibbs distribution*, represents the general form of the *density operators in thermodynamic equilibrium when the averages of the constants of the motion* $\widehat{A}_i$ *are given.* It will be the foundation for all applications to the physics of equilibrium phenomena in the remainder of the present book. We shall also find it useful to introduce time-dependent distributions of the form (4.6) for non-equilibrium systems (§ 14.3.4). When the energy $\langle H \rangle$ is amongst the quantities which are given on average, the multiplier relating to $\widehat{H}$ is traditionally written as β.

In § 1.3.2 we met already with an example of a Boltzmann-Gibbs distribution. The system consisted of a paramagnetic ion with Hamiltonian $\widehat{H}_1 = -(\boldsymbol{B} \cdot \boldsymbol{\mu}_1)$ in the Hilbert space of a spin-$\frac{1}{2}$ particle, and with a given average energy,

$$-\left(\boldsymbol{B} \cdot \langle \widehat{\boldsymbol{\mu}}_1 \rangle\right) = \frac{U}{N}.$$

According to (4.6) its density operator in thermal equilibrium has the form

$$\widehat{D} = \frac{1}{Z_1} e^{-\beta \widehat{H}_1},$$

where β is the Lagrangian multiplier related to the constraint on the energy and where Z_1 is a normalization factor. We find again expression (1.29) and we have seen that β is directly related to the temperature. The method used in Chap.1 to justify this distribution did not appeal to the maximum entropy principle, but was solely based upon the indifference principle. We shall show more generally in § 5.7.2 how the Boltzmann-Gibbs equilibrium distribution can be derived from the latter principle.

Historically the fact that the equilibrium distributions have the exponential form (4.6) was established gradually, in the case of classical statistical mechanics, during the second half of the nineteenth century. This discovery preceded that of the logarithmic form of the statistical entropy, that is, the order was the inverse of the one we are following in the present book (§ 3.4.2). The *Maxwell distribution* (1860) was the first occurrence of an exponential law. Maxwell showed by a heuristic argument (§ 7.2.2) that the distribution of the velocities $\boldsymbol{v}$ of the molecules of mass m in a perfect gas was proportional to $\exp(-\frac{1}{2}\beta m v^2)$. In 1868 he extended this exponential form to a gas subject to an *external potential*. However, the expressions that he wrote down referred to a *single* molecule taken from the population which makes up the gas, and not to the whole of the macroscopic system. It was left to *Boltzmann* to recognize in 1871 that the phase density of a more general classical system of *interacting* atoms, with a Hamiltonian H, should be proportional to

$e^{-\beta H}$, and to find in connection with this result expression (4.30) for the entropy which we shall derive below. Boltzmann justified the form $e^{-\beta H}$ mainly by using *dynamical* arguments, defining the equilibrium state as the one which is reached in an evolution during which the entropy increases (§§ 3.4.2 and 4.1.5); he also relied on agreement with the results from thermodynamics. The extension to *open systems*, where the number of particles can change thanks to exchanges with the outside world, is due to *Gibbs* who, in his book of 1902, introduced the distinction between micro-canonical, canonical, and grand canonical ensembles. In the last case he noted that the constraints (4.2) contain not only the given value of the average energy, but also that of the average number of particles, and he wrote down the corresponding equilibrium distribution (4.33). He only considered the equilibrium state itself and not the way the system evolves towards this state, and the justifications he gave for the equilibrium distributions were of a *statistical* kind (§ 5.7.2), anticipating the arguments given in § 4.1.3.

4.2.2 Variational Method

The above proof is not totally complete: we must, in fact, still prove that expression (4.6) makes the statistical entropy $S(\widehat{D})$ a *maximum* and not merely *stationary* as we have just shown. Moreover, the λ_i variables remain to be determined.

A possible method for checking that, in fact, (4.6) gives the maximum of $S(\widehat{D})$ starts from the remark that our calculation has produced *only one* stationary value whereas we expected at least one maximum and one minimum. However, we have not taken into account the condition that the operator $\widehat{D}$ must be positive. The domain in which the unknown quantities $\langle k|\widehat{D}|k'\rangle$ vary therefore has a *boundary*, defined by the vanishing of at least one of the eigenvalues $p_m \geq 0$ of $\widehat{D}$. On this boundary we may find other extrema of $S(\widehat{D})$; however, $S(\widehat{D})$ cannot be a maximum at a boundary point, as $-kp_m \ln p_m$ increases with an infinite slope near $p_m = 0$. The maximum of $S(\widehat{D})$ is thus necessarily the single stationary value that we have found and it lies inside the allowed domain for $\widehat{D}$.

One could also note that the second differential of (4.5) with respect to the matrix elements of $\widehat{D}$ is the same as that of $-\operatorname{Tr}\widehat{D}\ln\widehat{D}$. On the other hand, the concavity property (3.26) of the statistical entropy can be extended, if we take (3.18) into account, to operators $\widehat{D}$ which have arbitrary traces. This implies that the second differential of (4.5) is negative at all points where it is defined, that is, at all points where $\widehat{D}$ is positive. In particular, at the point (4.6) expression (4.5) takes the value $\operatorname{Tr}\widehat{D} = 1$, its first differential vanishes, and its second differential is negative. This point is therefore the absolute maximum of (4.5) for arbitrary variations of $\widehat{D}$. (One can use (3.18) and check, as an exercise, that (4.5) is equal to or less than 1). As a result, if we impose the constraints (4.2), the Boltzmann-Gibbs distribution is, indeed, the one which provides the maximum of S.

Below we shall give, without using the Lagrangian multiplier method, a direct proof of the fact that the maximum of S under the constraints (4.2) is reached for the Boltzmann-Gibbs density operator (4.6).

Let us consider the statistical entropy $S(\widehat{D})$ associated with the Boltzmann-Gibbs distribution (4.6). Starting from (3.17) we readily find

$$S(\widehat{D}) \;=\; -k\,\langle \ln \widehat{D}\,\rangle \;=\; k \ln Z + k \sum_i \lambda_i \,\langle A_i \rangle. \tag{4.7}$$

Let us compare this statistical entropy with the $S(\widehat{\mathcal{D}})$ which is associated with an arbitrary, normalized, density operator $\widehat{\mathcal{D}}$ different from $\widehat{D}$. To do this we apply the lemma (3.18), replacing $\widehat{X}$ by $\widehat{\mathcal{D}}$ and $\widehat{Y}$ by (4.6):

$$S(\widehat{\mathcal{D}}) \;<\; -k \operatorname{Tr} \widehat{\mathcal{D}} \ln \widehat{D} \;=\; k \ln Z + \sum_i k\lambda_i \operatorname{Tr} \mathcal{D}\widehat{A}_i. \tag{4.8}$$

Let us now assume that $\widehat{\mathcal{D}}$ gives the same expectation values for the given observables as $\widehat{D}$, namely, that

$$\operatorname{Tr} \widehat{\mathcal{D}}\widehat{A}_i \;=\; \langle A_i \rangle \;=\; \operatorname{Tr} \widehat{D}\widehat{A}_i. \tag{4.9}$$

Comparing (4.7), (4.8), and (4.9) we see immediately that in that case

$$S(\widehat{D}) \;>\; S(\widehat{\mathcal{D}}).$$

The Boltzmann-Gibbs distribution thus provides us, indeed, with a *statistical entropy which is larger* than that of any other density operator *satisfying the same constraints* (4.9) and it gives us the general formal solution for the macro-state resulting from the maximum entropy principle.

As a useful by-product of the proof that we have just given we can construct an *approximation method* which allows us to replace the exact Boltzmann-Gibbs density operator (4.6), when it is too complicated, by a more manageable approximate density operator $\widehat{\mathcal{D}}$. Let is consider $\widehat{\mathcal{D}}$ in (4.8) as a *trial density operator*, with unit trace, and let us rewrite this inequality in the form

$$\boxed{\frac{1}{k}\, S(\widehat{\mathcal{D}}) - \sum_i \lambda_i \operatorname{Tr} \widehat{\mathcal{D}}\widehat{A}_i \;<\; \ln Z}\;, \qquad \forall\;\; \widehat{\mathcal{D}} \neq \widehat{D}. \tag{4.10}$$

For given values of the multipliers λ_i the left-hand side of (4.10) appears to be a function of $\widehat{\mathcal{D}}$ which reaches its *maximum*, equal to $\ln Z$, when $\widehat{\mathcal{D}}$ is the Boltzmann-Gibbs equilibrium distribution (4.6). Let us assume that the latter is too complicated to allow us to evaluate physical quantities $\langle A_i \rangle$ or S and let us assume that we have chosen a class of density operators $\widehat{\mathcal{D}}$ which are sufficiently simple for us to be able to carry out these calculations. We are looking for a criterion to find from that class the *approximate density operator which provides the best possible approximation for* $\ln Z$; from the latter quantity we can derive the thermodynamic properties of the system at equilibrium, as we shall see in what follows. In the restricted class of trial density operators that we are considering the best choice for $\widehat{\mathcal{D}}$ is the one which *gives the maximum value for the left-hand side of* (4.10), the closest to $\ln Z$. Finding this maximum thus gives us an approximation which will be of interest when the Boltzmann-Gibbs distribution cannot be used for practical calculations, especially in the case of *interacting particles* or when we want to explain *phase transition phenomena* (§§ 9.3.1 and 11.2.1, Exercs.9a and 11f).

4.2.3 Partition Functions

We still must determine the Lagrangian multipliers λ_0, or $Z = e^{\lambda_0+1}$, and λ_i as functions of the data $\langle A_i \rangle$ by expressing that (4.6) satisfies the constraints (4.2) and (4.2′).

First of all, the normalization (4.2′) gives

$$\boxed{Z = \operatorname{Tr} e^{-\sum_i \lambda_i \widehat{A}_i}} \quad . \tag{4.11}$$

This equation enables us to consider the normalization constant Z of the equilibrium distribution (4.6) *as a function of the Lagrangian multipliers* λ_i *and of the data given with certainty* which include control parameters. The former dependence appears explicitly in (4.11), while the latter is implicit, through the definition of the trace over the Hilbert space and through the form of the operators $\widehat{A}_i$, in particular of the Hamiltonian. This function $Z\{\lambda_i\}$ is called the *partition function* associated with the equilibrium statistical ensemble where the values of the observables $\widehat{A}_i$ are given on average. It was introduced by Planck in 1921 and called "Zustandssumme" (state sum) whence the notation "Z". We have already seen an example of an application in Chap.1 (Eqs.(1.13), (1.30)). In the general case it is of great practical use to evaluate it.

Let us, indeed, write down the conditions (4.2) which should determine the multipliers λ_i as functions of the averages $\langle A_i \rangle$:

$$\begin{aligned} \langle A_i \rangle &= \operatorname{Tr} \widehat{D} \widehat{A}_i = \frac{1}{Z} \operatorname{Tr} e^{-\sum_j \lambda_j \widehat{A}_j} \widehat{A}_i \\ &= -\frac{1}{Z} \frac{\partial}{\partial \lambda_i} \operatorname{Tr} e^{-\sum_j \lambda_j \widehat{A}_j}, \end{aligned}$$

where we have used the form (2.19) of the derivative of a trace of an operator. Using the definition (4.11) we find now [2]

$$\boxed{\frac{\partial}{\partial \lambda_i} \ln Z\{\lambda_j\} = -\langle A_i \rangle} \quad . \tag{4.12}$$

[2] If the system is macroscopic and can undergo a phase transition, it can happen (§ 5.7.1) that the partial derivatives of $\ln Z$ are not defined for some values of the multipliers λ_i; this occurs, for instance, when we describe a solid-liquid transition by means of a grand canonical ensemble.

Determining the Lagrangian multipliers λ_i thus amounts to *inverting equations* (4.12), which we can write down immediately once we know the partition function Z.

The function $Z\{\lambda_i\}$ has special properties which imply that the solution of Eqs.(4.12) in terms of the $\{\lambda_i\}$, if it exists, is *unique.* Let us, in fact, assume that we have found two sets of values $\{\lambda_i\}$ and $\{\lambda'_i\}$ which are solutions of (4.12). The density operators $\widehat{D}$ and $\widehat{D}'$ which correspond to them would, by virtue of (4.7) and (4.8), be such that $S(\widehat{D}') < S(\widehat{D})$ and $S(\widehat{D}) < S(\widehat{D}')$ which is impossible. The maximum of the statistical entropy is thus unique. However, it is possible that (4.12) has no solutions, no density operator being able to satisfy the constraints (4.2); this would happen, for instance, if $\widehat{A}_i$ is the particle number operator $\widehat{N}$ and one requires that its average $\langle N \rangle$ be negative.

In practice, one usually avoids inverting Eqs.(4.12) by taking *the λ_i rather than the averages $\langle A_i \rangle$ as the parameters characterizing the equilibrium.* This is what we did in § 4.1 where we studied the equilibrium of a system of spins as a function of the parameter β, that is, of the temperature, rather than as a function of the associated variable, the energy U. We shall see more generally in Chap.5 that the Lagrangian multipliers, introduced here as a mathematical artifice to take the constraints imposed by the data (4.2) into account when we look for the maximum of $S(\widehat{D})$, have direct macroscopic physical interpretations.

We can also derive the *correlations* and the *statistical fluctuations* of the constants of motion $\widehat{A}_i$ from the partition function, in the most common case when the observables $\widehat{A}_i$ commute. In fact, we find that

$$\frac{\partial^2 Z}{\partial \lambda_i \partial \lambda_j} = \mathrm{Tr}\, \mathrm{e}^{-\sum_k \lambda_k \widehat{A}_k}\, \widehat{A}_i \widehat{A}_j,$$

and hence

$$\frac{\partial^2 \ln Z}{\partial \lambda_i \partial \lambda_j} = \langle A_i A_j \rangle - \langle A_i \rangle \langle A_j \rangle . \tag{4.13}$$

The matrix of the right-hand side of (4.13) is positive as, for arbitrary C_i,

$$\sum_{ij} C_i C_j^* \left[\langle A_i A_j \rangle - \langle A_i \rangle \langle A_j \rangle \right] = \left\langle \left| \sum_i C_i \left(A_i - \langle A_i \rangle \right) \right|^2 \right\rangle .$$

The function $\ln Z$ is thus a *convex* function of its variables λ_i, which directly explains why the solution of Eqs.(4.12) in terms of the λ_i is unique.

4.2.4 Equilibrium Entropy

Finally, using (4.7) and (4.12), we can easily express the equilibrium entropy associated with the Boltzmann-Gibbs distribution in terms of the partition function:

$$\boxed{\frac{S}{k} = \ln Z - \sum_i \lambda_i \frac{\partial}{\partial \lambda_i} \ln Z} \quad . \tag{4.14}$$

Let us consider a *shift in the equilibrium*, that is, an infinitesimal change in the parameters $\langle A_i \rangle$ which characterize this equilibrium, and hence in the parameters λ_i which are related to the $\langle A_i \rangle$. By virtue of (4.12) we have

$$d \ln Z = - \sum_i \langle A_i \rangle \, d\lambda_i, \tag{4.15}$$

whence, if we use (4.14),

$$dS = k \sum_i \lambda_i \, d\langle A_i \rangle. \tag{4.16}$$

These relations show that the *natural variables for* S *are the data* $\langle A_i \rangle$, whereas $\ln Z$ *is useful as a function of the* λ_i. Considering S as a function of the $\langle A_i \rangle$ we then find from (4.16) that

$$\frac{1}{k} \frac{\partial S}{\partial \langle A_i \rangle} = \lambda_i, \tag{4.17}$$

a relation which expresses the *Lagrangian multipliers as the changes in the statistical entropy when the equilibrium is shifted.*

The relations (4.7) and (4.17) make it possible to write the expression for $\ln Z$ in terms of S/k:

$$\ln Z = \frac{S}{k} - \sum_i \langle A_i \rangle \frac{\partial}{\partial \langle A_i \rangle} \frac{S}{k}. \tag{4.18}$$

Comparing (4.14) with (4.18) and (4.12) with (4.17) shows that there exists, apart from a sign, a complete *symmetry* between the function $\ln Z$ of the variables λ_i and the function S/k of the variables $\langle A_i \rangle$. The transformation (4.14) which allows us to go from one to the other is a *Legendre transformation.* We shall in § 6.3.1 give the general formalism of such transformations and some applications; they are very useful whenever one wants to change from one set of variables to the conjugate ones, which are defined as the partial derivatives of some function. Simultaneously with this change in variables, one should arrange a *change in the function* through the Legendre transformation, in order to be able to get again the old variables by simple differentiations, as in (4.12) and (4.17). It will be essential to make absolutely clear – especially for applications to thermodynamics – *which variables we consider to be independent* when calculating partial derivatives. For instance, the partial derivatives of S are simple only if all the variables are the $\langle A_i \rangle$.

The general theory of Legendre transformations (§ 6.3.1) makes it possible to connect the second derivatives of $\ln Z$ and S with respect to their natural variables with each other. In the present case we have, according to (4.13),

$$\Big\langle \big(A_i - \langle A_i \rangle\big)\big(A_j - \langle A_j \rangle\big)\Big\rangle = \frac{\partial^2 \ln Z}{\partial \lambda_i \partial \lambda_j} = -\frac{\partial \langle A_i \rangle}{\partial \lambda_j} = -\frac{\partial \langle A_j \rangle}{\partial \lambda_i}, \tag{4.19}$$

whereas by taking the derivative of (4.17) we get

$$\frac{1}{k}\frac{\partial^2 S}{\partial \langle A_i \rangle \partial \langle A_j \rangle} = \frac{\partial \lambda_i}{\partial \langle A_j \rangle} = \frac{\partial \lambda_j}{\partial \langle A_i \rangle}. \tag{4.20}$$

The matrices of the second derivatives of $\ln Z$ and of $-S/k$ are thus the inverse of each other. As a result S is a *concave* function of the variables $\langle A_i \rangle$ with its second derivatives forming a negative matrix.

4.2.5 Factorization of Partition Functions

The Hilbert space of a macroscopic system is characterized by a very large set of quantum numbers; this makes the evaluation of the trace (4.11) laborious and often impracticable. However, most applications which we shall be concerned with refer to systems consisting of *non-interacting entities*; these entities can be particles, or spins, or more abstract objects such as vibrational modes or defects in a solid. In such a case the Hilbert space $\mathcal{E}_\mathrm{H}$ in which the trace is calculated can be decomposed into a direct product (§ 2.1.1) of simple Hilbert spaces $\mathcal{E}_\mathrm{H}^q$ associated each with one entity, and the Hamiltonian $\widehat{H}$, one of the constants of motion $\widehat{A}_i$, can be written as a direct sum of operators $\widehat{H}_q$, each of which operates in an elementary Hilbert space $\mathcal{E}_\mathrm{H}^q$, while the action of $\widehat{H}_q$ in $\mathcal{E}_\mathrm{H}^q$ is defined by (2.12). The system is thus decomposed into subsystems, characterized by the index q, which do not interact with one another. The other constants of motion $\widehat{A}_i$, such as the numbers of particles or the momentum, are usually additive so that, for the system considered,

$$\sum_i \lambda_i \widehat{A}_i = \bigoplus_q \sum_i \lambda_i \widehat{A}_i^q \equiv \bigoplus_q \widehat{K}_q \tag{4.21}$$

has the form of a *direct sum of operators* $\widehat{K}_q$ which act, respectively, in the Hilbert space $\mathcal{E}_\mathrm{H}^q$ associated with the subsystem q.

Under those conditions the evaluation of the partition function (4.11) is greatly simplified. In fact,

$$Z = \mathrm{Tr}\, \mathrm{e}^{-\oplus_q \widehat{K}_q} = \mathrm{Tr}\left[\bigotimes_q \mathrm{e}^{-\widehat{K}_q}\right]$$

becomes the trace, in $\mathcal{E}_{\mathrm{H}}$, of a tensor product of operators. According to (2.11) this trace can be factorized as a *product of traces* tr_q, each of which is calculated in a space $\mathcal{E}_{\mathrm{H}}^q$:

$$\boxed{Z = \prod_q Z_q, \qquad Z_q = \mathrm{tr}_q\, \mathrm{e}^{-\widehat{K}_q}} \quad . \tag{4.22}$$

Because of its practical importance we shall give another proof of this result. Let us choose in each space $\mathcal{E}_{\mathrm{H}}^q$ a base $\{|m_q\rangle\}$ which diagonalizes $\widehat{K}_q$; in the most often encountered case where the observables $\widehat{A}_q$ commute, $|m_q\rangle$ is their common eigenbase. Let us denote by $K_q(m_q)$ the eigenvalue of $\widehat{K}_q$ associated with $|m_q\rangle$. The space $\mathcal{E}_{\mathrm{H}}$ is spanned by the base $\{|m_a, \ldots, m_q, \ldots\rangle\}$ and the eigenvalue of $\sum_i \lambda_i \widehat{A}_i$ corresponding to the micro-state $|m_a, \ldots, m_q, \ldots\rangle$ is $K_a(m_a) + K_b(m_b) + \ldots + K_q(m_q) + \ldots$. In this base the *trace* (4.11) may be written as a *multiple sum*,

$$\begin{aligned} Z &= \sum_{m_a, m_b, \ldots, m_q, \ldots} \mathrm{e}^{-K_a(m_a)}\, \mathrm{e}^{-K_b(m_b)} \ldots \mathrm{e}^{-K_q(m_q)} \ldots \\ &= \left[\sum_{m_a} \mathrm{e}^{-K_a(m_a)}\right] \left[\sum_{m_b} \mathrm{e}^{-K_b(m_b)}\right] \cdots \left[\sum_{m_q} \mathrm{e}^{-K_q(m_q)}\right] \cdots \\ &= Z_a\, Z_b \ldots Z_q \ldots , \end{aligned} \tag{4.22'}$$

which we can factorize immediately.

We note also that for the non-interacting systems which we are considering the Boltzmann-Gibbs density operator (4.6) can be factorized as a tensor product,

$$\widehat{D} = \widehat{D}_a \otimes \widehat{D}_b \cdots \otimes \widehat{D}_q \otimes \cdots , \qquad \widehat{D}_q = \frac{1}{Z_q}\, \mathrm{e}^{-\sum_i \lambda_i \widehat{A}_i^q}, \tag{4.23}$$

of *density operators relating to the various subsystems.* The absence of *interactions* thus implies *in thermodynamic equilibrium* the absence of *correlations* between subsystems.

The factorization (4.22) of Z implies that $\ln Z$ is a sum of terms $\ln Z_q$ relating to the various subsystems. All quantities such as (4.12) or (4.14) which are found by differentiation are also *additive.* In particular, we find again the additivity, $S = \sum_q S_q$, of the statistical entropy, in agreement with (3.20) and (4.23).

In this book we shall meet with many examples of factorizations like (4.22) and (4.23). However, one should in this respect note the importance of the choice of the equilibrium statistical ensemble. In Chap.1 the spins, situated at the sites $i = 1, \ldots, N$, did not interact, but because of the restriction (1.11) one could not consider them as independent entities in the *microcanonical* ensemble. One takes better advantage of the absence of interactions, if one works in the *canonical* ensemble (Exerc.4c). Even in the microcanonical ensemble, we evaluated the canonical partition function Z as a technical artifice in order to find the microcanonical partition function W, and its expression (1.13) was found just by using the factorization of the contributions from the various sites.

More generally, we shall often choose the statistical ensemble with a view to take advantage of a factorization of Z. Moreover, a comparison of the various approaches in different ensembles can enable us to obtain interesting results (Exercs.2b, 4f, 5a, and 9g).

4.2.6 Summary: Technique of Studying Systems at Equilibrium

In this section we shall collect the essential results which we have just derived from the maximum disorder principle and which provide us with a systematic method for studying equilibrium problems in statistical physics. Most applications considered in what follows will be treated using this method.

In order to study a system at equilibrium we start by *defining the Hilbert space* of its states and by *looking for the constants of the motion* and their associated observables $\widehat{A}_i$, such as the energy or the number of particles of each kind. The averages $\langle A_i \rangle$ of these observables are data which characterize equilibrium on the macroscopic scale; we associate with each of them a Lagrangian multiplier λ_i.

The density operator describing the system at equilibrium is then the *Boltzmann-Gibbs distribution*

$$\widehat{D} = \frac{1}{Z}\, \mathrm{e}^{-\sum_i \lambda_i \widehat{A}_i}. \tag{4.6}$$

At equilibrium, the expectation value of any observable $\widehat{A}$ can be calculated using the general relation $\langle A \rangle = \mathrm{Tr}\, \widehat{D}\widehat{A}$. However, for the particular observables $\widehat{A}_i$ entering (4.6), this calculation is simplified, if we first evaluate the normalization constant Z, that is, the *partition function*

$$Z\{\lambda_i\} = \mathrm{Tr}\, \mathrm{e}^{-\sum_i \lambda_i \widehat{A}_i}, \tag{4.11}$$

considered as function of the variables λ_i. In fact, we find the *average values* $\langle A_i \rangle$ by differentiation:

$$\langle A_i \rangle = -\frac{\partial}{\partial \lambda_i} \ln Z\{\lambda_j\}, \tag{4.12}$$

as well as the *correlations* and *fluctuations* (4.13).

More generally, let us assume that we are interested in the average value of an observable $\widehat{X}$ that can be produced, starting from one of the conserved quantities $\widehat{A}_i$, through differentiation with respect to a parameter ξ which occurs in the definition of the latter. For instance, the derivative of the Hamiltonian with respect to an external magnetic field is, apart from the sign, the magnetic moment operator. Knowing Z as function of ξ is sufficient to evaluate $\langle X\rangle$, since

$$\boxed{\langle X\rangle = \left\langle \frac{\partial A_i}{\partial \xi} \right\rangle = -\frac{1}{\lambda_i}\frac{\partial}{\partial \xi} \ln Z} \, . \tag{4.24}$$

This expression is proved in the same way as (4.12); the fact that some operators may not be commuting does not matter when we take the first derivative (2.19) of a trace.

The applications treated in the present book will, in general, be simple enough for the eigenvalues of the constants of the motion $\widehat{A}_i$ to be known. We shall therefore work in the representation in which the observables $\widehat{A}_i$ are diagonal and we shall calculate the trace (4.11) as a sum over eigenvalues, all the time taking advantage of *factorizations* such as (4.22). We must beware of possible degeneracies, as in the trace (4.11) each distinct eigenvalue is weighted by its multiplicity.

The logarithm of the partition function, ln Z, can also be obtained as the maximum of

$$\frac{1}{k}\, S(\widehat{\mathcal{D}}) - \sum_i \lambda_i \,\mathrm{Tr}\,\widehat{\mathcal{D}}\widehat{A}_i, \tag{4.10}$$

over the set of all density operators $\widehat{\mathcal{D}}$. By restricting this set to a class $\widehat{\mathcal{D}}$ which allows a manageable calculation of (4.10), for instance, by taking $\widehat{\mathcal{D}}$ as factorized, we can, by looking for the maximum of (4.10) in this class, find a *variational approximation* to ln Z.

The *equilibrium entropy* is the maximum of the statistical entropy over the set of density operators satisfying the constraints (4.2) on the given averages $\langle A_i\rangle$. As a function of $Z\{\lambda_i\}$ it can be expressed as

$$S = k \ln Z - \sum_i \lambda_i \frac{\partial}{\partial \lambda_i} k \ln Z. \tag{4.14}$$

The partition function $Z\{\lambda_i\}$ thus plays an essential rôle in the practical calculation of the properties of a system at equilibrium: the averages $\langle A_i\rangle$ or $\langle X\rangle$ can be found from it through differentiation, (4.12) or (4.24), and the statistical entropy, considered as a function of the variables $\langle A_i\rangle$, is connected with $k \ln Z$ through the Legendre transformation (4.14). The expressions $\mathrm{Tr}\,\widehat{D}\widehat{A}$ for the quantities $\langle A\rangle$ or $-k\,\mathrm{Tr}\,\widehat{D} \ln \widehat{D}$ for S will therefore be useless for systems at equilibrium after the preliminary calculation of $Z\{\lambda_i\}$. We shall, moreover, see (§ 5.6) that the parameters λ_i and the function Z can be interpreted in terms of commonly used macroscopic quantities. For instance, the Lagrangian multiplier β associated with the energy observable $\widehat{H}$ is directly related to the absolute temperature, $\beta = 1/kT$. To make the physical interpretation easier we shall be led to changes in the functions, using for instance $-kT \ln Z$ instead of Z, and in the variables, using T instead of β, in order to work directly with the traditional macroscopic quantities. Equations (4.12) and (4.14) will thus be replaced by equivalent formulæ which have an obvious thermodynamic interpretation (see the tables in § 5.6.6).

4.3 Canonical Ensembles

The previous section presented in a very general way the density operator formalism for thermodynamic equilibrium. To make these results look less abstract we shall show how they apply to a *fluid.* The nature of the fluid will appear through the form of the Hamiltonian $\widehat{H}_N$ which contains the kinetic energy of the N particles, their interaction potential, as in (2.65), and which depends possibly on their internal degrees of freedom, if they have a structure. The container that encloses the fluid is described by a box potential which is zero inside and infinite outside, and which also occurs in $\widehat{H}_N$. In agreement with the remarks in § 4.1.4, the data characterizing a thermodynamic equilibrium state are the *energy* U, the *number* N of particles, and the *volume* Ω of the box. Depending on whether these data are given exactly or only on average, we obtain different descriptions of the equilibrium state through different statistical ensembles which were defined in § 4.1.2. We shall show in § 5.5.3 that they are equivalent for the evaluation of most physical quantities, provided the system is macroscopic. In practice, it will thus be convenient, when we want to deal with a problem, to choose the ensemble which gives rise to the simplest calculations, taking into account the remark at the end of § 4.2.5.

4.3.1 The Canonical Ensemble

In the canonical ensemble proper, also called "petit canonical" ensemble, which we introduced in § 4.1.2, thermal equilibrium is characterized by giving the number of particles N and the volume Ω exactly and the *energy* U *on the average.* A Lagrangian multiplier, commonly denoted by β, is associated with the corresponding constraint

$$\mathrm{Tr}\,\widehat{D}\widehat{H}_N = U. \tag{4.25}$$

The Boltzmann-Gibbs distribution reduces to

$$\widehat{D} = \frac{1}{Z_\mathrm{C}}\,\mathrm{e}^{-\beta \widehat{H}_N}. \tag{4.26}$$

If we denote the eigenkets of $\widehat{H}_N$ by $|m\rangle$ and the eigenenergies by E_m, the *probability* p_m *for the eigenstate* $|m\rangle$ *is the exponential*

$$p_m = \frac{1}{Z_\mathrm{C}}\,\mathrm{e}^{-\beta E_m}. \tag{4.26$'$}$$

The *canonical partition function*, as function of the multiplier β and of the exact data N and Ω, is given by

$$Z_{\rm C}(\beta, N, \Omega) = {\rm Tr}\, {\rm e}^{-\beta \widehat{H}_N} = \sum_m {\rm e}^{-\beta E_m}, \tag{4.27}$$

where the summation is over all N-particle states. The parameter β is connected with the average energy U through

$$-\frac{\partial \ln Z_{\rm C}(\beta, N, \Omega)}{\partial \beta} = U. \tag{4.28}$$

It is important to note that the sum (4.27) is over the *eigenstates* $|m\rangle$, and not over the *energy values* E_m; when the energy levels are degenerate each of them appears as often as the *degree of degeneracy of the level* $d(E_m)$. One method of summation consists in grouping the terms which belong to the same energy value together and writing (4.27) in the form

$$\boxed{Z_{\rm C} = \sum_{E_m} d(E_m)\, {\rm e}^{-\beta E_m}} . \tag{4.29}$$

In this connection one should note that the probability distribution for the energy *is not an exponential one*: it contains as factor of the decreasing exponential (4.26′) a distribution, which in general increases rapidly, representing the *level density*, that is, the factor multiplying dE in the expression for the number of eigenstates of the Hamiltonian with energies lying between E and $E + dE$ (§ 5.5.3 and Exerc.9g).

Nevertheless, it is often difficult to calculate the degrees of degeneracy and this restricts the use of (4.29). For most practical applications the states m will be characterized by a set of independent quantum numbers, $m = (m_1, m_2, \ldots)$, and the energies E_m will have the special form $E_m = \varepsilon_1(m_1) + \varepsilon_2(m_2) + \ldots$; under those conditions the simplest way to evaluate (4.27) is to characterize the states m not by their energies but by the *set of quantum numbers* $m = (m_1, m_2, \ldots)$. Thanks to the factorization (4.22) the calculation is then simplified without there being any need to worry about degeneracies, as we have

$$Z_{\rm C} = \sum_m {\rm e}^{-\beta E_m} = \left(\sum_{m_1} {\rm e}^{-\beta \varepsilon(m_1)}\right) \left(\sum_{m_2} {\rm e}^{-\beta \varepsilon(m_2)}\right) \cdots .$$

Exercises 1b, 4c, and 4f give examples of this situation.

The equilibrium entropy $S(U)$ is given by

$$S(U) = k \ln Z_{\rm C} + k\beta U, \tag{4.30}$$

and the Lagrangian parameter $k\beta$ is its derivative with respect to U. Finally, the energy is not defined exactly, and its statistical fluctuations in canonical equilibrium can be found from (4.13) or (4.20),

$$\left\langle \left(\widehat{H}_N - U \right)^2 \right\rangle = \frac{\partial^2 \ln Z_{\mathrm{C}}(\beta)}{\partial \beta^2} = -k \bigg/ \frac{\partial^2 S(U)}{\partial U^2} . \tag{4.31}$$

Of course, Z_{C} depends also on the variables N and Ω. However, its second derivative with respect to N is, for instance, not connected with the statistical fluctuations in N, which are zero.

4.3.2 The Grand Canonical Ensemble

The grand canonical ensemble, also called "grand ensemble", was defined in § 4.1.2. It corresponds to situations where the number of particles N in the system, considered to be a random variable, and the energy U are *both given in the form of averages.* Technically speaking it is very useful, as it often leads to simpler calculations than the canonical ensemble, of which it is anyway the equivalent for a macroscopic system. In the case of a finite system these ensembles represent different physical stuations. We saw in § 4.1.2 that the canonical ensemble was suited for the description of a system which can exchange heat with a thermostat. The grand canonical ensemble similarly describes an *open system*, that is, a system which can exchange particles and heat with the outside. This is, for instance, the case for the part of the fluid lying in a fixed volume element, while the rest of the fluid plays the rôle of a particle and energy bath. Exercise 4b provides another example.

When the system is in a pure state with a well defined particle number N, its representative ket belongs to a Hilbert space $\mathcal{E}_{\mathrm{H}}^{(N)}$. In the case we are now considering N is not exactly known; we have indicated in § 2.3.6 that we must associate with the system a Hilbert space $\mathcal{E}_{\mathrm{H}}$, called *Fock space*, which is constructed as the *direct sum* $\bigoplus_{N=0}^{\infty} \mathcal{E}_{\mathrm{H}}^{(N)}$ of N-particle spaces (§ 2.1.1). In this space we introduce the *particle number observable* $\widehat{N}$ in order to treat the energy and the particle number on the same footing. Its eigenvalues are the integers N and its eigenvectors the kets for which the system has a well defined number of particles. Its eigensubspace corresponding to N is $\mathcal{E}_{\mathrm{H}}^{(N)}$; for $N = 0$, $\mathcal{E}_{\mathrm{H}}^{(0)}$ is a one-dimensional space which describes the vacuum. In the situation under consideration the eigenvalue N of $\widehat{N}$ is a random variable and only its expectation value $\langle N \rangle$ is known, as is the case for the energy which is the eigenvalue of the Hamiltonian. Observables such as $\widehat{N}$, $\widehat{H}$, or the total momentum, which do not change the number of particles of the ket on which they operate, commute with $\widehat{N}$ and must be considered as a set of observables each operating in a subspace $\mathcal{E}_{\mathrm{H}}^{(N)}$. More precisely, the Hamiltonian $\widehat{H}$ in Fock space is the direct sum of the Hamiltonians $\widehat{H}_N$ in the N-particle subspaces $\mathcal{E}_{\mathrm{H}}^{(N)}$. The density operator $\widehat{D}$ operates in the Fock space, and the constraints (4.2) on the average of the energy, U, and of the particle number, $\langle N \rangle$, can be written in the form

$$\mathrm{Tr}\, \widehat{D}\widehat{H} = U, \qquad \mathrm{Tr}\, \widehat{D}\widehat{N} = \langle N \rangle. \tag{4.32}$$

Note that the symbol Tr has no longer the same meaning as in § 4.3.1 as we are now dealing with a trace over $\mathcal{E}_{\mathrm{H}} = \bigoplus_N \mathcal{E}_{\mathrm{H}}^{(N)}$. Writing Tr_N for the trace used for the canonical ensemble with N particles and $\widehat{D}_N$ for the component of $\widehat{D}$ in $\mathcal{E}_{\mathrm{H}}^{(N)}$ obtained by projection, the constraints (4.32) and the normalization (4.2′) can be written as

$$\sum_N \mathrm{Tr}_N\, \widehat{D}_N \widehat{H}_N = U, \qquad \sum_N N\, \mathrm{Tr}_N\, \widehat{D}_N = \langle N \rangle,$$

$$\sum_N \mathrm{Tr}_N\, \widehat{D}_N = 1.$$

The quantity $\mathrm{Tr}_N\, \widehat{D}_N$ represents the *probability that the system consists of N particles.*

Traditionally the multipliers associated with the constraints (4.32) are denoted by β and $-\alpha$ so that we can write for the *grand canonical density operator* (4.6)

$$\widehat{D} = \frac{1}{Z_{\mathrm{G}}}\, \mathrm{e}^{-\beta\widehat{H}+\alpha\widehat{N}}. \tag{4.33}$$

Its normalization coefficient is the *grand partition function*

$$\boxed{\begin{aligned} Z_{\mathrm{G}}(\alpha,\beta) &= \mathrm{Tr}\, \mathrm{e}^{-\beta\widehat{H}+\alpha\widehat{N}} \\ &= \sum_N \mathrm{e}^{\alpha N} \sum_m \mathrm{e}^{-\beta E_m^{(N)}} = \sum_N \mathrm{e}^{\alpha N}\, Z_{\mathrm{C}}(\beta, N) \end{aligned}}\,, \tag{4.34}$$

where the $E_m^{(N)}$ are the energies of the N-particle micro-states. The grand partition function $Z_{\mathrm{G}}(\alpha)$ is thus a Laplace transform with respect to N of the canonical partition function $Z_{\mathrm{C}}(N)$.

Writing for the sake of simplicity N instead of $\langle N \rangle$, we find the following relations between the averages U and N and the Lagrangian multipliers β and α:

$$\frac{\partial \ln Z_{\mathrm{G}}}{\partial \beta} = -U, \qquad \frac{\partial \ln Z_{\mathrm{G}}}{\partial \alpha} = N, \tag{4.35}$$

and we have for the equilibrium entropy

$$S(U,N) = k \ln Z_{\mathrm{G}} + k\beta U - k\alpha N. \tag{4.36}$$

Inverting the relations (4.35) we have

$$\frac{1}{k}\frac{\partial S}{\partial U} = \beta, \qquad \frac{1}{k}\frac{\partial S}{\partial N} = -\alpha. \tag{4.37}$$

Finally, the statistical fluctuations in the energy and the particle number are

$$\left.\begin{aligned} \left\langle\left(\widehat{H}-U\right)^2\right\rangle &= \frac{\partial^2 \ln Z_{\rm G}}{\partial\beta^2} \\ &= -k\,\frac{\partial^2 S}{\partial N^2}\bigg/\left[\frac{\partial^2 S}{\partial N^2}\,\frac{\partial^2 S}{\partial U^2}-\left(\frac{\partial^2 S}{\partial U\partial N}\right)^2\right], \\ \left\langle\left(\widehat{N}-N\right)^2\right\rangle &= \frac{\partial^2 \ln Z_{\rm G}}{\partial\alpha^2} \\ &= -k\,\frac{\partial^2 S}{\partial U^2}\bigg/\left[\frac{\partial^2 S}{\partial N^2}\,\frac{\partial^2 S}{\partial U^2}-\left(\frac{\partial^2 S}{\partial U\partial N}\right)^2\right]. \end{aligned}\right\} \quad (4.38)$$

Expressions (4.31) and (4.38) for the equilibrium fluctuations as functions of the second derivatives of S can be obtained directly from the general theory of Legendre transformations (Eq.(6.16)).

4.3.3 Other Examples of Ensembles

The remainder of the book will provide us with a large number of substances where knowledge of the energy levels enables us to use either (4.26–30) or (4.33–37) to calculate the physical properties in thermal equilibrium, either in the (petit) canonical ensemble, or in the grand canonical ensemble. Sometimes, for specific applications, we shall use statistical ensembles based on other choices of the data (4.2) in order to benefit from the flexibility of the partition function technique.

For instance, in Chap.1 we worked in the *microcanonical ensemble*, assuming that *the energy U was given exactly* and not as an average. Because of the discrete nature of the eigenenergies of $\widehat{H}$ we must, however, in such a case ask that the energy is not exactly equal to U, but lies between U and $U + \Delta U$, where ΔU must be small as compared to the measuring accuracy of U, but large as compared to typical energy level spacings. These conditions are compatible for large systems as the level density tends rapidly – exponentially – to infinity with the volume. The constraint on the energy defines an allowed Hilbert space $\mathcal{E}_{\rm H}^{W}$ with a large dimensionality, W. Here we need only one Lagrangian multiplier which takes account of the normalization (4.2′) and the Boltzmann-Gibbs distribution reduces to the result (4.1), as expected:

$$\widehat{D} = \frac{1}{W}\,\widehat{I}_W. \quad (4.39)$$

The microcanonical partition function is nothing but the number of levels, W, and (4.14) reduces to the equiprobability entropy,

$$S = k \ln W. \quad (4.40)$$

One could reconstruct the theory of Chap.1 (Exercs.1b and 4c) by using a canonical ensemble. The calculations are simpler and expressions (4.26–30) lead to the same physical results as in § 1.4.

Returning to the grand canonical ensemble, let us now consider a fluid containing several species of molecules $a, b, \ldots$. We must introduce for each species a constraint expressing the *conservation of the number of molecules* $N_a, N_b, \ldots$. There thus appear several associated Lagrangian multipliers $\alpha_a, \alpha_b, \ldots$, and the grand canonical density operator is

$$\widehat{D} = \frac{1}{Z_{\mathrm{G}}} \,\mathrm{e}^{-\beta\widehat{H}+\alpha_a\widehat{N}_a+\alpha_b\widehat{N}_b+\cdots}. \tag{4.41}$$

The *grand partition function* (4.29) has been generalized to become

$$Z_{\mathrm{G}}(\alpha_a, \alpha_b, \ldots, \beta) = \mathrm{Tr}\,\mathrm{e}^{-\beta\widehat{H}+\alpha_a\widehat{N}_a+\alpha_b\widehat{N}_b+\cdots}. \tag{4.42}$$

However, if the molecules can change into one another through *chemical reactions*, the constants of the motion are no longer the numbers of molecules, but the *numbers of the constituent atoms.* In that case, the density operator has still the form (4.41), but now $N_a, N_b, \ldots$ represent the total number of the various atoms which combine to make up the molecules (§§ 6.6.3 and 8.2.2).

One can easily extend the method to other physical systems and to other observables; this enables us to introduce new conjugate variables. For instance, if the system, rather than being enclosed in a box, can freely move in space, we introduce three Lagrangian multipliers $-\boldsymbol{\lambda}$ which are associated with the three components of the total *momentum* $\boldsymbol{P}$. The density operator has the form

$$\widehat{D} = \frac{1}{Z} \,\mathrm{e}^{-\beta\widehat{H}+(\boldsymbol{\lambda}\cdot\widehat{\boldsymbol{P}})},$$

where $-\boldsymbol{\lambda}$ occurs as the conjugate of the constant of motion $\langle\boldsymbol{P}\rangle$ just as β is the conjugate of the energy. As always, the Lagrangian multipliers have an interesting physical meaning: $\boldsymbol{\lambda}/\beta$ can be interpreted as the *velocity of the whole* system, as one can see through a change in the frame of reference (Exerc.4e).

Let us give still a few other examples. Consider a sample of a magnetic substance with Hamiltonian $\widehat{H}$ when there is no field and assume that we try to study its properties as a function of the total *magnetic moment*. The data are $\langle H\rangle$ and $\langle\boldsymbol{M}\rangle$ and the Boltzmann-Gibbs distribution is

$$\frac{1}{Z} \,\mathrm{e}^{-\beta\widehat{H}+\beta(\boldsymbol{B}\cdot\widehat{\boldsymbol{M}})}.$$

The Lagrangian multiplier $-\beta\boldsymbol{B}$ associated with the magnetic moment can be identified, apart from the factor $-\beta$, as the external magnetic induction which would be necessary to produce the moment $\langle\boldsymbol{M}\rangle$. The same considerations can be applied to a *dielectric* for which the conjugate variables would be the dipole moment and the electric field.

Similarly, consider a fluid kept in a cylinder by a moving piston; the *volume* Ω can fluctuate around its given average value. The associated Lagrangian multiplier will be identified (§ 5.6.5) with $\beta\mathcal{P}$, where $\mathcal{P}$ is the pressure exerted on the piston. In the study of *thermal engines*, and also in *chemistry*, it is often useful to take as the state variable the pressure rather than the volume and thus to work in the so-called *isobaric-isothermal* ensemble where the natural variables include the temperature and the pressure (Exerc.5a).

Another example is provided by nuclear physics: a nucleus is a set of interacting protons and neutrons; if we want to study its moment of inertia, we must know how it behaves when it rotates, that is, when we require it to have an *angular momentum* $\boldsymbol{J}$. The Lagrangian multipliers associated with $\widehat{\boldsymbol{J}}$ are the components of the vector $-\beta\boldsymbol{\omega}$ and the equilibrium density operator can be written in the form

$$\frac{1}{Z}\,\mathrm{e}^{-\beta\widehat{H}+\beta(\boldsymbol{\omega}\cdot\widehat{\boldsymbol{J}})};$$

here we can identify $\boldsymbol{\omega}$ with an *angular velocity.* In this case, the three components of $\widehat{\boldsymbol{J}}$ are constants of the motion which *do not commute* with one another, but that hardly affects the general formalism. The same idea can be applied to the equilibrium of a fluid in a rotating cylinder (Exerc.7b).

4.3.4 Equilibrium Distributions in Classical Statistical Mechanics

There is little to be changed in the formalism of §§ 4.2 and 4.3 when we go over to the classical approximation. The density operators and the observables are replaced by functions defined in phase space. The trace is replaced by an integration over that space, possibly with a summation over N: $\mathrm{Tr} \Longrightarrow \sum_N \int d\tau_N$. When the particle number N can fluctuate, the density in phase D is a set of components D_N, each of which is a function in the phase space $\mathcal{E}_{\mathrm{P}}^{(N)}$. We change from the discrete to the continuum: whereas the statistical entropy $S(\widehat{D})$ is a function of the unknown matrix elements of $\widehat{D}$, in classical statistical mechanics $S(D)$ is a *functional* of the phase density D. The equation expressing the stationarity of (4.5) is replaced by

$$0 = \sum_N \int d\tau_N\, \delta D_N \left[\ln D_N + \sum_i \lambda_i A_i + \lambda_0 + 1\right], \tag{4.43}$$

where the variations δD_N are arbitrary functions; it leads to the vanishing of the expression inside the square brackets which is the functional derivative of (4.5). The classical Boltzmann-Gibbs distribution is thus

$$D_N = \frac{1}{Z}\,\mathrm{e}^{-\sum_i \lambda_i A_i}, \tag{4.44}$$

where the partition function is defined by

$$\boxed{Z = \sum_N \int d\tau_N \, \mathrm{e}^{-\sum_i \lambda_i A_i}} \, . \tag{4.45}$$

All relations between the average values $\langle A_i \rangle$ and the Lagrangian multipliers λ_i, the equilibrium entropy S and the partition function Z, remain unchanged.

For instance, the *classical microcanonical* ensemble is described by a constant phase density, equal to $1/W$, in the volume W of the N-particle phase space which lies between the surfaces $H_N(\boldsymbol{r}_1, \boldsymbol{p}_1, \dots, \boldsymbol{r}_N, \boldsymbol{p}_N) = U$ and $H_N(\boldsymbol{r}_1, \boldsymbol{p}_1, \dots, \boldsymbol{r}_N, \boldsymbol{p}_N) = U + \Delta U$.

The *classical canonical* ensemble has a phase density

$$D_N = \frac{1}{Z_{\mathrm{C}}} \, \mathrm{e}^{-\beta H_N}, \tag{4.46}$$

with as partition function (Exerc.4d)

$$Z_{\mathrm{C}}(N) = \int d\tau_N \, \mathrm{e}^{-\beta H_N}. \tag{4.47}$$

Finally, the *classical grand canonical* ensemble is described by a density in phase D with components D_N in each of the $6N$-dimensional phase spaces, which are equal to

$$D_N = \frac{1}{Z_{\mathrm{G}}} \, \mathrm{e}^{-\beta H_N + \alpha N}, \tag{4.48}$$

while the classical grand partition function is related to Z_{C} through the same Laplace transformation (4.34) as in quantum mechanics.

Summary

An equilibrium state, which is characterized by macroscopic data, is microscopically and statistically described as the state of maximum disorder compatible with these data. During the approach to equilibrium the density operator evolves in the allowed Hilbert space where the conservation laws are satisfied; information is lost and the statistical entropy increases. Looking for the maximum of the statistical entropy under the constraints that some constants of motion are given as averages, leads to the Boltzmann-Gibbs distribution (4.6) which corresponds to the maximum of (4.10).

The calculation of the equilibrium macroscopic quantities is made easier by using partition functions; this technique is summarized in § 4.2.6: knowing the partition function (4.11) is sufficient (i) to find the relations (4.12) between the data characterizing the equilibrium state which are, on the one hand, the average values of conserved observables and, on the other hand, their associated Lagrangian multipliers, and (ii) to calculate the statistical entropy (4.14), the expectation values (4.24), and the correlations or fluctuations (4.13). We describe two general methods, factorization and the variational method, to evaluate partition functions.

Depending on the number and nature of the data, we introduce different statistical equilibrium ensembles with their own natural variables, for instance, U and N for the microcanonical ensemble, β and N for the canonical ensemble, and β and α for the grand canonical ensemble.

Exercises

4a Relation Between Fluctuation and Response

1. Consider a system which is in equilibrium in a uniform magnetic field $\boldsymbol{B}$ in the z-direction. This field occurs in the Hamiltonian, $\widehat{H} = \widehat{H}_0 - B\widehat{M}$, through the second term. The Hamiltonian $\widehat{H}_0$ when there is no field present commutes with the operator $\widehat{M}$ which is the z-component of the total magnetic moment $\widehat{\boldsymbol{M}}$. We neglect the "diamagnetic" term which is quadratic in B and which comes from the term in A^2 from $(\boldsymbol{p}-e\boldsymbol{A})^2/2m$. Write down the general expression for the magnetic susceptibility, that is, the static linear response of the system to an infinitesimal perturbation B. Compare it with the statistical fluctuation in M. This kind of relation between fluctuations and linear responses is very general.

2. Application: Show that Curie's law is satisfied whenever the energy levels in the presence of the field B have the form $E_{n\lambda}(B) = -BM_n + E_\lambda$, where the M_n are the eigenvalues of $\widehat{M}$, where λ denotes the other quantum numbers, and where E_λ is independent of n and B. The models studied in Chap.1 and Exercs.1b and 1c were special cases of this general form. Check that the expression for the Curie constant in Exerc.1b satisfies the general relation between fluctuations and responses.

Solution:

1. Taking twice the derivative of ln Z with respect to B and using (4.24) and (4.13) we find

$$\langle M \rangle = \mathrm{Tr}\, \widehat{D}\widehat{M} = \frac{1}{\beta}\frac{\partial}{\partial B} \ln Z(\beta, B),$$

$$\Delta M^2 = \frac{1}{\beta^2}\frac{\partial^2}{\partial B^2} \ln Z.$$

On the other hand,

$$\chi = \frac{1}{\Omega}\frac{\partial \langle M \rangle}{\partial B}\bigg|_{B=0} = \frac{1}{\Omega kT}\,\Delta M^2\bigg|_{B=0}.$$

The linear response χ is proportional to the fluctuations in the associated observable M, evaluated for the case when the perturbation B is not there.

2. Curie's law is valid provided ΔM^2 is independent of T when $B = 0$. This is true for the models considered since, for any function f,

$$\langle f(M)\rangle\Big|_{B=0} = \frac{\sum_{n\lambda} f(M_n)\,\mathrm{e}^{-\beta E_\lambda}}{\sum_{n\lambda} \mathrm{e}^{-\beta E_\lambda}} = \frac{\sum_n f(M_n)}{\sum_n 1}.$$

The Curie constant of Exerc.1b is equal to $N\mu_{\mathrm{B}}^2 g^2 j(j+1)/3\Omega k$ and

$$3\Delta M^2\Big|_{B=0} = \langle M^2\rangle\Big|_{B=0} = \left\langle\left(\frac{-\sum_i g\mu_{\mathrm{B}}\boldsymbol{J}_i}{\hbar}\right)^2\right\rangle = N\mu_{\mathrm{B}}^2 g^2 j(j+1).$$

4b Adsorption

When a gas is in equilibrium with a solid wall, its molecules can become attached to that wall on special sites where they are trapped in a way which depends on the structure of the wall. Adsorption is thus an equilibrium between two phases, the gas and the system of bound molecules, similar to a chemical equilibrium.

To study this we consider a model: the wall has $\mathcal{N}$ sites on each of which a molecule can get bound with a binding energy $-u$. The molecules which are not adsorbed form a perfect gas. Calculate the grand partition function for the adsorbed molecules and hence, using the thermodynamic properties of gases, which we shall obtain in § 7.3.1, calculate the average number N of adsorbed molecules as function of the gas pressure and the temperature.

Solution:

The micro-states are characterized by stating for each of the $\mathcal{N}$ sites whether it is occupied or not by a molecule. The grand partition function can be factorized into contributions relating to each of the sites. If the site is occupied, the energy is $-u$ and the number of particles equals 1; if it is not occupied, both quantities are zero. Hence we have

$$Z_{\mathrm{G}}(\alpha,\beta) = \left(\mathrm{e}^{\beta u+\alpha}+1\right)^{\mathcal{N}}.$$

From (4.35) it follows that

$$N = \frac{\partial \ln Z_G}{\partial \alpha} = \frac{\mathcal{N}}{1+\mathrm{e}^{-(\beta u+\alpha)}}.$$

At equilibrium the values of $\beta = 1/kT$ and of α are the same in the wall and in the gas. In the gas, α is related to the pressure $\mathcal{P}$ through

$$\mathrm{e}^{\alpha} = \mathcal{P}h^3\left(2\pi m\right)^{-3/2}\left(kT\right)^{-5/2},$$

so that we have

$$\frac{N}{\mathcal{N}} = \frac{\mathcal{P}}{\mathcal{P}+\mathcal{P}_0}, \qquad \text{where}$$

$$\mathcal{P}_0 \equiv \left(kT\right)^{5/2} \left(\frac{2\pi m}{h^2}\right)^{3/2} \mathrm{e}^{-u/kT}.$$

This relation which is called the *Langmuir adsorption isotherm* gives for each value of T, which enters through $\mathcal{P}_0$, the number N of adsorbed molecules as a function of the pressure. This number decreases with pressure. When $T \to 0$, we find that $N \to \mathcal{N}$ and practically all sites are occupied; this effect resembles the condensation of vapours on cold walls; for instance, the drying of crockery in dishwashers is achieved by cooling the walls by a cold water jet. When $T \gg u/k$, we find that $N \to 0$ and the walls are degassed; this is the reason why one heats the walls of containers in which one wishes to make high vacua (Exerc.14c).

We can obtain the same results by treating the N adsorbed molecules on the $\mathcal{N}$ sites in a canonical ensemble. Their canonical partition function is equal to

$$Z_C(N,\beta) = \frac{\mathcal{N}!}{N!(\mathcal{N}-N)!}\,\mathrm{e}^{\beta N u},$$

which for macroscopic values of $\mathcal{N}$ and N gives

$$\ln Z_C \sim N \ln \frac{\mathcal{N}}{N} + (\mathcal{N}-N) \ln \frac{\mathcal{N}}{\mathcal{N}-N} + \beta N u.$$

In order to write down the condition that these molecules are in equilibrium with those in the gas, we must equate the chemical potentials $\mu = \alpha/\beta$ in the two systems. For the wall we find α in the canonical ensemble by using (5.48); this leads to

$$\alpha = -\frac{\partial}{\partial N} \ln Z_C = \ln \frac{N}{\mathcal{N}-N} - \beta u.$$

Langmuir's law then follows by again using the relation between α and $\mathcal{P}$ in the gas.

4c Free Energy of a Paramagnetic Solid

Rederive the results of § 1.4, starting from the canonical Boltzmann-Gibbs equilibrium distribution:

1. Write down the general expressions for the internal energy U, the statistical entropy S, the magnetic moment $\langle M \rangle$, the susceptibility χ, and the spin specific heat C, as functions of $\ln Z_C$. Show that $F \equiv -\beta^{-1} \ln Z_C$ can be identified with the free energy, $1/k\beta$ with the absolute temperature T, and S with the entropy.

2. Evaluate Z_C in the model of Chap.1 and hence derive the above thermodynamic quantities.

3. Compare for that model the canonical and microcanonical ensembles.

Hints:

1. The results follow from (4.30) and the relation

$$dF = -S\,d\frac{1}{k\beta} - \langle M \rangle\, dB.$$

2. Factorization leads to Z_{C} in the form $\left(2\cosh \beta\mu_{\mathrm{B}}B\right)^N$.

3. The macroscopic quantities are the same; only the fluctuations in the energy and the magnetic moment are different. The entropies differ by a non-extensive contribution of order $\ln N$.

4d Absence of Magnetism in Classical Mechanics

Show that if the particles, the electrons and nuclei, which make up a sample of matter were to obey the laws of classical mechanics, introducing an arbitrary magnetic field would, at thermal equilibrium, have no effect whatever (Bohr-van Leeuwen theorem). Write down first the Hamiltonian in a magnetic field and then the probability distribution in canonical equilibrium for the various observables, which are functions of the positions and the velocities of the particles. Remember that the momentum of a charged particle is not an observable (§ 2.1.3).

The existence of magnetic moments is thus a purely quantum effect, whether we are dealing with the intrinsic magnetic moments of the particles, which are proportional to their spins, or with magnetic moments related to currents. Langevin paramagnetism (Exerc.1c) is only apparently classical: the magnetic moment $\boldsymbol{\mu}$ which was *a priori* associated by Langevin with a paramagnetic ion has necessarily a quantum origin.

Solution. In the Hamiltonian which has the form (2.65) the magnetic field $\boldsymbol{B} = \operatorname{curl} \boldsymbol{A}$ only occurs through the kinetic energy term. The velocity of the i-th particle is

$$\boldsymbol{v}_i = \frac{d\boldsymbol{r}_i}{dt} = \frac{\partial H}{\partial \boldsymbol{p}_i} = \frac{1}{m_i}\left[\boldsymbol{p}_i - e_i\boldsymbol{A}(\boldsymbol{r}_i)\right].$$

If we use the volume element (2.55) we find that in canonical equilibrium (4.46) leads to a probability density for the particle positions $\boldsymbol{r}_i$ and velocities $\boldsymbol{v}_i$ which is equal to

$$\frac{1}{Z_{\mathrm{C}}}\frac{1}{N!}\prod_{i=1}^{N}\left(\frac{m_i}{h}\right)^3$$

$$\times \exp\left[-\beta\sum_{i=1}^{N}\left(\frac{1}{2}m_i\boldsymbol{v}_i^2 + V(\boldsymbol{r}_i)\right) - \beta\sum_{i,j=1;\,i<j}^{N} W\left(\left|\boldsymbol{r}_i - \boldsymbol{r}_j\right|\right)\right].$$

This probability density is independent of the field $\boldsymbol{B}$. The same is true for the expectation value of all the observables, which for identical particles are symmetric functions of the $\boldsymbol{r}_i$ and $\boldsymbol{v}_i$, such as the current density at a particular point.

4e Galilean Invariance

Show that the entropy of a system at equilibrium in the absence of external fields remains unchanged, if the system is subjected to a uniform translation (§ 4.3.3). In order to do this, express the Lagrangian multiplier associated with the momentum $\boldsymbol{P}$ as a function of the translational velocity and compare the thermodynamic quantities in the two frames of reference.

Solution. If the momentum $\boldsymbol{P}$ is given on average, the partition function is equal to

$$Z = \text{Tr}\, e^{-\beta\widehat{H}+(\boldsymbol{\lambda}\cdot\widehat{\boldsymbol{P}})} = \text{Tr}\, e^{-\beta\widehat{H}'},$$

where $\widehat{H}'$ differs from the Hamiltonian $\widehat{H}$ in that the particle momenta $\boldsymbol{p}_i$ are replaced by $\boldsymbol{p}_i' = \boldsymbol{p}_i - m_i\boldsymbol{u}$, with $\boldsymbol{\lambda} = \beta\boldsymbol{u}$, and by adding an extra term $\Delta = M\lambda^2/2\beta^2$, where M is the total mass. In the Galilean frame with velocity $\boldsymbol{u}$, $\boldsymbol{p}_i'$ is the momentum of the i-th particle and the interaction energy is the same as in the initial frame, so that $\widehat{H}' - \Delta$ has the same form as $\widehat{H}$. The calculation of Z in this frame shows that $\ln Z = \ln Z_0 + \beta\Delta$, where Z_0 is the partition function at rest. The term $\beta\Delta$ gives, as expected, the average momentum $\partial \ln Z/\partial\boldsymbol{\lambda} = M\boldsymbol{u}$ and a contribution $\frac{1}{2}Mu^2$ to the energy $-\partial \ln Z/\partial\beta$, but it does not contribute to the entropy (4.14).

4f Oscillators in Canonical Equilibrium

1. Write down the canonical partition function Z_1 of a quantum harmonic oscillator of frequency $\omega/2\pi$ ($\varepsilon = \hbar\omega$).

2. What happens in the case of an isotropic two- or three-dimensional oscillator? Write down the partition function Z_N for N oscillators with the same frequency.

3. Express the statistical entropy S as function of the energy U for N oscillators. Compare this result with Planck's in the form it was obtained in Exerc.3e.

4. Derive from Z_N the number of ways $W_N(P)$ to decompose an integer P into a sum $p_1 + p_2 + \ldots + p_N$, where the p_i are non-negative integers – the method is the same as the one used in § 1.2.3.

Solution:

1. The energy levels are $E_p = (p + \frac{1}{2})\hbar\omega$ with $p \geq 0$, and hence

$$Z_1 = \sum_p e^{-\beta E_p} = \frac{1}{2\sinh(\frac{1}{2}\beta\varepsilon)}.$$

2. In two dimensions the micro-states are characterized by two quantum numbers p_1 and p_2 with $E_{p_1p_2} = (p_1 + p_2 + 1)\hbar\omega$. The factorization of Z gives us $Z_2 = \left(Z_1\right)^2$. Note that using (4.29) to calculate the partition function would have been more complicated: the allowed energy levels are $E_m = m\hbar\omega$, with m a positive integer, but they have a multiplicity $d(E_m) = m$ which leads to

$$Z_2 = \sum_m m\,\mathrm{e}^{-\beta m\hbar\omega} = \left(Z_1\right)^2.$$

Similarly, factorization gives us $Z_3 = \left(Z_1\right)^3$, $Z_N = \left(Z_1\right)^N$.

3. We find from Z_1 and Z:

$$U = -N\,\frac{\partial}{\partial\beta}\,\ln Z_1 = \frac{1}{2}N\varepsilon\,\frac{\mathrm{e}^{\beta\varepsilon}+1}{\mathrm{e}^{\beta\varepsilon}-1},$$

$$S = k\left(\ln Z + \beta U\right) = \frac{1}{2}Nk\left[-\ln\left(\mathrm{e}^{\beta\varepsilon}+\mathrm{e}^{-\beta\varepsilon}-2\right)+\beta\varepsilon\,\frac{\mathrm{e}^{\beta\varepsilon}+1}{\mathrm{e}^{\beta\varepsilon}-1}\right],$$

$$\mathrm{e}^{\beta\varepsilon} = \frac{2U+N\varepsilon}{2U-N\varepsilon},$$

$$S = Nk\,\left[\left(\frac{U}{N\varepsilon}+\frac{1}{2}\right)\ln\left(\frac{U}{N\varepsilon}+\frac{1}{2}\right)-\left(\frac{U}{N\varepsilon}-\frac{1}{2}\right)\ln\left(\frac{U}{N\varepsilon}-\frac{1}{2}\right)\right].$$

Planck did not include the "zero-point" energy $\frac{1}{2}\hbar\omega$ in the expression for the oscillator levels so that his average energy per oscillator, E, is related to the internal energy by $U/N = E + \frac{1}{2}\varepsilon$. Replacing U as function of E enables us to identify the entropy of the canonical ensemble which we have calculated above with the entropy of a micro-canonical ensemble for $N \to \infty$ which was calculated in Exerc.3e.

4. The eigenvalues of $\widehat{H}$ are $(P + \frac{1}{2}N)\varepsilon$ with a multiplicity $W_N(P)$. We have thus

$$Z_N = \sum_{P=0}^{\infty} W_N(P)\,\exp\left[-\left(P+\frac{1}{2}N\right)\beta\varepsilon\right],$$

or, if we use $Z_N = \left(Z_1\right)^N$ and put $z = \mathrm{e}^{-\beta\varepsilon}$,

$$\sum_{P=0}^{\infty} W_N(P)\,z^P = \frac{1}{(1-z)^N}.$$

Comparing the expansions of the left- and the right-hand sides we find

$$W_N(P) = \frac{(P+N-1)!}{(N-1)!P!},$$

which is the same result as we found in Exerc.3e by a combinatorial analysis method. Using the partition function Z_3 is, for instance, the fastest way to find the multiplicities of the levels of a three-dimensional harmonic oscillator.

4g The Ehrenfests' Urn Model

In order to understand how the approach to equilibrium is compatible with the reversibility of the motion, P. and T. Ehrenfest (§ 3.4.3) considered the following process. Two urns A and B contain, respectively, $N+m$ and $N-m$ numbered balls. Every second, one draws a number at random and one moves the ball with that number from the one urn to the other, thus changing m

by +1 or by −1. Like the trajectory of a classical micro-state in phase space, the $m(t)$ curve, with t and m both integers, is reversible and recurrent.

1. Show that nevertheless, whatever the value of $m(0)$, $m(t)/N$ is practically always small for sufficiently large t. To do this, calculate the probability distribution $P(m)$ and the variance $\langle m^2(t)\rangle$ for $t \to \infty$.

2. Consider a rather improbable situation m_0, for instance, $N = 100$, $m_0 = 90$. The curve $m(t)$ intersects $m = m_0$ as often when ascending as when descending. At first sight this seems to indicate that m has as much chance of increasing as of decreasing, if it starts from an initial value $m(0) = m_0$, and this seems to contradict the approach to equilibrium, $m/N \to 0$. In order to resolve this paradox, calculate the probabilities for m to increase or to decrease, when it starts from m_0, and show that the most frequent configuration for the curve $m(t)$ in the neighbourhood of a point $m(t) = m_0$ is neither a descent, nor an ascent, but a maximum.

Answers:

1. We find

$$P(m) \;\to\; \frac{(2N)!}{2^{2N}(N+m)!(N-m)!} \;\sim\; \frac{1}{\sqrt{N\pi}}\,\mathrm{e}^{-m^2/N},$$

$$\left\langle m^2\right\rangle \;\to\; \tfrac{1}{2}\,N.$$

2. Notwithstanding the symmetry of the $m(t)$ curve and the fact that it intersects $m = m_0$ as often when ascending as when descending, the probability of increasing, when starting from m_0, is $(N - m_0)/2N$ and that for decreasing $(N + m_0)/2N$. However, the probability that m_0 is crossed either when ascending or when descending is $(N^2 - m_0^2)/4N^2$, whereas the probability that it is a maximum, $(N + m_0)^2/4N^2$, is very much larger.

4h Loaded Die

Observations on a badly balanced die have shown that 6 occurs twice as often as 1. Nothing peculiar was observed for the other faces. What are the probabilities p_m $(1 \le m \le 6)$ which according to the maximum statistical entropy criterion we should assign to the various faces?

Solution. The constraints are

$$\sum_{m=1}^{6} p_m \;=\; 1, \qquad 2p_1 - p_6 \;=\; 0.$$

Hence we find

$$p_m \;=\; \frac{1}{Z}\,\mathrm{e}^{-\lambda A_m},$$

with

$$
\begin{aligned}
A_1 &= 2, \quad A_m = 0, \quad \text{for } 2 \le m \le 5, \\
A_6 &= -1, \quad Z = \mathrm{e}^{-2\lambda} + 4 + \mathrm{e}^{\lambda},
\end{aligned}
$$

and $\partial Z / \partial \lambda = 0$ leads to

$$
p_1 = 0.107, \quad p_m = 0.170, \quad \text{for } 2 \le m \le 5, \quad p_6 = 0.214.
$$

5. Thermodynamics Revisited

"On a pensé que la chaleur pouvait être produite par un mouvement intestin et vibratoire des molécules des corps; mais ce système, qui paraît combattu par des observations indirectes, est maintenant généralement abandonné. Ce n'est plus qu'en Allemagne qu'il peut encore compter quelques partisans."

C. Bailly, Manuel de Physique, 1826

"– N'avez-vous point quelques principes, quelques commencements des sciences?
– Oh! oui, je sais lire et écrire."

Molière, Le Bourgeois Gentilhomme

"Là où je cherchais les grandes lois, on m'appellait fouilleur de détails."

Marcel Proust, Le Temps Retrouvé

In this chapter we shall travel the last stage of the journey which, for a physical system in equilibrium, leads us from its microscopic description to an understanding of its macroscopic properties. After having incorporated statistics in the microscopic description of the state of a system (Chap. 2), we learned how to calculate the uncertainty in that state (Chap. 3). This provided us with a criterion for assigning to a system, only known through a small number of data, the least biased probability law. For an equilibrium situation we obtained in this way the Boltzmann-Gibbs distribution (Chap. 4) which, at least in principle, enables us to evaluate the expectation value of any microscopic physical quantity. Our task, however, is not yet finished as we must still investigate the relevance of our theoretical results to our actual macroscopic experience, which is reflected in the elaborate results of thermodynamics. Let us first of all note that on a macroscopic scale the properties of the various substances do not seem to have *any probabilistic character*; we must thus explain how our microscopic theory, in which statistics plays such a fundamental rôle, can account for the determinism of macroscopic phenomena (§ 5.5).

Moreover, it is essential for us to recognize the counterparts in our microscopic approach of the many concepts introduced in thermodynamics. This is easy for such macroscopic quantities as the momentum, centre-of-mass position, magnetic moment, or charge density, which can be identified as the expectation values of the corresponding observables, provided their statistical fluctuations are relatively small. Similarly, the number of moles is directly connected with the number of molecules. Also, it is natural to equate the internal energy with the expectation value of the Hamiltonian. A macroscopic electromagnetic field applied to a sample enters directly the Hamiltonian, at least in the unquantized approximation (see § 13.1). Nevertheless, the microscopic description of thermodynamic equilibrium by a probability law with the maximum uncertainty is already less obvious, and the arguments given in § 4.1 will be completely justified only in the present chapter. Still worse, thermodynamics is based upon many, more or less intuitive, concepts which cannot be readily formulated mathematically and whose nature is far from clear on a microscopic scale: temperature, pressure, work, heat, entropy, ... For the particular physical system considered in Chap.1 we were able to find the probabilistic meaning of the temperature and the entropy. Before undertaking the study of other systems in Chaps.7 to 13, we shall show in general how to *identify microscopically the thermodynamic quantities*. We shall thus be able to interpret the results obtained by applying statistical physics to some substance or other.

Doing that, we shall also *prove the Laws of thermodynamics*. The latter discipline, gradually developed during the nineteenth century, distilled from empirical data general principles governing the behaviour of all equilibrium systems. We shall recall them here in their traditional form, numbered from zero to three (§§ 5.1 to 5.4), and we shall show that they are not just irreducible "laws", as their name suggests, but just consequences of the microscopic laws of matter and of the probabilistic approach expounded in Chap.4. We shall return to this proof in § 6.1.3 and again find the principles of equilibrium thermodynamics in a more modern form. In that sense, statistical mechanics shows up as a more "fundamental" discipline than thermodynamics, as the latter appears to follow from the former.

The proof of the Laws of thermodynamics in their conventional form helps us to understand the microscopic significance of quantities connected with energy exchanges: temperature (§§ 5.1.2 and 5.3.2), entropy (§ 5.3.2), heat (§ 5.2.2), work (§ 5.2.3), and pressure (§ 5.6.5). Moreover, we stress the importance of the *chemical potentials*, quantities associated with exchanges of particles of a given kind (§§ 5.1.4 and 5.6.3). In fact, the concepts of chemical potential and of temperature will appear at the same level in all what follows in the present book. We also emphasize the relations between the partition functions introduced through the microscopic approach of Chap.4 and the various thermodynamic potentials (§ 5.6). The tables in § 5.6.6 summarize these relations; they will be useful for us when we apply the methods of equilibrium statistical physics to actual problems.

Finally, in § 5.7 we discuss finite size systems; they give rise to observable fluctuations which can be predicted by statistical mechanics.

5.1 The Zeroth Law

5.1.1 Relative Temperatures in Thermodynamics

The first concept introduced in thermodynamics is that of relative temperature. It is based upon a, seemingly trivial, property of thermal contact between two bodies. When such a contact has no effect, the two bodies are in thermal equilibrium one with the other. One then observes that, if two bodies are *separately in thermal equilibrium with a third body, they remain in thermal equilibrium with each other* if they are put in contact with one another. It took some time to realize the importance of this property for the logical structure of thermodynamics. For this reason, it was considered to be a preliminary law, as fundamental as the other ones. Mathematically, the Zeroth Law reflects the existence of an *equivalence relation* between systems in thermal equilibrium, which defines their common temperature. The possibility to index the temperature by comparison, using its correspondence with some physical quantity directly measurable on a thermometer, rests on this law.

5.1.2 Thermal Contact in Statistical Physics

From the microscopic view-point of statistical mechanics, if two systems a and b are *thermally isolated*, their Hamiltonian $\widehat{H}$ is the sum of two terms $\widehat{H}_{\rm a}$ and $\widehat{H}_{\rm b}$ which only depend on the dynamical variables of the systems in question. As $\widehat{H}_{\rm a}$ and $\widehat{H}_{\rm b}$ commute, there are two constants of the motion, the energies $U_{\rm a}$ and $U_{\rm b}$ of the two systems, respectively. Let us write down the conditions that a and b are separately in thermal equilibrium *in a canonical ensemble. For each system* we must introduce a Lagrangian multiplier β, and this results in

$$\begin{aligned} \widehat{D}_{\rm a} &= \frac{1}{Z_{\rm a}}\, {\rm e}^{-\beta_{\rm a}\widehat{H}_{\rm a}}, & \widehat{D}_{\rm b} &= \frac{1}{Z_{\rm b}}\, {\rm e}^{-\beta_{\rm b}\widehat{H}_{\rm b}}, \\ Z_{\rm a}(\beta_{\rm a}) &= {\rm Tr}_{\rm a}\, {\rm e}^{-\beta_{\rm a}\widehat{H}_{\rm a}}, & Z_{\rm b}(\beta_{\rm b}) &= {\rm Tr}_{\rm b}\, {\rm e}^{-\beta_{\rm b}\widehat{H}_{\rm b}}, \\ U_{\rm a} &= -\frac{\partial}{\partial\beta_{\rm a}} \ln Z_{\rm a}(\beta_{\rm a}), & U_{\rm b} &= -\frac{\partial}{\partial\beta_{\rm b}} \ln Z_{\rm b}(\beta_{\rm b}). \end{aligned} \tag{5.1}$$

The global density operator of the two systems is the tensor product

$$\widehat{D} = \widehat{D}_{\rm a} \otimes \widehat{D}_{\rm b} = \frac{1}{Z_{\rm a}(\beta_{\rm a}) Z_{\rm b}(\beta_{\rm b})}\, {\rm e}^{-\beta_{\rm a}\widehat{H}_{\rm a} - \beta_{\rm b}\widehat{H}_{\rm b}}, \tag{5.2}$$

where $\widehat{H}_{\rm a}$ stands for $\widehat{H}_{\rm a} \otimes \widehat{I}_{\rm b}$. There is clearly no correlation between the two systems a and b in (5.2).

Bringing the two systems a and b in thermal contact means on the microscopic scale that they are coupled through a weak interaction. (In § 5.2 we shall see the form of the coupling between two systems which can exchange work rather than heat.) We must thus include in the Hamiltonian,

$$\widehat{H} = \widehat{H}_a + \widehat{H}_b + \widehat{V},$$

an interaction term $\widehat{V}$ depending on observables of the two parts a and b. The interaction $\widehat{V}$ is so *weak* that we can neglect it as compared to $\widehat{H}_a$ and $\widehat{H}_b$. Nevertheless, its presence is sufficient to allow slow *energy exchanges* between a and b, so that there remains only *a single constant of the motion*, the total energy U. Similarly, in Chap.1 the interactions between magnetic moments were necessary in order that the total available energy be divided up among the moments in a completely random way; however, they were negligible in expression (1.2) for the energy levels and in the probability distribution (1.20) which follows from it. The canonical density operator is now expressed in terms of *only one* Lagrangian multiplier β which is related to the total energy:

$$\widehat{D} = \frac{1}{Z} \mathrm{e}^{-\beta \widehat{H}}, \tag{5.3}$$

and it is legitimate, neglecting $\widehat{V}$ as compared to $\widehat{H}_a + \widehat{H}_b$, to approximate it by

$$\widehat{D} = \frac{1}{Z_a(\beta) Z_b(\beta)} \mathrm{e}^{-\beta(\widehat{H}_a + \widehat{H}_b)}. \tag{5.3'}$$

Comparison with (5.2) shows that *thermal contact has the effect of making the Lagrangian multipliers β_a and β_b equal.* If we then again separate the two systems which were brought into thermal contact, that is, if we suppress the interaction $\widehat{V}$ between them, their density operator remains unchanged and equal to (5.3′), and we have for each of the isolated parts a and b the expressions (5.1), but now with $\beta_a = \beta_b$. Inversely, if we bring two systems such that $\beta_a = \beta_b$ into thermal contact, their density operator will remain practically unchanged, as (5.2) is then practically the same as (5.3).

The Zeroth Law of thermodynamics follows immediately and we see that *the Lagrangian multipliers β provide a relative temperature scale*, since thermal equilibrium between two systems is reflected in the equality of their associated β's.

At the same time we have solved the general problem of *energy partition*, that is: How does the total available energy $U_a + U_b$ divide up when we bring two systems in thermal contact? Before the contact, equations (5.1) determine β_a and β_b as functions of U_a and U_b; after thermal contact, it is sufficient to write that β_a and β_b transform to $\beta'_a = \beta'_b = \beta$ and the total energy $U_a + U_b$ remains unchanged, so as to determine the final values U'_a and U'_b.

Since for each of the systems the function $U(\beta)$ is a decreasing function, a property following from the convexity (4.31) of $\ln Z$ and the form (4.28) of the relation between U and β, the final, common, value of the "temperature" β lies between the initial values β_a and β_b. The *"colder"* system, that is, the one which takes energy from the other, is, though, the one with the *larger value of the parameter* β; the β scale of relative temperatures is thus graded in a sense opposite to that of the ordinary scales. This simple remark shows at the same time one of the forms of the Second Law of thermodynamics: it is always the system with the higher temperature which gives up heat to the other system when they are brought into thermal contact.

We have noted (§ 4.1.6) that the parts a and b may well correspond to independent degrees of freedom of the same system, rather than to subsystems which are separated in space. For instance, in a paramagnetic solid, the part a may correspond to the electron spin degrees of freedom and b to the vibrational degrees of freedom of the atoms in the crystal lattice. The spin-spin interactions happen to be large enough to establish thermal equilibrium between the spins in a time of the order of 10^{-7} to 10^{-8} s. The spin-lattice interactions, though, are weak so that global equilibrium of the spin plus lattice degrees of freedom is established much more slowly, after times of the order of 10^{-2} to 10^{-4} s. For intermediate times, the subsystems a and b are separately in equilibrium, and we can talk about a *spin temperature* and a *lattice temperature*. These two temperatures tend to become equal due to the spin-lattice coupling in a time of the order of 10^{-2} to 10^{-4} s, but magnetic resonance experiments performed sufficiently fast show up metastable situations of the type (5.1). Further on we shall meet with other examples of substances in which the smallness of some couplings allows, at least over not too long periods, the coexistence of two temperatures (for gases, see § 8.4.5 and for semiconductors, see § 15.2.2).

As expression (4.27) for Z_C shows, the parameter β varies between 0 and ∞, provided the energy spectrum is bounded from below but not from above; it varies between $-\infty$ and $+\infty$ in the rarer cases where the spectrum is bounded both from below and from above (spin temperature). In the second case, the tendency to equalize the relative temperatures β is reflected in a behaviour which looks pathological in terms of the absolute temperature $T = 1/k\beta$. In particular, *negative* absolute temperatures must be regarded as *higher* than ordinary, positive, absolute temperatures (Exerc.1a).

The above argument was developed for a canonical ensemble. We shall show in § 5.5.3 that for macroscopic systems the results are independent of which statistical ensemble we use. Better to understand the significance of the relative temperature and of the Zeroth Law, we shall, however, as we did in Chap.1, treat the problem of thermal contact once again, but in a *microcanonical ensemble.* Before coupling, each of the systems is, for given values of the energies U_a and U_b (within a margin of ΔU) represented by the density operator (4.39) which maximizes the corresponding statistical entropy S_a or S_b. The latter are given by (4.40) as functions of U_a or U_b. After coupling, the total energy remains unchanged, but now the total statistical entropy must be a maximum. Let us assume that each of the two systems is in microcanonical equilibrium, with energies U'_a and U'_b that we must determine. The total entropy $S_a(U'_a) + S_b(U'_b)$ must be a maximum, with $U'_a + U'_b = U_a + U_b$ given (within a margin of $2\Delta U$). If one now *defines* a parameter β_a associated to the system a by

$$\beta_a = \frac{1}{k}\frac{\partial}{\partial U_a} S_a(U_a) = \frac{\partial}{\partial U_a} \ln W_a, \tag{5.4}$$

the condition for a maximum statistical entropy can be written as

$$\frac{\partial}{\partial U'_a}\left[S_a(U'_a) + S_b(U_a + U_b - U'_a)\right] = k(\beta_a - \beta_b) = 0,$$

so that the energies U'_a and U'_b are still determined by requiring that the relative temperatures β_a and β_b are equal. However, the temperature β is now no longer defined microscopically as a Lagrangian multiplier as in the case of the canonical equilibrium, but as the derivative (5.4) of the logarithm of the microcanonical partition function with respect to the energy; the latter is now given exactly, and no longer on average.

In fact, even if we neglect the interaction $\widehat{V}$ between the systems a and b, neither of them is any longer in microcanonical equilibrium contrary to our above assumption, even though the ensemble a + b is in such an equilibrium. The energy U_a can, at the expense of U_b, show fluctuations which exceed ΔU and which we have just neglected. Nevertheless, for macroscopic systems of the kind considered in thermodynamics, the calculations of § 1.2.4, which can easily be extended to the general case, show that the relative values of these fluctuations are negligible; this completely justifies the above results. At any rate, when we find in § 5.5.3 that in the limit of large systems all ensembles are equivalent, we shall identify the Lagrangian multiplier β_a and the average energy U_a of the canonical ensemble with the derivative variable β_a and the exact energy U_a of the microcanonical ensemble. On the contrary, in the case of finite systems, the Zeroth Law, which remains rigorous in the canonical ensemble, is not rigorous in the microcanonical ensemble where the shape of the probability distribution plays a rôle. The definition (5.4) of the microcanonical temperature then depends, in particular, on the choice of ΔU; together with the microcanonical entropy, it loses its meaning if ΔU tends to zero for fixed volume, due to the discrete nature of the spectrum. We cannot therefore apply without ambiguity the thermodynamic concepts to *small* systems unless we work with *canonical*, or grand canonical, ensembles.

5.1.3 Thermometers and Thermostats

When we bring two systems with different relative temperatures β_a and β_b into thermal contact, their density operator changes from (5.2) to (5.3), and we find the final temperature β by requiring that the sum of the energies U'_a and U'_b, given by (5.1) in terms of $\beta'_a = \beta'_b = \beta$, equals the sum of the initial energies U_a and U_b. If the system b is very small compared to a, the variation in the energy $U'_a - U_a = U_b - U'_b$ of the system a has the same order of magnitude as the energies of b, that is, it has a very small relative magnitude. As a result, the change in the temperature $\beta - \beta_{\rm a}$ of the larger system a is negligible: the state of the system a is *practically unchanged* by its interaction with b. On the other hand, the system b changes by *adjusting its temperature* to that of a.

A *thermometer* is thus a system b for which one observes some property, such as the volume, the resistivity, ..., depending on the energy $U_{\rm b}$, and hence on the common temperature $\beta_{\rm b} = \beta_{\rm a}$; the temperature $\beta_{\rm a}$ is therefore characterized by that property. The system b must be small compared to the system a whose temperature one wants to measure, but sufficiently large that the relative statistical fluctuations of the observed quantity are small (§ 5.7.3). On the other hand, a *thermostat* is a large system a which serves as a *heat source*; the temperatures of much smaller systems b which are brought into thermal contact with a adjust themselves to the practically constant temperature of a.

5.1.4 Extensions: Open Systems

The Zeroth Law of traditional thermodynamics only deals with thermal equilibrium between two systems which can exchange heat. Neither of these systems is *isolated*, but each one is *closed*, that is, it does not exchange matter with the outside. The extension of thermodynamics to *open* systems is essential in all those cases where a weak coupling enables not only heat exchanges, but also *particle exchanges*. There are numerous examples: adsorption (Exerc.4b), chemical equilibrium (§§ 6.6.3 and 8.2.2), phase equilibrium (§§ 6.4.6 and 9.3.3), osmosis (§ 6.6.2), electrostatic equilibrium (§ 11.3.3). The arguments of § 5.1.2, when extended to the grand canonical ensemble (§ 4.3.2), then show that the Lagrangian multiplier α associated with N plays, apart from its sign, exactly the same rôle with regards to particle exchanges as the temperature β with regards to heat exchanges. When two systems a and b can exchange particles (and energy) *their parameters* α (and β) *become equal.* A large system, which plays the rôle of a *particle reservoir, imposes its value of* α on systems which are brought into contact with it; this determines the number of particles which flow from one to the other. The variable β, the conjugate of the energy, characterizes the *tendency of a system to absorb heat*; the variable α, the conjugate of N, characterizes the *tendency to lose particles.* This concept is just as important as the temperature and can be used in the same way. Of course, one must introduce for each kind of particle which can be exchanged a different parameter α.

In the microcanonical ensemble α is defined as $-k^{-1}\partial S/\partial N$, and in the canonical ensemble as $-\partial \ln Z_C/\partial N$, expressions which are the analogues of (5.4). Here again, the system must be a macroscopic one in order that N can be treated as a continuous variable and that the equilibrium between two systems implies the equality of the variables α.

These ideas can be extended to *any other conservative quantity* A_i *which can be exchanged* between two systems, such as momentum, angular momentum (Exerc.7b), or, for two fluids separated by a mobile piston, volume. In all these cases, exchanges leading to equilibrium are determined by expressing that the final state is the most disordered state possible, as measured by the statistical entropy. This introduces a quantity λ_i which is the conjugate of A_i and which takes the same value in the two systems between which the exchange of A_i may take place. As in the case of β or α, we can microscopically introduce the variable λ_i in two different ways which are equivalent for a large system. If the statistical ensemble is characterized by giving the average $\langle A_i \rangle$, the corresponding Lagrangian multiplier λ_i is the conjugate variable. If, however, the statistical ensemble is characterized by an exactly given A_i, the variable λ_i which is the conjugate of A_i is the derivative with respect to A_i of the logarithm of the partition function of the ensemble considered – like (5.4) for β and the energy.

5.2 The First Law

After we have identified the macroscopic scales of relative temperatures with decreasing functions of β, we now turn to the microscopic interpretation of the concepts of internal energy, heat, work, and pressure.

5.2.1 Internal Energy

If a system which can exchange work and heat with the outside undergoes a change from an initial state 1 to a final state 2, the First Law of thermodynamics expresses that the sum of the work $\mathcal{W}$ and heat $\mathcal{Q}$ received equals the change in a *function U of the state of the system*, the internal energy:

$$U_2 - U_1 = \mathcal{W} + \mathcal{Q}. \tag{5.5}$$

The history of the discovery of the First Law and of the concept of the internal energy shows how difficult that was. It required bringing together ideas from many different fields. This can be seen from the interests of the dozen or so scientists who, more or less at the same time, between 1837 and 1847, and independently, established the equivalence between work and heat or who enounced the principle of the conservation of energy. Let us mention: Marc Seguin (Annonay 1786–1875), Gustave Hirn (Colmar 1815–1890), and William Rankine (Edinburgh 1820–Glasgow 1872), engineers who perfected the steam engine, Julius Robert Mayer (Heilbronn, Wurttemberg 1814–1878), a naval doctor, interested in metabolism, Justus von Liebig (Darmstadt 1803–Munich 1873), a chemist and physiologist, William Grove (Swansea 1811–London 1896), an electrochemist, Hermann von Helmholtz (Potsdam 1821–Berlin 1894), a young doctor starting in that period his work in physics, James Prescott Joule (Manchester 1818–Sale 1889), a rich brewer who had become a physicist and had already studied the thermal effect of electric current, Ludvig Colding (Holbaek 1815–Copenhagen 1888), motivated by metaphysical ideas. The equivalence between the two forms, kinetic and potential, of mechanical energy had been established during the eighteenth century and, starting from Lavoisier, the study of heat had become quantitative. At that time it used to be classified under the name "caloric", as one of the four "imponderable fluids", together with light, electricity, and magnetism. Between 1798 and 1804, Benjamin Thompson, an American officer, who had become minister of war in Bavaria under the name of Count Rumford, measured the production of heat starting from mechanical energy in the drilling of cannon, a first step towards the equivalence of heat and work. Nevertheless, the idea of *caloric* as a *conserved fluid* was still wide-spread. One can understand how in those circumstances Carnot enounced the Second Law in 1824 *before* the discovery of the First Law (§3.4.1): more or less consciously the flow of caloric from a hot source to a cold one was commonly compared with that of water producing work in the waterwheel of a mill. A few years later, Carnot realized that the work provided by a thermal engine was, in fact, due to the transformation of part of the heat transferred (§ 3.4.1). However, he died during the 1832 cholera epidemic and most of his furniture and papers were destroyed as a hygienic measure – even though at the present time one wonders whether he died from that illness. Some of his writings, recovered later on, were published, but only half a century later. One therefore had to wait for the important experimental and theoretical work of Mayer (1842), Joule (1843), Colding (1843), and Helmholtz (1847), before science was enriched both by the idea of the *quantitative equivalence of work and heat* and by the concept of *internal energy*. Very soon the latter was extended to include chemical and electric energies, even though a direct proof of the equivalence of the reaction heat and the electrical energy in electrochemistry was difficult to give. At the end of the nineteenth century the idea of energy as an uncreated and indestructible substance even gave birth to a philosophy tainted with mysticism: energetics. Its adherents, in particular the physical chemist Ostwald, were amongst the obstinate opponents of kinetic theory and of Boltzmann, for whom the ultimate reality in Nature lay in material particles, rather than in energy.

From the point of view of statistical mechanics, the principle of the *conservation of energy* is a simple consequence of the dynamical laws. The internal energy U is first identified with the mean value of the Hamiltonian $\widehat{H}$ in the state $\widehat{D}$:

$$\boxed{U = \mathrm{Tr}(\widehat{D}\widehat{H})} \quad . \tag{5.6}$$

For an *isolated system* (§ 2.2.6) the change with time of the internal energy can then be derived from Ehrenfest's equation (2.29) which gives

$$\frac{dU}{dt} = \frac{1}{i\hbar}\left\langle\left[\widehat{H},\widehat{H}\right]\right\rangle = 0, \tag{5.7}$$

so that the internal energy U is a constant. If we consider the system under study together with the one with which it exchanges work and heat as a single, isolated system, conservation of the total macroscopic energy of this composite system thus immediately follows from that of the microscopic energy (5.6).

5.2.2 Heat

Let us now consider a *non-isolated system* which may exchange energy with the outside. One can on the macroscopic scale distinguish the various forms of energy exchange, heat, on the one hand, and (mechanical, electrical, ...) work, on the other hand; this distinction is of considerable practical importance. At the moment, however, we have on the microscopic scale only available the internal energy concept (5.6). In order to interpret the concepts of heat and work in our microscopic and statistical formalism, we must analyse the energy exchanges of a system with the outside and discriminate between the reasons which may change its internal energy (5.6).

We have seen (§ 5.1.2) that heat exchange with the outside is due to a coupling potential $\widehat{V}$ which is not well defined and which is small compared to the Hamiltonian $\widehat{H}$ of the system itself. Because of the presence of this potential the evolution of the density operator $\widehat{D}$ of the system is coupled with that of the outside, and $\widehat{D}$ varies by $d\widehat{D}$ during a time dt; the Liouville-von Neumann equation $i\hbar d\widehat{D}/dt = [\widehat{H}, \widehat{D}]$, valid when the system was isolated, is no longer suitable for describing such changes in $\widehat{D}$. Hence, the internal energy (5.6) can now change under the influence of the thermal contact, by an amount dU equal to

$$\boxed{\delta Q = \mathrm{Tr}(d\widehat{D}\ \widehat{H})} \tag{5.8}$$

during a time dt. One can interpret this change as a heat exchange with the outside.

For instance, in § 5.1.2 the system a changes from an equilibrium state (5.1) with temperature β_a to the state (5.3') with temperature β, thus gaining an amount of heat,

$$\begin{aligned} Q &= \mathrm{Tr}\left[\frac{1}{Z_a(\beta)}\ e^{-\beta\widehat{H}_a}\ \widehat{H}_a\right] - \mathrm{Tr}\left[\frac{1}{Z_a(\beta_a)}\ e^{-\beta_a\widehat{H}_a}\ \widehat{H}_a\right] \\ &= -\frac{\partial}{\partial\beta}\ \ln\ Z_a(\beta) + \frac{\partial}{\partial\beta_a}\ \ln\ Z_a(\beta_a). \end{aligned} \tag{5.9}$$

The gain of heat by the system b in the same transformation is clearly the opposite of (5.9), because of the conservation of the total energy $U = U_a + U_b \simeq \langle \widehat{H}_a + \widehat{H}_b \rangle$.

If, as in (5.9), the density operator $\widehat{D}$ commutes with the Hamiltonian $\widehat{H}$, it is characterized by the probabilities p_m for the eigenstates $|m\rangle$ of $\widehat{H}$ with energies E_m:

$$\widehat{D} = \sum_m |m\rangle\, p_m\, \langle m|.$$

The internal energy (5.6) is then equal to $U = \sum_m p_m E_m$, so that the gain of heat,

$$\delta \mathcal{Q} = \sum_m dp_m \, E_m, \tag{5.10}$$

corresponds to a *redistribution of the probabilities* of the various states $|m\rangle$: the system *heats up when the relative probabilities of the states with the higher energies increase.* The concept of heat is thus intimately associated with the probabilistic nature of the macro-states.

5.2.3 Work and Forces

In the preceding subsection we have given a microscopic interpretation to the concept of heat received by a system. In order to interpret the exchanges of work we must now consider couplings which, in contrast to coupling through thermal contact, are not infinitesimal. As a first approach we assume that the system studied does not exchange heat with the outside, but that it is in contact with external sources of work which impose on some macroscopic variables ξ_α a given time dependence. This situation can be modelled by assigning to the system a Hamiltonian $\widehat{H}(t)$ which is *time-dependent* through parameters ξ_α changing in a *controlled way* (§ 2.1.5). If, for instance, the system is the fluid of a thermal engine enclosed in a cylinder, its Hamiltonian depends on the position of the piston or, what amounts to the same, on the available volume Ω which is there the variable ξ_α; the displacement of the piston, assumed to be given, is associated with exchange of work at the macroscopic scale and is microscopically reflected in the time-dependence of $\widehat{H}$ through Ω. Similarly, the work done by an external (electric, magnetic, or gravitational) field is associated with parameters ξ_α in the Hamiltonian that characterize this field. Generally speaking, macroscopic physics defines the *work received* by the system as

$$\delta \mathcal{W} = \sum_\alpha X_\alpha \, d\xi_\alpha, \tag{5.11}$$

for an infinitesimal change $d\xi_\alpha$ in the *"position" variables* ξ_α. The coefficients X_α play the rôle of the associated *"forces"*. If $d\xi_\alpha$ denotes the displacement of the piston, X_α is the force applied to the piston; if $d\xi_\alpha = d\Omega$ is the expansion, $-X_\alpha = \mathcal{P}$ is the applied pressure. If ξ_α is an induction field $\boldsymbol{B}$, in which the system has been placed, the outside source is the set of coils producing the field, and the work received has again the form (5.11), as we saw in (1.31); nevertheless, here the magnetic moment $\boldsymbol{M} = -X_\alpha$ plays the rôle of the "force" while the "displacement" $d\xi_\alpha$ is the change $d\boldsymbol{B}$ of the field. As we shall see many times, the traditional nomenclature of thermodynamics is not always suitable, and we must be careful to avoid confusions.

Returning to the microscopic scale, we want to study the effect of a change $d\xi_\alpha$ in the parameters ξ_α which takes place during the time dt. The Hamiltonian is changed from $\widehat{H} = \widehat{H}(\{\xi_\alpha\})$ to

$$\widehat{H} + d\widehat{H} = \widehat{H}(\{\xi_\alpha + d\xi_\alpha\}) = \widehat{H} + \sum_\alpha \widehat{X}_\alpha \, d\xi_\alpha. \tag{5.12}$$

The operators

$$\widehat{X}_\alpha \equiv \frac{\partial \widehat{H}}{\partial \xi_\alpha}, \tag{5.13}$$

which we shall interpret as the *"force" observables* of the system, conjugate to the "position" variables ξ_α, characterize the *way the system reacts to the coupling* with the outside. If, for instance, ξ_α is the field of gravity $\boldsymbol{g}$, which occurs in $\widehat{H}$ in the form $mg \sum_i \widehat{z}_i$, where $\widehat{z}_i$ is the height of particle i, the conjugate observable $\widehat{X}_\alpha$, coupled to the field strength, is $\widehat{X}_\alpha = m \sum_i \widehat{z}_i$ which is proportional to the height of the centre of mass.[1] Similarly, an external electric potential at a point $\boldsymbol{r}$ is coupled to the charge density observable at that point. In the case of a uniform magnetic induction $\boldsymbol{B}$, the observable $-\widehat{X}_\alpha$ is the magnetic moment of the system in the direction of $\boldsymbol{B}$. Finally, if ξ_α is an ordinary position variable, such as that of a piston, $\widehat{X}_\alpha$ is the observable describing the ordinary force applied by the piston on the system.

The slow change (5.12) in $\widehat{H}$ gives rise to a change in the energy of the system, which we shall interpret as an *exchange of work* with the outside *in an adiabatic transformation.* Let us, first of all, note that while the Hamiltonian changes the *statistical entropy remains constant* since (3.29) is valid even when $\widehat{H}$ is time-dependent; this is in agreement with identifying $S(\widehat{D})$ with the thermodynamic entropy. To evaluate the change in the internal energy we use Ehrenfest's equation (2.29) extended to statistical mixtures (§2.2.6). The observable $\widehat{H}$, whose expectation value we are studying, depends explicitly on time through the parameters ξ_α so that the second term in (2.29) makes a contribution; the first term vanishes, as in (5.7). Hence we find, according to (5.12),

$$\frac{dU}{dt} = \frac{d}{dt}\langle \widehat{H} \rangle = \sum_\alpha \left\langle \frac{\partial \widehat{H}}{\partial \xi_\alpha} \right\rangle \frac{d\xi_\alpha}{dt},$$

and the infinitesimal increase in U is

$$\boxed{\delta W = \sum_\alpha \langle \widehat{X}_\alpha \rangle \, d\xi_\alpha} \,. \tag{5.14}$$

This equation enables us to interpret the change in the microscopic energy for the transformation studied as the macroscopic work received (5.11), and to identify the *mean values* of the observables (5.13) in the state $\widehat{D}$,

[1] It might shock people to find $m\langle \sum \widehat{z}_i \rangle dg$ as the work done by the field of gravity. We are, in fact, dealing here with the energy received by the system if the gravitational acceleration g were to change. More generally, the definition of work gives rise to subtleties when there are long-range forces or fields which one either may or may not include when defining the system studied. We shall discuss this point in the context of electromagnetism (§ 6.6.5).

$$\langle \widehat{X}_\alpha \rangle = \left\langle \frac{\partial \widehat{H}}{\partial \xi_\alpha} \right\rangle = X_\alpha, \tag{5.15}$$

with the *"forces"* occurring in expression (5.11) for this work.

The word *"adiabatic"* has here been used in its macroscopic thermodynamic meaning, a reversible transformation without exchange of heat with the outside. In quantum mechanics, one uses the same word to denote the limit where the Hamiltonian depends very weakly on time, that is, where the parameters ξ_α vary very slowly. The *"adiabatic theorem"* which is valid in this limit states the following. Let $|m(t)\rangle$ and $E_m(t)$ be the eigenstates and eigenvalues of $\widehat{H}(t)$ at time t. Assume that at the initial time t_0 the system is in the state $|m(t_0)\rangle$, and let us find out how this state changes according to the Schrödinger equation (2.24). One can show that the solution of this equation tends, apart from a time-dependent phase factor, to $|m(t)\rangle$ in the adiabatic limit as $d\xi_\alpha/dt \to 0$; more precisely, the condition for the validity of the adiabatic approximation $|\psi(t)\rangle \propto |m(t)\rangle$ is

$$\left| \left\langle m' \left| \frac{\partial \widehat{H}}{\partial t} \right| m \right\rangle \right| \ll \frac{\left(E_m - E_{m'}\right)^2}{\hbar}, \tag{5.16}$$

at all times and for all other energy levels m'. It is tempting to adapt this result to statistical physics by combining the two meanings of the word adiabatic. If the macro-state at time t is

$$\widehat{D}(t) = \sum_m |m(t)\rangle\, p_m\, \langle m(t)|, \tag{5.17}$$

and if the adiabatic theorem is valid, we have at time $t + dt$

$$\widehat{D}(t+dt) = \sum_m |m(t+dt)\rangle\, p_m\, \langle m(t+dt)|, \tag{5.18}$$

where the probabilities p_m remain constant in time. To calculate the change in the internal energy,

$$U(t) = \mathrm{Tr}\, \widehat{D}(t)\, \widehat{H}(t) = \sum_m p_m\, E_m(t), \tag{5.19}$$

we use

$$\begin{aligned} dE_m &= \langle m + dm|\widehat{H} + d\widehat{H}|m + dm\rangle - \langle m|\widehat{H}|m\rangle \\ &= \langle m|d\widehat{H}|m\rangle - E_m d(\langle m|m\rangle) = \langle m|d\widehat{H}|m\rangle, \end{aligned}$$

whence

$$dU = \sum_m p_m\, dE_m = \mathrm{Tr}\, \widehat{D}\, d\widehat{H}. \tag{5.20}$$

We thus recover expression (5.14) for the work received, which here appears associated with the *shift in the energy levels* of the system resulting from the change in the external parameters.

Although this reasoning is common and although its final consequences (5.19) and (5.20) are correct, it is fallacious and rests upon a semantic confusion. The adiabatic theorem does not allow us to derive (5.18) from (5.17) unless the condition (5.16) is satisfied for all pairs m, m' of levels. However, even if there are no accidental degeneracies or quasi-degeneracies, the energy levels of a macroscopic system, even a small one, lie *extraordinarily densely.* A typical numerical estimate was given in § 1.2.3. The right-hand side of (5.16) is extremely small, and the characteristic time which follows from this for the quantum adiabaticity is enormously huge – much, much longer than the age of the universe! On the other hand, the characteristic times for thermodynamic adiabaticity are, although long compared to the times for microscopic evolution, on our scale. Expression (5.18) is therefore not valid for a macroscopic system.

Moreover, an adiabatic evolution in the thermodynamic sense must at all times pass through macro-states which lie very closely to thermal equilibrium. If, therefore, in (5.17) p_m is proportional to $\exp[-\beta E_m(t)]$, the canonical equilibrium expression, we expect the eigenvalues of $\widehat{D}(t + dt)$ to be proportional to $\exp[-\beta' E_m(t + dt)]$ with a temperature $\beta' = \beta + d\beta$ which may have changed. However, expression (5.18) gives as eigenvalues the p_m, which, in general, are not proportional to an expression like $\exp[-\beta'(E_m + dE_m)]$ to first order in dt. In order that this would occur, the spectrum $E_m(t)$ should undergo only a simple dilatation and translation as t changes. In the case where levels cross during the evolution of $\widehat{H}(t)$ there might even occur a population inversion, with a lower level having a smaller probability than a higher level. Retaining at all times a canonical equilibrium therefore makes it necessary that the probabilities for the micro-states $|m(t)\rangle$ do not remain constant, as in (5.18), but change from $p_m \propto \exp(-\beta E_m)$ to $p'_m \propto \exp[-\beta'(E_m + dE_m)]$. The macro-state resulting from a thermodynamic *adiabatic transformation differs* thus in the case of canonical equilibria from the one resulting from the *adiabatic theorem.* The situation looks better for microcanonical equilibria, where $p_m = 1/W = p'_m$ remains constant, so that the adiabatic theorem associates the equiprobability in the range ΔE at time t with the equiprobability for the resulting levels at time $t + dt$; however, for a large system one may expect crossing, or quasi-crossing, of levels, which prevents either the adiabatic theorem to hold or the state (5.18) to be microcanonical.

Anyhow, for a study of thermodynamic adiabatic transformations, one still should prove that an evolution of $\widehat{H}(t)$, which is slow on the microscopic scale but *violates condition* (5.16), will lead $\widehat{D}(t)$ approximately through *equilibrium states.* Assuming this result, one can prove as an exercise, using the properties (5.14) or $dS = 0$ found above, that the change in temperature $d\beta = \beta' - \beta$ is for a canonical ensemble given by

$$\frac{d\beta}{\beta} = \frac{\langle \widehat{H} d\widehat{H} \rangle - \langle \widehat{H} \rangle \langle d\widehat{H} \rangle}{\langle \widehat{H}^2 \rangle - \langle \widehat{H} \rangle^2}. \tag{5.21}$$

5.2.4 Exchanges Between Systems and Sources

We have assumed above that a source of work could be represented by a time-dependent Hamiltonian. Let us return to the microscopic study of the work concept, using a more fundamental method similar to the one of § 5.1.2. We shall thus treat the ensemble formed by *the system of interest and the sources* with which it is interacting as a *single system.* This will enable us (i) to take heat and work exchanges into account at the same time; (ii) to replace the parameters ξ_α whose time-dependence was assumed to be known by *dynamical variables* governed by their own equations of motion; (iii) to show explicitly the microscopic conditions for the validity of the work concept; (iv) to consider the possibility of feedback of the system on the work source and on the dynamics of the ξ_α variables. This last possibility must necessarily be taken into account when an exchange of work takes place between two systems of comparable size rather than between a system and a much larger source which imposes on the ξ_α their changes.

Let us assume that the system a can exchange heat with the system b and work with the system c, without coupling between b and c. The Hamiltonian $\widehat{H}$ of the system a+b+c can be written as

$$\widehat{H} = \widehat{H}_{\mathrm{a}} + \widehat{H}_{\mathrm{b}} + \widehat{V} + \widehat{H}_{\mathrm{c}}, \tag{5.22}$$

where $\widehat{H}_{\mathrm{b}}$ and $\widehat{H}_{\mathrm{c}}$ contain only variables referring to b and c, respectively, where $\widehat{V}$ is a small term which describes the thermal coupling between a and b, and where the interaction responsible for the work done by c on a is included in $\widehat{H}_{\mathrm{a}}$. This last operator contains therefore not only observables pertaining to a, but also some *observables* $\widehat{\xi}_\alpha$ *of the system* c. The latter commute with one another and with the operators of the system a; this enables us to identify $\widehat{H}_{\mathrm{a}}$ with the Hamiltonian $\widehat{H}$ of § 5.2.3 and to define, as in (5.13), operators $\widehat{X}_\alpha$ which can, however, still involve the operators $\widehat{\xi}_\alpha$ of the system c. Let us note that including into $\widehat{H}_{\mathrm{a}}$ the term which describes the coupling between a and c introduces between these systems an *asymmetry* on which the definition of work itself depends (§ 6.6.5).

The internal energies of a, of b, and of c are, respectively, the expectation values of $\widehat{H}_{\mathrm{a}}$, $\widehat{H}_{\mathrm{b}}$, and $\widehat{H}_{\mathrm{c}}$ and $\langle H_{\mathrm{a}}\rangle$ includes the energy of the coupling between a and c. We can again write down their changes using Ehrenfest's equation (2.29). The heat given up by b to a during the time interval dt is thus found in the form

$$\delta Q = -dU_{\mathrm{b}} = -\frac{dt}{i\hbar}\left\langle \left[\widehat{H}_{\mathrm{b}}, \widehat{H}\right]\right\rangle = -\frac{dt}{i\hbar}\left\langle \left[\widehat{H}_{\mathrm{b}}, \widehat{V}\right]\right\rangle, \tag{5.23}$$

where we have used (5.22). Similarly, the work done by c on a equals

$$\delta W = -dU_{\mathrm{c}} = -\frac{dt}{i\hbar}\left\langle \left[\widehat{H}_{\mathrm{c}}, \widehat{H}\right]\right\rangle = -\frac{dt}{i\hbar}\left\langle \left[\widehat{H}_{\mathrm{c}}, \widehat{H}_{\mathrm{a}}\right]\right\rangle. \tag{5.24}$$

The thermal coupling $\widehat{V}$ is sufficiently weak that its average value can be neglected when compared with U_{a}, U_{b}, and U_{c}. The total system a+b+c being isolated, we have

$$dU_{\rm a} = \delta Q + \delta \mathcal{W} = \frac{dt}{i\hbar}\left\langle\left[\widehat{H}_{\rm a}, \widehat{H}\right]\right\rangle = \frac{dt}{i\hbar}\left\langle\left[\widehat{H}_{\rm a}, \widehat{V} + \widehat{H}_{\rm c}\right]\right\rangle. \tag{5.25}$$

We must still rewrite expression (5.24) so that we can recognize it as the macroscopic work (5.11). The non-commutativity of $\widehat{H}_{\rm a}$ and $\widehat{H}_{\rm c}$ comes solely from the presence of the operators $\widehat{\xi}_\alpha$ in $\widehat{H}_{\rm a}$. Let us, to begin with, assume that this is dependence is *linear*. In that case we have

$$\begin{aligned} \delta \mathcal{W} &= -\frac{dt}{i\hbar}\left\langle\left[\widehat{H}_{\rm c}, \widehat{H}_{\rm a}\right]\right\rangle = \frac{dt}{i\hbar}\sum_\alpha \left\langle \widehat{X}_\alpha \left[\widehat{\xi}_\alpha, \widehat{H}_{\rm c}\right]\right\rangle \\ &= \left\langle \widehat{X}_\alpha \widehat{\frac{d\xi_\alpha}{dt}}\right\rangle dt, \end{aligned} \tag{5.26}$$

where

$$\widehat{\frac{d\xi_\alpha}{dt}} = \frac{1}{i\hbar}\left[\widehat{\xi}_\alpha, \widehat{H}_{\rm c}\right] = \frac{1}{i\hbar}\left[\widehat{\xi}_\alpha, \widehat{H}\right] \tag{5.27}$$

is the velocity operator of $\widehat{\xi}_\alpha$ in the Heisenberg picture (§ 2.1.5). If we *neglect the correlations between the "force" and the "displacement"*, we get from (5.26)

$$\delta \mathcal{W} = \sum_\alpha \left\langle \widehat{X}_\alpha\right\rangle d\left\langle\widehat{\xi}_\alpha\right\rangle, \tag{5.28}$$

which enables us to identify $\left\langle \widehat{X}_\alpha\right\rangle = X_\alpha$ and $\left\langle\widehat{\xi}_\alpha\right\rangle = \xi_\alpha$ with the macroscopic "force" and "position" in (5.11), respectively.

We obtain the same result under more general conditions by assuming that the *statistical fluctuations in* $\widehat{\xi}_\alpha$ *are negligible*, that is, that the quantities ξ_α *behave classically*. In fact, in this case $\widehat{X}_\alpha$ practically commutes with $\left[\widehat{\xi}_\alpha, \widehat{H}_{\rm c}\right]$, although it may contain the operators $\widehat{\xi}_\alpha$ which do not commute with $\widehat{H}_{\rm c}$. Equation (5.26) remains valid, and (5.27) is, according to (2.78), equal to the Poisson bracket $\{\xi_\alpha, H\}$ which determines the classical velocity of ξ_α. The absence of correlations, which we used to get to (5.28), is implied by the absence of fluctuations in ξ_α, and hence in $d\xi_\alpha$. The dynamics of the ξ_α, assumed in § 5.2.3 to be given, are now provided by the equations of motion (5.27).

The heat received δQ is, according to (5.23), (5.24), (5.25), given by the two equivalent expressions:

$$\delta Q = -\frac{dt}{i\hbar}\left\langle\left[\widehat{H}_{\rm b}, \widehat{V}\right]\right\rangle = \frac{dt}{i\hbar}\left\langle\left[\widehat{H}_{\rm a}, \widehat{V}\right]\right\rangle. \tag{5.29}$$

We can also, by eliminating the systems b and c, see that the work (5.26) is nothing but $\left\langle d\widehat{H}_{\rm a}\right\rangle$, where the operator $d\widehat{H}_{\rm a}$ is calculated through the change in the variables ξ_α, as in (5.20). By subtraction we thus recover again in the general case expression (5.8) for the heat received by the system a.

From the above we shall remember that when a closed system can exchange work and heat with the outside, the change in its internal energy (5.6) during an infinitesimal transformation consists of two parts:

$$dU = \mathrm{Tr}(\widehat{D}\, d\widehat{H}) + \mathrm{Tr}(d\widehat{D}\, \widehat{H}). \tag{5.30}$$

The first term is identified with the mechanical, electric, magnetic, ... *work* received during the transformation and can, according to (5.20) or (5.14), be expressed in terms of the shift in the energy levels of the system or of the change in the coupling parameters ξ_α. The second term vanishes when there is no infinitesimal coupling $\widehat{V}$ with the outside, as then $i\hbar\, d\widehat{D} = [\widehat{H}, \widehat{D}]dt$. It is the *heat* received, (5.8) or (5.10), associated with the *change in the state of disorder* of the system.

The microscopic expression (5.30) for the change in the internal energy exhibits the *reversibility* of the work exchanges, established by macroscopic experiments. In fact, the work $\delta\mathcal{W} = \mathrm{Tr}(\widehat{D}\, d\widehat{H})$ received during a given transformation $d\xi_\alpha$ is the opposite of that associated with the inverse transformation $-d\xi_\alpha$. In contrast, there is no reason why the changes $d\widehat{D}$ should change sign with $d\xi_\alpha$, since they may occur even when the ξ_α remain fixed: the heat exchanges $\delta\mathcal{Q} = \mathrm{Tr}(d\widehat{D}\, \widehat{H})$ *cannot be controlled* like the work exchanges by letting the parameters change. This fact is now found at the microscopic level and we shall make it more rigorous by proving the Second Law.

5.3 The Second Law

5.3.1 Energy Downgrading in Thermodynamics

The Second Law of thermodynamics can be stated in various ways which all reflect the *irreversibility of the evolution of macroscopic systems.* In § 5.1.2 we recalled Clausius's statement (1854): spontaneous exchanges of heat between two systems can only take place in one direction, *from hot to cold*; mathematically this makes temperatures have an ordered structure. In another form the Second Law establishes a fundamental distinction between work and heat: work, whether mechanical, electric, magnetic, chemical, ..., can always be downgraded into heat whereas it is impossible to construct a perpetual motion of the second kind which would produce work *in a closed cycle* taking heat from a *single source* at a uniform temperature (Kelvin 1854). Similarly, we know Carnot's statement (1824) which allows us to define the thermodynamic absolute temperature starting from the *maximum yield of thermal engines* functioning with two heat sources at different temperatures, and which has many important technical applications. Carathéodory (1909) gave a form which led to interesting mathematical developments: there exist in the neighbourhood of any initial state other states which *cannot be reached* by an adiabatic transformation. In § 3.4.1 we saw how the Second Law of thermodynamics led to the introduction of the entropy concept, in a form which was not yet statistical.

All these statements are, as one can show through macroscopic considerations, equivalent to one another and to Clausius's statement (1865) which expresses the Second Law in *analytic form* as follows. If a system is in thermodynamic *equilibrium*, one can assign to it two quantities, the *absolute temperature* T, which is a special case of the relative temperatures defined by the Zeroth Law, and the *entropy* S, which is a function of the variables characterizing the state of the system. In a *quasi-static*, but not necessarily reversible transformation, during which the system at all times passes through equilibrium states while exchanging heat and work with the outside, one has

$$\boxed{dS = \frac{\delta Q}{T}} \quad . \tag{5.31}$$

In mathematical terms the heat δQ received during the time interval dt, which is not an exact total differential, has T as an integrating factor. Any other transformation leading from an equilibrium state 1 to another equilibrium state 2 is *irreversible*, and the change in entropy satisfies the *inequality*

$$S_2 - S_1 > \sum_j \int \frac{\delta Q_j}{T_j^{(\mathrm{S})}}, \tag{5.32}$$

where $T_j^{(\mathrm{S})}$ is the temperature of the source j providing the amount of heat δQ_j. The system, which passes through non-equilibrium states, may not have a well defined temperature, in contrast to the sources.

We note that the Second Law defines the absolute temperature and the thermodynamic entropy up to a possible multiplicative constant depending on units; moreover, it defines the thermodynamic entropy up to an additive constant. When a system consists of several independent subsystems, the addition of the equalities (5.31) shows that the *thermodynamic entropy is additive.*

5.3.2 Entropy and Absolute Temperature

Generalizing the results of § 1.3.3 we shall now recover the Second Law as a *consequence of statistical mechanics.* Let us first of all consider a *quasi-static transformation* from the microscopic point of view. At any time the system, in canonical thermal equilibrium, is described by a Boltzmann-Gibbs distribution, but the exchanges of work and heat with the outside are reflected in changes with time in β and in $\widehat{H}$ through the parameters ξ_α:

$$\widehat{D} = \frac{1}{Z_{\mathrm{C}}(t)}\, \mathrm{e}^{-\beta(t)\widehat{H}(t)}, \qquad Z_{\mathrm{C}}(t) = \mathrm{Tr}\, \mathrm{e}^{-\beta(t)\widehat{H}(t)}. \tag{5.33}$$

We note that the time-dependence of $\widehat{D}$ does not follow Eq.(2.49) which would be valid for an isolated system with a Hamiltonian $\widehat{H}(t)$: the system is, in fact, in thermal contact with the outside and we assume that this coupling leads it at all times to the state with maximum disorder.

In §§ 5.2.2 and 5.2.4 we have identified the amount of heat received by the system during the time dt with

$$\delta Q = \operatorname{Tr} d\widehat{D}\,\widehat{H},$$

where we can use (5.33) to express $d\widehat{D}$ as a function of the changes $d\beta$, $d\xi_\alpha$. We have also shown in (3.28) that for an arbitrary infinitesimal change $d\widehat{D}$ of the density operator $\widehat{D}$ the statistical entropy changes by

$$dS = -k \operatorname{Tr} d\widehat{D} \ln \widehat{D}.$$

In the present case we can, if we use (5.33) and the relation $\operatorname{Tr} d\widehat{D} = 0$, write this quantity in the form

$$\begin{aligned} dS &= k \operatorname{Tr} d\widehat{D} \ln Z_\mathrm{C} + k \operatorname{Tr} d\widehat{D}\,\beta\widehat{H} \\ &= k\,\beta \operatorname{Tr} d\widehat{D}\,\widehat{H}, \end{aligned}$$

which, finally, gives by comparison with (5.8)

$$\delta Q = \operatorname{Tr} d\widehat{D}\,\widehat{H} = \frac{1}{k\beta}\,dS, \qquad \forall\ d\beta,\ d\xi_\alpha. \tag{5.34}$$

We saw in § 5.1 that the Lagrangian multiplier β, introduced to take into account that the energy is a constant of the motion, defined a relative temperature scale. We must identify (5.34) with the expression $T\,dS_\mathrm{th}$, where the thermodynamic entropy S_th is a function of the various variables β, ξ_α which characterize the equilibrium , and where T depends only on β. There is a unique solution to this problem. Apart from a multiplying constant, β is *the inverse of the thermodynamic absolute temperature*:

$$\boxed{\beta = \frac{1}{kT}}\,, \tag{5.35}$$

and the *statistical entropy*

$$S = -k \operatorname{Tr}(\widehat{D} \ln \widehat{D}) = k \ln Z_\mathrm{C} + k\,\beta U, \tag{5.36}$$

which was introduced at the microscopic and statistical level as a measure of the disorder, can be identified with the thermodynamic entropy S_th defined by Clausius's equation (5.31), apart from an additive constant.

To prove this result we rewrite the identity

$$dS = k\,\beta T\,dS_\mathrm{th}$$

in the form

$$\frac{\partial S}{\partial \beta} = k\,\beta T\,\frac{\partial S_\mathrm{th}}{\partial \beta}, \tag{5.37a}$$

$$\frac{\partial S}{\partial \xi_\alpha} = k\,\beta T\,\frac{\partial S_\mathrm{th}}{\partial \xi_\alpha}. \tag{5.37b}$$

Taking the derivative of (5.37a) with respect to ξ_α and of (5.37b) with respect to β, and using the fact that T is not a function of the ξ_α, we get

$$\frac{\partial^2 S}{\partial\beta\partial\xi_\alpha} = k\,\beta T \frac{\partial^2 S_{\rm th}}{\partial\beta\partial\xi_\alpha} = k\,\beta T \frac{\partial^2 S_{\rm th}}{\partial\beta\partial\xi_\alpha} + \frac{d(k\beta T)}{d\beta}\frac{\partial S_{\rm th}}{\partial\xi_\alpha},$$

which implies that βT is a constant. Integration of (5.37), after we have made the choice $k\,\beta T = 1$, then shows that $S_{\rm th} - S$ is a constant.

In (5.35) and (5.36) we meet again with the arbitrary multiplicative constant k which we introduced in Chap.3 into the definition of the statistical entropy. If we want to take for the unit of the absolute temperature T the kelvin, the most commonly used scale in practice, which is defined by putting the temperature of the triple point of water equal to 273.16 K, we must identify k with the *Boltzmann constant* (1.35), as in § 1.3.3. Another choice which follows naturally from statistical physics would be to put $k = 1$; the quantity $1/\beta$ then defines the *absolute temperature measured in energy units* and the entropy S is *dimensionless.* In order to remember the magnitude of k, bear in mind that room temperature $T = 300$ K corresponds to an energy kT equal to $\frac{1}{40}$ eV.

If the system consists of several non-interacting subsystems, each at thermal equilibrium, the *additivity of the entropy* follows immediately from the statistical independence of the various parts (§ 3.2.2c). We have also seen that in this case (§ 4.2.5) the partition functions are multiplied, and the entropies are added.

Statistical mechanics enables us to prove simply a few inequalities satisfied by the absolute temperature and the entropy. For most systems the Hamiltonian has no upper bound and we have seen that then $\beta > 0$, and hence $T \geq 0$. It then follows from (5.34) that *S is an increasing function of the internal energy U*, for a given Hamiltonian $\widehat{H}$ ($\delta\mathcal{W} = 0$). For systems where β can vary between $+\infty$, corresponding to zero absolute temperature T, and $-\infty$, the entropy $S(U)$ has a maximum at $T = \pm\infty$ (Exerc.1a). On the other hand, it follows from (4.31) that

$$T^2\,\frac{\partial U}{\partial T} = -\,\frac{1}{\partial^2 S/\partial U^2} = \frac{1}{k}\,\left\langle\left(\widehat{H} - U\right)^2\right\rangle, \tag{5.38}$$

so that, for given $\widehat{H}$, *T is an increasing function, and S a concave function of U* (Exerc.5d).

Above we have used a canonical ensemble. In a *grand canonical ensemble* we have

$$dS = -\,k\,{\rm Tr}\,d\widehat{D}\,\ln\,\widehat{D} = k\,\beta\,{\rm Tr}\,d\widehat{D}\,\widehat{H} - k\,\alpha\,{\rm Tr}\,d\widehat{D}\,\widehat{N}. \tag{5.39}$$

The second term, $-k\alpha d\langle N\rangle$, vanishes in a transformation (5.30) of a closed system, and the first term can be identified, as above, with $\delta Q/T$. In a *microcanonical* ensemble the temperature β is defined by (5.4) for transformations which keep the parameters ξ_α fixed, that is, where $dU = \delta Q$, and we find again $dS = k\beta\,\delta Q$.

The microscopic interpretations of the absolute temperature and of the entropy, and also that of the chemical potential, as we shall see, are basically *statistical* by nature and are associated with a Boltzmann-Gibbs *equilibrium*: the temperature is a parameter occurring in the canonical probability distribution, and the entropy represents lack of information. On the other hand, for the interpretation of the various quantities, such as heat, work, or forces, associated with energy, we did not need assume that $\widehat{D}$ represented an equilibrium macro-state. In Chaps.14 and 15 we shall see that the *local* temperature concept retains a meaning for some non-equilibrium states, and that the dynamical analysis of irreversibilities leads to the introduction of *several* entropies, all following from the lack of information defined in Chap.3.

5.3.3 Irreversibility

We have just proved the equivalence between the microscopic *degree of disorder* and the thermodynamic *entropy* for the *equilibrium states*, or the states close to equilibrium, which a system passes through during a quasi-static transformation. The second part (5.32) of the Carnot-Clausius principle refers to irreversible transformations in which the intermediate states may be close to equilibrium – a situation which we shall study in Chap.14 – but also quite far from equilibrium, like the sudden mixing of two fluids with different temperatures or explosive chemical reactions. In such states thermodynamics does not define an entropy whereas statistical mechanics allows us at any time to associate a statistical entropy with the density operator given by theory. We can, however, continue to identify the two entropies in the initial or final states which are, by assumption, in equilibrium. We saw in §§ 3.2.3 and 4.1.5 that on the microscopic scale the entropy of an isolated system increases for statistical reasons, such as badly known coupling between the various parts of the system, loss of information to inaccessible degrees of freedom, or increase in microscopic disorder. It follows that the thermodynamic entropy of an *isolated system* which undergoes an arbitrary transformation from one *equilibrium* state to another *cannot decrease*. In § 6.1 we shall come back to this point from both the macroscopic and the microscopic points of view.

This conclusion remains valid if the system, although *thermally isolated*, can exchange work with the outside. In fact, the proof of § 3.2.3 did not assume the Hamiltonian to be time-independent; it only used the unitarity of the evolution operator $\widehat{U}$, which holds for a time-dependent Hamiltonian, even though the specific form (2.27) for $\widehat{U}$ is no longer valid.

On the other hand, the entropy may decrease for a system *in thermal contact with the outside*. For instance, a system in thermal equilibrium which cools off loses entropy whereas the outside gains more entropy; similarly, biological organisms become ordered while the neighbouring disorder increases. In such a case we shall again find the irreversibility property (5.32), postulated in the framework of macroscopic thermodynamics, by regarding the system under study and those with which it exchanges heat as a single thermally isolated system, whose total entropy increases as we have just seen. The changes in entropy of the heat sources j at temperatures $T_j^{(\mathrm{S})}$ can easily be calculated, as these sources are large systems undergoing quasi-static transformations and receiving, respectively, amounts of heat $-\delta \mathcal{Q}_j$ during the time dt. The change in the statistical entropy of the system is $S_2 - S_1$, and altogether we have thus

$$S_2 - S_1 - \sum_j \int \frac{\delta \mathcal{Q}_j}{T_j^{(\mathrm{S})}} \geq 0,$$

which completes the proof of the Second Law.

We shall study in greater detail in Chap.14 the irreversibility of processes for which the various parts of the system are at all times in a state close to equilibrium.

In our discussions we have neglected correlations between the system and the sources; these are absent, by assumption, in the initial state, but may appear in the final state due to the interactions describing thermal contact. The entropy of the final state is, because of these correlations, smaller than the sum of the entropies of the system and the sources (Eq.(3.21)) so that the appearance of correlations has no effect other than strengthening inequality (5.32).

As an exercise, let us analyse the irreversibility of a transformation of two systems, initially in thermal equilibrium at different temperatures, which leads from the state $\widehat{D}$ given by (5.2) to another, not necessarily equilibrium, state $\widehat{D}_t$. Due to the random coupling $\widehat{V}$ between a and b, the total statistical entropy increases (§3.2.3). The systems a and b exchange heat; their internal energies become

$$\mathrm{Tr}\, \widehat{D}_t \widehat{H}_{\mathrm{a}} = U_{\mathrm{a}} + \mathcal{Q}, \qquad \mathrm{Tr}\, \widehat{D}_t \widehat{H}_{\mathrm{b}} = U_{\mathrm{b}} - \mathcal{Q},$$

as we can neglect $\mathrm{Tr}\, \widehat{D}_t \widehat{V}$. Combining the irreversibility inequality (3.31) with the lemma (3.18), we find

$$-k\, \mathrm{Tr}\, \widehat{D} \ln \widehat{D} = S \leq S_t \leq -k\, \mathrm{Tr}\, \widehat{D}_t \ln \widehat{D},$$

and hence, using (5.2),

$$\left(\beta_{\mathrm{a}} - \beta_{\mathrm{b}}\right) \mathcal{Q} \geq 0. \tag{5.40}$$

The sign of the heat $\mathcal{Q}$ received by a is that of the difference $T_{\mathrm{b}} - T_{\mathrm{a}}$ of the initial temperatures.

This example shows that statistical mechanics can express irreversibility in a more general way than thermodynamics since, in contrast to the thermodynamic entropy, the *statistical entropy is defined for non-equilibrium states.*

5.4 The Third Law or Nernst's Law

5.4.1 Macroscopic Statement

The Third Law (1906), due to Walther Nernst, expresses the *impossibility to reach the absolute zero.* In analytical form this impossibility is reflected by the fact that at zero temperature the entropy is independent of the parameters ξ_α which characterize the equilibrium state of the system: otherwise one could reach the absolute zero by an adiabatic transformation, varying some of the ξ_α.

The thermodynamic entropy was defined apart from an additive constant. The Third Law shows that this constant is independent of the system; one can therefore choose it in such a way that *the entropy vanishes at the absolute zero* (Planck).

5.4.2 Statistical Entropy at the Absolute Zero

In contrast to the thermodynamic entropy defined by the Second Law, the statistical entropy associated with a macro-state does not contain an arbitrary additive constant. When we identified in (5.36) those two entropies with one another we implicitly fixed the additive constant of the thermodynamic entropy. We shall see that this adjustment of the constant corresponds exactly to Nernst's law. In other words, the statistical entropy should vanish at the absolute zero.

When $T \to 0$ or $\beta \to \infty$, the probabilities

$$p_m = \frac{e^{-\beta E_m}}{\sum_n e^{-\beta E_n}} = \frac{e^{-\beta(E_m - E_0)}}{\sum_n e^{-\beta(E_n - E_0)}} \tag{5.41}$$

tend exponentially to zero for all excited states $|m\rangle$ with energies larger than that of the ground state, E_0. The density operator $\widehat{D}$ reduces to the *projection onto the ground state*, provided it is not degenerate, and thus describes a pure state with an entropy $S = -k \sum_m p_m \ln p_m$ equal to *zero.* The Third Law therefore follows from the statement that the *ground state* of most systems *is not degenerate.*

In fact, this condition is too restrictive. For large systems such as are considered by thermodynamics, the Third Law means that the entropy *per unit volume* tends to zero. To derive it from microscopic physics, it is thus sufficient to prove that

$$\lim_{\beta\to\infty} \lim_{\Omega\to\infty} \frac{1}{\Omega} S \to 0, \tag{5.42}$$

where the volume Ω tends to infinity *before* the temperature tends to zero. At low temperatures the entropy, whether canonical or microcanonical, is of the order of $k \ln W$, where W is the number of eigenstates of $\widehat{H}$ with energies below $E_0 + \Delta E$, the interval ΔE remaining finite as $\Omega \to \infty$. This number

W thus represents the *multiplicity of the ground state E_0 together with the weakly excited states*; it depends on the volume of the system. Nernst's law only requires $\ln W/\Omega \to 0$; it therefore holds as long as the *multiplicity W grows less rapidly than an exponential of the volume*. In practice, W is always large, as the ground state of a macroscopic system is always degenerate or nearly degenerate; however, experiments show that in most cases $\ln W/\Omega$ tends to zero in the large volume limit, and this explains why the vanishing of S at $T = 0$ has been given the status of a Law.

One could imagine systems with a pathological behaviour, $\ln W \propto \Omega$, but it doesn't look like they exist in nature. For instance, when there are no applied fields (§ 1.2.1), all states of the independent spin system, which is a model for a paramagnetic solid, have the same energy, and $W = 2^N$. The entropy per unit volume is, at all temperatures, equal to $(kN/\Omega) \ln 2$ and it violates Nernst's law. Nevertheless, in a real solid, the interactions between the magnetic moments cannot be neglected at very low temperatures. They tend to correlate these moments, that is, to increase the order; in general, below a certain temperature the solid becomes ferromagnetic or antiferromagnetic, and the entropy of this new phase tends to zero at zero temperature, as should be the case. We shall find a similar effect (§ 12.2.3) in solid helium 3 where each atomic nucleus has spin $\frac{1}{2}$. The only interactions felt by these N spins, due to the magnetic moments of the nuclei, are very weak, of the order of 10^{-7} eV. As a result, when the temperature is lowered, but stays above a few mK, everything behaves as if the ground state were 2^N-fold degenerate. The entropy tends to the value $kN \ln 2$: it looks as if Nernst's law is violated. Only below 1 mK does the entropy fall towards zero because of the establishment of nuclear magnetic ordering.

Another, more common, effect where Nernst's law is violated in practice occurs when matter, cooled down to very low temperatures, remains during a nearly infinite time in a *non-equilibrium* disordered state, so that the experimentally measured entropy remains positive and does not tend to zero. For instance, *glass* is in a metastable amorphous state, even at the lowest temperatures that can be reached which, nevertheless, are much lower than its crystallization temperature; only the entropy of the equilibrium crystalline state would tend to zero (§ 4.1.6).

Let us finally note that the Third Law is *quantum mechanical* in origin. Whereas the quantum mechanical statistical entropy is always positive or zero, we have seen (§ 3.3.2) that there is no lower bound for the classical statistical entropy. The classical approximation would lead to negative, non-physical values for the entropy and it is therefore certainly violated at the low temperatures where Nernst's law is relevant.

5.5 The Thermodynamic Limit

5.5.1 Extensive and Intensive Variables

Thermodynamics is usually interested in systems which are, at least locally, *homogeneous* on a macroscopic scale. The dimensions of such systems can be considered to be infinite as compared to the distances between the elementary microscopic constituents; the number of the latter is very large, and the relative magnitude of surface effects becomes negligible. This is called the *thermodynamic limit.*

In this limit macroscopic experiments show that the various quantities of interest can be classified either as extensive – varying proportionally to the volume of nested subsystems of the system under study – or intensive – remaining invariant under such a subdivision. Amongst the *extensive* quantities we have the volume Ω, the *entropy* S, the *constants of motion* such as the internal energy U, the number of particles N of each kind, and the total momentum, as well as other additive quantities such as the mass or the magnetic moment. Amongst the *intensive* quantities we have the *variables which are the conjugates of the constants of motion*, such as the reciprocal β of the absolute temperature or variables λ_i of the same kind (§ 5.1.4); let us also mention the pressure, the magnetic field, or the average energy per particle. These various quantities are connected with one another through thermodynamic relations which make their extensive or intensive nature obvious, as soon as one *postulates*, for instance, for a fluid, that the *entropy*, considered as a function of the volume Ω and of the constants of motion such as U and N, is *homogeneous* of degree 1:

$$S(x\Omega, xU, xN) \;=\; x\, S(\Omega, U, N). \tag{5.43}$$

Moreover, the thermodynamic quantities are *independent of the shape of the sample.*

These properties present such a body of evidence that, notwithstanding their importance, one did not feel compelled to state them as a separate Law of thermodynamics. They reflect, in fact, our intuition about what is matter at our scale: a homogeneous continuum, the characteristics of which are not altered by subdivision, except by scaling of extensive quantities; a stable substance which remains locally unchanged when one brings fragments together. Two counter-examples will help us to feel why extensivity is less trivial than it looks. (i) A metal at equilibrium, carrying a negative *charge* which is proportional to its mass, is not extensive. In fact, the excess electrons position themselves in this case near the surface of the material (§ 11.3.3). Their (Coulomb) contribution to the internal energy, although macroscopic, is not proportional to the volume, and it depends on the shape of the sample; the metal is not homogeneous and if it is cut into pieces at least its electrostatic properties are changed. (ii) *Stellar* matter cannot reach an equilibrium state where it is stable and homogeneous; a star evolves perpetually while contracting under the influence of gravitation, which violates extensivity. Even if they have exactly the same composition, two stars with different masses can differ considerably (Exerc.6e, 15f).

5.5.2 Extensivity from the Microscopic Point of View

A complete justification of the Laws of thermodynamics, starting from statistical physics, requires a *proof of the extensivity* (5.43), a property which was postulated in macroscopic physics. This proof is difficult and appeals to special conditions which must be satisfied by the interactions between the particles. In particular, they must repel one another at small distances apart, in order that matter doesn't collapse, and they must not have too long a range. Using these conditions one can prove in one or other of the canonical ensembles that $\ln Z/\Omega$ has a finite limit, independent of the shape of the container, when the volume of the latter tends to infinity while U/Ω, N/Ω or β, α remain fixed. This property is the same as (5.43) since the entropy is the Legendre transform (4.14) of $k \ln Z$ and both therefore have the same degree of homogeneity.

In order to get a feel for the proof one may imagine that one cuts up a large volume of matter at equilibrium into elementary cubes with macroscopic dimensions. To begin with one neglects the correlations between the cubes. At this stage the extensivity is a clear consequence of the factorization property (4.22). If the interatomic forces have a finite range, the contribution of the interactions between one cube and another comes from near their interface. It is thus expected to be negligible compared to the contributions from the interior of the cubes, provided the size of the cubes is much larger than the range of the forces. However, it is not simple to make this argument rigorous.

The proof is even more complicated if, instead of modelling the system as a combination of atoms which interact through short-range forces, one represents it in a more fundamental way by electrons and nuclei, which is the structure of most substances; their interactions are long-range Coulomb forces. The proof of the extensivity in this case requires that *the total charge be zero* and it is crucial to use quantum mechanics. In several instances we shall see that the *indistinguishability* of quantum particles plays an essential rôle in the extensivity: Gibbs paradox (§ 8.2.1), stability of neutron stars and white dwarves (§ 10.1.4).

Extensivity is violated and the system collapses if the potential $V(r)$ between two particles is attractive at short distances apart (end of § 9.3.1). In fact, if $V(0) = -a$ is negative, the energy is dominated by a term $-\frac{1}{2}aN^2$ in the configurations where the particles stick together; hence at equilibrium the internal energy per particle U/N cannot tend to a finite limit as $N \to \infty$. This is the mechanism which leads to the instability of stellar matter through gravitation. In fact, the absence of short-range attractions is not sufficient to guarantee extensivity: the latter can be violated even when there are no interactions (Exerc.12d).

We note that at the microscopic level the extensivity of the microcanonical entropy $S = k \ln W$ means that the energy level density $W/\Delta E$ of a macroscopic system is large, of the order of magnitude of an exponential of the volume. For extensive substances one can also derive a general concavity property (Exerc.9g).

The extensivity of physical systems in the thermodynamic limit is an important property. It expresses the *stability of matter* and it is thanks to it that we can speak of samples of some substance or other without having to specify the shape or size. It has many consequences both at the theoretical level: *the vanishing of statistical fluctuations* in the thermodynamic limit (§ 5.7.1) and the pairing off of intensive and extensive variables as *conjugate pairs* (§ 5.6), and also at the technical, mathematical level: *the equivalence of various ensembles* (§ 5.5.3), the Gibbs-Duhem relation (§ 5.6.5), and the necessity to use the canonical ensemble for a finite subsystem (§ 5.7.2).

5.5.3 Equivalence of Ensembles

We shall show that, if a system is extensive, we obtain for all, extensive or intensive, macroscopic physical quantities the same value whatever statistical ensemble we are using, be it a microcanonical, canonical, grand canonical, or any other one. In other words, in the thermodynamic limit *giving the constants of motion exactly or only as averages leads to the same results.*

As thus the thermodynamic conclusions are, for a macroscopic system, independent of our choice of ensemble, we shall put this latitude to our advantage in the remainder of this book and work in the statistical ensemble that leads to the *simplest calculations*, taking into account, in particular, the possibilities of factorizing the partition function (§ 4.2.5). Most often it will be the canonical or grand canonical ensemble. For instance, one can check that in the example of Chap.1 for a paramagnetic salt (Exerc.4c) the canonical ensemble leads to the same results as the microcanonical one, but more easily.

To establish this equivalence we only need compare the partition functions associated with two different ensembles, as all macroscopic quantities follow from them (§ 4.2.6). To fix ideas, let us consider the same system in canonical equilibrium (§ 4.3.1) and in grand canonical equilibrium (§ 4.3.2); the number of particles is fixed exactly to be N in the first case and as an expectation value in the second case. The partition functions are related to one another through the Laplace transform (4.34):

$$Z_{\mathrm{G}}(\beta, \Omega, \alpha) \;=\; \sum_{N} \mathrm{e}^{\alpha N} \, Z_{\mathrm{C}}(\beta, \Omega, N). \tag{5.44}$$

Let us assume that extensivity has been established for the canonical ensemble, so that

$$\frac{1}{\Omega} \ln \, Z_{\mathrm{C}}(\beta, \Omega, N) \;=\; f(\beta, x) \tag{5.45}$$

is a function of $x \equiv N/\Omega$, and is practically independent of Ω. The intensive variable x can be treated as continuous in the limit as $\Omega \to \infty$ and we can write (5.44) as

$$Z_{\mathrm{G}}(\beta, \Omega, \alpha) = \Omega \int_0^\infty dx\, \mathrm{e}^{\Omega[\alpha x + f(\beta, x)]}. \tag{5.46}$$

As in §1.2.4 we note that in the vicinity of the maximum of $\alpha x + f(\beta, x)$ with respect to $x \equiv N/\Omega$ the integrand in (5.46) has a *very pronounced maximum* in the neighbourhood of which it behaves as a Gaussian with a narrow peak; the width of this peak is small as $\Omega^{-1/2}$. The integral (5.46), which is completely dominated by this maximum (Exerc.5b), gives as $\Omega \to \infty$ the dominant behaviour of $\ln Z_{\mathrm{G}}$ in the form

$$\ln Z_{\mathrm{G}}(\beta, \Omega, \alpha) \sim \max_N \left[\alpha N + \ln Z_{\mathrm{C}}(\beta, \Omega, N)\right]. \tag{5.47}$$

The relation (5.47) between the canonical and the grand canonical partition functions in the thermodynamic limit is just a *Legendre transformation* (§§ 4.2.4 and 6.3.1). The variables α and N are in that relation connected through

$$\alpha = -\frac{\partial}{\partial N} \ln Z_{\mathrm{C}}(\beta, \Omega, N), \tag{5.48}$$

which, taking the derivative of (5.47) with respect to α, implies that

$$N = \frac{\partial}{\partial \alpha} \ln Z_{\mathrm{G}}(\beta, \Omega, \alpha). \tag{5.49}$$

The number of particles N in the canonical ensemble which dominates the sum (5.44) is thus the same as the average number $\langle N \rangle$ in the grand canonical ensemble, as (5.49) and (4.35) are the same. The variable α, the conjugate of N, which in the canonical ensemble is defined by the derivative (5.48), just as the temperature β is defined in the microcanonical ensemble by (5.4), can be identified with the Lagrangian multiplier α of the grand canonical ensemble. Even though Z_{G} and Z_{C} are not functions of the same variables and though they take on different values, their partial derivatives with respect to $-\beta$ (the energy), with respect to Ω, which is proportional to the pressure (§ 5.6.1), or with respect to other variables occurring in the Hamiltonian, are equal, as one can see by differentiating (5.47). This completes the identification of the quantities α, N, β, U, and so on, in the canonical and grand canonical ensembles. We note that we had implicitly anticipated this result by using the same notation in the two ensembles. Finally, the entropy, given by (4.30) in the canonical ensemble and by (4.36) in the grand canonical ensemble, also takes asymptotically the same value when Z_{G} and Z_{C} are related through (5.47). Of course, the statistical fluctuations are not the same; for instance, ΔN vanishes in the canonical ensemble, but is large, as $\sqrt{N}$, in the grand canonical ensemble. We shall show, however, that the relative size of fluctuations is small in any ensemble, except at a critical point or when the system splits into phases (§ 5.7.1).

Using the *saddle-point method* (Exerc.5b) enables us to establish (5.47) in a more rigorous manner and to calculate the corrections in lower orders in Ω. Formula (5.94) thus yields, if we assume that the solution of (5.48) is unique and we denote it by $N_0(\beta, \Omega, \alpha)$,

$$\begin{aligned} \ln Z_G(\beta, \Omega, \alpha) &= \alpha N_0 + \ln Z_C(\beta, \Omega, N_0) \\ &\quad - \frac{1}{2} \ln \left| \frac{1}{2\pi} \frac{\partial^2 \ln Z_C}{\partial N_0^2} \right| + \mathcal{O}\left(\frac{1}{\Omega}\right). \end{aligned} \tag{5.50}$$

From (5.50), (4.30), (4.36), and (5.48) it follows that the *two entropies differ* by

$$S_G - S_C = \frac{k}{2} \ln \frac{\partial N_0}{\partial \alpha} + \mathcal{O}(1). \tag{5.51}$$

Although this quantity is large, of order $\ln \Omega$, it has a small relative size, of order $\ln \Omega / \Omega$. The fact that S_G is larger is the result of the definition itself of the grand canonical ensemble, which we found by looking for the maximum of the entropy for given $\langle N \rangle$, a weaker constraint than in the canonical ensemble where we have the additional condition $\Delta N = 0$. Expression (5.51), which is equal to $k \ln \Delta N$, where ΔN^2 is the variance (4.38), (4.35) in the grand canonical ensemble, measures the *uncertainty about the value of* N in that ensemble.

We have here derived Z_G from Z_C. *Inversely*, if the extensivity has been established in the grand canonical ensemble, one must start from the Laplace transform

$$Z_C(\beta, \Omega, N) = \frac{1}{2\pi i} \int_{\alpha_0 - i\pi}^{\alpha_0 + i\pi} d\alpha \, e^{-\alpha N} Z_G(\beta, \Omega, \alpha), \tag{5.52}$$

which is the inverse of (5.44) and where α_0 is arbitrary. The argument is analogous to the earlier one, interchanging the rôles of the variables α and x. Nevertheless we work here in the plane of the complex variable α and we should deform the contour of integration so that it passes through the saddle point (Exerc.5b), that is, through the point α_0 where $-\alpha N + \ln Z_G(\beta, \Omega, \alpha)$ is stationary. This point is a minimum for real α, but a maximum along the contour $\alpha_0 - i\pi$, $\alpha_0 + i\pi$. We get in this way

$$\ln Z_C(\beta, \Omega, N) \sim \min_{\alpha} \left[-\alpha N + \ln Z_G(\beta, \Omega, \alpha) \right]. \tag{5.53}$$

Using the fact that $\ln Z_C$ and $\ln Z_G$ are concave functions, we see that this relation is the inverse Legendre transform of (5.47), which proves again the equivalence of the two ensembles.

Either the calculation leading from (5.44) to (5.47) or the one leading from (5.52) to (5.53) can be adapted to any other pair of ensembles; for instance, N and α are replaced by U and β to show the equivalence between the microcanonical and the canonical ensembles.

The canonical distribution depends *exponentially* on the energy of the microstates whereas the microcanonical distribution of those energies is *localized at a point* U, with a margin ΔU. It may seem surprising that such different behaviours lead to equivalent results in the thermodynamic limit. In fact, the distribution $p(E)$ of the *probability for the energy* in the canonical ensemble is proportional to the product of the Boltzmann factor $\mathrm{e}^{-\beta E}$ and the *level density* (§4.3.1). The latter is itself proportional to $\mathrm{e}^{S(E)/k}$, where $S(E)$ is the microcanonical entropy, so that we have

$$p(E) \;\propto\; \mathrm{e}^{-\beta E+S(E)/k}. \tag{5.54}$$

The extensivity of $S(E)$ then entails that $p(E)$ is close to the Gaussian (Exerc.5b)

$$p(E) \;\propto\; \exp\left[-\frac{1}{2k}\left|\frac{\partial^2 S}{\partial U^2}\right|\left(E-U\right)^2\right]. \tag{5.54'}$$

This probability is centred around the point U determined by the standard relation between energy and temperature,

$$-\beta+\frac{1}{k}\,\frac{\partial S(U)}{\partial U} \;=\; 0,$$

with statistical fluctuations small as $\Omega^{-1/2}$ in relative magnitude. The *exponential decrease* $\mathrm{e}^{-\beta E}$ is thus combined with the *very rapid increase in the level density* to give rise to a *sharp peak* which differs little from the microcanonical energy distribution.

5.6 Thermodynamic Potentials

5.6.1 Entropy and Massieu Functions

The use of thermodynamic potentials is one of the important tools of thermodynamics. A *thermodynamic potential* is a *function of certain variables* characterizing the state of the system, which has the property of being a *maximum* or a *minimum at equilibrium*, and whose *partial derivatives* with respect to these variables have a simple interpretation. Having available such a function allows us in practice to calculate heat capacities, equations of state, and, more generally, all equilibrium properties of the system. We shall systematically use this approach for most applications that we shall treat.

The prototype of a thermodynamic potential is the equilibrium *entropy as function of conserved quantities* such as the energy U or the particle number N, and *of the parameters* ξ_α which occur in the Hamiltonian such as the volume Ω or an external field. Its partial derivative with respect to U is the inverse $k\beta$ of the absolute temperature; its derivative with respect to N is a variable, $-k\alpha$, of the same kind which governs the exchange of particles (§ 5.1.4). To find the physical meaning of its derivatives with respect to the "positions" ξ_α we bear in mind the general expression for the differential of the internal energy

$$dU = T\,dS + \mu\,dN + \sum_\alpha X_\alpha\,d\xi_\alpha. \tag{5.55}$$

In § 5.6.3 we shall discuss the term associated with the variation of N; the last terms represent the *work received* (5.11). From (5.55) we get the differential of S:

$$dS = \frac{1}{T}\,dU - \frac{\mu}{T}\,dN - \sum_\alpha \frac{X_\alpha}{T}\,d\xi_\alpha. \tag{5.56}$$

It assigns to each of the *natural variables* U, N, ξ_α of S its *conjugate variable* with respect to S, namely,

$$\frac{1}{T} = k\beta, \qquad -\frac{\mu}{T} = -k\alpha, \qquad -\frac{X_\alpha}{T}. \tag{5.57}$$

Hence, the partial derivatives of S with respect to the ξ_α are directly related to the "force" variables X_α.

From the microscopic point of view the natural variables U, N, ξ_α occurring in S are exactly those which characterize a *microcanonical equilibrium.* The entropy *as a thermodynamic potential* can thus be identified with

$$S(U, N, \xi_\alpha) = k \ln W(U, N, \xi_\alpha), \tag{5.58}$$

where the weight W is the *microcanonical partition function.*

It is important to note that the entropy, if we consider it to be a function of *other variables*, for instance, T, N, ξ_α instead of U, N, ξ_α, *does not have simple partial derivatives*; it is no longer a thermodynamic potential. If one wants to change variables while conserving the duality between conjugated variables, we must perform a Legendre transformation (§§ 4.2.4 and 6.3.1), which consists in changing the function at the same time we change the variables. In particular, substituting $1/T$ for U leads to the introduction of the *Massieu function*

$$\Psi_C\left(\frac{1}{T}, N, \xi_\alpha\right) = S(U, N, \xi_\alpha) - \frac{1}{T}\,U, \tag{5.59}$$

where U and T are related to one another through $\partial S/\partial U = 1/T$. This function is the *thermodynamic potential* for the variables $1/T$, N, ξ_α, as its differential is equal to

$$d\Psi_C = -U\,d\left(\frac{1}{T}\right) - \frac{\mu}{T}\,dN - \sum_\alpha \frac{X_\alpha}{T}\,d\xi_\alpha. \tag{5.60}$$

Through its derivatives Ψ_C provides the same relations as $S(U, N, \xi_\alpha)$; the first pair $(U, 1/T)$ of conjugate variables has become $(1/T, -U)$, and the others remained the same.

The natural variables of (5.59) are in statistical mechanics those which characterize a *canonical equilibrium* state, apart from a change in units by k

in $1/T = k\beta$. Comparing (5.59) with (4.30) enables us to identify the Massieu potential (5.59) with the *logarithm of the partition function*,

$$\Psi_{\mathrm{C}}\left(\frac{1}{T}, N, \xi_\alpha\right) = k \ln Z_{\mathrm{C}}(\beta, N, \xi_\alpha), \tag{5.61}$$

and Eqs.(5.59), (5.60) repeat the results of §§4.2.6 and 4.3.1, while giving a physical interpretation to the various quantities which occur as partial derivatives.

One can in the same way introduce another Massieu potential,

$$\begin{aligned}\Psi_{\mathrm{G}}\left(\frac{1}{T}, -\frac{\mu}{T}, \xi_\alpha\right) &= k \ln Z_{\mathrm{G}}(\beta, \alpha, \xi_\alpha) \\ &= S(U, N, \xi_\alpha) - \frac{1}{T} U + \frac{\mu}{T} N,\end{aligned} \tag{5.62}$$

suited to the variables $1/T = k\beta$, $-\mu/T = -k\alpha$, ξ_α which characterize the *grand canonical equilibrium.* In (5.62) U and N are determined by $\partial S/\partial U = 1/T$ and $\partial S/\partial N = -\mu/T$, and the Massieu function Ψ_{G} appears as the Legendre transform of (5.58) with respect to U and N. Here again, the differential

$$d\Psi_{\mathrm{G}} = -U\, d\left(\frac{1}{T}\right) + N\, d\left(\frac{\mu}{T}\right) - \sum_\alpha \frac{X_\alpha}{T}\, d\xi_\alpha \tag{5.63}$$

shows that Ψ_{G} is, indeed, a thermodynamic potential.

We can readily generalize this to any other canonical ensemble. This enables us to interpret $k \ln Z$ as a thermodynamic Massieu potential, provided we take as variables the natural parameters characterizing the equilibrium macro-state for that ensemble. Its partial derivative with respect to each of the variables is the conjugate variable, apart possibly from a sign; this sign is easily determined by bearing in mind Eqs.(5.56) or (5.57) and the fact that each Legendre transformation changes one sign. We have implicitly assumed that the system is extensive, by assuming that the entropy has the same value in each ensemble. For each pair of *conjugate variables*, one is then *extensive* and the other one *intensive.* The extensive variables are the conserved quantities (U, N) together with some of the parameters ξ_α occurring in the Hamiltonian, like the volume Ω; on the other hand, the other ξ_α parameters, such as the magnetic field $\boldsymbol{B}$, are intensive.

For many decades thermodynamics put more emphasis on the *energy* concept than on the *entropy* concept. As a result the differential form (5.55) has been the foundation for fixing the traditional nomenclature and formalism. Nonetheless, it would have been more satisfactory to use (5.56) systematically. In fact, with that expression it is easier to connect thermodynamics and statistical mechanics, as we have just seen. Moreover, (5.56) puts the quantities U and N which are of the same, conservative, nature on a par, whereas the entropy S, which plays a very different rôle from that of all other variables, occurs on the left-hand side. In the case of an irreversible transformation, the equation (5.56) is replaced by an inequality which shows up the increase in entropy of an isolated system. The use as thermodynamic potentials of the entropy and of the Massieu functions, which follow from it, fits in well with these remarks. An extra simplification would follow if we chose our units such that $k = 1$, with $T = 1/\beta$ being measured in joules and S being dimensionless. Nevertheless, we have given in to the weight of tradition and mainly used in what follows the thermodynamic potentials following from the energy, which are the ones used in most of the literature. In fact, the basic quantities (5.57) which occur in connection with the Massieu potentials, in particular $1/T = k\beta$, $-\mu/T = -k\alpha$, $\mathcal{P}/T$ do not even have a name, notwithstanding the important rôle they play in thermodynamics, and we shall have to talk about the temperature T, the chemical potential μ, and the pressure $\mathcal{P}$, following the common usage. We shall work with the variables (5.57) more natural albeit less usual, only in Chaps. 6 and 14 where we discuss non-equilibrium thermodynamics.

François Massieu, engineer, geologist, physicist, and mathematician (Vatteville 1832–Paris 1896) introduced in 1870 the above kind of thermodynamic potentials, which are Legendre transforms of the entropy, and which he called "characteristic functions". His aim was to determine indirectly the specific heat of a vapour, starting from its equation of state which can be found more easily experimentally, with applications to steam engines in view (Exerc. 6a). Notwithstanding their immediate success, the Massieu functions soon became forgotten. Gibbs's (1875) and Helmholtz's thermodynamic potentials became in final reckoning the popular ones, in particular, through Planck's treatise (1897) on thermodynamics; nevertheless, besides these standard thermodynamic potentials, Planck had rediscovered and used one of Massieu's potentials!

5.6.2 Free Energy

Expression (5.55) for the change in the *internal energy* for a reversible transformation shows that this quantity is a thermodynamic potential, provided one expresses it as function of the variables S, N, and ξ_α. We have just emphasized the drawbacks of this traditional starting point. Moreover, from a statistical mechanics point of view, the function $U(S, N, \xi_\alpha)$ is not directly given in any ensemble, in contrast to $S(U, N, \xi_\alpha)$; furthermore, it is inconvenient to treat S as a variable.

If, nevertheless, one wishes to take the *temperature* T, instead of S, as state variable side by side with N and the ξ_α, it is convenient to introduce the (Helmholtz) *free energy* as thermodynamic potential:

$$F(T, N, \xi_\alpha) = U(S, N, \xi_\alpha) - TS. \qquad (5.64)$$

This is the Legendre transform of the internal energy with respect to the entropy, with $\partial U/\partial S = T$. In fact, the differential of F which follows from (5.55) is

$$dF = -S\,dT + \mu\,dN + \sum_\alpha X_\alpha\,d\xi_\alpha. \tag{5.65}$$

The pairs of *conjugate variables with respect to the free energy* are $(T, -S)$, (N, μ), (ξ_α, X_α), and the partial derivatives of F with respect to its natural variables T, N, ξ_α give, at equilibrium, the values of the other variables in each pair.

In contrast to the internal energy, the free energy as thermodynamic potential can easily be calculated in the framework of statistical mechanics. In fact, the natural variables T, N, ξ_α of F are directly connected with those of the *canonical ensemble* and comparing (5.64) with (5.59) and (5.61) shows that the free energy can be simply expressed in terms of the *canonical partition function* $Z_C(\beta, N, \xi_\alpha)$ through

$$F(T, N, \xi_\alpha) = -kT \ln Z_C\left(\frac{1}{kT}, N, \xi_\alpha\right). \tag{5.66}$$

Therefore, in order to obtain the macroscopic properties of a system from the canonical formalism, it amounts to the same to work with the Massieu function (5.61) or with the free energy (5.66), and to use the partial derivatives (5.60) or (5.65), respectively.

The free energy is used when one is interested in exchanges of work of a system with the outside during quasi-static *isothermal transformations*: the work is, in fact, in that case given by the change (5.65) in F. On the other hand, if one is considering adiabatic transformations, the internal energy U, whose change (5.55) gives the work, is more advantageous.

In § 6.3.3 we shall see that the free energy is also introduced in the study of a system, possibly in a non-equilibrium state, which is *coupled to a thermostat.* In that case, T denotes the temperature of the latter. In *equilibrium* the free energy is a *minimum*; this explains why it is called "thermodynamic *potential*".

For a system, in communication not only with a thermostat, but also with a manostat which maintains its pressure $\mathcal{P}$ at a constant value, the proper thermodynamic potential is the *free enthalpy*, or *Gibbs potential*,

$$G(T, N, \mathcal{P}) = U - TS + \mathcal{P}\Omega = -kT \ln Z_{\rm i}, \tag{5.67}$$

the double Legendre transform of U with respect to S and Ω. It is suited to the natural variables T, N, and $\mathcal{P}$, since

$$dG = -S\,dT + \mu\,dN + \Omega\,d\mathcal{P}, \tag{5.68}$$

and it can, according to (5.67), be derived directly from the *isobaric-isothermal* ensemble.

The *enthalpy*

$$H(S, N, \mathcal{P}) = U + \mathcal{P}\Omega, \tag{5.69}$$

the thermodynamic potential suitable for a system in communication with a manostat, but thermally insulated, cannot be simply derived in statistical mechanics.

5.6.3 Chemical Potentials

In § 5.1.4 we saw that the Lagrangian multiplier α associated with N in the grand canonical ensemble measures the *tendency of a system to give up particles.* The relation (5.62) between the grand partition function and the entropy enables us also to express α in terms of the variable μ, introduced in thermodynamics through (5.55), that is,

$$\boxed{\mu = \frac{\alpha}{\beta} = \left.\frac{\partial U}{\partial N}\right|_{S,\Omega} = \left.\frac{\partial F}{\partial N}\right|_{T,\Omega}} \quad . \tag{5.70}$$

One gets the expression for μ in terms of the free energy from (5.65).

The variable μ is called "*chemical potential*". Notwithstanding this name, it is a quantity which is useful in many other fields than just chemistry; moreover, the term "potential" does not refer to a thermodynamic potential like the entropy, the Massieu functions, the internal energy, the free energy, or the enthalpy, since μ is the conjugate variable of N with respect to U. Again, it is necessary here to beware of semantic confusions which are easily introduced through the nomenclature inherited from history. The word "potential" applied to μ has another meaning: at a given temperature, if two systems have different μ, or α, the one with the higher chemical potential will give up particles to the other one. In principle, it would be better to work with the variable α, but its replacement by μ is legitimate in all circumstances where the temperature is uniform, which is often the case. A notable exception is connected with thermoelectric effects, which are phenomena mixing particle and heat exchanges (see § 14.4.3).

We have implicitly assumed that the system consisted of N particles which were all of the same kind. Often we shall encounter circumstances where the existence of several kinds of particles entails the introduction of *several chemical potentials* $\mu_1, \mu_2, \ldots$ which are, respectively, associated with each of the constituent species (§§ 4.1.4 and 4.1.6). For instance, in a *gas mixture* (§ 8.2), the grand canonical equilibrium is characterized by several multipliers, $\alpha_1 = \beta\mu_1$, $\alpha_2 = \beta\mu_2$, ... which are conjugate to the conserved variables N_1, N_2, ... Nevertheless, if this gas is subject to *chemical reactions*, N_1, N_2, ... denote numbers of molecules which can be transformed into one another, and the above grand canonical distribution describes a quasi-equilibrium state where the numbers of molecules are frozen in. The final chemical equilibrium, after reactions, is obtained by writing down relations between the chemical potentials μ_1, μ_2, ... (§§ 6.6.3 and 8.2.2). This is the context in which historically the variables μ were introduced, and this explains their name.

However, the chemical potential concept is fruitful in all circumstances where exchange of particles between two systems is allowed. To determine the *equilibrium* one writes down that *the chemical potentials* of those particles of which the total number is conserved, but which can be exchanged – they may be molecules, atoms, ions, radicals, electrons, elementary particles, depending on the case considered – *must be equal* in the two systems (Exerc.8e). If the chemical potentials are unequal, knowing them enables us to *determine the direction of the exchanges*, just as the sign of the temperature difference determines the direction of the heat exchange. For instance, a liquid in the presence of its vapour *evaporates* if its chemical potential is higher. *Equilibrium between phases* (§ 6.4.6) requires that the chemical potentials, as well as the temperatures and the pressures, are equal. Similarly, two fluids of the same kind separated by a porous partition exchange molecules through the pores: *effusion* (§ 7.4.1). In the case of *osmosis* (§ 6.6.2) the semi-permeable wall can let through the solvent molecules, so that their chemical potentials become the same on the two sides, but it cannot let through the molecules in solution; the numbers of molecules in solution are separately conserved on both sides, which, in general, will lead to different chemical potentials for them. This asymmetry produces a difference in *osmotic pressure. Adsorption* and desorption of a gas on the wall of the vessel that encloses it result from the exchange of molecules between the gas and some sites on the wall, in particular, microscopic irregularities where they can possibly be trapped; the two systems of which we must compare the chemical potentials are in this case the gas and the collection of sites that may be occupied (Exerc.4b). It is useful to compare this situation with that of a gas of *photons* (Chap.13) in an enclosure. In that case the number of photons is not conserved by the interactions with the walls, so that we should not introduce a Lagrangian multiplier α; it comes to the same to put $\alpha = 0$, and hence the *chemical potential vanishes* (§ 10.5.2).

The chemical potential is similarly a major tool in electrostatics and electrodynamics, where it is often called the "*electrochemical potential*". We are dealing in that case with the chemical potential μ of the mobile charge carriers, ions in *ionic solutions*, electrons in *metals*, or electrons and holes in *semiconductors. Electrostatic equilibrium* (§ 11.3.3) is determined by requiring that the *chemical potential is uniform.* If the latter varies from one point to another, the charged particles *tend to move* in the direction of decreasing μ. In fact, if there is a difference $\Delta\mu$ between the chemical potential for particles with charge q at two points, on a macroscopic scale the quantity $\Delta\mu/q$ can be identified with the *electromotive force* felt by these particles between the two points in question (§ 14.4.2). We must again beware of the nomenclature: in general, although μ is called the electrochemical potential, $\Delta\mu/q$ *is not* equal to the difference in the macroscopic *electric potential,* since the latter contains also a contribution from the charge density (§ 11.3.3). In particular, if charged substances are in electrostatic equilibrium, their electric potential, in contrast to μ, is not uniform. A characteristic example is the spontaneous appearance of an electric potential between two metals or semiconductors which are brought into contact. The chemical potential of the electrons would not be the same in each of them if they were taken in isolation. As soon as they are brought into contact this produces a transfer of electrons from one substance to the other in order to equalize the chemical potentials. The double layer of positive and negative charges which is produced in this way at the interface gives rise to an electric potential (§ 11.3).

In the above examples the systems a and b which can exchange particles are either *separated objects*, such as the gas and the wall in the case of adsorption, or *different parts* of the same, continuous object, for instance, volume elements of a substance which is not uniformly charged; they may even be entities *superposed upon one another in space*. In fact, in a mixture of molecules in the gaseous phase or of ions in a dilute solution, it is useful to treat the sets of molecules of each species as different systems a, b, ..., and then to assume that one brings a, b, ... into chemical contact, allowing reactions involving exchanges of atoms, ions, or radicals. Similarly, in a semiconductor it will be fruitful to regard the conduction electrons and the holes as two independent systems which can annihilate one another, and then to require that the difference of their numbers be conserved, but not the numbers of the charge carriers of each kind (§ 11.3.4).

Along the lines of § 5.1.4 we have interpreted the chemical potential as the quantity *governing the exchange* of particles. Its expression (5.70) gives us another interpretation, namely, that of the *marginal energy of an extra particle.* Let us, in fact, assume that we add a particle to an open system while keeping the entropy and the ξ_α parameters, such as the volume, fixed; the transformation is reversible so that both the initial and the final state are in equilibrium. The change dU in the internal energy is then equal to the chemical potential μ. The latter is, according to (5.65), also equal to the change in the free energy when, for a given temperature, one reversibly adds a particle to the system. This interpretation of μ as the energy of an extra particle is particularly useful for systems of non-interacting fermions at zero temperature (§ 10.4.2), where μ is called the *Fermi energy.* More generally, if the numbers of constituent particles change by dN_j during a *reversible* infinitesimal shift in equilibrium, the change (5.55) in the internal energy contains a contribution $\sum_j \mu_j dN_j$ which corresponds, for instance, to the chemical energy of a reaction (§ 6.6.3).

The distinction between work and heat was established for transformations of closed systems and the change in energy $\mu\, dN$ in (5.55) has a different status from both work and heat. Even from the point of view of statistical physics, the splitting of dU into the two terms (5.30), which are, respectively, interpreted as work and heat when $dN = 0$, cannot unambiguously be extended to transformations of open systems. In fact, in the grand canonical ensemble $\widehat{H}$ must be treated as a quantity which is independent of N, but has components for the different values of N; when there is no work done, (5.30) then reduces to its second term $\mathrm{Tr}\, d\widehat{D}\, \widehat{H}$. However, (5.39) shows that this term equals $T\, dS + \mu\, dN$ so that it seems legitimate to consider $\mu\, dN$ as a contribution to the heat, added to the usual contribution. On the other hand, in the microcanonical or canonical ensembles, N appears through the Hamiltonian and plays the same rôle as the ξ_α parameters; the energy $\mu\, dN$ is thus analogous to a work term $X\, d\xi$, even though the two terms of (5.30) have in this case hardly any meaning. It is thus preferable to regard $\mu\, dN$ neither as heat nor as work, but as another form of energy.

5.6.4 Grand Potential

Suppose we want to replace at the same time in the characterization of the state of a system the entropy S and the number of particles N by their conjugate variables with respect to the internal energy, that is, by the *temperature* T and the *chemical potential* μ. We must introduce the *thermodynamic potential* A suited to these new variables, which is defined as the double Legendre transform of U,

$$A(T, \mu, \xi_\alpha) = U - TS - \mu N = F - \mu N. \tag{5.71}$$

This function is called the *grand potential*. Its differential

$$dA = -S\,dT - N\,d\mu + \sum_\alpha X_\alpha\,d\xi_\alpha \tag{5.72}$$

shows that one obtains the various thermodynamic quantities of interest and their mutual interrelations by taking derivatives of A with respect to the natural variables T, μ, and ξ_α.

Just as the free energy is suitable for the description of a system coupled to a thermostat, the grand potential is particularly convenient whenever one studies an open system coupled both to a thermostat and to a particle reservoir, which fix the values of T and of μ, respectively. The grand potential is also useful when one studies the equilibrium of two systems which can exchange energy and particles, as in that case it is sufficient just to require that the values of T and of μ are the same in the two systems.

From a microscopic point of view a comparison of (5.71) and (4.36) shows that the grand potential is expressed in terms of the *grand partition function* $Z_G(\beta, \alpha, \xi_\alpha)$ through

$$A(T, \mu, \xi_\alpha) = -kT \ln Z_G\left(\frac{1}{kT}, \frac{\mu}{kT}, \xi_\alpha\right). \tag{5.73}$$

This result was expected, since the natural variables T, μ, of the grand potential are simply related to those which characterize a state in the *grand canonical ensemble*, that is, β, α. In most applications that we shall consider, the latter ensemble will be the most convenient. We shall thus evaluate (5.73), and after that use (5.71) and (5.72) to find the properties of the system. It would have been slightly simpler to rely on the Massieu potential Ψ_G defined by (5.62) and on its differential (5.63) in terms of β and α, but we shall conform to tradition and rather introduce the grand potential

$$A = -T\Psi_G. \tag{5.74}$$

The correspondence (5.47) between different ensembles in the thermodynamic limit implies the macroscopic relation (5.71) between A and F.

5.6.5 Pressure; The Gibbs-Duhem Relation

The macroscopic state of a fluid consisting of a single kind of particles is characterized by three variables, for instance, the extensive quantities U, N, and Ω. The volume Ω is a ξ_α type variable and its conjugate variable $X_\alpha = -\mathcal{P}$ defines the pressure. The "*equation of state*" which connects pressure, temperature, and particle density N/Ω is obtained in the *canonical ensemble* by using (5.60) or (5.65). We find thus

$$\mathcal{P} = kT \frac{\partial}{\partial \Omega} \ln Z_{\mathrm{C}} = - \frac{\partial F(T, \mu, \Omega)}{\partial \Omega}. \tag{5.75}$$

In the *grand canonical ensemble* the equation of state is given in a *parametric form*, involving μ, through the equations

$$N = - \frac{\partial A}{\partial \mu}, \qquad \mathcal{P} = - \frac{\partial A}{\partial \Omega}. \tag{5.76}$$

However, $\ln Z_{\mathrm{G}}$ or A are functions of one extensive variable Ω only, plus two intensive variables. The extensivity of $\ln Z_{\mathrm{G}}$ for a fluid (§5.5.2) therefore results in A being proportional to Ω for fixed T and μ, or in A/Ω being a function of T and μ only, which can be expressed as

$$\frac{\partial A(T, \mu, \Omega)}{\partial \Omega} = \frac{A}{\Omega}. \tag{5.77}$$

It then follows from (5.76) that

$$\boxed{-\mathcal{P}\Omega = A = -\frac{1}{\beta} \ln Z_{\mathrm{G}} = U - TS - \mu N}, \tag{5.78}$$

which shows that, apart from the sign, *the grand potential per unit volume is equal to the pressure.*

Taking the differential of (5.78) and using (5.55) yields

$$S\,dT - \Omega\,d\mathcal{P} + N\,d\mu = 0,$$

an equation which means that the intensive variables T, μ, and $\mathcal{P}$ are not independent. This identity is known in thermodynamics as the *Gibbs-Duhem relation.* It implies that, if one wants to characterize a fluid through three independent variables, at least one of them must be extensive (§ 6.2.3).

5.6.6 Summary: Tables of the Thermodynamic Potentials

We shall now complete the arguments of §§ 4.2.6 and 4.3 for a practical solution of problems of statistical physics at equilibrium by gathering in two tables the various statistical ensembles and the corresponding thermodynamic potentials for a fluid consisting of a single kind of particle.

Table 5.1. Canonical ensembles and their Massieu potentials $\Psi = k \ln Z$

Natural variables	Ensemble	Partition function	Entropy	Differential
U, N, Ω	Microcanonical	W	$S = k \ln W$	$k^{-1}dS =$ $\beta dU - \alpha dN + \beta \mathcal{P} d\Omega$
β, N, Ω	Canonical	$Z_C = \mathrm{Tr}_N\, e^{-\beta \widehat{H}_N}$	$\Psi_C + k\beta U$	$k^{-1}d\Psi_C =$ $-U d\beta - \alpha dN + \beta \mathcal{P} d\Omega$
β, α, Ω	Grand canonical	$Z_G = \mathrm{Tr}\, e^{-\beta \widehat{H} + \alpha \widehat{N}}$	$\Psi_G + k\beta U - k\alpha N$	$k^{-1}d\Psi_G =$ $-U d\beta + N d\alpha + \beta \mathcal{P} d\Omega$
$\beta, N, \beta\mathcal{P}$	Isobaric-isothermal	$Z_i = \int d\Omega \mathrm{Tr}\, e^{-\beta \widehat{H} - \beta \mathcal{P} \Omega}$	$\Psi_i + k\beta U + k\beta \mathcal{P} \Omega$	$k^{-1}d\Psi_i =$ $-U d\beta - \alpha dN - \Omega d(\beta \mathcal{P})$

Table 5.2. Thermodynamic potentials. The partition functions are defined in Table 5.1

Natural variables	Ensemble	Thermodynamic potential	Relations	Differential
S, N, Ω		Internal energy U		$dU =$ $T dS + \mu dN - \mathcal{P} d\Omega$
T, N, Ω	Canonical	Free energy $F = -kT \ln Z_C$	$F = U - TS$	$dF =$ $-S dT + \mu dN - \mathcal{P} d\Omega$
T, μ, Ω	Grand canonical	Grand potential $A = -kT \ln Z_G$	$A = U - TS - \mu N$	$dA =$ $-S dT - N d\mu - \mathcal{P} d\Omega$
$T, N, \mathcal{P}$	Isobaric-isothermal	Free enthalpy $G = -kT \ln Z_i$	$G = U - TS + \mathcal{P}\Omega$	$dG =$ $-S dT + \mu dN + \Omega d\mathcal{P}$
$S, N, \mathcal{P}$		Enthalpy H	$H = U + \mathcal{P}\Omega$	$dH =$ $T dS + \mu dN + \Omega d\mathcal{P}$

The various ensembles differ in the *choice of the three natural variables* used to describe the equilibrium state. In all cases one starts by evaluating the *partition function* and from that one derives at once a *thermodynamic potential.* The expression for the differential of the latter then provides the various thermodynamic quantities *through simple differentiations* with respect to the natural variables. It will hardly ever be useful for this purpose to use the Boltzmann-Gibbs density operator itself.

Table 5.1 summarizes the *Massieu potential* formalism, which is simpler and more directly connected with statistical mechanics. The pairs of conjugated variables with respect to S are $(U, k\beta)$, $(N, -k\alpha)$, $(\Omega, k\beta\mathcal{P})$. In all cases, the Massieu functions are given by $k \ln Z$, which in the case of the microcanonical ensemble reduces to $S = k \ln W$.

However, one usually rather employs the *potentials connected with the energy* which are summarized in Table 5.2. The pairs of conjugated variables with respect to U are (S, T), (N, μ), $(\Omega, -\mathcal{P})$, and their thermodynamic interpretations are obvious. We shall usually work in the *canonical* ensemble or in the *grand canonical* ensemble which lead to the simplest calculations of the partition function. The other rows in the table are especially useful when one is concerned with macroscopic thermodynamics. For instance, a heat engineer, interested in a gas the pressure of which is fixed from the outside, will use the enthalpy; a chemist studying reactions at given temperature and pressure will use the free enthalpy. Note that the inclusion of S amongst the natural variables prevents us from connecting the thermodynamic potential, that is, the energy or the enthalpy, to a canonical partition function.

5.7 Finite Systems

5.7.1 Statistical Fluctuations

If we want to identify the results of statistical mechanics with our everyday experience, we must still understand why the predictions of macroscopic physics *appear to be deterministic* notwithstanding the *probabilistic* nature of the underlying theory. To explain this, at least for equilibrium situations, we must prove that the various statistical quantities, the mean values of which have been identified with the corresponding macroscopic variables, show *negligible relative statistical fluctuations* in any of the canonical ensembles. For the example of Chap.1 we have already shown this kind of property in § 1.2. In the general case, we shall see that the existence of a *thermodynamic limit*, expressed by the extensivity of $\ln Z$, guarantees that the relative fluctuations in macroscopic systems are extremely small.

Let us, in fact, consider the variance (4.31) of the *energy*, ΔU^2, in the canonical ensemble. It is the second derivative of $\ln Z_C$ (an extensive quantity) with respect to β (an intensive quantity) so that it is proportional to the volume or to the number of particles, N. The energy U is itself proportional to N, with $U/N > 0$ if we choose the origin suitably. As a result, the *relative fluctuation* of the energy, $\Delta U/U$, *tends to zero as* $N^{-1/2}$ in the thermodynamic limit. The dimensionless coefficient has no reason for being large; hence we expect that, for instance, for one mole where $N = 6 \times 10^{23}$, the energy will be defined better than to one in 10^{10}. The statistical uncertainty in U is thus much smaller than the experimental errors.

It is remarkable that this result follows from our principle of *maximum uncertainty* which has been the foundation for obtaining the canonical probability distribution. The energy had been left free to fluctuate around the mean value that we had assigned to it; nevertheless, it hardly ever strays far from it, as its probability distribution in the limit of a large system tends to the Gaussian (5.54) with a small width of relative order $N^{-1/2}$. Even though the uncertainty S *about the whole set* of degrees of freedom of the system is a maximum, the energy is known *practically exactly.*

The situation is the same when one brings two *macroscopic objects into thermal contact.* We saw in § 5.1.2 that the equalizing of the temperatures determined the mean values U_a and U_b of the energies of the two systems. We did not discuss the statistical partition of the total energy between a and b. Nevertheless, the answer to this question can be found in § 1.2.4. In the microcanonical ensemble all configurations where the total available energy $U_a + U_b$ is split arbitrarily between a and b have the same probability. However, by far the most of the micro-states correspond to situations where the energy of a is close to U_a, with a relative spread of order $N^{-1/2}$. Statistical mechanics explains in this way why the total energy practically always splits up in the same way between two systems which are in thermal contact, even though all partitions are possible, in principle.

These considerations can immediately be extended to energy or particle number fluctuations in a grand canonical ensemble, which are given by (4.38). The smallness of these fluctuations, when the thermodynamic limit exists, enables us to understand intuitively the equivalence between the various ensembles which may describe the equilibrium of a macroscopic system (§ 5.5.3). Indeed, it comes to the same whether we give a constant of the motion such as U or N *exactly* or only give *its mean value*, as in the second case its relative statistical fluctuations turn out to be negligible anyway.

Nevertheless, there is one important exception. When a system can undergo a *phase transition* (§§ 6.4.6 and 9.3.3), the characterization of its state on the macroscopic scale is not always unique if the number of intensive data is too large. For instance, if we give the temperature, the chemical potential, and the volume of a fluid, the proportion of the two phases remains undetermined along the liquid-vapour coexistence curve in the T, μ plane. The microscopic counterpart of this situation corresponds to the grand canonical ensemble, where the macro-state is characterized by the same variables. It is thus not surprising to find that the Boltzmann-Gibbs distribution provides energy and particle number fluctuations, ΔU and ΔN, which are *proportional* *to Ω rather than to* $\sqrt{\Omega}$ when the values of α and β correspond to an equilibrium between the two phases. This pathological behaviour results from the fact that in the thermodynamic limit $\ln Z_G/\Omega$ is not an analytical function of α and β: its first derivatives are discontinuous and its second derivatives diverge along the coexistence curve. Moreover, we shall see (§§ 6.4.5 and 6.4.6) that the phase transition is characterized by the vanishing of the denominator

$$\frac{\partial^2 S}{\partial N^2}\frac{\partial^2 S}{\partial U^2} - \left(\frac{\partial^2 S}{\partial U \partial N}\right)^2$$

of (4.38). Note finally that the various ensembles *are not equivalent as regards fluctuations* when there is a phase transition (§ 5.7.3, Exercs.9g, 12b, 12c). For instance, ΔU and ΔN remain always zero in the microcanonical ensemble, even if it describes several phases, whereas we have just seen that they can be macroscopic in the grand canonical ensemble.

As regards the thermodynamic variables ξ_α, we saw in § 5.2.4 when we interpreted them on the microscopic scale as dynamic variables of the source c of work, that the *definition itself of work* makes it necessary that their fluctuations are negligible. This implies, in particular, that the source of work must be a macroscopic system for which the thermodynamic limit is valid.

We now turn to the fluctuations of the conjugated variables X_α, to be evaluated in the system a. For the sake of simplicity, let us assume that the latter is in canonical equilibrium, that its Hamiltonian $\widehat{H}_a$ is linear in the ξ_α, of the form

$$\widehat{H}_a = \widehat{H}_0 + \sum_\alpha \xi_\alpha \widehat{X}_\alpha, \tag{5.80}$$

and that the operators $\widehat{H}_0$ and $\widehat{X}_\alpha$ commute with one another. The mean values and the fluctuations of the X_α are then given by

$$\langle \widehat{X}_\alpha \rangle = -\frac{1}{\beta}\frac{\partial}{\partial \xi_\alpha} \ln Z_C, \qquad \Delta X_\alpha^2 = \frac{1}{\beta^2}\frac{\partial^2}{\partial \xi_\alpha^2} \ln Z_C. \tag{5.81}$$

The existence of a thermodynamic limit for $\ln Z_C$ again implies that the relative fluctuations of the X_α are small as $N^{-1/2}$, whether the ξ_α be extensive or intensive variables.

The *temperature* $T = 1/k\beta$ is defined *exactly* in the canonical and grand canonical ensembles, but its definition (5.4) in the microcanonical ensemble, where, in contrast, the energy is given practically exactly, within a margin ΔU, is imprecise for a finite system: it depends on the way the smoothing ΔU is done. Similarly, α is badly defined in the microcanonical and canonical ensembles, since we are dealing with a "derivative" with respect to the discrete variable N. In order to estimate, for instance, the precision of α in the canonical ensemble, let us compare it with the grand canonical ensemble where α does not fluctuate at all. Expression (5.52) shows Z_C as an integral of Z_G over α, and the dominant contribution to this integral gives us information about the distribution of α-values in the canonical ensemble. We have seen that, as $N \to \infty$, only the saddle-point $\alpha_0(N)$ contributes so that $\alpha = \alpha_0$ is defined exactly in the thermodynamic limit. If N is large, but finite, the weight of (5.52) is concentrated in a region $|\alpha - \alpha_0|$ which is small as $N^{-1/2}$, so that the *absence of fluctuations in* N is compensated by a small *uncertainty* in α. Nevertheless, we are not dealing here with a true statistical fluctuation, as α in this case is a complex variable.

Statistical mechanics allows us not only to calculate thermodynamic quantities, but also quantities like statistical fluctuations which on a macroscopic scale are not relevant. For instance, the statistical fluctuation of the *energy of a small system* in canonical equilibrium, which is given by (4.31), equals

$$\Delta U = \sqrt{kT^2C}, \tag{5.82}$$

where C denotes the specific heat of the sample.

More generally (Exerc.2b, 4a, 5c, 5d) the *theoretical predictions for a finite system are statistical by nature.* The first observation of a random phenomenon of this kind goes back to 1826; it was the motion of grains of pollen in suspension, observed through a microscope by Robert Brown, hence the name "Brownian" motion (§ 15.3.5). We shall see in § 5.7.3 how measurements on a small system enable us to confirm the probabilistic nature of thermal equilibrium.

5.7.2 Finite Part of an Infinite System

It is immaterial which ensemble describes an *infinite* system in equilibrium, but one may wonder what probability distribution one should attribute to a finite part of an infinite system. Let us as an example consider a *finite system* a *weakly coupled to a thermostat* b, that is, an energy bath of infinite dimensions. These two systems in thermal contact form an infinite system which, using the canonical ensemble, we can legitimately describe by the density operator (5.3′). We can directly eliminate the energy bath b through a partial trace and we then get the density operator of the system a:

$$\widehat{D}_{\rm a} = \frac{1}{Z_{\rm a}}\, {\rm e}^{-\beta \widehat{H}_{\rm a}}. \tag{5.83}$$

This result, obtained by assuming that the system a+b was in canonical equilibrium, is a general one, since for the infinite system a+b all ensembles are equivalent; we shall check it below for a microcanonical equilibrium of a+b.

A finite system, interacting weakly – or having interacted in the past – with a thermostat is thus described by a *canonical Boltzmann-Gibbs distribution.* Similarly, a small system a which can exchange – or which in the past has exchanged – *particles* with the reservoir b is in *grand canonical equilibrium.*

The probability distribution of the finite system, which has thus been determined theoretically, can be measured experimentally, for instance, by measuring statistical fluctuations. Statistical mechanics can thus be tested experimentally, not only through its thermodynamic predictions, but also through its statistical consequences for the equilibrium of small systems (§5.7.3).

Let us give an alternative justification for the canonical Boltzmann-Gibbs distribution for a *finite* part of an infinite system. The proof, due to Gibbs, will be solely based on the statistical *equiprobability* hypothesis of § 4.1.1. We assume thus that the system a+b, with Hamiltonian $\widehat{H}_{\mathrm{a}} + \widehat{H}_{\mathrm{b}}$ plus a small coupling which we can legitimately neglect, is described by a *microcanonical* distribution. We want to prove that a is described by a *canonical* distribution, provided b is large. Let us denote the eigenstates of the Hamiltonian $\widehat{H}_{\mathrm{a}} + \widehat{H}_{\mathrm{b}}$ in the Hilbert space $\mathcal{E}_{\mathrm{H}}^{\mathrm{a}} \otimes \mathcal{E}_{\mathrm{H}}^{\mathrm{b}}$ of the global system a+b by $|k_{\mathrm{a}} l_{\mathrm{b}}\rangle$ and the corresponding eigenenergies by $E_k^{\mathrm{a}} + E_l^{\mathrm{b}}$. The microcanonical density operator (4.39) can be written as

$$\widehat{D}_{\mathrm{a+b}} = \sum_{(U<E_k^{\mathrm{a}}+E_l^{\mathrm{b}}<U+\Delta U)} |k_{\mathrm{a}} l_{\mathrm{b}}\rangle \frac{1}{W} \langle k_{\mathrm{a}} l_{\mathrm{b}}|.$$

The density operator of the subsystem a in the Hilbert space $\mathcal{E}_{\mathrm{H}}^{\mathrm{a}}$ can, according to (2.39), be obtained by taking the partial trace over the states of $\mathcal{E}_{\mathrm{H}}^{\mathrm{b}}$:

$$\widehat{D}_{\mathrm{a}} = \sum_k |k_{\mathrm{a}}\rangle \, p_k \, \langle k_{\mathrm{a}}|, \tag{5.84}$$

$$p_k = \frac{1}{W} \sum_l 1, \qquad \text{for } l \text{ such that } U - E_k^{\mathrm{a}} < E_l^{\mathrm{b}} < U - E_k^{\mathrm{a}} + \Delta U. \tag{5.85}$$

The probability p_k is thus proportional to the number of states of the system b with energies between $U - E_k^{\mathrm{a}}$ and $U - E_k^{\mathrm{a}} + \Delta U$, that is, to the microcanonical partition function W^{b} of the system b, calculated at the energy $U - E_k^{\mathrm{a}}$. Let us assume that the thermodynamic limit is valid for the system b; we shall return to this point. The quantity $S^{\mathrm{b}} = k \ln W^{\mathrm{b}}$ is then extensive, like U, whereas E_k^{a} is finite. The E_k^{a} dependence of

$$W p_k = W^{\mathrm{b}}(U - E_k^{\mathrm{a}}) = \exp\left[S^{\mathrm{b}}(U - E_k^{\mathrm{a}})/k\right] \tag{5.86}$$

is obtained by expanding S^{b} in (5.86). However, as S^{b} is extensive, the successive terms in $\partial^2 S^{\mathrm{b}}/\partial U^2$, $\partial^3 S^{\mathrm{b}}/\partial U^3$, ... are, respectively, negligible as Ω^{-1}, Ω^{-2}, ..., where Ω is the size of the system b. Defining the temperature of b through

$$\beta \equiv \frac{1}{k} \frac{\partial S^{\mathrm{b}}(U)}{\partial U},$$

and introducing a new normalization constant,

$$Z_{\mathrm{a}} \equiv W \, \mathrm{e}^{-S^{\mathrm{b}}(U)/k},$$

we obtain in the limit as $\Omega \to \infty$ the required result:

$$p_k = \frac{1}{Z_a} e^{-\beta E_k^a}, \qquad \widehat{D}_a = \frac{1}{Z_a} e^{-\beta \widehat{H}_a}. \tag{5.87}$$

The extensivity of the system b, which was necessary for the above proof, is ensured if we assume that b has the structure of a "*Gibbs ensemble*", that is, if it consists of $\mathcal{N}$ subsystems, $b_1, b_2, \ldots, b_{\mathcal{N}}$, all identical and weakly coupled, with $\mathcal{N} \to \infty$. The canonical partition function of b is in this case $(Z_1)^{\mathcal{N}}$, where Z_1 is the one of b_1. Its behaviour for large $\mathcal{N}$ ensures that the canonical and microcanonical entropies are the same (§ 5.5.3), and hence, it implies the extensivity of S^b, which is proportional to $\mathcal{N}$.

Reinterpreting these arguments, we see that they give us, interestingly enough, a *justification for the maximum entropy criterion* that we introduced as a postulate in § 4.1.3. Here, we only assume the *indifference principle* of § 4.1.1, which looks completely natural. Let a be the finite system with which we want to associate a probability law, or, in quantum mechanics, a density operator, in order to make predictions from the sole knowledge that the expectation value of its energy is u. Let us introduce a statistical ensemble of $\mathcal{N}$ copies $b_1, b_2, \ldots, b_{\mathcal{N}}$ of the system a; this Gibbs ensemble, with $\mathcal{N} \to \infty$, which can be either a real system, or a thought experiment, describes a *collection of experiments* performed on the system a. Each of the samples a, $b_1, b_2, \ldots, b_{\mathcal{N}}$ is the result of the same macroscopic preparation which uniquely determines the expectation value u of the energy, but not its exact value, nor that of other variables. Let us now regard a+ $b_1+ \ldots b_{\mathcal{N}}$ as a single "supersystem". We identify the *expectation value* u for the energy of a with the *arithmetic mean* of the energies of the $\mathcal{N}+1$ samples, so that the supersystem is constrained to have the energy $U = (\mathcal{N}+1)u$. In contrast to what happened above when $b_1+\ldots+b_{\mathcal{N}}$ was a thermostat, the systems a, $b_1, \ldots, b_{\mathcal{N}}$ are in this case not coupled. The distribution of energy between them is no longer governed by random exchanges produced by the coupling, but arises solely from statistical considerations. According to the indifference principle we assume that *all possible results of experiments* performed on a or its copies are *equiprobable*. This amounts to assigning to the supersystem the microcanonical probability distribution characterized by the energy U, which is the only existing datum. As we are solely interested in the system a itself, we eliminate its copies by taking a partial trace over $b_1, \ldots, b_{\mathcal{N}}$. According to the above proof, this implies that we must assign to a *a canonical distribution* when the only datum is the expectation value u of its energy. This result is the same as what we established in Chap.4 by looking for the maximum of the statistical entropy of a, but that principle is now by-passed. The same method, with technical complications due to the non-commutation of observables, can be employed more generally[2] in the case where an arbitrary number of statistical data for a are given; that enables one to derive the maximum entropy "principle" from the indifference principle, and hence, to *construct von Neumann's expression* (3.17) *for the quantum entropy* starting solely from the *equiprobability concept.*

[2] R.Balian and N.Balazs, Ann. Phys. **179**, 97 (1987).

5.7.3 Observation of Statistical Distributions

Starting at the beginning of the twentieth century the combination of thermodynamics and kinetic theory made it possible to elucidate phenomena where one can observe directly the statistical nature of the microscopic physics: Einstein's theory of Brownian motion, which had remained ununderstood for decades (end of § 5.7.1), Smoluchowski's theory of fluctuations, the observation and explanation of "critical opalescence". This last phenomenon is the milky appearance of a fluid in the presence of its saturated vapour in the vicinity of the critical point, where the statistical fluctuations in the density become important and show up over distances of the order of μm; the medium loses its transparency because light is scattered by irregularities which are all the time present (Exerc.6d).

The *theory of the fluctuations* in an equilibrium system results directly from the use of the appropriate Boltzmann-Gibbs distribution and the elimination of the microscopic variables which are not observed. In this way we obtained in (5.54) the probability distribution $p(E)$ for the energy E of a finite system in canonical equilibrium. The Boltzmann factor $\mathrm{e}^{-\beta E}$ is weighted in that case by the number of configurations with energy E, $W = \mathrm{e}^{S(E)/k}$, and the concavity of S provides for $p(E)$ a shape which shows a maximum near the equilibrium value $E = U$. A similar calculation (§ 1.2.4) provided the probability that the energy of one system in thermal contact with another has a given value; this probability is an exponential of the sum of the entropies of the two systems.

More generally, let $\lambda = \{\lambda_\alpha\}$ be a set of macroscopic variables, including the energy, of which we want to know the probability distribution; the system, which is finite and maintained by a thermostat at a temperature T, is therefore in canonical equilibrium (§ 5.7.2). As the variables λ_α are macroscopic, the number $W(\lambda)\, d\lambda$ of microscopic configurations for which they lie between λ_α and $\lambda_\alpha + d\lambda_\alpha$ is large; we have written here $d\lambda$ for $\prod_\alpha d\lambda_\alpha$. This number defines $S(\lambda) \equiv k \ln W(\lambda)$, an *extension of the microcanonical entropy to the state close to equilibrium defined by putting constraints on the* λ_α. Denoting the energy of this state λ by $E(\lambda)$ the required probability equals

$$p(\lambda)\, d\lambda \;\propto\; \mathrm{e}^{-\beta E(\lambda)+S(\lambda)/k}\, d\lambda. \tag{5.88}$$

It contains the Boltzmann factor multiplied by an exponential of the entropy $S(\lambda)$, which produces the exponential of a Massieu potential of the type (5.59). If the finite system considered here can also exchange particles with an external reservoir, we would similarly obtain $p(\lambda)$ as the exponential of a Massieu potential of the type (5.62). The *probability that macroscopic variables* λ *deviate from their equilibrium values* is thus directly expressed in terms of the *entropy* $S(\lambda)$ *of the state* λ, or one of its *Legendre transforms* depending on the nature of the equilibrium studied.

Expression (5.88) shows that, as we saw in § 5.7.1, the statistical fluctuations are the larger, the smaller the system. The expectation values $\langle\lambda_\alpha\rangle$, the only quantities considered in macroscopic thermodynamics, are close to the values for which the Massieu function $S(\lambda) - E(\lambda)/T$ is a maximum. The fluctuations in λ_α around $\langle\lambda_\alpha\rangle$ can become large if the maximum is not very pronounced, which is the case near a critical point; that is the reason why critical opalescence can be observed on a macroscopic scale. In the case of a large system displaying phase separation the maximum of $S - E/T$ may even be reached for some range of λ-values (§§ 5.7.1 and 6.4.6, Exerc.9g). For a finite sample, $p(\lambda)$ is then spread over this range, but its shape is sensitive to detailed effects, such as surface phenomena or a gravitational field.

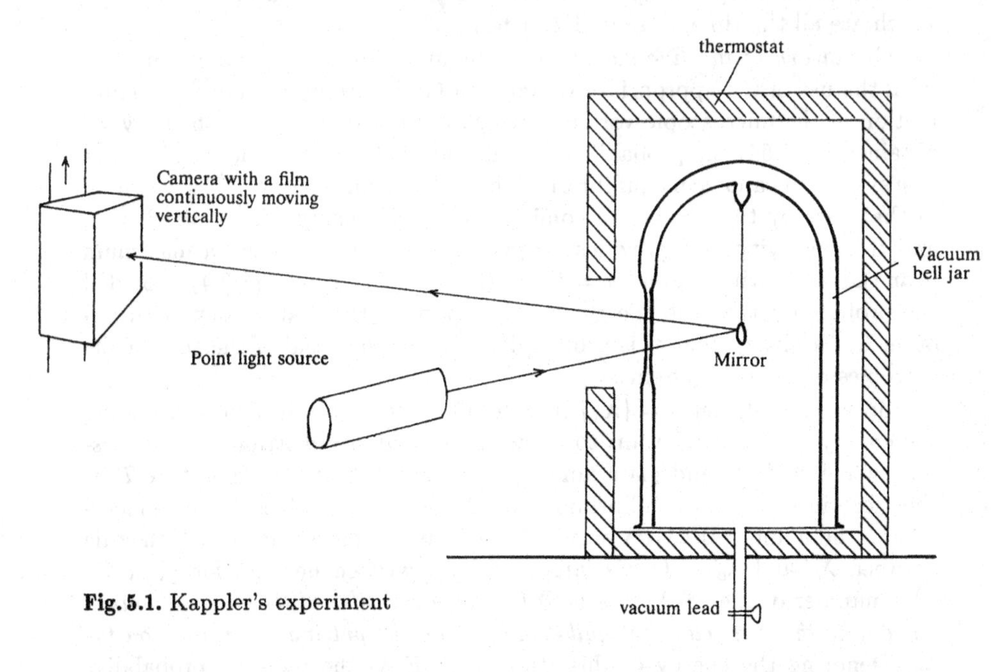

Fig. 5.1. Kappler's experiment

The *direct observation of the probability law* for a macroscopic object at equilibrium in a thermostat was the object of Kappler's experiment (1931). Using the displacement of a reflected light beam, he registered the rotational motion of a very light and small (1 to 2 mm^2) vertical mirror, suspended by a torsion wire of quartz with a diameter of the order of μm, and placed in an isothermal enclosure (Fig.5.1). The temperature and the pressure of the gas surrounding the mirror could be varied at will (Exerc.5e).

The mirror is a system with one pair of degrees of freedom, the angle of rotation, θ, and its conjugate momentum. One observes that the angle of rotation θ does not remain zero with respect to its equilibrium position. The Hamiltonian describing the motion of the mirror itself as if it were isolated is that of a classical harmonic oscillator,

$$H = \frac{p_\theta^2}{2\mathcal{I}} + \frac{1}{2}C\theta^2, \tag{5.89}$$

where $\mathcal{I}$ is the moment of inertia of the mirror and $C\theta$ the restoring couple, previously measured. The momentum p_θ is proportional to the angular velocity $\omega = d\theta/dt = p_\theta/\mathcal{I}$. Apart from these two collective variables, the micro-state of the mirror depends on a large number of microscopic variables which do not play any rôle here, since the energy and the entropy associated with them are independent of θ and p_θ. The mirror is a finite system, weakly coupled to the gas and to the thermostat which surrounds it. According to § 5.7.2 it must be described by the (classical) canonical distribution law at the temperature of the thermostat. We should thus check experimentally that the probability distribution for the variables θ and ω is the Gaussian law

$$p(\theta, \omega)\, d\theta\, d\omega = \frac{\sqrt{\mathcal{I}C}}{2\pi kT}\, \mathrm{e}^{-(\mathcal{I}\omega^2 + C\theta^2)/2kT}\, d\theta\, d\omega. \tag{5.90}$$

We used the fact that the volume element $d\tau = d\theta\, dp_\theta/h$ of classical statistical mechanics is proportional to $d\theta\, d\omega$.

At each time the ordinate of the registration curve gives the position θ of the mirror and its slope gives ω. One can clearly not predict the values of these random variables at a given time. A study of the system over a very long time nevertheless enables one to determine directly the probability law $p(\theta, \omega)$. *Successive measurements*, in fact, are a *repeated experiment* and they provide samples governed by the Boltzmann-Gibbs distribution. Hence the relative length of time during which the values of θ and ω lie within the range $d\theta\, d\omega$ should be equal to the probability $p(\theta, \omega)\, d\theta\, d\omega$ given by (5.90).

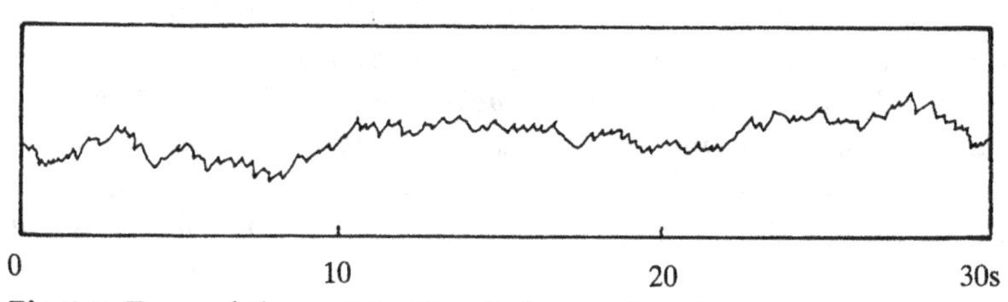

Fig. 5.2. Trace of the spot in Kappler's experiment

This was confirmed by the analysis of more than a hundred hours of registration (Fig.5.2) thus giving direct experimental support for the theoretical ideas which led to the Boltzmann-Gibbs distribution (5.90). These measurements also made it possible to check that the mean values of the kinetic and potential energies,

$$\tfrac{1}{2}\, \mathcal{I}\, \langle \omega^2 \rangle = \tfrac{1}{2}\, C\, \langle \theta^2 \rangle = \tfrac{1}{2}\, kT, \tag{5.91}$$

satisfy the equipartition theorem (§ 8.4.2). Finally, the proportionality constant measured through (5.91) is Boltzmann's constant $k = R/N_\mathrm{A}$, where R is the molar gas constant. Kappler's experiment is thus an *experimental determination of Avogadro's number* N_A and this is done with a 1% accuracy.

Let us note, to end with, that this experiment illustrates an inherent difficulty for high accuracy measurements. Expression (5.91) shows, in fact, that the angle of the mirror of a *galvanometer* at a temperature T shows fluctuations of the order of $\sqrt{kT/C}$ so that it is impossible to measure very weak electric currents which would give rise to deflections below the size of the thermal fluctuations. These *thermal noise* effects are very general and are commonly found in electronics: for instance, there exists at the ends of an open resistance a random potential difference, due to the fluctuations in the electron velocities, with a zero mean but with fluctuations proportional to the square root of the temperature. Each signal with an amplitude smaller than those fluctuations is drowned in the noise background which one cannot make disappear completely. The reduction of thermal noise, which is *indispensable in order to carry out high precision measurements*, makes it necessary to use apparatus maintained at *low temperatures*, as this is the only means of decreasing the statistical fluctuations of thermal origin.

Summary

All thermodynamic laws and concepts – such as relative temperatures, thermal equilibrium, conservation and downgrading of energy, heat, work, entropy, absolute temperature, vanishing of entropy at the absolute zero, chemical potential, pressure – can be derived from the microscopic statistical approach, involving density operators, averages of observables, statistical entropy, and Boltzmann-Gibbs equilibrium ensembles. Just as the temperature governs heat exchanges, the chemical potentials govern particle exchanges.

In the thermodynamic limit, which is reached for most large systems, the variables can be classified as intensive or extensive, the canonical ensembles are all equivalent, and the relative statistical fluctuations become negligible. A finite system placed in a thermostat is in canonical equilibrium. The statistical distribution of macroscopic variables is obtained by extending the entropy to states close to equilibrium.

To calculate in practice the macroscopic properties of a system in thermal equilibrium one uses a thermodynamic potential, such as the free energy or the grand potential, which one evaluates through the partition function in the natural variables of the potential. We summarized the useful formulae in § 5.6.6.

Exercises

5a Elasticity of a Fibre

The object of this exercise is to understand how the elasticity of organic materials such as wool or rubber is governed by an entropy mechanism. We shall use here a very rough model, schematising a wool fibre; Prob.2 discusses a slightly more realistic model, better suited to rubber. Characteristic of these models is that we *neglect the energy.* We shall see that nevertheless the thermodynamic concepts can still be applied.

On a microscopic scale a wool fibre consists of long polymer chains. Each chain is itself formed by a sequence of identical links, which are more or less large protein radicals and which are connected with one another. Each of the links can, nevertheless, be in various quantum states, to each of which corresponds a certain length of the link. Transitions between these states change the total length of the fibre and they are the origin of the elasticity. We shall therefore represent a chain by the following model. It consists of N elements, where N is large, each of which can occur in two micro-states, a short state of length $l - a$ and a long state of length $l + a$. To simplify we assume that the energies of the two states are the same and that the energy associated with the hooking up of the links is independent of their states. Choosing a suitable energy origin one can thus assume that the total energy of the chain equals zero in each of the 2^N possible micro-states. The chain is placed in a thermostat at temperature T. Its total length L is determined by applying a force f at its end.

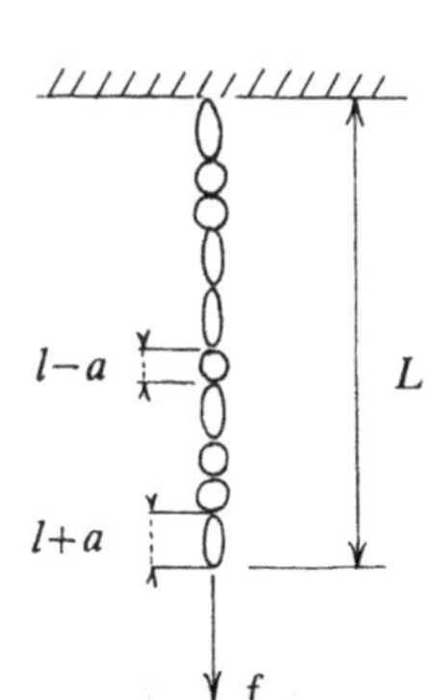

Fig. 5.3. Model of a wool fibre

1. Write down the equation of state of the chain, that is, the relation between f, L, and T, and evaluate its entropy. To do this one may consider that the total length is given as a constraint and introduce a Lagragian multiplier, the physical meaning of which must be found, or one may start from the entropy as function of the length of the chain, or one may write down the canonical equilibrium for the total system, that is, the fibre together with the weight providing the tension.

2. Study the behaviour of the length as function of the tension at the given temperature. How does the elasticity modulus vary with temperature?

3. What happens if one heats the fibre for a given tension? What, if one releases the tension adiabatically? These results can be observed, for instance, by using a hair-drier to heat a strong elastic band under tension by a weight. The shortening produced in this way is due to the origin of the elasticity of the fibre: for a metallic wire the term dominating in $f = \partial F/\partial L$ is the energy; for a fibre, as well as for rubber consisting of an entanglement of long chains without intrinsic elasticity, the dominant term is the entropy in $F = U - TS$, and its contribution has the opposite sign to U. In our model the entropy term is even the only one.

Entropic elasticity was first observed for rubber by Gough (1805) and studied quantitatively by Joule (1859).

Solution:

1a. *Constraint on L method* (see Chap.4). Let $l_i = l + a\sigma_i$, with $\sigma_i = \pm 1$, be the random length of the ith link. A macro-state is characterized by the probability $D(\sigma_1, \ldots, \sigma_N)$. In order to take the constraint

$$\left\langle \sum_i l_i \right\rangle = L$$

on the average length of the chain into account we introduce a Lagrangian multiplier φ. Thermal equilibrium is found by looking for the maximum of $S(D)$ under this constraint. Chapter 4 shows that the result is the Boltzmann-Gibbs distribution

$$D = \frac{1}{Z} \exp\left[\varphi \sum_i (l + a\sigma_i)\right]$$

$$Z = \sum_{\{\sigma_i\}} \exp\left[\varphi \sum_i (l + a\sigma_i)\right] = \left(2 \cosh \varphi a\right)^N e^{\varphi N l},$$

and that the relation between L and φ is

$$L = \frac{1}{Z} \frac{\partial Z}{\partial \varphi}.$$

We work here in an isobaric-isothermal kind of ensemble (§§ 4.3.3 and 5.6.6), with a Hamiltonian which vanishes identically so that D is independent of the β multiplier. The product $-\mathcal{P}\Omega$ is replaced by fl, and the φ multiplier plays the rôle of $-\mathcal{P}/kT$.

In order to interpret these results macroscopically we introduce the thermodynamic potential, identified with the free enthalpy of §§ 5.6.2 and 5.6.6,

$$G(T, f) = -kT \ln Z = -fNl - NkT \ln\left(2 \cosh \frac{af}{kT}\right),$$

as function of the variable $f = kT\varphi$. This change in variable is similar to the introduction of the chemical potential $\mu = kT\alpha$ as variable instead of α in the grand potential A; it is also exhibited in (5.57) which, in terms of the forces X_α, expresses the variables which are the conjugates of the displacements $d\xi_\alpha$, here dl, with respect to the entropy. We can thus identify G and dG with

$$G = U - TS - fL, \qquad dG = -S\,dT - L\,df,$$

and f can be interpreted as the tension. From this we get the equation of state

$$L(T,f) = -\frac{\partial G}{\partial f} = Nl + Na \tanh \frac{af}{kT}$$

and the entropy

$$S = -\frac{\partial G}{\partial T} = Nk \ln\left(2 \cosh \frac{af}{kT}\right) - \frac{Naf}{T} \tanh \frac{af}{kT}.$$

One can check that the internal energy $U = G + TS + fL$ vanishes.

1b. *Microcanonical ensemble method* (see Chap.1). The length of the chain can have the values

$$L = Nl + 2na - Na, \qquad 0 \leq n \leq N.$$

The equilibrium macro-state corresponding to the length L of the fibre is characterized by stating that all micro-states corresponding to a length L, within a margin ΔL such that $2a \ll \Delta L \ll 2Na$, are equiprobable. The – statistical or thermodynamic – entropy is

$$\begin{aligned} S(L) &= k \ln W = k \frac{\Delta L}{2a} \ln \frac{N!}{n!(N-n)!} \\ &\simeq kN \left[-\left(\frac{1}{2} + \frac{L - Nl}{2Na}\right) \ln\left(\frac{1}{2} + \frac{L-Nl}{2Na}\right) \right. \\ &\quad \left. - \left(\frac{1}{2} - \frac{L-Nl}{2Na}\right) \ln\left(\frac{1}{2} - \frac{L-Nl}{2Na}\right)\right], \end{aligned}$$

where we have used Stirling's formula. The length L varies from $Nl - Na$ to $Nl + Na$ and the variation of S is given by the curve in § 1.2.3 with a maximum of $kN \ln 2$ at $L = Nl$.

The use of the standard thermodynamic formulæ, such as $T = \partial S/\partial U$, is made here difficult by the fact that the energy U, on which S normally depends, only takes on a single value $U = 0$. Nevertheless, like the volume of a fluid, the length is here imposed from outside by the constraint f, and can change in reversible transformations when f and T change. The chain in such a transformation takes up an amount of heat $T\,dS$ and an amount of work $f\,dL$; however, the internal energy remains always equal to zero so that we find $T\,dS + f\,dL = 0$, that is,

$$f = -T \frac{dS}{dL} = \frac{kT}{2a} \ln \frac{Na - Nl + L}{Na + Nl - L}.$$

This expression is equivalent to the equation of state $L(T,f)$ that we found earlier.

Another way to find the meaning of f consists in using the free energy $F = U - TS$ which here reduces to

$$F(T,L) = -TS(L),$$

and has the differential

$$dF = -S\,dT + f\,dL.$$

The relation $f = \partial F/\partial L$ again gives $f = -T\,dS/dL$.

By expressing S as a function of f we can check that the results are the same as under a).

1c. *Canonical ensemble method* (see Exerc.4c). We change the definition of the system by including in it not only the fibre, but also the weight f suspended from it. The energy of each microstate of this new system is now

$$E(\sigma_1,\ldots,\sigma_N) = -f\sum_i \left(l + a\sigma_i\right).$$

We write down the expression for the canonical equilibrium of this system:

$$D = \frac{1}{Z}\,\exp\left[\beta f\sum_i (l + a\sigma_i)\right],$$

which gives the same results as under a) when we identify φ with βf. The thermodynamic potential is interpreted here as a free energy:

$$F'(T,f) = U' - TS,$$

where U' is the internal energy of the fibre plus weight system, which equals $-fL$. Here f appears as one of the ξ_α variables entering the Hamiltonian, whereas in G we had $U = 0$ while f/kT was the Lagrangian multiplier associated with the length L. Notwithstanding the differences in the definition of the systems and in the interpretation, we have $F'(T,f) = G(T,f)$. We can also compare F' with the free energy of method b): $F' = F - fL$, where the last term is the free energy of the weight in the gravitational field.

We notice the formal analogy with paramagnetism (Chap.1 and Exerc.4c). The lengths $l \pm a$ correspond to the two spin values, f to B, L to M, and the energy of the fibre plus weight system has the same form as the energy of the N magnetic moments placed in an external field B.

Note. The assignment of an interaction energy of the kind $-fL$ to one or other part of a composite system is arbitrary. It is useful in the methods a) and b) to assign the interaction $-fL$ to the *exterior* of the fibre, which allows exchanges of free energy between the fibre and the outside; the free energy of the fibre is in that case of a pure entropy nature. In method c) we include the energy $-fL$ in the system, as in § 5.2.4; if f remains constant, there is no exchange of work between the system and the exterior, and conservation of energy can in a reversible transformation be expressed as $dU = -f\,dL = \delta Q = T\,dS$.

We also find this kind of situation, for instance, in magnetism (see § 6.6.5). If the system studied is a magnetic moment M, placed in a magnetic field regarded, as in Chap.1, to be external, it receives an amount of work $W_1 = -\int_0^B M\,dB$ during a change in B, which also changes M; the interaction energy $-MB$ between the matter and the field is counted as part of the system. If, on the other hand, we include the field itself in the system, the work received $W_2 = \int_0^M B\,dM =$

$W_1 + BM$ includes the extra energy given up by the windings to produce the field B.

2. If the tension is not too large, the elongation is proportional to it: *Hooke's law*. In this linear regime we have

$$\frac{f}{L - Nl} \sim \frac{kT}{Na^2},$$

and the elasticity modulus is proportional to the temperature, a property similar to Curie's law. The rigidity increases with the temperature.

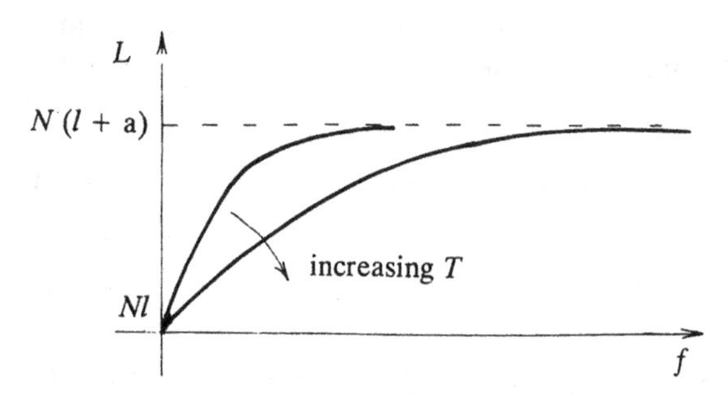

Fig. 5.4. Elongation of a fibre as function of the applied force

3. The fibre shortens from its maximum length $N(l + a)$ at $T = 0$ to Nl as $T \to \infty$, whatever the given value of f, except for $f = 0$ when the dilatation coefficient vanishes. This behaviour is the opposite of that of a metallic wire, which lengthens if one heats it under constant tension.

During an adiabatic release, when

$$dS = \frac{\partial S}{\partial T}\, dT + \frac{\partial S}{\partial f}\, df = 0,$$

the fibre cools down according to

$$\frac{dT}{T} = \frac{df}{f}.$$

The length remains constant since the release of the tension (which for fixed T makes the fibre shorten itself) is compensated by a cooling (which by itself would make the fibre lengthen itself). The cooling through adiabatic release resembles adiabatic demagnetization (§ 1.4.4). In that case also, if the paramagnetic salt were not coupled to the outside, an adiabatic lifting of the field B would cool the sample without changing M.

Thanks to the behaviour of its isotherms and adiabats in the L, f plane, rubber can evolve along a Carnot cycle between a cold and a hot source, like water vapour in a steam engine. The design, practical realization, and theoretical and experimental study of the yield of small thermal engines working on that principle is of considerable pedagogical interest.

5b Saddle-point or Steepest Descent Method

In statistical mechanics one often must evaluate integrals of the kind $\int e^{Af(x)}\,dx$ in the limit as $A \to \infty$. The symbol x may represent a real or a complex variable, or several variables, or even a discrete variable over which one must sum. One also meets with integrals of the form $\int e^{Af(x)}\,g(x)\,dx$ (see §§ 1.2.4, 1.2.6, 5.5.3, Exerc.9b). The integration limits may be either finite or infinite. One obtains the final result in all cases by noting that the exponential greatly exaggerates the variations of $f(x)$. Therefore, if the maximum of $f(x)$ is reached for $x = x_0$, a point either on the integration path or at one of its limits, the weight $e^{Af(x)}$ is *extremely strongly peaked* around x_0, and the integral is completely dominated by the vicinity of x_0. To lowest order one thus finds

$$\ln \int e^{Af(x)}\,dx \sim Af(x_0), \qquad f(x_0) = \max f(x), \tag{5.92}$$

$$\frac{\int e^{Af(x)}\,g(x)\,dx}{\int e^{Af(x)}\,dx} \sim g(x_0). \tag{5.93}$$

The proof of these results as well as the calculation of correction terms are based on expanding $f(x)$ and $g(x)$ around x_0. If, for instance, $f(x)$ is a real function of a real variable, and if x_0 lies within the integration limits, with $f'(x_0) = 0$, $f''(x_0) < 0$, we have

$$\begin{aligned} \int e^{Af(x)}\,dx &\approx e^{Af(x_0)} \int e^{Af''(x_0)(x-x_0)^2/2} \\ &\quad \times \left[1 + Af'''(x_0)\frac{(x-x_0)^3}{3!} + \ldots\right] dx \\ &\sim e^{Af(x_0)} \left[\frac{-2\pi}{Af''(x_0)}\right]^{1/2}. \end{aligned} \tag{5.94}$$

If x_0 is the lower integration limit, with $f'(x_0) < 0$, we find

$$\begin{aligned} \int e^{Af(x)}\,dx &\approx e^{Af(x_0)} \int e^{Af'(x_0)(x-x_0)} \\ &\quad \times \left[1 + Af''(x_0)\frac{(x-x_0)^2}{2} + \ldots\right] dx \\ &\sim e^{Af(x_0)} \left[-Af'(x_0)\right]^{-1}. \end{aligned} \tag{5.95}$$

In the case of a function $f(x)$ of a complex variable, the dominant point x_0 is the one where $\mathrm{Re} f(x)$ is a maximum. If this point does not lie at an endpoint of the integration contour, it satisfies $f'(x_0) = 0$ and we are dealing with a *saddle-point* on the map which represents the relief of the surface $\mathrm{Re} f(x)$ in the complex x plane. We must thus first deform the contour such that it passes through this saddle-point, following the path of the *steepest descent* – hence the name of this method. One proceeds in the same way, if x is real and f purely imaginary, in which case the condition $f'(x_0) = 0$ means that x_0 is the point where the phase of the integrand is stationary; this provides the so-called *stationary phase method.*

1. Prove the Stirling formula (see the list of formulae at the end of this volume).

2. If $S(N, P)$ denotes the microcanonical entropy of a system of oscillators, which was calculated in Exerc.3e, explain why it is the same as the canonical entropy of Exerc.4f in the limit as $N \to \infty$, P/N finite.

3. Consider a system of N spins 1 in a magnetic field. As in Chap.1, its energy levels are given by

$$E = \varepsilon P = \varepsilon \sum_{i=1}^{N} \sigma_i, \qquad |P| < N,$$

where now each σ_i can take on the values $+1, -1, 0$, and P is an integer. Evaluate the multiplicity $W(N, P)$ of the levels in the limit $N \gg 1$, $N - |P| \gg 1$, starting from the canonical partition function found in Exerc.1b.

4. Adapt the arguments of § 5.5.3 to prove the equivalence of the canonical and the microcanonical ensembles for extensive systems, starting from either the one or the other. Evaluate the difference between the entropies.

Hints:

2. Using the extensivity of $S(N, P)$ one finds from

$$Z_N = \sum_P \exp\left[\frac{1}{k} S(N, P) - \left(P + \frac{N}{2}\right)\beta\varepsilon\right]$$

the asymptotic form

$$k \ln Z_N \sim \max_P \left[S(N, P) - k\beta\varepsilon\left(P + \frac{N}{2}\right)\right].$$

As a result, the canonical entropy for that value of P is equal to

$$S = k\left(1 - \beta\frac{\partial}{\partial\beta}\right) \ln Z_N \sim S(N, P).$$

3. Starting from

$$Z_N = \left(1 + e^{\beta\varepsilon} + e^{-\beta\varepsilon}\right)^N = \sum_P W(N, P)\, e^{-\beta\varepsilon P},$$

we find

$$W = \frac{1}{2\pi i} \oint \frac{dz}{z}\, z^P \left(1 + z + \frac{1}{z}\right)^N.$$

The contour encircling the origin can be deformed in such a way that it passes through the two saddle-points, on the real axis, of which only the highest,

$$z_0 = \frac{1}{2(N+P)} \left[\sqrt{4N^2 - 3P^2} - P\right]$$

(when $P \geq 0$) contributes. After some calculations we find from Eq.(5.94)

$$W \sim \frac{1}{\sqrt{2\pi}(4N^2 - 3P^2)^{1/4}} z_0^P \left[\frac{N\left(N + \sqrt{4N^2 - 3P^2}\right)}{N^2 - P^2}\right]^{N+1/2} .$$

5c Electric Shot Noise

An electric current always shows statistical fluctuations because of the discrete nature of the carriers. For instance, a heated cathode in a radio valve emits electrons at random: the *thermionic effect.* The result is a mean current $\langle I \rangle = e\nu$, where ν is the average number of electrons emitted per unit time. Calculate the statistical fluctuation ΔI in the thermionic current during an interval t. Evaluate that fluctuation for a current of 1 μA and a measuring time of 1 s.

Answer. The number of electrons n emitted during the time t is a random variable like the variable n in Exerc.2b in the limit as $\Omega \to \infty$; here t is similar to v, ν to N/Ω. Its fluctuation is thus $\Delta n^2 = \langle n \rangle = \nu t$. The probability p_n is the Poisson law,

$$p_n = \frac{(\nu t)^n}{n!} \mathrm{e}^{-\nu t}.$$

Numerical Application: $\Delta I = 0.4 \times 10^{-12}$ A. The resulting fluctuation ΔV of 0.4 μV at the ends of a resistance of 1 MΩ can be detected by a sensitive amplifier.

5d Energy Fluctuations and Heat Capacity

1. The energy of a system weakly coupled to a thermostat with which it exchanges heat is a random quantity. Express its statistical fluctuations as a function of the heat capacity. This relation is yet another example of the relation between fluctuations and response (Exerc.4a). Numerical example: a water drop of 1 μm diameter.

2. Consider a waterdrop with a well defined energy U in microcanonical thermal equilibrium. The energy E_1 of a part λ (in mass) of the drop is a random variable which can fluctuate around its mean value $U_1 = \lambda U$. What is the number W of levels of the drop such that the part λ has an energy situated within a small range δ around E_1? What is the probability law for the relative change $x = (E_1 - U_1)/U_1$ in E_1?

Hints:

1. We have

$$\Delta U^2 = \frac{\partial^2 \ln Z_{\rm C}}{\partial \beta^2} = kT^2 \frac{\partial U}{\partial T} = kT^2 C.$$

As the specific heat of water is 4.18 J K^{-1} g^{-1}, we have $C = 2.2 \times 10^{-12}$ J K^{-1}, and hence $\Delta U = 1.6 \times 10^{-15}$ J, which is comparable with the change of 2.2×10^{-15} in U produced by a heating up by 1 mK.

2. As in §§ 1.2.4 and 5.7.3 we find

$$p(x) \propto W \propto \exp\left[\frac{1}{k} S(E_1, \lambda N) + \frac{1}{k} S(E - E_1, N - \lambda N)\right],$$

which reduces to a Gaussian with variance

$$\Delta x^2 = \frac{kT^2 C}{U^2} \frac{1-\lambda}{\lambda},$$

when we expand around $x = 0$.

5e Kappler's Experiment

1. In Kappler's experiment (§ 5.7.3) one does not know the equilibrium position of the pendulum, as it moves without ceasing. How can one measure $\langle \theta^2 \rangle$?

2. Numerical application: evaluate $\mathcal{I}$ for an aluminium mirror (density 2.7×10^3 kg m^{-3}) of thickness 0.1 mm and with a surface area of 2 mm^2. To evaluate $\mathcal{C}$ one measures the period τ of the oscillations of the pendulum. What is the order of the displacement of the spot at 1 m for $T = 300$ K and $\tau = 60$ s?

3. What happens, if one evacuates the air from the bell jar?

Answers:

1. It is sufficient to measure the angle θ_1 by taking an arbitrary origin for directions, since $\langle \theta^2 \rangle = \langle \theta_1^2 \rangle - \langle \theta_1 \rangle^2$.

2. $\mathcal{I} \simeq 10^{-13}$ kg m^2. As $\mathcal{C} = 4\pi^2 \mathcal{I}/\tau^2$, we find for the displacement of the spot about 2 mm.

3. The probability law for θ is independent of the gas pressure, as long as the collisions between the gas and the mirror occur sufficiently often. If the gas is too rarefied, however, the mirror is no longer at thermal equilibrium at the temperature T of the surrounding gas. Neverthelesss, it might still be brought to equilibrium through the suspension wire which transmits thermally excited mechanical vibrations, and also through the equilibrium electromagnetic radiation under the bell jar (Chap.13). This radiation acts upon the mirror through the random radiation pressure. However, in both cases it takes a long time to establish equilibrium, as the couplings are extremely weak, and measurements over shorter periods are not significant.

6. On the Proper Use of Equilibrium Thermodynamics

"Abandonnant les théories ambitieuses d'il y a quarante ans, encombrées d'hypothèses moléculaires, nous cherchons aujourd'hui à élever sur la Thermodynamique seule l'édifice tout entier de la physique mathématique. Les deux principes de Meyer et de Clausius lui assureront-ils des fondations assez solides pour qu'il dure quelque temps? Personne n'en doute; mais d'où nous vient cette confiance?"

H. Poincaré, La Science et l'Hypothèse, 1906

"Je ne sais ce que c'est des principes, sinon des règles qu'on prescrit aux autres pour soi."

D. Diderot, Jacques le Fataliste

"Quelque loi qu'il vous dicte, il faut vous y soumettre."

Racine, Phèdre

"Voulez-vous de bonnes lois; brûlez les vôtres, et faites-en de nouvelles."

Voltaire, Dictionnaire Philosophique

Like the other two traditional branches of macroscopic physics, mechanics and electromagnetism, thermodynamics was constructed progressively by induction. To start with, experimental observations were synthesized by *laws*, such as Newton's, Coulomb's or Gay-Lussac's laws; this process, which started very early for mechanics (Archimedes), accelerated from the seventeenth to the nineteenth century for the whole of macroscopic physics. During the nineteenth century one realized through a new induction that these laws could be derived from a few unifying *principles* which were more abstract, but very general: Lagrangian or Hamiltonian analytical mechanics, the Maxwell equations, and the Laws of thermodynamics. Finally, the first third of the twentieth century has brought a new understanding and unification by basing all these principles upon the new, microscopic, *quantum and statistical* physics. Apart from its philosophical interest, each stage of the unification has extended the possibilities for predictions by allowing one to proceed henceforth by deduction. From fundamental principles one derives new laws, on which one can confidently build even before experimental checks; in particular we have seen and we shall see how statistical physics allows us to calculate equations of state which thermodynamics introduces as empirical data. One can thus consider, at the present level of knowledge, that thermodynamics is *incomplete*: statistical mechanics must intercede when one tries either to get a better *understanding* of the significance of thermodynamic quantities, or to *calculate* them for some substance or other starting from its microscopic structure. The preceding chapters have shown us in a general way how to realize this programme; the following chapters will present us with many applications to simple substances.

Nevertheless, for many practical or technical applications where one is dealing with more complicated objects, it is often more efficient *not to start from the most fundamental level* possible and not to try and calculate everything. Macroscopic thermodynamics then provides a framework which, while rigorous and general, is economic, and which suffices to establish many relations and inequalities between different quantities. It is therefore essential, especially for the engineer, to be able to apply thermodynamics autonomously by appealing only to its own foundations. In Chap.5 we have reminded ourselves of the traditional formulation of these foundations, which we inherited from the nineteenth century. Nevertheless there exists a more modern and more synthetic presentation, due to Callen, which enables us better to master the subject and which has the advantage of being sufficiently close to statistical physics to benefit from contributions from it. To this aim we shall draw inspiration from Callen's book,[1] and again discuss the basic principles in the unified form that he has given where the *entropy* plays the dominant rôle (§§ 6.1 and 6.2). We shall then indicate the powerful *techniques* used when applying thermodynamics, especially those dealing with changes of variables (§ 6.3) and with general properties derived from the Laws of thermodynamics (§§ 6.4 and 6.5). Finally we discuss various *examples* (§ 6.6).

At the price of sometimes repeating ourselves, we have written the present chapter in such a way that it can be read and used independently of the remainder of the book. On the other hand, the contents of Chaps.2 to 5 suffice for an understanding of the applications of statistical physics which we shall consider, starting from Chap.7.

Below we shall restrict ourselves to macroscopic, stable or metastable, *equilibrium* states, reserving until Chap.14 the study of the temporal development of macroscopic processes. The term "thermodynamics" is therefore twofold inadequate: the Laws deal not with "*dynamics*", but with "*statics*". Moreover, thermodynamics is not solely a theory of "*thermal*" phenomena: it covers *all exchanges*, of energy and of heat, of course, but also of *particles*, of momentum and of any other conserved quantities. We shall often use in this chapter the term "thermostatics" in order to contrast it to the true "thermodynamics" of Chap.14.

---

[1] H.B.Callen, *Thermodynamics*, Wiley, New York, 1960.

6.1 Return to the Foundations of Thermostatics

6.1.1 The Object of Thermodynamics

The complete characterization of a physical system would involve the knowledge of a huge number of microscopic data. Macroscopic physics, on the other hand, restricts itself to the study of a reduced number of *collective variables*: shapes, densities, mean positions or velocities for mechanics – and the disciplines connected with it, such as acoustics, fluid dynamics, elasticity, or the strength of materials – charges, currents, or electric and magnetic polarizations for electromagnetism, molecular concentrations for chemistry.

Nevertheless, there persist at the macroscopic level some consequences of the hidden microscopic degrees of freedom. For instance, energy can be transferred to them from the collective degrees of freedom in the form of *heat.* Thermodynamics enables us, without explicitly introducing these microscopic coordinates on which, in principle, the mechanical, electromagnetic, or chemical collective coordinates depend, to take into account their residual macroscopic effects, such as heat or electric resistivity.

Thermostatics deals with *thermal*, and also *osmotic, electric, or chemical*, *equilibrium* states which remain unchanged with time and which are independent of the history of the system. These equilibrium states often are metastable, since the time for the establishing of absolute equilibrium, where all physical, chemical, or nuclear reactions have come to an end, can be huge (see the end of § 4.1.5). On the other hand, statistical mechanics and thermodynamics of irreversible non-equilibrium processes enable one to explain a large number of phenomena, but they do not constitute a discipline which is as coherent or systematic as the theory of thermostatic equilibria, to which we shall restrict ourselves in the present chapter. We shall embark upon the study of the dynamics near equilibrium in Chap.14.

6.1.2 The Maximum Entropy Principle

Instead of basing thermodynamics on the traditional principles reviewed in Chap.5, we start from Callen's formulation which is directly based upon a postulate about the existence and the maximum property of the entropy. We shall first state this principle in a general and abstract form, and then clarify its meaning through comments. Later on we shall see that it encompasses the standard Laws.

The equilibrium states of a system are characterized on the macroscopic scale by a set of *extensive variables* A_i and by a function of these variables, the entropy, S, which is *continuously differentiable, positive, and additive*: the entropy of a composite system is the sum of those of its parts, the entropy of a homogeneous substance is extensive. In an *isolated composite* system the lifting of some constraints may allow exchanges between subsystems, which are reflected in changes in the A_i; the domain $\mathcal{A}$ that is allowed for the A_i variables is restricted by the remaining constraints and by the *conservation laws*. In the final equilibrium state that the system reaches, the value of the A_i variables is determined by looking for the *maximum of the entropy in the domain* $\mathcal{A}$.

Amongst the extensive A_i variables we have the *internal energy*; the state of a fluid is further characterized by its *volume* Ω and its *number* N *of molecules*. Mixtures and chemical equilibria may involve several numbers of the various kinds of particles. When the system consists of several homogeneous fragments, the A_i comprise the variables, such as energies, volumes, numbers of moles, ..., relating to *each of the fragments*, and the index i denotes both the nature of the variable and the subsystem. It is also possible that the properties of the system at equilibrium, such as an electrically charged substance or a solid under constraints, vary *continuously* from point to point, on a scale which is large compared to microscopic distances. The A_i variables then refer to each volume element, and the index i includes the point coordinates. These variables may also denote in electromagnetism the *charge* or the magnetic or electric *dipole moment* of each volume element, or for an elastic solid the product of the 6 components of the *deformation tensor* with the volume elements. Note that the extensive A_i variables can include, depending on the circumstances, either the ξ_α or the X_α variables introduced in the definition (5.11) of work; this will be discussed in §§ 6.2.1 and 6.6.5.

Most of the A_i variables are *conservative*, their change being compensated by an opposite change of the corresponding variable for another part of the system. This is the case for the *internal energy*, for the *numbers of molecules* of each type when there are no chemical reactions, or for the numbers of moles of atoms, ions, or radicals in the case of chemical equilibrium. The *volume* can also be considered to be conservative if the system consists of two parts separated by a moving piston, the volumes of which occur amongst the A_i. The conservation laws and in particular the First Law appear indirectly in the above principle, through the definition of the domain $\mathcal{A}$.

The present formulation is directly adapted to the determination of the macroscopic equilibrium state of a composite system with parts which are more or less partially in communication with one another, once we know the expressions for the entropies of these parts: the partition of the *energy* between systems in thermal contact, of *molecules* in osmotic equilbrium, of the *charge* density in electrostatic equilibrium, of *constituent atoms* between the various molecules which they can form in chemical equilibrium, or even the position of a moving *piston* separating two fluids. We shall see that the generality of exchange situations makes it actually possible for the maximum entropy principle to encompass the whole of thermostatics.

In particular, even though we are dealing with equilibrium states, it also applies to *quasi-static* processes which are sufficiently slow that the system at each moment can be considered to consist of parts all nearly in equilibrium; in this case it implies that the total entropy is a non-decreasing function of time. If, however, we are dealing with an *irreversible* process, such as a Joule expansion, a sudden mixing, or a chemical reaction, the principle refers only to the initial and the final time; it states that the total entropy of the system must increase from an initial equilibrium state to a final equilibrium state, even though the intermediate states are arbitrary. For a non-isolated system, *any decrease in the entropy must be accompanied by an increase, at least as great, in the entropy of the systems to which it is coupled.*

We may under certain circumstances partly abandon the *additivity* condition on the entropy. For instance, if we want to use a semi-macroscopic theory to determine the density variation at a *liquid-vapour interface* near the critical point, we divide space into volume elements where the fluid is practically uniform. The A_i variables are the energy, the number of particles, and the volume of each element. The additive part of the entropy is the sum of the entropies of these volume elements; however, experiments show that one must include an extra contribution which is not there for a uniform fluid and which tends to restore uniformity. A simple empirical model, valid for slow spatial variations in the density $n(\boldsymbol{r})$, consists in taking for that contribution the expression $-K \int (\nabla n)^2 \, d^3\boldsymbol{r}$. Looking for the maximum of the total entropy, under the constraint that the density changes in space from that of the liquid to that of the vapour, gives us the structure of the interface. The latter must have a minimum area to reduce the effects of the extra entropy term. In this way we understand the origin of *capillary forces* which are proportional to K (Exerc.6c and 6d).

Similarly, in electromagnetism or gravitational theory (Exerc.6e), *the energy is no longer extensive* because of the existence of long-range potentials, but one can easily extend the principle of maximum entropy to such cases.

The additivity of the entropy is somewhat subtle in the case of *chemical equilibria*. In the gaseous phase, it is useful to analyze the system as consisting of subsystems, not separated in space, consisting of the molecules of the various species. For instance, for a H, H_2 mixture, the state variables A_i are the volume Ω, the energy U_1 and number N_1 of unbound H atoms, and the energy U_2 and number N_2 of H_2 molecules. The $H_2 \leftrightarrows 2H$ equilibrium is determined by looking for the maximum of the total entropy under the constraints that $U_1 + U_2$ and $N_1 + 2N_2$ are fixed. This entropy is, indeed, the sum of the entropies $S_1(U_1, \Omega, N_1)$ and $S_2(U_2, \Omega, N_2)$ of each chemical species, calculated as if the molecules occupied the whole of the volume Ω. However, if one assigns to each species a volume Ω_1 or Ω_2, proportional to the number of particles, the entropy contains besides the sum $S_1(U_1, \Omega_1, N_1) + S_2(U_2, \Omega_2, N_2)$ a contribution which is the *mixing entropy* (§ 8.2).

We have already stressed (§§ 4.1.4 and 4.1.5) the frequent occurrence of *quasi-equilibrium* situations where there is not enough time to establish a true equilibrium. Such situations can be treated in the general framework defined above. Amongst the extensive A_i variables we must include certain quantities which enable us to distinguish a metastable state from a stable state. For instance, a metastable supersaturated vapour and the stable liquid phase are distinguished by the volume they occupy. It is sufficient in order to determine a final metastable state to constrain these quantities in such a way that *only the metastable region is accessible*; the quasi-equilibrium that we are looking for is provided by the maximum of the entropy in the domain $\mathcal{A}$ thus demarcated. The true equilibrium, on the other hand, corresponds to the absolute maximum.

These remarks show that, notwithstanding its power and its generality, the principle of thermostatics cannot be used without precautions. The choice of the A_i variables which characterize the macroscopic state is essential, but not always obvious. For instance, when there is no applied magnetic field, one is tempted to forget to include the magnetic moment amongst the A_i; this omission, in general without any consequences, is nevertheless unfortunate for a ferromagnetic substance below the Curie temperature. Similarly, when we want to describe certain plastic substances or materials which have shape memory, it may be necessary to introduce "*hidden variables*", the macroscopic meaning of which is not evident. The choice of such variables can sometimes be guided by statistical physics; by default we must resort to empiricism. We have just seen that a certain amount of empiricism was also necessary to introduce adequate constraints in the case of metastability.

6.1.3 Connection with Statistical Physics

The formal analogy between the maximum *entropy* principle of thermostatics and the maximum *statistical entropy* principle (§§ 4.1.3 and 5.7.2), on which the microscopic study of equilibrium systems is based, suggests that identifying these two entropies would enable us to have the former principle based upon the latter one. Things are, however, not quite that simple, as the statistical entropy $S(\widehat{D})$ is related to the *microscopic description* of the macro-state of the system by means of the density operator, whereas the entropy $S(A_i)$ of thermostatics depends solely on the *macroscopic A_i variables*.

Before the various parts of the system are put into contact, the A_i can take on arbitrary values. Thermostatics determines their final value in the equilibrium state reached after interaction. We also know that the latter is on the microscopic scale described by the density operator which makes the statistical entropy a maximum under the constraints defining the domain $\mathcal{A}$. We are thus led to justify the maximum entropy principle of thermostatics by *proceeding in two stages*, just as we did in § 5.1.2 where the A_i variables were the energies U_a and U_b of the two subsystems.

In the *first stage* we find out what is the *least biased* density operator $\widehat{D}_{A_i}$ describing the situation where *all A_i variables are frozen* into a given arbitrary value. Assuming, for instance, that these constraints all relate to the expectation values of the observables $\widehat{A}_i$, the procedures of § 4.2.1 show that $\widehat{D}_{A_i}$ has the *exponential form* (4.6), that is,

$$\widehat{D}_{A_i} = \frac{1}{Z} \exp\left(- \sum_i \lambda_i \widehat{A}_i \right), \tag{6.1}$$

where the Lagrangian multipliers λ_i are adjusted in such a way that

$$\frac{\partial \ln Z}{\partial \lambda_i} = - A_i. \tag{6.2}$$

We are then led to identify the *entropy $S(A_i)$ of thermostatics*, which is a function of the A_i variables, with the *statistical entropy of the distribution* (6.1), which by means of (4.7) and (6.2) can be expressed in terms of the λ_i parameters and hence of the A_i parameters. The entropy of thermostatics is thus the *maximum of the statistical microscopic entropy* $S(\widehat{D})$ over the set of distributions $\widehat{D}$ compatible with the constraints $\langle \widehat{A}_i \rangle = A_i$:

$$\boxed{S(A_i) \equiv \max_{\widehat{D}} S(\widehat{D}), \qquad \text{for} \qquad \mathrm{Tr}\, \widehat{D}\widehat{A}_i = A_i} \, . \tag{6.3}$$

More generally, if some data A_i, such as the volumes of the subsystems for a canonical ensemble, are given exactly and not as expectation values, they enter directly into $\widehat{D}$, either through the definition of the Hilbert space or through the Hamiltonian, and there is no need to introduce the conjugate λ_i variables (§ 4.1.2). The relation (6.3) still holds, some constraints on the A_i variables now being implemented directly and not through the equations $\mathrm{Tr}\, \widehat{D}\widehat{A}_i = A_i$. As in the case of the canonical equilibrium ensembles (§ 5.5.3), here also the specific procedure followed is immaterial: the extensivity of each subsystem implies that the same function $S(A_i)$ is obtained, whether microscopically the A_i are treated as expectation values or as exact data.

The interpretation of $S(A_i)$ in the framework of information theory is clear. We are dealing with the *uncertainty in the macro-state* of the system when the A_i are the *only quantities which are known.* In fact, any density operator $\widehat{D}$ different from (6.1) and satisfying the constraints on the A_i yields $S(\widehat{D}) < S(A_i)$; being biased it contains not only the relevant information on the A_i, but also a certain amount, $S(A_i) - S(\widehat{D})$, of irrelevant information about other quantities (§ 4.1.3). The entropy $S(A_i)$ of thermostatics can thus be interpreted as the *relevant statistical entropy relating to only the A_i variables.*

The *second stage* of statistical physics, looking for the *maximum maximorum* of S, when some of the constraints on the A_i have been lifted, then coincides with the maximum entropy principle of thermostatics. The latter is therefore just a *restriction on the macroscopic variables* of the maximum statistical entropy principle of § 4.1.3, which itself is a consequence of the indifference principle (§ 5.7.2).

Thus, even if a substance is too complex for us to be able to evaluate (6.3) from its microscopic structure, statistical physics allows us to base the maximum entropy postulate on information theory. When one relies on experiments to find $S(A_i)$, one determines, actually, the *degree of disorder prevailing at the microscopic scale for given values of the macroscopic variables A_i*. The additivity and the extensivity of the entropy, postulated at the macroscopic scale, are related to the sub-additivity of the statistical entropy (§ 3.2.2) and appear as approximations, justified if the thermodynamic limit exists: the difference between the global entropy and the sum of those of the parts is, in general, a negligible surface effect. The interface and capillarity phenomena evoked in § 6.1.2 provide a notable exception.

6.1.4 Entropy and Disorder

In the statement of § 6.1.2 the entropy appeared as a quantity which is clearly fundamental, but abstract and rather mysterious. Its significance as a measure of the microscopic disorder, or, what amounts to the same, of the lack of information, enables us to understand its maximum property. The *unavoidable increase* in entropy reflects simply the *increase in disorder* which takes place when certain constraints preventing exchanges between subsystems are lifted. The amount by which it increases itself measures how much information has been lost at the microscopic scale when the A_i variables change spontaneously from the value to which they were constrained initially to their final value which allows a larger microscopic disorder.

It may be illuminating to analyse the thermodynamic processes in this light. For instance, the existence of a *latent heat* of melting, $L = T\Delta S$, means that the crystallization of a substance *increases its order suddenly*, by an amount measured by ΔS. In a *refrigerator* the lowering of the temperature is accompanied by an increase in order; we must pay a price for this, namely, we must *create elsewhere a larger disorder*, for instance, through the chemical reaction of burning fuel in an electrical plant which feeds the grid. Similarly, *living organisms* are more ordered than the inert matter from which they emerge; they can develop only thanks to subtle physico-chemical mechanisms which increase the disorder in their surroundings.

The rôle of the hot source in a *thermal engine* is clear, namely, it provides energy. However, in cooling off, this source becomes more ordered, which is impossible without creating disorder elsewhere. The *cold source* plays the indispensable rôle of a *source of order*. The total disorder increases when the temperatures of the two sources approach one another, and heat thus tends to flow from one to the other. We take advantage of this spontaneous flux to deflect part of it for our purposes in the form of work in the engine (§ 6.6.1).

These examples involve the energy, but conceptually the latter is not an essential ingredient: one should not restrict *dissipation*, that is, irreversible increase in entropy, to the transformation of work into heat. For instance, a *mixture* of coffee and milk, at the same temperature, will entail an increase in entropy without involving *any thermal effect*.

It took a century before it was recognized that *entropy* and *lack of information* or *disorder* were equivalent (§ 3.4.5). The fact that this equivalence is far from being intuitive occurs because Avogadro's number, and hence Boltzmann's constant, are large in SI units. To be sure, one can use information to make the entropy decrease, as in the Maxwell demon paradox. One must, however, at the same time, in a closed cycle, increase the entropy of an apparatus in order to gain that information. Nevertheless, the thermodynamic entropy of a macroscopic body is always equivalent to a huge amount of information, since the natural thermodynamic unit of 1 J K^{-1} is equivalent to 1.05×10^{23} bits. Hence such transformations cannot easily be studied experimentally.

6.2 Thermodynamic Identities

Important identities between various physical quantities can be derived from the very existence of a fundamental function $S(A_i)$.

6.2.1 Intensive Quantities

Given that we are going to reduce the problems of thermostatics to the quest, for an isolated system, of the maximum of S in a domain $\mathcal{A}$ for the A_i variables, it is natural to introduce and to interpret the partial derivatives of S with respect to these variables. Separating the total energy $U = A_0$ from the other A_i variables, we write

$$S = S(U, \{A_i\}), \tag{6.4}$$

$$dS = \frac{1}{T}\, dU + \sum_{i \geq 1} \gamma_i \, dA_i. \tag{6.5}$$

We assume here that only one of the A_i variables is an energy; if not, one introduces several temperatures.

The partial derivatives γ_i, like T, are homogeneous functions of degree 0 of the extensive A_i variables: they are the *intensive variables* which are conjugate to the A_i with respect to the entropy. Their relations with the A_i,

$$\boxed{\frac{1}{T} = \frac{\partial S}{\partial U}, \qquad \gamma_i = \frac{\partial S}{\partial A_i}}, \tag{6.6}$$

define the *equations of state* of the system. These equations of state are not independent, since they should be derivable from a *single* function $S(U, \{A_i\})$; inversely, giving one of them is not sufficient to characterize the system, as integrating it introduces arbitrary functions.

Identifying (6.4) with (6.3) enables us to connect the *intensive variables*, through $1/T = k\beta$, $\gamma_i = k\lambda_i$, with the Lagrangian multipliers λ_i which appear in (6.1) and which take the constraints $\langle \widehat{A}_i \rangle = A_i$ into account.

On the other hand, if the "position" variables ξ_α which are involved in the empirical definition (5.11) of the work are all extensive, they are, with U and N, part of the macroscopic variables that we have here called A_i. The differential (6.5) is thus the same as (5.56), so that the intensive variables,

$$\left.\begin{aligned} \gamma_i &= -\frac{\mu}{T} \qquad \text{when } A_i = N, \\ \gamma_i &= -\frac{X_\alpha}{T} \qquad \text{when } A_i = \xi_\alpha, \end{aligned}\right\} \tag{6.7}$$

can directly be interpreted in terms of the *chemical potential* μ and the *forces* X_α. Those include, in particular, $-\mathcal{P}$ if $A_i = \Omega$, or the components of the *stress* tensor if A_i denotes the corresponding component of the deformation tensor of an elastic solid. Recall that one usually inverts the relation (6.4) between entropy and energy as $U(S, \{A_i\})$ and that one uses the differential form (5.55) instead of (6.5).

We shall not reconsider here the discussions of Chap.5 which enabled us to find the meaning of the diverse quantities ξ_α, X_α, μ, now replaced by the A_i and the γ_i, and to identify work, chemical energy, and heat. Let us, nevertheless, note that here the A_i variables are *all extensive* whereas in Chap.5 for the sake of greater generality we allowed some of the "position" variables ξ_α to be intensive. For instance, we introduced magnetic work (1.31) of the form $-M\,dB$ for a quasi-static transformation of a paramagnetic system; in § 5.2.3 (footnote 1) we have also considered work done by gravitational forces where the ξ_α variable was intensive. In fact, such situations occur in electromagnetism and in gravitational theory because the long-range nature of the interactions makes the definition of the physical system, and at the same time that of *work*, *ambiguous*. In § 6.6.5 we shall discuss in detail the example of work in a dielectric where the ξ_α and X_α variables can exchange their rôles, depending on the point of view. In what follows we shall restrict ourselves to circumstances where the variable $\xi_\alpha = A_i$ is extensive, for instance, a magnetic or electric moment; the conjugate variable $X_\alpha = -T\gamma_i$ is then the $\boldsymbol{B}$ or $\boldsymbol{E}$ field, whereas in § 1.3.3 we had the opposite situation.

6.2.2 Conditions for Equilibrium Between Two Systems

When conservative quantities can be exchanged between various parts of a system, equilibrium is reached when the *corresponding intensive variables are equal.* If, in fact, for two subsystems, A_1 and A_2 denote the same conservative quantity which can be exchanged, equilibrium is determined by looking for the maximum of $S(A_1, A_2, \ldots)$ for a fixed value of $A_1 + A_2$. In the domain $\mathcal{A}$ defined in this way we have $dA_1 + dA_2 = 0$, so that $dS = 0$ implies $\gamma_1 = \gamma_2$.

We have thus found again the *Zeroth Law* and its extensions (§ 5.1), as a consequence of the maximum entropy principle: equalling of temperatures, of the variables $-\mu/T$, or $\mathcal{P}/T$ in the case of exchanges of, respectively, energy, particles, or volume.

In the case of chemical equilibrium (§ 6.6.3), the relations which express the conservation laws in terms of the extensive A_i variables are not as simple as $A_1 + A_2 =$ const; in that case the equilibrium conditions take a different form, but can again be written in terms of the intensive variables. Finally, if an A_i variable is allowed to take on arbitrary values, *without any constraint* imposed by a conservation law, its *conjugate intensive variable vanishes* at equilibrium, since $\partial S/\partial A_i = 0$. For instance, photons can be created or absorbed by the wall of the vessel in which they are enclosed; their number is not conserved, and the *chemical potential vanishes* for a gas of photons at equilibrium (§ 10.5.2 and Chap.13).

6.2.3 Gibbs-Duhem Relations

These relations, which we have already written down in § 5.6.5 for a fluid, express mathematically the extensivity of S *for each homogeneous part* of a composite system. Changing the volume Ω by $\lambda\Omega$ multiplies all extensive quantities by a factor λ, so that S satisfies the identity

$$S(\lambda U, \{\lambda A_i\}) \;=\; \lambda\, S(U, \{A_i\}) \tag{6.8}$$

for an extensive system. Differentiating (6.8) with respect to λ and using (6.6) for $\lambda = 1$, we get the *identity*

$$\frac{U}{T} + \sum_{i\geq 1} \gamma_i A_i \;\equiv\; S(U, \{A_i\}) \tag{6.9}$$

between the various intensive quantities $T, \gamma_i, U/S$, and A_i/S.

The Gibbs-Duhem relation is derived from (6.9) by differentiation and use of (6.5), which gives us

$$U\, d\left(\frac{1}{T}\right) + \sum_{i\geq 1} A_i\, d\gamma_i \;\equiv\; 0, \tag{6.10}$$

again expressing the fact that the *intensive variables* T, γ_i *are not independent.*

If the system consists of *several homogeneous phases* one can take the *additivity* of the entropy into account and write a Gibbs-Duhem relation *for each part*; the number of independent intensive variables is reduced accordingly (§ 6.4.6).

6.2.4 Mixed Second Derivatives Identity

A large number of experimental physical data are *response coefficients*, that is, ratios between variations of two extensive or intensive variables. All those quantities can be expressed in terms of the second derivatives of the fundamental function $S = S(U, \{A_i\})$, or of $U = U(S, \{A_i\})$.

For instance, the specific heat at constant volume of a fluid, which is defined by

$$C_v \equiv T \left(\frac{\partial S}{\partial T}\right)_{\Omega,N}, \tag{6.11}$$

equals

$$C_v = T \bigg/ \frac{\partial^2 U(S,\Omega,N)}{\partial S^2}. \tag{6.11'}$$

As an exercise one can prove that the expansion coefficient,

$$\alpha \equiv \frac{1}{\Omega}\left(\frac{\partial \Omega}{\partial T}\right)_{\mathcal{P},N}, \tag{6.12}$$

is given by

$$\frac{1}{\alpha} = \Omega \frac{\partial^2 U}{\partial S \partial \Omega} - \Omega \frac{\partial^2 U}{\partial S^2}\frac{\partial^2 U}{\partial \Omega^2} \bigg/ \frac{\partial^2 U}{\partial S \partial \Omega}. \tag{6.12'}$$

Quantities such as the isothermal compressibility or the specific heat at constant pressure can similarly be expressed in terms of second derivatives of the fundamental function, using elementary differential calculus.

Whereas the Gibbs-Duhem identity used the extensivity, one can find other relations between various physical quantities by simply using the fact that a mixed second derivative can be evaluated in two ways:

$$\frac{\partial}{\partial x_j}\left(\frac{\partial f}{\partial x_i}\right) = \frac{\partial}{\partial x_i}\left(\frac{\partial f}{\partial x_j}\right). \tag{6.13}$$

For instance, if we write, for $U(S, \Omega, N)$, that

$$\frac{\partial}{\partial \Omega}\left(\frac{\partial U}{\partial S}\right)_{\Omega,N} = \frac{\partial}{\partial S}\left(\frac{\partial U}{\partial \Omega}\right)_{S,N}$$

we find the identity

$$\left(\frac{\partial T}{\partial \Omega}\right)_{S,N} = -\left(\frac{\partial \mathcal{P}}{\partial S}\right)_{\Omega,N},$$

or

$$\left[\frac{1}{\Omega}\left(\frac{\partial \Omega}{\partial T}\right)_{S,N}\right]\left[\frac{1}{T}\left(\frac{\partial \mathcal{P}}{\partial S}\right)_{\Omega,N}\right] = -\frac{1}{\Omega T}, \tag{6.14}$$

which connects the expansion coefficient in an adiabatic transformation with the change in pressure produced by supplying heat.

This example illustrates a field of useful applications of thermodynamics: from a mathematically trivial identity one deduces *non-trivial identities between response coefficients*, which can be checked experimentally, or which enable us to predict one from the knowledge of another. More generally, in order to establish identities involving quantities, such as the expansion coefficient (6.12), which have more complicated expressions in terms of second derivatives, it will be useful to combine the above idea with a change of variables technique (§ 6.3.5). Let us remark right now that when the entropy of a homogeneous substance depends on n extensive variables, the number of its independent second derivatives is through (6.10) and (6.13) reduced to $\frac{1}{2}n(n-1)$ so that there exist only $\frac{1}{2}n(n-1)$ *independent response coefficients.* For a fluid, we have $n = 3$ which provides 3 independent coefficients. Thus, though the existence of the fundamental function $S(U, \{A_i\})$ is difficult to demonstrate experimentally, it has many implications for the response coefficients, which are readily measured and tabulated.

We also note that identities such as (6.14) are constraints on the equations of state (6.6). If one wants to characterize a system by its equations of state, it is necessary that they have equal mixed derivatives; moreover, they must satisfy the Gibbs-Duhem relation which reduces the number of equations of state by one. Conversely, these constraints suffice to determine the fundamental function S, which one obtains by integrating the differential system (6.6), with the integration constant being fixed by Nernst's Law.

6.3 Changes of Variables

The whole complexity, but also the richness, of thermodynamics arises from the multiplicity of the variables which may be brought into play for one actual problem or another. The present section gives the appropriate techniques for changing the variables.

6.3.1 Legendre Transformations

The fundamental function $S(U, \{A_i\})$ is well suited to a study of situations governed by the maximum entropy principle where the system is *isolated* and where the data refer to its extensive variables. Nevertheless, one often is dealing with certain features of the equilibrium state under study, which involve the associated intensive variables. For instance, if the system studied *is not isolated*, but maintained at a temperature T through thermal exchanges with a much larger system which plays the rôle of a thermostat, we must not only let the A_i $(i \geq 1)$ vary in a certain domain $\mathcal{A}$, but also let U vary in such a way that $1/T = \partial S/\partial U$ remains constant. Of course, the values of U

and the A_i for this kind of equilibrium can be obtained (§ 6.3.3) by applying the maximum entropy principle to the composite system including *both* the thermostat and the system proper; however, looking for the maximum of the entropy of the system *by itself* over the manifold defined by $\partial S/\partial U = 1/T$ and over the domain $\mathcal{A}$ for the A_i would lead to a *wrong result*, as energy exchanges also modify the entropy of the thermostat.

In fact, we have already stressed (§§ 2.3.3 and 4.2.4) that when we change variables, changing, for instance, from U to its conjugate variable $1/T$ with respect to S, we must also *change the function*. The fundamental function $S(U, \{A_i\})$ contains all the thermostatic information about the system; eliminating U between S and the definition $1/T = \partial S/\partial U$ allows us to consider S as a function of the temperature and the A_i; however, the expression obtained in this way would no longer suffice to characterize completely the properties of the system, since the two different entropies

$$S_1 = S(U, \{A_i\}) \qquad \text{and} \qquad S_2 = S(U + \varphi(\{A_i\}), \{A_i\}),$$

where φ is an arbitrary function, lead to the same function S of T and the A_i. We have already indicated that in order after a change of variables to conserve the information contained in the function $S(U, \{A_i\})$ we should perform a Legendre transformation.

The mathematical theory of Legendre transformations is simple. Let

$$f = f(\{x_i\}, \{t_j\})$$

be a differentiable function of the $n + p$ variables $x_1, \ldots, x_n, t_1, \ldots, t_p$, and let

$$y_i = \frac{\partial f}{\partial x_i} \tag{6.15}$$

be the conjugate variables of the x_i with respect to f. We want to use the y_i and the t_j as the new variables. The Legendre transform $g(\{y_i\}, \{t_j\})$ of f is the new function obtained by eliminating the $\{x_i\}$ between (6.15) and

$$g = f - \sum_{i=1}^{n} x_i y_i. \tag{6.16}$$

It is essential to note that g is of interest only as a *function of the* $\{y_i\}$, $\{t_j\}$, whereas f must always be considered as a *function of the* $\{x_i\}$, $\{t_j\}$. To go back from the y_i variables to the x_i variables and from $g(\{y_i\}, \{t_j\})$ to $f(\{x_i\}, \{t_j\})$ we differentiate the function f and (6.16), which gives

$$\left.\begin{aligned} df &= \sum_i y_i \, dx_i + \sum_j u_j \, dt_j, \\ dg &= -\sum_i x_i \, dy_i + \sum_j u_j \, dt_j, \end{aligned}\right\} \tag{6.17}$$

where the $u_j = \partial f/\partial t_j$ are the conjugate variables of the t_j. As a result the x_i variables can be expressed as functions of the y_i and the t_j, as follows,

$$x_i = -\frac{\partial g}{\partial y_i}, \tag{6.18}$$

and $f(\{x_i\},\{t_j\})$ is obtained by eliminating the y_i between (6.18) and

$$f = g + \sum_i x_i y_i. \tag{6.19}$$

Apart from a few signs, the Legendre transformation is *symmetric*; the inverse transformation is obtained *without integration*. The relations (6.17) show also that the u_j, the conjugate variables of the t_j with respect to f, are also the conjugate of the t_j with respect to g:

$$u_j = \frac{\partial f}{\partial t_j} = \frac{\partial g}{\partial t_j}. \tag{6.20}$$

Finally, we get the relations between the *second derivatives* of f and g with respect to their natural variables by taking the derivatives of (6.18) and (6.20) with respect to x_i, t_j:

$$\left.\begin{aligned}
\delta_{il} &= -\sum_k \frac{\partial^2 g}{\partial y_i \partial y_k}\frac{\partial^2 f}{\partial x_k \partial x_l}, \\
\frac{\partial^2 g}{\partial y_i \partial t_j} &= -\sum_k \frac{\partial^2 g}{\partial y_i \partial y_k}\frac{\partial^2 f}{\partial x_k \partial t_j}, \\
\frac{\partial^2 g}{\partial t_j \partial t_k} &= \frac{\partial^2 f}{\partial t_j \partial t_k} + \sum_{il} \frac{\partial^2 f}{\partial t_j \partial x_i}\frac{\partial^2 g}{\partial y_i \partial y_l}\frac{\partial^2 f}{\partial x_l \partial t_k}.
\end{aligned}\right\} \tag{6.21}$$

In particular, the matrices of the second derivatives of f with respect to the x_i and those of $-g$ with respect to the y_i are each other's inverse.

The mathematicians prefer to define $-g$ as the Legendre transform of f, which leads to a simplification of the signs in the formulae and to a greater symmetry. That convention is used in analytical mechanics to connect the *Hamiltonian* with the *Lagrangian*, according to (2.63); the conjugated variables are then the *momenta* p_k and the *velocities* $\dot{q}_k$, whereas the positions q_k play the rôle of the t_j variables above. Nevertheless, the sign convention (6.16) is more convenient in thermodynamics where one often wishes to be able to perform successive Legendre transformations on several variables.

If the function f is arbitrary, its Legendre transform g is not necessarily single-valued. Nevertheless, if f is *convex* (or *concave*), its matrix of second derivatives with respect to the x_i is positive (or negative), and its Legendre transform is not only *single-valued*, but also *concave* (or *convex*). We shall see in § 6.4.1 that the thermostatic entropy $S(U,\{A_i\})$ is concave and that $U(S,\{A_i\})$ is convex so that the thermodynamic potentials which follow from them by Legendre transformations are single-valued.

6.3.2 Massieu Functions and Thermodynamic Potentials

Let us remind ourselves that the Massieu functions (§ 5.6.1) are the thermodynamic potentials adapted to the replacement of some A_i variables ($i \geq 0$) by their conjugates γ_i with respect to the entropy: they are the Legendre transforms of the entropy with respect to those A_i variables. Similarly, the usual thermodynamic potentials are the Legendre transforms of $U(S, \{A_i\})$ with respect to some of its natural variables (§ 5.6).

Thermodynamic potentials are not just purely technical means for *changing from one variable to its conjugate.* We have seen (§§ 5.6.1 and 5.6.6) that the Massieu functions *can in statistical mechanics be calculated* directly in terms of the partition functions. We have also seen (§ 5.7.3) that their exponential can be identified with the *probability distribution of the macroscopic A_i variables* in one canonical equilibrium or other. When searching for *the equilibrium of non-isolated systems* in § 6.3.3, we shall meet them again as "potentials", in the sense that their *maximum* enables us to determine that equilibrium. Finally, knowing the thermodynamic potentials of a system is convenient for an evaluation of its *exchanges of work and heat* when it is in communication with sources (§ 6.3.4).

It follows from the Gibbs-Duhem identity that one cannot Legendre transform S or U with respect to all their extensive variables: for instance, for a fluid this transform would vanish, since $U - TS + \mathcal{P}\Omega - \mu N \equiv 0$. More generally, one must retain at least *one extensive variable for each homogeneous phase.*

6.3.3 Equilibrium in the Presence of Sources

Let us return to the problem of the equilibrium of a system in communication with a *thermostat at temperature* T. We must look for the maximum of the total entropy $S + S_{\text{th}}$ of the isolated combined system consisting of the system proper plus thermostat; the A_i ($i \geq 1$) parameters of the system vary over the domain $\mathcal{A}$, while the energies U and U_{th} of the system and the thermostat also can change under the constraint that $dU + dU_{\text{th}} = 0$. To simplify the discussion we assume that U is the only energy occurring amongst the A_i ($i \geq 0$) variables of the system; one can, however, readily extend the discussion to a composite system in which internal exchanges of energy may take place. Changes in the entropy around its equilibrium value must satisfy

$$d(S + S_{\text{th}}) = 0, \qquad d^2(S + S_{\text{th}}) < 0. \tag{6.22}$$

As the thermostat is at a temperature T, the variations of its entropy and its energy satisfy

$$dS_{\text{th}} = \frac{1}{T} dU_{\text{th}} = -\frac{1}{T}\, dU;$$

moreover, the thermostat is assumed to be so large that its temperature does not change, which means that

$$d^2 S_{\text{th}} = -\frac{1}{T}\, d^2U + \frac{dT}{T^2}\, dU = -\frac{1}{T}\, d^2U.$$

We can thus get rid of the thermostat in (6.22) and write down the conditions for the *equilibrium of the system in contact with the thermostat* in the form

$$d\left(S - \frac{1}{T}U\right) = 0, \qquad d^2\left(S - \frac{1}{T}U\right) < 0, \tag{6.23}$$

for all changes of U and of the A_i in the domain $\mathcal{A}$.

The form $S - U/T$ of the expression, the maximum of which we must find, *reflects the tendencies* to which the system is subject. The *first term* expresses that the evolution towards disorder tends to make *the entropy S increase.* The *second term* expresses that this effect is opposed and controlled by the coupling to the thermostat. The latter appears in (6.23) solely through its temperature T. The *energy U* of the system also tends to *decrease*, as U enters (6.23) with a minus sign, and this the more strongly, the lower the temperature T of the thermostat, as the coefficient of U is $1/T$.

From the point of view of *statistical physics* we have seen (Eq.(4.10)) that the Boltzmann-Gibbs canonical equilibrium distribution is obtained when we look for the maximum of

$$S(\widehat{\mathcal{D}}) - k\beta \operatorname{Tr} \widehat{\mathcal{D}}\widehat{H} \tag{6.24}$$

over all possible trial density operators $\widehat{\mathcal{D}}$. We also know (§ 5.7.2) that a system in contact with a thermostat at temperature T reaches *canonical equilibrium* with as multiplier $\beta = 1/kT$. Comparing (6.24) with (6.23) shows that we can directly justify (6.23) by, as in § 6.1.3, looking for the maximum of (6.24) *in two stages*: the first stage, under the constraints $\operatorname{Tr} \widehat{\mathcal{D}}\widehat{H} = U$ and $\operatorname{Tr} \widehat{\mathcal{D}}\widehat{A}_i = A_i$ for $i \geq 1$, gives for $\widehat{\mathcal{D}}$ the form (6.1): expression (6.24) then becomes the same as (6.23) and the second stage gives again the macroscopic equilibrium condition for a system coupled to a thermostat. This condition (6.23) is thus a *restriction on the macroscopic variables U* and A_i ($i \geq 1$) of the *microscopic variational principle* introduced in § 4.2.2.

One of the equilibrium conditions (6.23), $\partial S/\partial U = 1/T$, simply expresses that the temperature of the system adjusts itself to T. This enables us to eliminate U as function of the A_i and T. This elimination is nothing but the *Legendre transform* of S with respect to U. It leads to the *Massieu function* Ψ_{C} defined by (5.59) and we must now look for its maximum with respect to the A_i variables, while the value of T is imposed by the thermostat. It is traditional to state this result in terms of the *free energy*:

> When a system is maintained at a temperature T and when certain internal exchanges allow the A_i parameters, other than the total energy,

to change in a domain $\mathcal{A}$, the equilibrium values of these parameters are obtained by looking for the *minimum of the free energy* on $\mathcal{A}$ for the given value of T.

More generally, let us assume that we are trying to determine the equilibrium state of a system which, instead of being isolated as in § 6.1.2, is *in contact with a reservoir* which forces some of the intensive variables γ_i to be fixed. We proceed as in the case of the thermostat – which is an energy reservoir. As variables we take, on the one hand, the intensive variables fixed by the reservoir and, on the other hand, the extensive variables pertaining to the other degrees of freedom; the domain $\mathcal{A}$ then occurs as a constraint on the latter. We perform the appropriate Legendre transform on S, which introduces a Massieu function depending on the intensive variables. The equilibrium state is obtained by looking for the *maximum of this Massieu function* over $\mathcal{A}$ for fixed values of its intensive variables.

A more usual procedure is based upon the remark that for an isolated system the maximum entropy postulate is, in the general case of systems at a positive temperature, equivalent to the search for the *minimum energy for a given entropy*. Indeed, the relation

$$dS = \frac{1}{T}\,dU + \sum_i \gamma_i\,dA_i$$

shows that in equilibrium we have $\sum_i \gamma_i\,dA_i = 0$ on the domain $\mathcal{A}$, which means that, if U is expressed in terms of S and the A_i variables, it is stationary for S fixed. At equilibrium the second differential,

$$d^2S = \frac{1}{T}\,d^2U + \frac{\partial}{\partial U}\left(\frac{1}{T}\right)\,dU^2 + 2\sum_i \frac{\partial^2 S}{\partial U \partial A_i}\,dU\,dA_i + \sum_i \gamma_i\,d^2A_i + \sum_{ij} \frac{\partial^2 S}{\partial A_i \partial A_j}\,dA_i\,dA_j \qquad (6.25)$$

is negative when $dU = d^2U = 0$ and for variations of the A_i in $\mathcal{A}$. As a result d^2U is positive for variations of the A_i in $\mathcal{A}$, if $dS = d^2S = dU = 0$. Thus, for an isolated system, the maximum of $S(U, \{A_i\})$ for U fixed and the A_i in $\mathcal{A}$ coincides with the minimum of $U(S, \{A_i\})$ for S fixed and the A_i in $\mathcal{A}$. This discussion shows that in order to find the equilibrium of a system in contact with a *reservoir* we must look for the *minimum of the thermodynamic potential* which is the Legendre transform of $U(S, \{A_i\})$ with respect to the variables whose conjugates are fixed externally.

Thus, we must look for the minimum of the *enthalpy* if a system is maintained at a given *pressure*; if it is maintained at *given temperature and given pressure*, we must find the minimum of the *free enthalpy*; and if it is in communication with a *thermostat and a particle reservoir*, we must look for the minimum of the *grand potential.*

6.3.4 Exchanges Between Systems in the Presence of Sources

Work done on a system during an infinitesimal *reversible* transformation, that is, a quasi-static transformation such that the total entropy of the system and the sources with which it interacts remains constant, is, according to (6.5), defined by

$$\delta \mathcal{W} = dU - T\,dS = -T \sum_{i \geq 1} \gamma_i\,dA_i. \tag{6.26}$$

In an adiabatic transformation, that is, a reversible transformation without heat exchange, the work received is equal to the change in the internal energy U.

Let us assume that the system a is *in communication with a thermostat and with another system* b to which it, reversibly, supplies useful work $\delta \mathcal{W}_{\mathrm{b}}$. The work $\delta \mathcal{W} = -\delta \mathcal{W}_{\mathrm{b}}$ done on the system a is given by (6.26), where T remains fixed, or

$$-\,\delta \mathcal{W}_{\mathrm{b}} = \delta \mathcal{W} = d(U - TS) = dF. \tag{6.27}$$

It is thus equal to the *change in its free energy* at the constant temperature fixed by the thermostat. As the latter can provide or absorb heat, the work $\delta \mathcal{W}_{\mathrm{b}}$ supplied to b is not equal to the decrease in the energy U of the system itself, but to the decrease in F. The existence of irreversibilities changes (6.27) into the inequality $\delta \mathcal{W}_{\mathrm{b}} < -dF$. This justifies the name of "free energy", as $-dF$ is the work which, at best, one can get from the system, taking into account that the latter is itself coupled to the thermostat.

Let us similarly consider a system a which is in contact with a *volume reservoir* and which *adiabatically* provides an amount of work $\delta \mathcal{W}_{\mathrm{b}}$ to another system b. This time, we must, if we want to find $\delta \mathcal{W}_{\mathrm{b}}$, subtract from the total work $-\delta \mathcal{W}$ provided by the system a, the work, $-\mathcal{P}\,d\Omega$, that it provides to the reservoir maintaining the constant pressure $\mathcal{P}$. In the case of a reversible transformation we have thus

$$\delta \mathcal{W}_{\mathrm{b}} = -\delta \mathcal{W} - \mathcal{P}\,d\Omega = -\,d(U + \mathcal{P}\Omega) = -\,dH. \tag{6.28}$$

Here, the available work is given by the *decrease in the enthalpy* H at the constant pressure imposed by the reservoir, and at constant entropy since the transformation is adiabatic.

The enthalpy is also useful to determine *heat exchanges* undergone by a homogeneous substance maintained at constant pressure, since

$$dH = T\,dS + \Omega\,d\mathcal{P} + \mu\,dN$$

then reduces to its first term. In technical applications thermal exchanges often occur in the open air and the atmosphere plays the rôle of the pressure reservoir. This is why engineering handbooks often give the enthalpy of substances at atmospheric pressure per mole or per unit mass as function of the temperature. We must note that the latter variable is not one of the natural variables of H in its rôle of thermodynamic potential, since one can extract all the properties of the material only through differentiation of $H(S,\mathcal{P},N)$; however $H(T,\mathcal{P},N)$ is adapted to the required aim: heat exchange determination.

Finally, if the system a is in contact both with a thermostat at a temperature T and with a volume reservoir at a pressure $\mathcal{P}$, we see similarly that the useful work that it can supply to another system b in a reversible transformation is the *decrease in its free enthalpy*

$$\delta\mathcal{W}_{\mathrm{b}} \;=\; -\,d(U-TS+\mathcal{P}\Omega) \;=\; -\,dG, \tag{6.29}$$

at given temperature and pressure. The use of the free enthalpy G is particularly useful for the study of transformations in the open air where the atmosphere plays the rôles both of an energy and a volume reservoir, for instance, for chemical reactions which often start and finish at atmospheric temperature and pressure. In that case, by virtue of the Gibbs-Duhem relation, G equals $\sum \mu_j N_j$, where the μ_j and the N_j denote the chemical potentials and the number of atoms of the various species.

These considerations stress the practical and technical use of the thermodynamic potentials as means of evaluating exchanges of work and of heat with a system, some intensive variables of which are fixed through contact with a reservoir. This is just the reason why the functions F, G, or H were originally introduced, and they are called thermodynamic "potentials" because their changes can be interpreted as work.

6.3.5 Calculational Techniques

(a) *Maxwell Relations.* The relation (6.13) between the mixed second derivatives, written out for either one of the fundamental functions $S(U,\{A_i\})$ or $U(S,\{A_i\})$, provided us with interesting identities between some response coefficients. On the other hand, we have seen that a convenient way to change variables consists in performing a Legendre transform on one of the fundamental functions, by introducing the appropriate Massieu function or thermodynamic potential. Writing down that the mixed derivatives of any of those functions are equal leads to new identities – the Maxwell relations. If we call the pairs of *conjugate variables*, either with respect to S or with respect to U, (x_1,y_1), (x_2,y_2), ..., (x_n,y_n), where for each couple (x,y) one variable is extensive and the other one intensive, the *Maxwell identities* have the form

$$\left(\frac{\partial y_1}{\partial x_2}\right)_{x_1 x_3 \ldots x_n} = \pm\left(\frac{\partial y_2}{\partial x_1}\right)_{x_2 x_3 \ldots x_n}; \tag{6.30}$$

the + sign corresponds to the case where the variables y_1 and y_2 are of the same kind, and the – sign to the case where one is intensive and the other one extensive.

The interest of the Maxwell relations for establishing relations between response coefficients lies in the fact that they encompass the equations (6.21) which follow from Legendre transforms and the identities (6.13).

(b) *Use of Jacobians.* The foregoing calculation techniques, Legendre transforms and thermodynamic potentials, Gibbs-Duhem relations, Maxwell relations, have a drawback. The n variables which were used to characterize a state are, in fact, *not arbitrary*: each of them should be taken to be one or other of each of the n pairs of conjugated variables, (A_i, γ_i) with respect to S, or (ξ_α, X_α) with respect to U. We cannot as yet deal with cases where, for instance, we need include both $\mathcal{P}$ and Ω among the state variabes.

Nevertheless, a response coefficient has the general form of a partial derivative $\partial y/\partial x_1$ of a function y of n *arbitrary state variables* $x_1, x_2, \ldots, x_n$; each of these is a function of the n original variables $A_1, A_2, \ldots, A_n$. It is thus useful to know how simply to carry out arbitrary changes of variables. If we have n functions $y_1, y_2, \ldots, y_n$ of n variables $x_1, x_2, \ldots, x_n$, the *Jacobian*

$$\frac{d(y_1, y_2, \ldots, y_n)}{d(x_1, x_2, \ldots, x_n)} \equiv \det \frac{\partial y_i}{\partial x_j} \tag{6.31}$$

is defined as the determinant of the first derivatives of the y's with respect to the x's. The Jacobians are *antisymmetric* with respect to the x and the y variables. They occur in the calculations of *response coeffcients* through the identity

$$\left(\frac{\partial y}{\partial x_1}\right)_{x_2,\ldots,x_n} = \frac{d(y, x_2, \ldots, x_n)}{d(x_1, x_2, \ldots, x_n)}. \tag{6.32}$$

Under changes of variables, they have simple properties which result from those of determinants, and which are reflected in the important *group laws*

$$\begin{aligned}\frac{d(y_1, y_2, \ldots, y_n)}{d(x_1, x_2, \ldots, x_n)} &= \left[\frac{d(x_1, x_2, \ldots, x_n)}{d(y_1, y_2, \ldots, y_n)}\right]^{-1} \\ &= \frac{d(y_1, y_2, \ldots, y_n)}{d(z_1, z_2, \ldots, z_n)}\,\frac{d(z_1, z_2, \ldots, z_n)}{d(x_1, x_2, \ldots, x_n)}.\end{aligned} \tag{6.33}$$

Manipulating the relations (6.33) is an efficient method for finding relations between response coefficients, once these are written in the form (6.32).

(c) *Exterior Algebra.* A still more powerful method, but slightly less elementary, uses exterior differential calculus; we shall here review a few elements of it which may be useful for thermodynamics. Consider a manifold characterized by n variables $x_1, x_2, \ldots, x_n$, with differentials $dx_1, dx_2, \ldots, dx_n$. A 0-form is identified with a function $f(x_1, x_2, \ldots, x_n)$. A 1-form,

$$\omega^{(1)} \equiv \sum_i f_i \, dx_i, \tag{6.34}$$

is an element of the n-dimensional vector space spanned by the base of differentials $dx_1, dx_2, \ldots, dx_n$; the f_i are arbitrary functions of the $x_1, x_2, \ldots, x_n$. To introduce the 2-forms, we define *exterior products*, also called *Grassmann products*, $dx_i \wedge dx_j \equiv -dx_j \wedge dx_i$ which are a set of $\frac{1}{2}n(n-1)$ *antisymmetric objects* deduced from the differentials. Now, a 2-form,

$$\omega^{(2)} \equiv \sum_{i,j} f_{ij}\, dx_i \wedge dx_j, \tag{6.35}$$

is an element of the $\frac{1}{2}n(n-1)$-dimensional vector space spanned by the $dx_i \wedge dx_j$; this space is generated as the antisymmetrized tensor product of the space of the 1-forms with itself. Only the antisymmetric part of f_{ij} with respect to its indices is involved in (6.35). Similarly one constructs by recurrence up to n-forms; the antisymmetrization implies that there exists for n-forms only a single base element $dx_1 \wedge dx_2 \wedge \ldots dx_n$ and hence that all n-forms can be expressed as $f(x_1, x_2, \ldots, x_n)\, dx_1 \wedge dx_2 \wedge \ldots dx_n$. The operations on the forms obey the rules of ordinary algebra, except for the fact that products of differentials are always understood as being completely antisymmetrized, and that the order in which they are written matters. For instance, the product of the 2-form $dx_1 \wedge dx_2$ with the 1-form dx_3 is $dx_1 \wedge dx_2 \wedge dx_3$; the product of $dx_1 \wedge dx_2$ with dx_1 vanishes.

One can easily *change variables* using the equations

$$dx_i = \sum_k \frac{\partial x_i}{\partial y_k}\, dy_k, \tag{6.36a}$$

together with their exterior products. In this way we find for the 2-forms

$$dx_i \wedge dx_j = \sum_{k<l} \frac{d(x_i, x_j)}{d(y_k, y_l)}\, dy_k \wedge dy_l, \tag{6.36b}$$

and for the n-forms

$$dx_1 \wedge dx_2 \wedge \ldots\; dx_n = \frac{d(x_1, x_2, \ldots, x_n)}{d(y_1, y_2, \ldots, y_n)}\, dy_1 \wedge dy_2 \wedge \ldots dy_n. \tag{6.36c}$$

The *Jacobians* thus occur here naturally.

The *exterior derivative of some form* $\omega^{(k)}$ is a form $\omega^{(k+1)}$, obtained by differentiating the coefficients according to the ordinary rules, and by considering the products of differentials generated in this way as exterior products. This amounts to combining the normal rules of differential calculus with the antisymmetrization rule for exterior products, and also with the convention $ddx_i = 0$. The exterior derivative of a 0-form f is the same as its ordinary differential: it is a 1-form (6.34) with $f_i = \partial f/\partial x_i$. The exterior derivative of the 1-form (6.34) is the 2-form

$$d\omega^{(1)} = \sum_{i,j} \frac{\partial f_i}{\partial x_j}\, dx_j \wedge dx_i = \sum_{i<j} \left(\frac{\partial f_j}{\partial x_i} - \frac{\partial f_i}{\partial x_j}\right) dx_i \wedge dx_j. \tag{6.37}$$

The essential property on which the use of exterior calculus in thermodynamics relies is the fact that *differentiating any k-form twice always leads to zero*:

$$dd\omega^{(k)} = 0. \tag{6.38}$$

For a 0-form this reduces to the identity (6.13) of the mixed derivatives; the rule $ddx_i = 0$ is just a special case of (6.38). Exterior calculus thus enables us to systematize, to generalize, and to combine conveniently several tools introduced above: the use of *Jacobians* in *arbitrary* changes of variables and the *mixed derivatives* identity.

As an example we derive for a fluid directly from $ddU = 0$ the identity

$$dT \wedge dS - d\mathcal{P} \wedge d\Omega + d\mu \wedge dN = 0. \tag{6.39}$$

By taking $\mathcal{P}$ and Ω as variables in (6.39) we get for N fixed the relation

$$\frac{d(T,S)}{d(\mathcal{P},\Omega)} \equiv \left(\frac{\partial T}{\partial \mathcal{P}}\right)_\Omega \left(\frac{\partial S}{\partial \Omega}\right)_\mathcal{P} - \left(\frac{\partial T}{\partial \Omega}\right)_\mathcal{P} \left(\frac{\partial S}{\partial \mathcal{P}}\right)_\Omega = 1 \tag{6.40}$$

between the changes in temperature and exchanges of heat during expansions or compressions.

(d) *Examples.* We have seen in § 6.2.4 that for a fluid only three response coefficients are independent. As an exercise one could apply the above techniques to prove the identity

$$C_\mathrm{p} - C_\mathrm{v} = \frac{T\Omega\alpha^2}{\kappa_T} \tag{6.41}$$

between the specific heats at constant pressure and constant volume (6.11), the expansion coefficient (6.12), and the isothermal compressibility

$$\kappa_T = -\frac{1}{\Omega}\left(\frac{\partial \Omega}{\partial \mathcal{P}}\right)_{T,N}. \tag{6.42}$$

One could also express in terms of these coefficients, which are the simplest ones to measure, the change of pressure through heating,

$$\left(\frac{\partial \mathcal{P}}{\partial T}\right)_{\Omega,N} = \frac{\alpha}{\kappa_T}, \tag{6.43}$$

the adiabatic compressibility,

$$\kappa_S = \frac{\kappa_T C_\mathrm{v}}{C_\mathrm{p}}, \tag{6.44}$$

the changes in the temperature and in the chemical potential through adiabatic compression,

$$\left(\frac{\partial T}{\partial \mathcal{P}}\right)_{S,N} = \frac{T\Omega\alpha}{C_\mathrm{p}}, \qquad \left(\frac{\partial \mu}{\partial \mathcal{P}}\right)_{S,N} = \frac{\Omega}{N}\left(1 - \frac{ST\alpha}{C_\mathrm{p}}\right), \tag{6.45}$$

the change in temperature through a *Joule expansion*, that is, a sudden irreversible expansion without exchange of work or heat with the outside,

$$\left(\frac{\partial T}{\partial \Omega}\right)_{U,N} = \frac{1}{C_{\rm v}}\left(\mathcal{P} - \frac{T\alpha}{\kappa_T}\right), \tag{6.46}$$

or through a *Joule-Thomson expansion*,

$$\left(\frac{\partial T}{\partial \mathcal{P}}\right)_{H,N} = \frac{\Omega}{C_{\rm p}}\,(\alpha T - 1). \tag{6.47}$$

The latter is an irreversible expansion without heat exchange with the outside, but where the initial and the final pressures are fixed. We shall show in § 9.2.5 that in that kind of expansion the enthalpy remains constant. The inversion temperature (§ 9.2.5), below which the Joule-Thomson expansion is an efficient means of cooling, is given by $\alpha T = 1$ according to (6.47), as αT increases when T decreases.

6.4 Stability and Phase Transitions

In this section we use the fact that S is not only stationary, but maximum at equilibrium.

6.4.1 Concavity of the Entropy

We shall now see that the maximum entropy principle itself implies that, for a homogeneous extensive substance, the *entropy* $S(\{A_i\})$ is a *concave function of the* A_i *coordinates* ($i \geq 0$), including the energy. Let us mentally split the system into two equal parts; the number of A_i variables is doubled. If the function $S(\{A_i\})$ were not concave, we could find changes ε_i in the A_i such that

$$2S(\{\tfrac{1}{2}A_i\}) < S(\{\tfrac{1}{2}A_i + \varepsilon_i\}) + S(\{\tfrac{1}{2}A_i - \varepsilon_i\}).$$

Due to the extensivity of S, the left-hand side equals $S(\{A_i\})$, that is, the entropy of the whole of the homogeneous system; the formation of heterogeneity associated with non-zero ε_i could thus make the entropy increase, so that the substance would not be homogeneous at equilibrium in contradiction to our original assumption.

Experiments show that the entropy is practically everywhere twice differentiable; its second derivatives can be discontinuous, but they remain bounded. The concavity of the entropy is thus equivalent to the fact that its second derivatives form a non-positive matrix. However, the extensivity implies that the *determinant of this matrix vanishes*. In fact, the Gibbs-Duhem relation (6.10), which we can also write in the form

$$\sum_{i\geq 0} A_i \,\frac{\partial^2 S}{\partial A_i \partial A_j} = 0, \qquad \forall\ j \geq 0,$$

entails the existence of an eigenvector A_i associated with the eigenvalue 0 of the $\partial^2 S/\partial A_i \partial A_j$ matrix.

It is advisable to check for any approximate theory or any empirical determination of the entropy that it is, indeed, a concave function of the extensive variables.

It follows from (6.25) that the second differential d^2U of the function $U(S,\{A_i\})$, when expressed in terms of dS and the dA_i $(i \geq 1)$, is equal to the product of $-T$ with the second differential d^2S of $S(U,\{A_i\})$, expressed in terms of dU and the dA_i $(i \geq 1)$. As a result, the *energy is a convex function of the extensive S and A_i variables.*

6.4.2 Thermodynamic Inequalities

The response coefficients that are most often the subject of measurements and of practical applications are functions of the second derivatives of S (§ 6.3.5), which are the elements of a *negative matrix.* From this we can derive inequalities necessarily satisfied by these coefficients, which enable us to check the consistency of an approximate theory, to detect experimental errors, or to make predictions.

In particular, the second derivative of S with respect to the variable U must be negative or zero. Hence, T is an increasing function of U, and the specific heat at constant volume is positive. The second derivatives of $-S$ and of U with respect to the other A_i variables must similarly be positive or zero. For instance, the adiabatic compressibility of a fluid,

$$\kappa_S = -\frac{1}{\Omega}\left(\frac{\partial \Omega}{\partial \mathcal{P}}\right)_{S,N} = \frac{1}{\Omega}\Big/\frac{\partial^2 U}{\partial \Omega^2}, \tag{6.48}$$

is positive.

More generally, all the diagonal minors of a positive matrix are positive, which leads to new inequalities. Conversely, an $n \times n$ matrix for which a set of *n nested diagonal minors* of ranks 1, 2, $\ldots, n$ have positive determinants is positive. Here, the determinant of the $\partial^2 S/\partial A_i \partial A_j$ matrix is zero due to the extensivity, so that $n-1$ conditions are necessary and sufficient to express that it is a non-positive matrix. The $\frac{1}{2}n(n-1)$ *independent response coefficients* are thus constrained to *satisfy $n-1$ fundamental inequalities.*

Even without writing those down, it is easy to find some of their consequences by manipulating the thermodynamic potentials. Let us, in fact, consider the Legendre transform $g(\{y_i\},\{t_j\})$ of a convex function: the matrix of the second derivatives of $f(\{x_i\},\{t_j\})$ is positive. The relations (6.21) show that the matrix of the $\partial^2 g/\partial y_i \partial y_k$ is negative, whereas the matrix of the $\partial^2 g/\partial t_j \partial t_l$ is positive. One can see this by evaluating $d^2 f$ for

$$dx_i = \sum_{l,j} \frac{\partial^2 g}{\partial y_i \partial y_l}\,\frac{\partial^2 f}{\partial x_l \partial t_j}\,dt_j,$$

and using (6.21). The second derivatives of a thermodynamic potential with respect to its *extensive* variables are thus *positive* or zero; with respect to its *intensive* variables, they are *negative.* By considering, for instance, the free energy and the Ω variable, we see that the isothermal compressibility,

$$\kappa_T = -\frac{1}{\Omega}\left(\frac{\partial\Omega}{\partial\mathcal{P}}\right)_{T,N} = \frac{1}{\Omega}\Big/\frac{\partial^2 F}{\partial\Omega^2}, \tag{6.49}$$

is positive. From (6.41) it follows that $C_\mathrm{p} > C_\mathrm{v}$.

6.4.3 The Le Chatelier Principle

Let us consider a fluid, the thermostatic properties of which are characterized by its internal energy $U = U(S, \Omega, N)$. The matrix of the second derivatives of U is non-negative. One of its eigenvalues is zero, expressing the extensivity, and one can eliminate it by letting N be constant. We shall write down the two fundamental inequalities which we have shown to exist in § 6.4.2; in the present case $n = 3$. To do that we express the conditions that two nested minors are positive:

$$\frac{\partial^2 U}{\partial S^2} \geq 0, \qquad \frac{\partial^2 U}{\partial S^2}\,\frac{\partial^2 U}{\partial\Omega^2} - \left(\frac{\partial^2 U}{\partial S\,\partial\Omega}\right)^2 \geq 0.$$

The first condition can be rewritten as $T/C_\mathrm{v} \geq 0$, or

$$C_\mathrm{v} > 0. \tag{6.50}$$

If we use the properties of Jacobians and the definitions (6.11) and (6.49), we see that the second condition is equivalent to

$$\begin{aligned} 0 \leq \frac{d(T,-\mathcal{P},N)}{d(S,\Omega,N)} &= \frac{d(T,-\mathcal{P},N)}{d(T,\Omega,N)}\,\frac{d(T,\Omega,N)}{d(S,\Omega,N)} \\ &= -\left(\frac{\partial\mathcal{P}}{\partial\Omega}\right)_{T,N}\left(\frac{\partial T}{\partial S}\right)_{\Omega,N} = \frac{T}{\Omega C_\mathrm{v}\kappa_T}, \end{aligned}$$

or

$$\kappa_T > 0. \tag{6.51}$$

The two conditions (6.50) and (6.51), which are necessary and sufficient for the *stability of the equilibrium of a homogeneous fluid*, are called *Le Chatelier's principle*: a system is stable if heating it increases its temperature, and if an isothermal expansion decreases its pressure. In another form, along the lines of § 6.4.1, Le Chatelier's principle expresses the fact that if we create a heterogeneity in the system, the reaction produced by this tends to restore the equilibrium. Indeed, if two parts of the system are brought to different temperatures $T_1 > T_2$, the resulting heat flux, directed from 1 to 2, has as a consequence that the temperature difference diminishes (Eq.(6.50)); if they are brought to different pressures, $\mathcal{P}_1 > \mathcal{P}_2$, the transfer of matter from 1 to 2 which tends to shift the densities towards the initial value has as a consequence that the pressure difference diminishes (Eq.(6.51)).

The additivity of the entropy implies that the stability of the equilibrium of a system consisting of several parts is guaranteed by the stability of its homogeneous components.

6.4.4 The Le Chatelier-Braun Principle

The Le Chatelier-Braun principle is concerned with deviations from equilibrium which involve no longer one, but two variables. To fix ideas, let A_1 be the energy and A_2 the volume. We assume that initially their conjugate intensive variables with respect to the entropy, γ_1 (equal to $1/T$) and γ_2 (equal to $\mathcal{P}/T$), have been determined through contact with two reservoirs, one of which can exchange energy and the other volume; after that the system has been isolated. First of all, the equilibrium is shifted by changing A_1 to $A_1 + dA_1$, while A_2 is fixed: transfer of energy without change in volume. The conjugated variable γ_1 changes by

$$d\gamma_1 = \left(\frac{\partial \gamma_1}{\partial A_1}\right)_{A_2} dA_1, \tag{6.52}$$

in the opposite sense to A_1 according to Le Chatelier's principle. Nevertheless, the other intensive variables are also, indirectly, changed by

$$d\gamma_2 = \left(\frac{\partial \gamma_2}{\partial A_1}\right)_{A_2} dA_1; \tag{6.53}$$

for instance, heating changes not only the temperature, but also the pressure. Secondly, we bring the system in communication not with the thermostat for γ_1, which would merely bring it back to its initial state, but with the second reservoir for γ_2 which can change the volume A_2. In the final equilibrium state, γ_2 will thus return to its original value, but A_1 (the energy) remains fixed at the perturbed value $A_1 + dA_1$. During this process, γ_1 (the temperature) changes indirectly by $d'\gamma_1$. Let us compare $d'\gamma_1$ with the initial change $d\gamma_1$. As γ_2 changes by $-d\gamma_2$, given by (6.53), we obtain

$$d'\gamma_1 = -\left(\frac{\partial \gamma_1}{\partial \gamma_2}\right)_{A_1} d\gamma_2 = -\left(\frac{\partial \gamma_1}{\partial \gamma_2}\right)_{A_1} \left(\frac{\partial \gamma_2}{\partial A_1}\right)_{A_2} dA_1. \tag{6.54}$$

Taking the ratio of (6.54) and (6.52) we find

$$\begin{aligned} -\frac{d'\gamma_1}{d\gamma_1} &= \left(\frac{\partial \gamma_1}{\partial \gamma_2}\right)_{A_1} \left(\frac{\partial \gamma_2}{\partial A_1}\right)_{A_2} \Big/ \left(\frac{\partial \gamma_1}{\partial A_1}\right)_{A_2} \\ &= \frac{d(\gamma_1, A_1)}{d(A_2, A_1)} \, \frac{d(A_2, A_1)}{d(\gamma_2, A_1)} \, \frac{\partial^2 S}{\partial A_1 \partial A_2} \Big/ \frac{\partial^2 S}{\partial A_1^2} \\ &= \left(\frac{\partial^2 S}{\partial A_1 \partial A_2}\right)^2 \Big/ \frac{\partial^2 S}{\partial A_1^2} \, \frac{\partial^2 S}{\partial A_2^2}. \end{aligned}$$

It thus follows from the concavity of S that

$$0 \leq -\frac{d'\gamma_1}{d\gamma_1} \leq 1. \tag{6.55}$$

The Le Chatelier-Braun principle reflects these inequalities: Let us consider a system, the equilibrium of which is shifted by a transfer referring to A_1; the conjugated intensive variable γ_1 has been perturbed, as well as another variable γ_2. A later process which brings γ_2 back to its initial value, thanks to a transfer referring to the quantity A_2, has *indirectly the effect of reducing the intial perturbation of* γ_1.

6.4.5 Critical Points

The eigenvalues of the matrix of the second derivatives of $S(U, \{A_i\})$ are negative or zero. We saw in § 6.4.1 that one of them, associated with the extensivity property, vanishes everywhere. The critical points correspond to particular values of the variables where at least one other eigenvalue vanishes. One can classify these points as critical, tricritical, ... points according to the rank of the determinant of the second derivatives of S.

For instance, for a fluid characterized by $S(U, \Omega, N)$, the rank of this determinant is 2 in an arbitrary point but falls to 1 in a critical point. This can be expressed by stating that the Jacobian

$$J = \begin{vmatrix} \dfrac{\partial^2 S}{\partial U^2} & \dfrac{\partial^2 S}{\partial U \partial \Omega} \\ \dfrac{\partial^2 S}{\partial \Omega \partial U} & \dfrac{\partial^2 S}{\partial \Omega^2} \end{vmatrix} = \frac{d(1/T, \mathcal{P}/T, N)}{d(U, \Omega, N)} \tag{6.56}$$

vanishes, for fixed N. As a result several *response coefficients diverge*: in fact, the expressions for the expansion coefficient,

$$\alpha = \frac{1}{J\Omega T^2}\left(-\mathcal{P}\frac{\partial^2 S}{\partial U^2} + \frac{\partial^2 S}{\partial U \partial \Omega}\right), \tag{6.57a}$$

the specific heat at constant pressure,

$$C_{\mathrm{p}} = \frac{1}{JT^2}\left(-\mathcal{P}^2\frac{\partial^2 S}{\partial U^2} + 2\mathcal{P}\frac{\partial^2 S}{\partial U \partial \Omega} - \frac{\partial^2 S}{\partial \Omega^2}\right), \tag{6.57b}$$

or the isothermal compressibility,

$$\kappa_T = \frac{1}{J\Omega T}\left(-\frac{\partial^2 S}{\partial U^2}\right), \tag{6.57c}$$

all have the Jacobian J in the denominator, whereas the specific heat at constant volume,

$$C_{\mathrm{v}} = -\left(T^2\frac{\partial^2 S}{\partial U^2}\right)^{-1}, \tag{6.57d}$$

may remain finite.

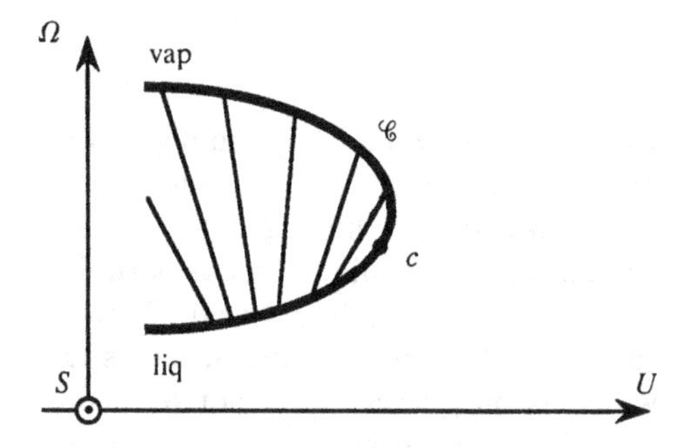

Fig. 6.1. The U, Ω plane for a liquid-vapour system

In general, the vanishing of a function of two variables U and Ω determines in the U, Ω plane a curve. Nevertheless, the function $J(U, \Omega)$ has special properties imposed on it by the concavity of the entropy, which restrict the possibilities. Assuming that $S(U, \Omega)$ can be differentiated at least thrice, one might imagine isolated critical points, as in the case when $S \propto -\left[(U - U_0)^2 + K(\Omega - \Omega_0)^2\right]^2$, or lines of critical points, as in the case when $S \propto -(U - U_0)^4 - K(\Omega - \Omega)^2$. However, such points have not been observed in Nature, as they are *not generic.* One expects, in fact, that in a real system the phenomena are not qualitatively sensitive to a small change in the physical parameters such as the particle masses or interactions. In the above-mentioned cases, however, the behaviour near the point $U = U_0$, $\Omega = \Omega_0$ or near the line $U = U_0$ would be changed by a small change in the parameters, which either would make the critical points disappear or would destroy the concavity of S. The real critical points, the properties of which remain *stable* when such a change takes place, have as their prototype the *liquid-vapour* transition. The structure of the function $S(U, \Omega)$ is then the following (Fig.6.1). A curve C where the second derivatives of S are discontinuous bounds a region in the U, Ω plane inside which J is identically equal to zero. In that region the $S(U, \Omega)$ surface is developable,[2] and it describes the coexistence between the two, liquid and gas, fluid phases (§ 6.4.6). Outside the C curve the function S is concave with $J > 0$ and describes a homogeneous fluid. The critical point c is the apex of the C curve. The second derivatives of S are continuous in that point, J is zero, but $S(U, \Omega)$ behaves, as experiments have shown, *non-analytically.* The theory of this *critical behaviour*, which had for a long time remained an unsolved problem, was seriously tackled only in the 1970s, inspired by Kenneth G. Wilson. Landau's theory which came earlier provides us with many useful results as a first approximation (Exerc.6d), even though it is not satisfactory in the immediate vicinity of the critical point.

[2] A developable surface in a 3-dimensional space is a surface which can be mapped onto a plane by means of a length-preserving transformation. This property is characterized by the relation $J \equiv 0$. It entails the existence on the surface of a set of straight lines, all tangent to some space curve.

6.4.6 Phase Separation

Experiments show the existence of systems which, while being homogeneous for certain values of the extensive parameters, for other values of these parameters split up spontaneously into two or more distinct homogeneous phases. The latter can either be qualitatively different (solid-fluid, or solids with different crystalline structures) or be qualitatively similar (liquid-gas). To show how this phenomenon enters the framework of thermodynamics, let us assume that an approximate theory which is only valid for a homogeneous system, or that an extrapolation of experimental measurements carried out on a homogeneous sample, have led to an expression for the entropy S_0 which, as function of the extensive variables, is *not concave*. Let S_1 be the convex envelope of S_0; it is defined, for each set of values of the extensive variables, as the upper bound of the set of points which can be regarded as barycentres of the points of S_0. Part of the surface that represents S_1 is the same as certain sheets a, b, ... of S_0; in such points, such as 4 (see Fig.6.2), the system is stable. The remainder of S_1 consists of a set of segments such as 2–3 in the figure – or, more generally, triangles, tetrahedra, ..., depending on the number of state variables – which are bitangent to S_0 at their ends; the corresponding sheet of S_1 is thus a fragment of a ruled surface which connects the sheets a, b, ... where $S_1 = S_0$. If S, as in § 6.4.5, depends on two state variables, S_1 is a developable surface. In the example of the liquid-vapour equilibrium (Fig.6.1), the part $S_1 = S_0$ describing homogeneous phases lies outside $\mathcal{C}$ and is connected; however, S_0 includes two separate sheets when the phases a and b differ qualitatively.

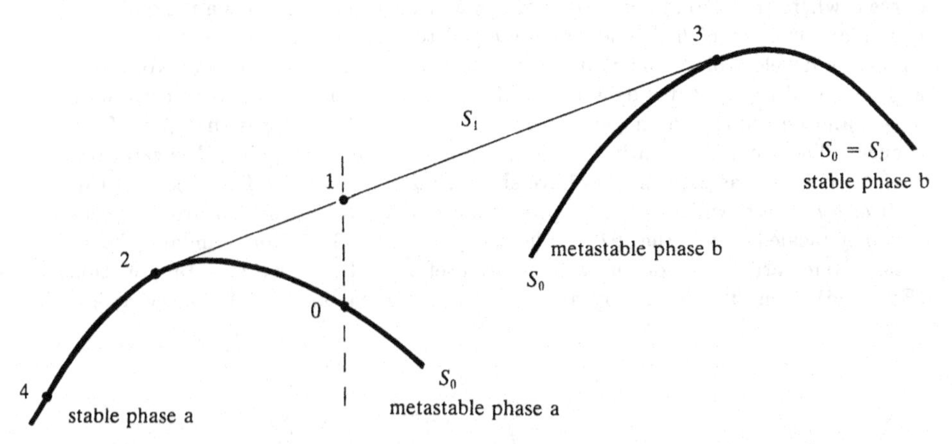

Fig. 6.2. Vertical section of the entropy surface for a system with two phases

Let us consider a point 0 of S_0 such that $S_0 < S_1$, situated on the continuation of the sheet a where $S_1 = S_0$. If 0 is sufficiently close to the region which represents stable homogeneous states such as 4 (Fig.6.2), it describes a state which is also *homogeneous, but unstable.* Such extrapolations of stable states to unstable homogeneous states are often observed when the lifetime of the latter is rather long: we are dealing with *metastable* states – supercooling of liquids or gases, superheating of solids or liquids, allotropic metastabilities as in diamond, hysteresis, and so on.

Nevertheless, in the strict framework of thermostatics the state 0 must be excluded. Let us, indeed, show by the reasoning of § 6.4.1 that the entropy can be increased by separating the system into two (or more) phases. The point 1 of S_1 that is associated with the same extensive variables as the point 0 is the barycentre of points 2 and 3, with weights λ_2 and λ_3 ($\lambda_2 + \lambda_3 = 1$), respectively. The two points 2 and 3 describe stable homogeneous phases a and b; the point 1 describes an inhomogeneous system, consisting of two parts which are similar to 2 and 3, but for which all extensive variables are weighted by λ_2 and λ_3. Thus, through transfer of the various conserved quantities from one region to another, the entropy can increase from the homogeneous state 0 to the heterogeneous state 1: the maximum entropy principle therefore implies that the system *spontaneously splits into two (or more) phases.* At equilibrium the *proportions of each phase* are λ_2 and λ_3, and the entropy is equal to

$$S_1 = \lambda_2 S_2 + \lambda_3 S_3. \tag{6.58}$$

In the region of coexistence of the phases we may characterize the system by its global extensive variables as above; since the representative surface of the entropy S_1 contains in those regions straight lines (or planes), the rank of the matrix of the second derivatives of S_1 decreases by one whenever a new phase appears. Nevertheless, it is better to characterize the system by the overabundant set of extensive variables for *each* of the m possible phases, rather than by its *global* extensive variables. The entropy in that representation is the sum of the contributions from the various phases; the extensivity of the phases entails the existence of m Gibbs-Duhem relations, so that the rank of the matrix of the second derivatives of S is decreased by m in the coexistence region. However, the extensive variables of the various phases are *not independent.* We must write down the *equilibrium conditions between the phases*, expressing that the total entropy is a maximum, that is, stationary against the transfer from one phase to another of any conservative quantity – energy, volume, or matter. As in § 6.2.2 these conditions can be found by writing down that *all intensive variables* calculated in one phase are *equal* to the corresponding variables in the other phases.

For instance, the conditions for equilibrium between the phases a and b of a *pure substance*, where one is a solid and the other a fluid, or one a liquid and the other a saturated vapour, can be expressed in terms of the three equations of state of each phase, by writing down that the *temperatures*, the *pressures*, and the *chemical potentials* must be equal:

$$\left.\begin{aligned} T_a(U_a/N_a, \Omega_a/N_a) &= T_b(U_b/N_b, \Omega_b/N_b), \\ \mathcal{P}_a(U_a/N_a, \Omega_a/N_a) &= \mathcal{P}_b(U_b/N_b, \Omega_b/N_b), \\ \mu_a(U_a/N_a, \Omega_a/N_a) &= \mu_b(U_b/N_b, \Omega_b/N_b). \end{aligned}\right\} \tag{6.59}$$

Maxwell's rule for constructing the horizontal part of the isotherms (§ 9.3.3) follows from this. The equality of the intensive variables entails the existence of 3 relations (6.59) between the 2×2 independent ratios U_a/N_a, Ω_a/N_a, U_b/N_b, and Ω_b/N_b of extensive variables. If, for instance, we consider the temperature as an independent variable, all other variables – pressure, chemical potential, energy, or entropy per particle for each of the phases, density for each of the phases – are determined when the two phases coexist. It is customary to represent the states of a substance by a *phase diagram*: for instance, in the $T, \mathcal{P}$ plane, the coexistence of two phases is possible along curves, outside which the stable equilibrium state is homogeneous. On such a coexistence curve the relative amount N_a/N_b of the two phases still remains undetermined. If 3 phases coexist, we obtain, instead of (6.59), 6 equations for 3×2 ratios U/Ω and Ω/N: instead of a coexistence curve, we have a *triple point*, as for the solid-liquid-vapour equilibrium.

These considerations can be generalized for arbitrary phase equilibria, involving, for example, mixtures of fluids; in that case it is customary to draw the phase diagrams in terms of relative concentrations rather than in terms of chemical potentials. The number of independent intensive variables, obtained by writing down relations similar to (6.59), is, for a mixture of r pure substances in m phases, equal to $r+2-m$. This count, called the *Gibbs phase rule*, enables us to classify the various kinds of equilibrium between phases and the corresponding shapes of phase diagrams. We still must specify an extensive variable for each of the phases in order completely to characterize the state of the system.

Equations (6.59) show that the *energy per mole* and the *density* are *discontinuous* when we pass from one phase to another. This is why such a change is called a *first-order phase transition*, in contrast to critical points which are considered to be second-order, as the above quantities are there continuous. Actually, for a liquid-gas transition the area lying inside the curve C in Fig.6.1, which describes the coexistence region, is represented in the $T, \mathcal{P}$ phase diagram by the first-order transition line $\mathcal{P}_s(T)$, also called the *saturation curve*. This line ends in a critical point (§ 6.4.5), the equivalent of the point c in Fig.6.1. Along it the discontinuities of U/N and Ω/N decrease, until they vanish at the critical point. Nevertheless, there remains in that point a trace of these discontinuities, as the derivatives of the functions U/N and Ω/N at the critical point are infinite. The divergences of the specific heat, of the compressibility, and of the expansion coefficient appear as the limit of the discontinuities associated with the first-order transition.

The discontinuity of the entropy across the coexistence line is reflected by the existence of a *latent transition heat* per mole,

$$L_{a\to b} = T\left(\frac{S_b}{N_b} - \frac{S_a}{N_a}\right), \tag{6.60}$$

where S_a/N_a and S_b/N_b are the entropies per mole of each of the phases. The transition from one phase to another, which occurs at constant temperature and pressure, needs an amount of heat proportional to the amount of matter transformed. It is easy to relate this latent heat to the density difference between the two phases and to the slope of the coexistence curve. To do this, we first of all write down that the free enthalpies (5.67) of the two phases are proportional to one another along the saturation line $\mathcal{P}_s(T)$ by virtue of the Gibbs-Duhem identity (5.78) and of the fact that the chemical potentials are equal,

$$\mu = \frac{G_a(T,\mathcal{P}_s,N_a)}{N_a} = \frac{G_b(T,\mathcal{P}_s,N_b)}{N_b}.$$

By taking the derivative with respect to T along the curve $\mathcal{P}_s(T)$ we then get

$$\frac{1}{N_a}\left(\frac{\partial G_a}{\partial T} + \frac{\partial G_a}{\partial \mathcal{P}}\frac{d\mathcal{P}_s}{dT}\right) = \frac{1}{N_b}\left(\frac{\partial G_b}{\partial T} + \frac{\partial G_b}{\partial \mathcal{P}}\frac{d\mathcal{P}_s}{dT}\right),$$

which gives us the *Clapeyron relation*

$$L_{a\to b} = T\left(\frac{\Omega_b}{N_b} - \frac{\Omega_a}{N_a}\right)\frac{d\mathcal{P}_s}{dT}. \tag{6.61}$$

6.5 Low Temperatures

6.5.1 Nernst's Law

The maximum entropy principle of § 6.1.2 encompasses all Laws of thermostatics in the standard formulation of Chap.5, except the Third Law. For the continuity of our discussion we remind ourselves here of the statement of this law (§ 5.4.1): the *macroscopic entropy* S of any system *tends to zero*, as well as the *temperature*, when the energy tends to its *minimum value.*

6.5.2 Vanishing of Some Response Coefficients

A specific heat, for instance at constant volume or at constant pressure, is related to the entropy through

$$C_Y = T\left(\frac{\partial S}{\partial T}\right)_Y,$$

where Y denotes the set of all variables which are kept constant. As $S(0) = 0$, the integral

$$S(T,Y) = \int_0^T \frac{C_Y}{T}\, dT$$

converges at $T = 0$, so that all *specific heats vanish* in the low temperature limit.

The vanishing of S implies also that every partial derivative of S with respect to an extensive variable A_i or an intensive variable γ_i tends to 0, when the temperature and a set Y of other, extensive or intensive, variables are kept constant:

$$\left(\frac{\partial S}{\partial A_i}\right)_{T,Y} \underset{T\to 0}{\longrightarrow} 0, \qquad \left(\frac{\partial S}{\partial \gamma_i}\right)_{T,Y} \underset{T\to 0}{\longrightarrow} 0. \tag{6.62}$$

The *changes in the entropy* associated with any shift in the equilibrium at constant temperature thus *vanish* with the temperature.

It follows from (6.62) and the Maxwell relations (6.30) for a thermodynamic potential of the free energy type that

$$\left(\frac{\partial \gamma_i}{\partial T}\right)_{A_i,Y} \underset{T\to 0}{\longrightarrow} 0, \qquad \left(\frac{\partial A_i}{\partial T}\right)_{\gamma_i,Y} \underset{T\to 0}{\longrightarrow} 0. \tag{6.63}$$

The *changes with temperature* of any, intensive or extensive, quantity *vanish* at low temperatures. For example, *expansion coefficients*, or increases of pressure with temperature, vanish as $T \to 0$.

6.6 Other Examples

The domain of applications of thermodynamic methods which we surveyed in §§ 6.2 to 6.5 is huge, even if we restrict ourselves, as we have done, to equilibria. We shall limit ourselves here to mentioning a few significant problems.

6.6.1 Thermal Engines

Even though the processes involved in thermal engines are irreversible, thermostatics enables us to strike a balance by comparing an initial state with a final state, both of which are at equilibrium. The aim of a thermal engine is either to provide a given system with a certain amount of work, or to reduce its entropy. This would clearly be impossible if the system were isolated, as it would evolve towards a "dead" state with maximum entropy, while its energy would remain constant. The principle of the thermal engines (§ 6.1.4) consists just in *taking advantage of the increase in entropy* of a *composite* isolated system, one part of which is the system of interest. When the intensive variables (temperatures or pressures) of the various parts are different, spontaneous (heat or work) transfers occur from one part to another in order to raise the total entropy. One *exploits these transfers* to arrive at the required result, which is to provide work or to decrease the entropy locally.

The analysis of the exchanges occurring in a thermal engine which provides work, or a thermal power station which produces electricity, then involves three elements:

(i) A subsystem a, the engine itself, consisting of a heat source – which can be of chemical origin, involving combustion, or of nuclear, solar origin – together with its fuel supplies, and various mechanical, magnetic, or electric elements (cylinders, turbines, blastpipes, alternators, ...) which help to transform various kinds of energy one into another and to exchange them with the rest of the system;
(ii) A subsystem b which is meant to receive, from a, some (mechanical or electrical) work $\mathcal{W}_b$;
(iii) A cold source c which can exchange heat with a; it consists, for instance, of the atmosphere or of cool water, and may include various devices, such as radiators, circulators, evaporators.

When the engine operates, the subsystem a provides altogether some (positive) energy $-\Delta U_a$. Present technology does not allow us to use chemical or nuclear energies other than by first transforming them into heat; the energy provided, $-\Delta U_a$, will thus be of thermal origin so that the change in entropy ΔS_a is negative. As the entropy of the complete system a+b+c cannot decrease, the decrease in the entropy of a must be compensated by an increase of the entropy of c:

$$\Delta S_c \geq -\Delta S_a. \tag{6.64}$$

This increase ΔS_c occurs thanks to the supply of a certain amount of heat,

$$\mathcal{Q}_c = T_c \, \Delta S_c, \tag{6.65}$$

which must be deducted from the total energy,

$$-\Delta U_a = \mathcal{W}_b + \mathcal{Q}_c, \tag{6.66}$$

provided by a. Only part of this energy is finally transformed into useful work, the remainder having been given up as heat to the subsystem c. Thus, because the *energy source* a cannot give up energy without a decrease in entropy, the *double balance*, of *energy*, which is conserved, and of *entropy*, which must increase, makes it necessary that there are exchanges with a system c, which is the *source of negentropy*. In § 6.1.4 we stressed this essential rôle played by the subsystem c for transforming the heat generated by a, which is disordered energy attached to the microscopic degrees of freedom, into work, which is ordered energy attached to some collective degrees of freedom. The "*thermal pollution*" through giving up heat to c *is the price to be paid* for this necessary increase in order.

It is the aim of the engineer to extract the maximum useful work $\mathcal{W}_b$ for a given change in the state of a – a given fuel consumption – and for a given temperature of the cold source. Equations (6.64) to (6.66) show that we have, in agreement with (6.27),

$$\mathcal{W}_\mathrm{b} \le -\Delta U_\mathrm{a} + T_\mathrm{c}\,\Delta S_\mathrm{a}; \tag{6.67}$$

the upper bound is attained when the total entropy remains constant during the process. In fact, inequality (6.67) is valid between two equilibrium states for each elementary stage of the operation of the thermal engine, even when the signs of the exchanges are arbitrary, and even when ΔU_a contains both heat and work. In particular, it is possible that during some elementary process, such as compression in a petrol engine, the engine recovers from the subsystem b part of the work that it gave up to it during another stage of the cycle; $\mathcal{W}_\mathrm{b}$ is negative in that case. It follows from (6.67) that *the efficiency is optimum for reversible transformations.* By letting the total entropy increase, any irreversibility in final reckoning has the effect of *dissipating* part of the available energy $-\Delta U_\mathrm{a}$ in the form of extra heat given up at a total loss to the cold source c. The *elimination of causes for irreversibilities* should thus be a primary concern in the designing of thermal engines. Nevertheless a difficulty arises; a quasi-reversible transformation is on principle slow, as the system must at all times be close to equilibrium. To produce the wanted power one must realize a compromise, allowing the smallest amount of irreversibility, while providing the energy $\mathcal{W}_\mathrm{b}$ in the required delay. There is still much to be done in this direction, since the actual efficiency of thermal engines rarely reaches more than 30 to 40 % of the theoretical maximum efficiency.

When one applies inequality (6.67) to each of the elementary stages of the operation of the engine, ΔU_a consists both of work and of heat. Nevertheless, these stages are, in general, organised as *closed cycles* which return all variables of the system a, except the energy and the entropy, to their initial values. If we apply the above reasoning to the whole of a closed cycle, and if the hot and the cold sources have fixed temperatures T_a and T_c, the change in energy ΔU_a of a reduces to the heat $\Delta U_\mathrm{a} = \mathcal{Q}_\mathrm{a} = T_\mathrm{a}\Delta S_\mathrm{a}$, and the theoretical maximum efficiency, given by (6.67), can be written in the form

$$\left(\frac{\mathcal{W}_\mathrm{b}}{-\mathcal{Q}_\mathrm{a}}\right)_\mathrm{max} = \frac{T_\mathrm{a} - T_\mathrm{c}}{T_\mathrm{c}}. \tag{6.68}$$

The maximum efficiency only depends on the temperatures of the hot and the cold sources(Carnot's theorem). Apart from reducing the irreversibilities, the only means of improving the efficiency of thermal engines consists thus of *increasing the difference between the temperatures of the sources.*

A *refrigerator* or a *heat pump* contains the same theoretical elements, a, b, and c, as an engine, but the signs of the exchanges are the opposite. The goal we are now aiming at is, in both cases, to extract heat $-\mathcal{Q}_\mathrm{c} > 0$ from the cold subsystem c and to give it to the hot subsystem a. Of course, this process would be forbidden if the subsystem b were not present, as it would entail a decrease in the entropy of the system a+c. The price to be paid to make this possible is *to supply work* $-\mathcal{W}_{\rm b}$. *The double balance of energy and entropy* of Eqs.(6.64) to (6.66) remains valid; thanks to the work done $-\mathcal{W}_{\rm b}$ on the subsystem a, its energy increases by more than the amount $-\mathcal{Q}_{\rm c}$ received from c, and its entropy increase is larger than the decrease $-\Delta S_{\rm c}$ in the entropy of the cold source c.

The *efficiency of a refrigerator* is the ratio $(-\mathcal{Q}_{\rm c})/(-\mathcal{W}_{\rm b})$ of the heat $-\mathcal{Q}_{\rm c}$ extracted from the subsystem c of interest to the work supplied $-\mathcal{W}_{\rm b}$. From (6.67), which is valid whatever the signs of the energy exchanges, it follows that the efficiency again is optimum for reversible transformations and that for a closed cycle it then equals

$$\left(\frac{-\mathcal{Q}_{\rm c}}{-\mathcal{W}_{\rm b}}\right)_{\max} = \frac{T_{\rm c}}{T_{\rm a}-T_{\rm c}}. \tag{6.69}$$

This theoretical efficiency becomes infinite as $T_{\rm a} - T_{\rm c} \to 0$: it needs very little work to cool down a system c, when its temperature is just below that of the surrounding air, which usually plays the rôle of the hot source a. This efficiency decreases as the temperature difference increases: one needs to supply an infinite amount of work to extract heat from a system close to the absolute zero, in agreement with the Third Law.

For a heat pump the system of interest is the room a to be heated, while the cold source c is the atmosphere, and the work supplied, $\mathcal{W}_{\rm b}$, is electric energy. One could, of course, transform this into heat $\mathcal{Q}_{\rm a}$ using an electric radiator; however, it is much more efficient to use it to transfer an extra amount of heat $-\mathcal{Q}_{\rm c} > 0$ from the surrounding air to a. The *efficiency of a heat pump*, the ratio of the total amount of heat received by a to the electric energy consumed, $-\mathcal{W}_{\rm b}$, has for a closed cycle a theoretical maximum equal to

$$\left(\frac{\mathcal{Q}_{\rm a}}{-\mathcal{W}_{\rm b}}\right)_{\max} = \frac{T_{\rm a}}{T_{\rm a}-T_{\rm c}}. \tag{6.70}$$

For $T_{\rm a} = 20^\circ$ C, $T_{\rm c} = -10^\circ$ C, this ratio equals 10 so that one could save up to 90 % electricity by heating houses by heat pumps instead of radiators.

Extremely varied elementary processes are involved in thermal engines. During each of the stages of the operating cycle, the subsystems which together make up the engine evolve in relation to other parts which fix some of the variables. For instance, in a steam engine, the cycle is close to a Carnot cycle: isothermal expansion at a temperature $T_{\rm a}$, adiabatic expansion with cooling, isothermal compression at a temperature $T_{\rm c}$, adiabatic compression with heating. Mechanical energy is exchanged with b in those 4 stages, the total balance for $\mathcal{W}_{\rm b}$ being positive; heat is absorbed and given up during the isothermal transformations. The operating of a petrol engine, for which the irreversibilities are large, can be idealized by an Otto cycle, where the heat exchanges do not take place at constant temperature, but at constant volume; the ratio between the minimum volume (ignition) and the maximum volume (exhaust) is the compression ratio. In a diesel cycle the combustion occurs at constant pressure. The theoretical analysis of each of these elementary processes is made easier if one uses the appropriate thermodynamic potential (§ 6.3.5).

6.6.2 Osmosis

Osmosis effects are made possible by the existence of *semi-permeable* membranes that let some kinds of molecules through, but not other kinds. Such membranes play, in particular, an important rôle in biology. The system contains two solutions a and b, separated by a semi-permeable membrane; to simplify the discussion we assume that there are only two kinds of molecules, where the molecules 1 can pass through the partition, but the molecules 2 cannot. The possibility of transfers of energy and of particles 1 between a and b implies (§ 6.2.2) that the temperatures are equal,

$$T_a = T_b,$$

and also the chemical potentials of 1,

$$\mu_{1a} = \mu_{1b}. \tag{6.71}$$

On the other hand, the numbers N_{2a} and N_{2b} are fixed. There remain two independent parameters, the volumes of the two parts a and b; if, for instance, these two volumes are fixed, the pressures on the two sides of the wall will, in general, be unequal. The presence of a semi-permeable membrane will, in producing concentration differences, also produce a pressure difference, the so-called *osmotic pressure.* For instance, in blood the pressure is the same on the two sides of the wall of the red cells. However, if we replace the plasma by pure water, which can freely pass through the membrane, this produces an osmotic pressure because there are extra molecules inside the red cells that cannot pass the wall, and the red cells explode.

In order to study osmosis, it is convenient to take as variables the temperature and the pressures $\mathcal{P}_a$ and $\mathcal{P}_b$, supposed to be fixed by external reservoirs. The appropriate thermodynamic potential is the Gibbs function, or free enthalpy (§ 5.6.2),

$$\begin{aligned} G &= U - TS + \mathcal{P}_a\Omega_a + \mathcal{P}_b\Omega_b \\ &= G(T, \mathcal{P}_a, N_{1a}, N_{2a}) + G(T, \mathcal{P}_b, N_{1b}, N_{2b}), \end{aligned}$$

the minimum of which, as function of N_{1a} and N_{1b} again gives the equilibrium condition (6.71) in the form

$$\frac{\partial}{\partial N_{1a}} G(T, \mathcal{P}_a, N_{1a}, N_{2a}) = \frac{\partial}{\partial N_{1b}} G(T, \mathcal{P}_b, N_{1b}, N_{2b}). \tag{6.72}$$

This relation enables us to find N_{1a} and N_{1b}, and to express the osmotic pressure $\mathcal{P}_a - \mathcal{P}_b$ as function of the different concentrations, the temperature, and $\mathcal{P}_b$. If, on the other hand, we let the pressures vary, not only the volumes but also the *concentrations vary indirectly* (as in § 6.4.4) by the transfer of molecules of the kind 1 through the membrane. This is the principle of a method for desalination of sea water.

6.6.3 Thermochemistry

The study of chemical equilibria, like that of osmotic equilibria, enters the framework of the maximum entropy principle through replacing energy exchanges which lead to thermal equilibria by particle exchanges. The difference in the nature of these two conserved quantities, though, leads to the appearance of some new aspects.

The extensive variables that characterize the macroscopic state of a system in which chemical reactions can take place include not only the energy and the volume, but also *the numbers of molecules* of each species, or of *ions* in the case of ionic solutions. If the system contains several phases, for instance, crystals in contact with solutions, we must take into consideration the extensive variables of each phase. An initial metastable state, where all chemical reactions are blocked, is described by giving all these extensive variables arbitrary values. The allowed exchanges and chemical reactions establish between these variables constraints defining the domain $\mathcal{A}$ of § 6.1.2, and we find their equilibrium values by looking for the maximum of the entropy in the domain $\mathcal{A}$, provided the system is *isolated.* Transformations of particles into one another are automatically accompanied by energy exchanges so that the temperature T is uniform in the state of chemical equilibrium. Let us suppose that the volumes of the various phases have been determined by the maximum entropy principle; the total volume is fixed because we have assumed that the system exchanges neither work nor heat with the outside. We still must write down the conditions which ensure that S is a maximum for all allowed changes in the numbers of particles. Each partial derivative (6.7) of S with respect to a number of molecules or of ions equals $\gamma = -\mu/T = -k\alpha$, where μ is the appropriate chemical potential; because T is uniform, any relations between the γ variables are reflected by the same relations *between the chemical potentials* μ. In chemistry one usually works with numbers of *moles* rather than with numbers of *particles*; this multiplies the chemical potentials with Avogadro's number. We prefer in the present book to measure the chemical potentials in energy units, J or eV, rather than in J mol^{-1}.

Let us denote the numbers of the various particles X_j, which may be molecules or ions, by N_j. For greater generality we use a different index to indicate particles of the same kind which are part of different phases, and we treat their passing from one phase into another as a reaction. We thus use the notation

$$\mu_j \equiv -T \frac{\partial S}{\partial N_j} = kT\alpha_j \tag{6.73}$$

for the chemical potential of X_j, that is, a particle of a given kind in a given phase. We can then write down all possible reactions which lead to equilibrium in the general form

$$\sum_j \nu_j^{(k)} \mathrm{X}_j \leftrightarrows 0, \tag{6.74}$$

where, for each reaction k, the numbers $\nu_j^{(k)}$ are positive, negative, or zero integers. For instance, for the dissociation equilibrium of water

$$2\mathrm{H_2O} \leftrightarrows 2\mathrm{H_2} + \mathrm{O_2}, \tag{6.75}$$

at high temperatures, the ν_j are, respectively, -2 for the H_2O molecules, $+2$ for the H_2 molecules, and $+1$ for the O_2 molecules. For the equilibrium of carbon dioxide gas with water one must, besides the chemical reaction proper,

$$(\mathrm{CO_2})_{\mathrm{dis}} + \mathrm{H_2O} \leftrightarrows \mathrm{CO_3H^-} + \mathrm{H^+}, \tag{6.76a}$$

of water with dissolved CO_2, take into account the equation

$$(\mathrm{CO_2})_{\mathrm{gas}} \leftrightarrows (\mathrm{CO_2})_{\mathrm{dis}} \tag{6.76b}$$

for the equilibrium between gaseous CO_2 and CO_2 in solution. In the first reaction the $\nu_j^{(1)}$ are 0 for gaseous CO_2, -1 for dissolved CO_2, -1 for H_2O, $+1$ for the CO_3H^- ion, and $+1$ for the H^+ ion; in the second reaction the $\nu_j^{(2)}$ are -1 for gaseous CO_2, $+1$ for dissolved CO_2, and 0 for the other components. We look for the maximum of S when all reactions (6.74) are allowed. The variations dN_j in the numbers N_j of the X_j particles are then no longer arbitrary, but subject to constraints determined by these reactions. If we denote by $dM^{(k)}$ the infinitesimal shifts associated with each of the reactions (6.74), the general form of the allowed dN_j is

$$dN_j = -\sum_k \nu_j^{(k)}\, dM^{(k)}. \tag{6.77}$$

As the $dM^{(k)}$ are independent differentials, the conditions for chemical equilibrium can be written as

$$\frac{\partial S}{\partial M^{(k)}} = 0, \qquad \forall\, k,$$

that is,

$$\sum_j \frac{\partial S}{\partial N_j} \frac{\partial N_j}{\partial M^{(k)}} = 0,$$

or, if we use (6.73) and (6.77),

$$\boxed{\sum_j \nu_j^{(k)} \mu_j = 0}\;. \tag{6.78}$$

Chemical equilibrium is thus determined by writing down that *the chemical potentials are conserved for each of the allowed reactions* with a *weight* corresponding to the number of particles involved. For instance, for the equilibrium (6.75) we must have

$$2\mu(H_2) + \mu(O_2) = 2\mu(H_2O);$$

for the equilibrium (6.76) the chemical potentials of CO_2 are the same in the two phases and we must also have

$$\mu(CO_2) + \mu(H_2O) = \mu(CO_3H^-) + \mu(H^+).$$

The whole theory of chemical equilibria rests on the above result. In § 8.2.2 we shall use it to derive the laws for chemical equilibria in the gas phase, using the explicit formula, which we find from statistical physics, for the chemical potentials in a gas containing several kinds of particles.

Another point of view can help us to see intuitively the meaning of the relations (6.78). Let us use the example of the reaction (6.75). According to a remark in § 6.1.2 we can regard each of the *sets of* H_2O, H_2, *and* O_2 *molecules* as a *subsystem* of the, isolated, system consisting of the whole gas. By assigning to each subsystem the whole volume Ω, the total entropy is the sum of the three entropies of the H_2O, H_2, and O_2 gases. These three subsystems are thus superimposed upon one another in space. If we argue in terms of the *elementary constituents*, the H and O *atoms* – for other problems we might have to consider radicals or ions – the extensive state variables are the numbers $N(H/H_2O)$ and $N(H/H_2)$ of H atoms in the H_2O and H_2 subsystems, and the numbers $N(O/H_2O)$ and $N(O/O_2)$ of O atoms in the H_2O and O_2 subsystems. They have their own atomic chemical potentials $\mu(H/H_2O)$, $\mu(H/H_2)$, $\mu(O/H_2O)$, and $\mu(O/O_2)$. The *numbers of atoms are conserved* and they can be exchanged between the subsystems for chemical equilibium, in the same way as energy is exchanged for thermal equilibrium. As a result the atomic chemical potentials are *equal*:

$$\mu(H/H_2O) = \mu(H/H_2), \qquad \mu(O/H_2O) = \mu(O/O_2). \tag{6.79}$$

Moreover, the numbers of atoms and molecules are in each subsystem related through $N(H/H_2O) = 2N(O/H_2O) = 2N(H_2O)$, $N(H/H_2) = 2N(H_2)$, $N(O/O_2) = 2N(O_2)$. It follows then from the variation (6.73) of S under changes in the number of particles, for instance, from

$$\begin{aligned} -T\,dS(H_2O) &= \mu(H_2O)\,dN(H_2O) \\ &= \mu(H/H_2O)\,dN(H/H_2O) + \mu(O/H_2O)\,dN(O/H_2O) \\ &= [2\mu(H/H_2O) + \mu(O/H_2O)]\;dN(H_2O), \end{aligned}$$

that a *molecular chemical potential is the sum of the chemical potentials of the constituent atoms*:

$$\left.\begin{aligned}\mu(\mathrm{H_2O}) &= 2\mu(\mathrm{H/H_2O}) + \mu(\mathrm{O/H_2O}),\\ \mu(\mathrm{H_2}) &= 2\mu(\mathrm{H/H_2}),\\ \mu(\mathrm{O_2}) &= 2\mu(\mathrm{O/O_2}).\end{aligned}\right\} \tag{6.80}$$

Transferring a molecule amounts to transferring the set of atoms which make it up, and Eqs.(6.80) mean simply that the *tendency* of a system, such as water, to *transfer molecules* is the sum of the tendencies to *transfer the constituent atoms*, H, H, and O. Eliminating the atomic chemical potentials from (6.79) and (6.80) gives us the required relation $2\mu(\mathrm{H_2O}) = 2\mu(\mathrm{H_2}) + \mu(\mathrm{O_2})$.

On the microscopic scale we find this additivity of the chemical potentials by the same reasoning as in § 6.1.3, when we analyse the form of the grand canonical equilibrium distribution. To fix ideas, let us again take the example of the equilibrium (6.75) for which the grand partition function is equal to

$$Z_{\mathrm{G}}[\alpha(\mathrm{H}), \alpha(\mathrm{O})] = \mathrm{Tr}\, \exp\left[-\beta\widehat{H} + \alpha(\mathrm{H})\widehat{N}(\mathrm{H}) + \alpha(\mathrm{O})\widehat{N}(\mathrm{O})\right]. \tag{6.81}$$

Given that we are here writing down the formalism of a fundamental theory, the basic objects occurring in (6.81) are the H and O *atoms* which automatically have at equilibrium a unique chemical potential $kT\alpha(\mathrm{H})$ or $kT\alpha(\mathrm{O})$ in the whole of the system. The Hamiltonian $\widehat{H}$ includes the interactions which are responsible for binding the atoms into molecules. In order to evaluate (6.81) let us imagine that we have classified the resulting configurations according to the number of $\mathrm{H_2O}$, $\mathrm{H_2}$, and $\mathrm{O_2}$ molecules formed, and let us denote by $\widehat{H}'$ the effective Hamiltonian which describes the kinetic energies and the residual interactions between these molecules, as well as the internal dynamics of the molecules assumed to be indivisible. Let us define the molecular grand partition function by

$$\begin{aligned}&Z'_{\mathrm{G}}[\alpha(\mathrm{H_2O}), \alpha(\mathrm{H_2}), \alpha(\mathrm{O_2})]\\ &\quad\equiv \mathrm{Tr}\, \exp\Big[-\beta\widehat{H}' + \alpha(\mathrm{H_2O})\widehat{N}(\mathrm{H_2O})\\ &\quad\quad + \alpha(\mathrm{H_2})\widehat{N}(\mathrm{H_2}) + \alpha(\mathrm{O_2})\widehat{N}(\mathrm{O_2})\Big]\,;\end{aligned} \tag{6.82}$$

it describes a metastable macroscopic state where the average numbers of molecules, $\langle\widehat{N}(\mathrm{H_2O})\rangle$, $\langle\widehat{N}(\mathrm{H_2})\rangle$, and $\langle\widehat{N}(\mathrm{O_2})\rangle$, independently given, are taken into account through Lagrangian multipliers. We can then identify (6.81) with (6.82) provided the molecular chemical potentials satisfy the relation

$$\begin{aligned}&\alpha(\mathrm{H})\widehat{N}(\mathrm{H}) + \alpha(\mathrm{O})\widehat{N}(\mathrm{O})\\ &\quad\equiv \alpha(\mathrm{H_2O})\widehat{N}(\mathrm{H_2O}) + \alpha(\mathrm{H_2})\widehat{N}(\mathrm{H_2}) + \alpha(\mathrm{O_2})\widehat{N}(\mathrm{O_2}).\end{aligned}$$

This equation expresses that the system is at equilibrium; it is equivalent to (6.79) and (6.80), which microscopically proves (6.78). We see here again that the chemical potentials of the atoms add up, *whether the atoms are bound or not.*

So far we have assumed that the reactive system was *isolated*: its volume remained fixed and it did not exchange heat with the outside. In practice, its *temperature* and its *pressure* are often fixed from the outside, so that neither its internal energy, nor its volume are given. The equilibrium is then determined by looking not for the maximum of the entropy, but of the isobaric-isothermal Massieu function (§§ 5.6.6 and 6.3.3). This amounts to looking for the minimum of the *free enthalpy* G, where T and $\mathcal{P}$ are fixed, while the N_j vary under constraints as above. We find again the conditions (6.78), with $\mu_j = \partial G/\partial N_j$.

The *reaction heat* is defined, for fixed temperature and pressure, as the heat supplied to the outside thermostat by the system when the reaction considered takes place. More precisely, let us assume that near equilibrium a shift occurs for the kth reaction (6.74). By referring this shift to one mole, the numbers of molecules or ions of each species change by

$$dN_j = \nu_j^{(k)} N_{\mathrm{A}}, \tag{6.83}$$

quantities which, notwithstanding the presence of Avogadro's number N_{A}, we must treat as infinitesimal. According to § 6.3.4 the reaction heat, supplied to the outside at constant pressure, is given by the *change in enthalpy*

$$\begin{aligned} -\,dH &= -\,d(U + \mathcal{P}\Omega) \\ &= -T\,dS - \Omega\,d\mathcal{P} - \sum_j \mu_j\,dN_j = -T\,dS \end{aligned} \tag{6.84}$$

at constant T and $\mathcal{P}$ for the given changes (6.83) in the dN_j.

When a system consisting of m phases can undergo p independent chemical reactions between r molecular species, *Gibbs' phase rule*, which results from (6.78) and generalizes the remarks of § 6.4.6, gives $r + 2 - p - m$ as the number of independent intensive equilibrium variables.

Finally, Le Chatelier's principle (§ 6.4.3) takes on its most useful form in the case of chemical equilibria: it indicates how outside parameters affect the concentrations. When *the temperature is increased* at constant pressure the chemical equilibrium shifts in the *endothermic* direction. *Increasing the pressure* at constant temperature shifts the chemical equilibrium in the direction corresponding to a *decrease of volume.*

6.6.4 Elasticity

The mechanics of deformable media is tightly bound up with thermodynamics since any realistic description of a fluid or solid substance needs a simultaneous discussion of its mechanical and thermal behaviour. We shall restrict ourselves here to a few remarks about the *equilibrium of a homogeneous elastic solid* in the limit of small deformations.

The points in the solid are characterized by their coordinates $\boldsymbol{r}$ in a state which we take as the reference state. They become $\boldsymbol{s}$ after deformation, and the tensor $\partial s_\alpha/\partial r_\beta$ is independent of $\boldsymbol{r}$ for homogeneous deformations. Its antisymmetric part describes a rotation; its symmetric part, the *deformation tensor*, describes the dilatations and shears and contains 6 independent elements. In particular, its trace represents the relative increase of volume. The entropy, which is independent of the position and the orientation of the solid, is a function of 8 extensive variables, the energy, the number of particles, and the product of the 6 elements of the deformation tensor with the volume of the solid in its reference state. All *mechanical and thermal* equilibrium properties of the substance can be derived from this function. Expression (5.11) for the work done in an adiabatic transformation defines the intensive variables which are conjugate to the deformation tensor; they also form a symmetric tensor, the *stress tensor*, the 6 elements of which can be interpreted as forces per unit area, either parallel or at right angles to such an area.

The Maxwell relations reduce the number of independent response coefficients. For instance, for small deformations, the *isothermal elasticity coefficients*, defined as the partial derivatives, for constant T, of the 6 elements of the deformation tensor with respect to the 6 elements of the stress tensor depend solely on the temperature and the substance (*Hooke's law*). Their number is 21, rather than 36, for an arbitrary substance, as the 6×6 matrix that they form is symmetric as a consequence of (6.30).

Depending on the nature of the substance one can obtain other relations between the response coefficients from *symmetry* considerations. We shall come back to this in § 14.2.3 in the more general framework of non-equilibrium thermodynamics (*Curie's principle*). Let us note, for instance, that the number of elasticity coefficients drops from 21 to 2 for an amorphous, isotropic substance, and to 3 for a crystal with all the cubic symmetries. More generally, the techniques of §§ 6.3 and 6.4, combined with the exploitation of the symmetries of the substance, enable us to establish many relations and inequalities, mixing mechanical and thermal properties, between the various response coefficients, such as the dilatation coefficients, the elasticity coefficients, or the specific heats.

6.6.5 Magnetic and Dielectric Substances

Like the mechanical properties, the electromagnetic properties of substances are not independent of their thermal properties. We must thus introduce a fundamental function, the entropy or the internal energy, which depends on electromagnetic variables to characterize completely all the electromagnetic and thermal equilibrium properties. This synthesis, the subject of many treatises on electromagnetism of matter, goes beyond the framework of the present book.

Nevertheless we shall address the subtle problem of the *choice of thermodynamic variables* which often leads to confusion. The difficulty lies in the fact that the charges and currents interact *at a distance* through the intermediary of fields, so that the separation of the various parts of the system cannot be carried out in a unique way; this affects even the definition of work (§§ 5.2.3, 5.2.4, and 6.2.1). Moreover, a substance placed in an electric or magnetic field is a *non-isolated system* and this makes it necessary to use the proper thermodynamic potential. As a consequence, several procedures exist for the definition of the system and of its pairs of conjugate variables; they differ by the way in which the fields are taken into account. Different books on thermodynamics and electromagnetism use different procedures and, for that reason, often seem to contradict one another. We thus find it useful to compare here the various methods used in order to make access to the existing literature easier.

In § 11.3.3 we shall elucidate the meaning of the macroscopic electrostatic quantities by studying a solid at the microscopic scale. Let us just note now that the fields as well as the microscopic charge and current densities vary wildly over distances of the order of atomic sizes. The quantities characterizing the electromagnetic properties of a substance at the macroscopic scale are *spatial averages* which allow us to get rid of these variations on the atomic scale. We thus define a first category of quantities varying slowly on our scale, the averages $\boldsymbol{E}$ and $\boldsymbol{B}$ of the microscopic *electric field* and *magnetic induction*, and the total average *charge* and *current* densities ϱ_{tot} and $\boldsymbol{j}_{\text{tot}}$. Nevertheless, if the medium is polarizable or magnetizable, its macroscopic properties depend on the charge and current distributions on the atomic scale in two ways. On the one hand, there may exist microscopic itinerant charges that are free to travel through the medium. On the other hand, microscopic localized charges and currents, q_i and $q_i\boldsymbol{v}_i$, give rise to the *polarization*

$$\boldsymbol{P} = \frac{1}{\omega} \sum \boldsymbol{r}_i q_i, \tag{6.85a}$$

the average dipole moment per unit volume, and the *magnetization*

$$\boldsymbol{M} = \frac{1}{2\omega} \sum [\boldsymbol{r}_i \times q_i \boldsymbol{v}_i], \tag{6.85b}$$

the average magnetic moment per unit volume. The sums in (6.85) are over all charges situated inside a volume ω which contains a sufficiently large number of crystal cells for a solid or of molecules for a fluid; the coordinate origin for $\boldsymbol{r}_i$ is chosen to be at the centre of this volume. The vectors $\boldsymbol{P}$ and $\boldsymbol{M}$ vary slowly on our scale and the total average charge and current densities,

$$\varrho_{\text{tot}} = \varrho - \text{div}\, \boldsymbol{P}, \qquad \boldsymbol{j}_{\text{tot}} = \boldsymbol{j} + \frac{\partial \boldsymbol{P}}{\partial t} + \text{curl}\, \boldsymbol{M}, \tag{6.86}$$

can thus be split into the average contributions, ϱ and $\boldsymbol{j}$, from the *free charges* and those from the *electric and magnetic dipoles*. The various average quantities, $\boldsymbol{E}$, $\boldsymbol{B}$, ϱ, $\boldsymbol{j}$, $\boldsymbol{P}$, $\boldsymbol{M}$ are related to one another through the *charge conservation law*, $\text{div}\, \boldsymbol{j} + \partial\varrho/\partial t = 0$ and through the *macroscopic Maxwell equations*, which are obtained by taking a spatial average of the proper, microscopic, Maxwell equations, using Eqs.(6.86). We find for slow space-time variations and fixed substances

$$\left.\begin{aligned}
\operatorname{curl} \boldsymbol{E} &= -\frac{\partial \boldsymbol{B}}{\partial t}, & \operatorname{div} \boldsymbol{B} &= 0, \\
\operatorname{div} \boldsymbol{D} &= \varrho, & \operatorname{curl} \boldsymbol{H} &= \frac{\partial \boldsymbol{D}}{\partial t} + \boldsymbol{j}; \\
\boldsymbol{D} &\equiv \varepsilon_0 \boldsymbol{E} + \boldsymbol{P}, & \boldsymbol{H} &\equiv \frac{1}{\mu_0} \boldsymbol{B} - \boldsymbol{M}; \\
\boldsymbol{E} &= -\nabla \Phi - \frac{\partial \boldsymbol{A}}{\partial t}, & \boldsymbol{B} &= \operatorname{curl} \boldsymbol{A}.
\end{aligned}\right\} \tag{6.87}$$

The last two equations are equivalent to the first two and define, up to gauge transformations, the scalar and vector potentials Φ and $\boldsymbol{A}$. Here, as everywhere in the present book, we use SI units, a table of which is given at the end of each of the two volumes. It now remains for us to embed these equations into the framework of thermostatics, and, in particular, to introduce thermodynamic potentials and work. We shall show how such thermodynamic potentials can be constructed through an integration of the *equations of state*, which express here the polarization and the magnetization at equilibrium in terms of the fields.

Let us consider an uncharged *dielectric* at thermal and electrostatic equilibrium. In order to construct a thermodynamic potential which will characterize its properties we try to evaluate the work received during a transformation which changes its electric variables $\boldsymbol{P}$, $\boldsymbol{E}$, and $\boldsymbol{D}$. To do this, let us assume that the material is subjected to the action of a set of charges q_α carried by conductors α at potentials Φ_α, for instance, that it lies between the two plates of a capacitor. The potential $\Phi(\boldsymbol{r})$ is chosen so as to vanish at infinity. If there were no dielectric, these same charges q_α would produce a field $\boldsymbol{E}_0(\boldsymbol{r})$; when the dielectric is present, it produces an extra field by becoming polarized. This replaces $\boldsymbol{E}_0$ by the field $\boldsymbol{E}$ which in the whole of space is given by the Maxwell equations (6.87) in terms of the q_α. To shift the equilibrium state we must modify the charges q_α carried by the conductors. Let us imagine that infinitesimal charges δq_α are dragged for that purpose from infinity onto the conductors. The *total work* received during this *adiabatic transformation* by the total system consisting of the external charges, the field, and the dielectric, is

$$\delta \mathcal{W}_{\text{tot}} = \sum_\alpha \Phi_\alpha \, \delta q_\alpha. \tag{6.88}$$

Using the Maxwell equations (6.87) we can also write this expression in terms of the field as follows:

$$\delta \mathcal{W}_{\text{tot}} = \int (\boldsymbol{E} \cdot \delta \boldsymbol{D}) \, d^3 \boldsymbol{r}, \tag{6.89}$$

where the integral is over the whole of space. The field $\boldsymbol{E}$ appears as an intensive variable or a "force", and the product of the electric induction $\boldsymbol{D}$ with the volume as the associated extensive variable or "position", for the system consisting of the *ensemble* of the dielectric, the fields, and the external conductors.

(i) So far there is no ambiguity. The difficulties arise when one tries to split the work (6.89) into contributions from one or other part of the system. Actually, this work contains not only the energy supplied to polarize the dielectric, but also some energy used to produce the field. When there is no dielectric, the field generated by the charges q_α equals $\boldsymbol{E}_0 = -\nabla\Phi_0$, and the work provided by the change considered in the charges is

$$\sum_\alpha \Phi_{0\alpha}\,\delta q_\alpha = \varepsilon_0 \int (\boldsymbol{E}_0 \cdot \delta \boldsymbol{E}_0)\, d^3\boldsymbol{r} = \delta\left[\frac{1}{2}\varepsilon_0 \int \boldsymbol{E}_0^2\, d^3\boldsymbol{r}\right].$$

A natural procedure in this context consists in splitting the energy into two parts, one relating to the sources or to the field when there is no dielectric, which equals

$$\frac{1}{2}\varepsilon_0 \int \boldsymbol{E}_0^2\, d^3\boldsymbol{r}, \tag{6.90}$$

and another relating to the dielectric and its interactions with the field. The change of the latter during an adiabatic transformation is the work

$$\delta\mathcal{W}^{(0)} = \int (\boldsymbol{E} \cdot \delta\boldsymbol{D})\, d^3\boldsymbol{r} - \varepsilon_0 \int (\boldsymbol{E}_0 \cdot \delta\boldsymbol{E}_0)\, d^3\boldsymbol{r},$$

which, if we use (6.87), can be transformed into

$$\delta\mathcal{W}^{(0)} = \int \{(\boldsymbol{E}\cdot[\delta\boldsymbol{D} - \delta\boldsymbol{D}_0]) + ([\boldsymbol{D} - \boldsymbol{D}_0]\cdot\delta\boldsymbol{E}_0) - (\boldsymbol{P}\cdot\delta\boldsymbol{E}_0)\}\, d^3\boldsymbol{r}.$$

The first term can, if we once again use (6.87), be written as

$$\begin{aligned} -\int (\nabla\Phi \cdot [\delta\boldsymbol{D} - \delta\boldsymbol{D}_0])\, d^3\boldsymbol{r} &= \int \Phi\, \mathrm{div}\,[\delta\boldsymbol{D} - \delta\boldsymbol{D}_0]\, d^3\boldsymbol{r} \\ &= \int \Phi[\delta\varrho - \delta\varrho_0]\, d^3\boldsymbol{r} \\ &= \sum_\alpha \Phi_\alpha[\delta q_\alpha - \delta q_{0\alpha}] = 0; \end{aligned}$$

it vanishes because the free charges carried by each conductor are the same and are changed in the same way whether or not there is a dielectric. Similarly, the second term, equal to

$$\int \left([\boldsymbol{D} - \boldsymbol{D}_0]\cdot\boldsymbol{E}_0\right) d^3\boldsymbol{r} = \sum_\alpha \left(q_\alpha - q_{0\alpha}\right)\delta\Phi_{0\alpha},$$

vanishes, and the work $\delta\mathcal{W}^{(0)}$ equals

$$\delta\mathcal{W}^{(0)} = -\int (\boldsymbol{P}\cdot\delta\boldsymbol{E}_0)\, d^3\boldsymbol{r}. \tag{6.91}$$

The two conjugate variables appearing in this definition of the work received by the dielectric are the *applied field* $\boldsymbol{E}_0$, calculated from the charges q_α as if there were no dielectric, and the *polarization* of the latter. In contrast to the work (6.89) which involves the whole space, the work (6.91) received by the dielectric is expressed as an integral over the volume of the dielectric only.

If we know, theoretically or experimentally, the equation of state which gives the polarization $\boldsymbol{P}$ as function of the field $\boldsymbol{E}$, or, what amounts to the same, of $\boldsymbol{E}_0$, we can find a thermodynamic potential by integrating (6.91). Since usually $\boldsymbol{P}$ is expressed as a function of the applied field $\boldsymbol{E}_0$ and the temperature, this naturally introduces a *free* energy $F^{(0)}(T, \boldsymbol{E}_0)$; $F^{(0)}$ is the thermodynamic potential suitable for the description of the dielectic in the field produced by given charges, which would produce $\boldsymbol{E}_0$ if there were no dielectric. Expression (6.91) for $\delta\mathcal{W}^{(0)}$ enables us to express $F^{(0)}$ as

$$F^{(0)}\big(T, \boldsymbol{E}_0\big) - F^{(0)}(T, 0) \;=\; -\int_0^1 d\lambda \int \left(\boldsymbol{P}(T, \lambda\boldsymbol{E}_0)\cdot\boldsymbol{E}_0\right) d^3\boldsymbol{r}. \tag{6.92}$$

The difference (6.92) can be interpreted as the work done on the dielectric *if one brings it from infinity to the region where the field is* $\boldsymbol{E}_0$ when the dielectric is not there, this transformation taking place at constant temperature and for *fixed values of the outside charges*. The Legendre transform

$$U^{(0)}(S, \boldsymbol{E}_0) \;=\; F^{(0)}(T, \boldsymbol{E}_0) + TS \tag{6.93}$$

of (6.92) with respect to T is the *internal energy of the system consisting of the dielectric and of its interactions with the applied field* $\boldsymbol{E}_0$ *which would be created by the external charges alone.*

For a study of the dielectric at the *microscopic scale*, rather than writing down a global Hamiltonian of the dielectric and the outside charges, it is useful to restrict oneself to a partial Hamiltonian describing the dielectric in the field $\boldsymbol{E}_0$ and forgetting about the charges; this Hamiltonian depends on $\boldsymbol{E}_0$ only through the term

$$-\int \left(\boldsymbol{E}_0\cdot\widehat{\boldsymbol{P}}\right) d^3\boldsymbol{r}, \tag{6.94}$$

where $\int \widehat{\boldsymbol{P}}\, d^3\boldsymbol{r}$ is the sum of the microscopic electric dipoles; it contains other terms accounting for interactions between the latter. This approach is the microscopic counterpart of the preceding one. In fact, according to (5.14) and (5.15), if $\boldsymbol{E}_0$ is varied slowly and if the dielectric does not exchange heat with the outside, the change $\delta\langle\widehat{H}\rangle$ can be identified with (6.91) where $\boldsymbol{P}$ is interpreted as the statistical average of $\widehat{\boldsymbol{P}}$. By regarding this energy variation as work in an adiabatic transformation, we can identify $\langle\widehat{H}\rangle$ with the macroscopic internal energy (6.93) that we have just defined. The replacement of the applied field $\boldsymbol{E}_0$ in the macroscopic Maxwell equations (6.87) by the total field $\boldsymbol{E}$ is the result of the interactions between the dipoles (§ 11.3.3).

(ii) We have so far included in the definition of our system not only the dielectric, but also its interaction with the field $\boldsymbol{E}_0$. Nevertheless, we can completely get rid of the latter as follows. We no longer include in the Hamiltonian the interaction term (6.94); to fix the polarization, now to be considered an independent variable characterizing the state of the system, we must introduce a *constraint*. In the resulting Boltzmann-Gibbs distribution, expression (6.94) appears as the constraint term, and $\beta \boldsymbol{E}_0$ can now be interpreted as the Lagrangian multiplier.[3] On the macroscopic scale the fundamental function corresponding to the new variables,

$$U(S,\boldsymbol{P}) = F^{(0)} + \int (\boldsymbol{E}_0 \cdot \boldsymbol{P})\, d^3\boldsymbol{r} + TS, \tag{6.95}$$

is obtained through a Legendre transformation of $U^{(0)}$ with respect to $\boldsymbol{E}_0$. As it is also the expectation value of the new Hamiltonian, we can interpret it as the internal energy of *just the dielectric without the field.* Its differential,

$$dU = T\,dS + \int (\boldsymbol{E}_0 \cdot \delta\boldsymbol{P})\, d^3\boldsymbol{r},$$

allows us to consider

$$\delta\mathcal{W} = \int (\boldsymbol{E}_0 \cdot \delta\boldsymbol{P})\, d^3\boldsymbol{r} \tag{6.96}$$

as work received by the dielectric itself, whereas $\delta\mathcal{W}^{(0)}$, given by (6.91), includes the interaction of the field with the dielectric.

(iii) There is, however, another way to split the total work (6.89) into two parts, one to be related to the field without the dielectric and the other to the dielectric itself. Let us assume for this purpose that the system is controlled, no longer by fixing the charges q_α carried by the external conductors, but by *fixing their potentials* Φ_α, for instance, through a system of generators. If there is no dielectric, the potentials Φ_α, which we assume to be invariable, produce the field $\boldsymbol{E}_1$. It is essential to note that this new applied field differs from $\boldsymbol{E}_0$; in fact, when the dielectric is removed, the generators deliver currents and change the external charges q_α, while producing some electric work. The energy of the field when there is no dielectric is now

$$\frac{1}{2}\,\varepsilon_0 \int \boldsymbol{E}_1^2\, d^3\boldsymbol{r},$$

and we are led to define work received by the dielectric during an adiabatic transformation as the difference :

$$\delta\mathcal{W}^{(1)} = \int (\boldsymbol{E} \cdot \delta\boldsymbol{D})\, d^3\boldsymbol{r} - \varepsilon_0 \int (\boldsymbol{E}_1 \cdot \delta\boldsymbol{E}_1)\, d^3\boldsymbol{r}.$$

This work is generated by a change $\delta\Phi_\alpha$ in the potentials, and the fields $\boldsymbol{E}_1$ and $\boldsymbol{E}$ are this time associated with the same set of *potentials* Φ_α, whereas $\boldsymbol{E}_0$ and $\boldsymbol{E}$ were associated in $\delta\mathcal{W}^{(0)}$ with the same charges q_α. Using the Maxwell equations we can transform $\delta\mathcal{W}^{(1)}$ into

$$\delta\mathcal{W}^{(1)} = \int \{([\boldsymbol{E} - \boldsymbol{E}_1] \cdot \delta\boldsymbol{D}) + (\boldsymbol{D}_1 \cdot [\delta\boldsymbol{E} - \delta\boldsymbol{E}_1]) + (\boldsymbol{E}_1 \cdot \delta\boldsymbol{P})\}\, d^3\boldsymbol{r}.$$

---

[3] Similarly, in Exerc.5a when we described the elasticity of a fiber, we could in the Hamiltonian include or not the term coupling the length of the fiber with the stretching weight, which played the rôle of the field.

The first term now equals

$$-\int (\nabla[\Phi - \Phi_1] \cdot \delta \boldsymbol{D})\, d^3 r = \int (\Phi - \Phi_1) \operatorname{div} \delta \boldsymbol{D}\, d^3 r$$
$$= \int (\Phi - \Phi_1)\, \delta \varrho\, d^3 r$$
$$= \sum_\alpha (\Phi_\alpha - \Phi_{1\alpha})\ \delta q_\alpha = 0,$$

as the values of the Φ_α are now kept fixed. The second term vanishes similarly, and hence $\delta \mathcal{W}^{(1)}$ is equal to

$$\delta \mathcal{W}^{(1)} = \int (\boldsymbol{E}_1 \cdot \delta \boldsymbol{P})\, d^3 r. \tag{6.97}$$

Even though the work $\delta \mathcal{W}^{(1)}$ differs from $\delta \mathcal{W}^{(0)}$ and from $\delta \mathcal{W}$, because the applied fields $\boldsymbol{E}_0$ and $\boldsymbol{E}_1$, which are, respectively, related to fixed charges and fixed potentials, differ, (6.97) is still expressed as an integral over the inside of the dielectric only. It can be interpreted again as work received by the dielectric and by its interaction with the field but for fixed external potentials. The conjugate variables are now the *polarization* and the *applied field* $\boldsymbol{E}_1$, *calculated from the potentials* Φ_α as if there were no dielectric. Through integration we introduce a new free energy,

$$F^{(1)}(T, \boldsymbol{P}) - F^{(1)}(T, 0) = \int_0^1 d\lambda \int \big(\boldsymbol{E}_1(T, \lambda \boldsymbol{P}) \cdot \boldsymbol{P}\big)\, d^3 r, \tag{6.98}$$

which represents the *total* work supplied to *bring the dielectric from infinity to the region where it gets the polarization* $\boldsymbol{P}$, when this transformation is carried out at constant temperature and for *fixed values of the external potentials.*

(iv) The work (6.98) includes an electrostatic contribution

$$\sum_\alpha \Phi_\alpha (q_\alpha - q_{1\alpha}) = \int [(\boldsymbol{E}_1 \cdot \boldsymbol{D}) - (\boldsymbol{E} \cdot \boldsymbol{D}_1)]\ d^3 r$$
$$= \int (\boldsymbol{E}_1 \cdot \boldsymbol{P})\, d^3 r, \tag{6.99}$$

delivered by the generators. If, as in § 6.3.4, we want to eliminate this extra work, we are led again to introduce a new thermodynamic potential

$$F^{(2)}(T, \boldsymbol{E}_1) = F^{(1)}(T, \boldsymbol{P}) - \int (\boldsymbol{E}_1 \cdot \boldsymbol{P})\, d^3 r \tag{6.100}$$

through a Legendre transformation. The variation of $F^{(2)}$,

$$F^{(2)}(T, \boldsymbol{E}_1) - F^{(2)}(T, 0) = \int \delta \mathcal{W}^{(2)}$$
$$= -\int_0^1 d\lambda \int \big(\boldsymbol{P}(T, \lambda \boldsymbol{E}_1) \cdot \boldsymbol{E}_1\big)\, d^3 r, \tag{6.101}$$

then represents the work supplied to *bring the dielectric from infinity to the region where the field* $\boldsymbol{E}_1$ was produced by the *fixed potentials* Φ_α, excluding the work done by the generators. The free energy $F^{(2)}$ is the thermodynamic potential describing the dielectric itself, in the field produced by fixed potentials and equal to $\boldsymbol{E}_1$ when there is no dielectric. Notwithstanding their similarity, one should not confuse the free energies $F^{(0)}$ and $F^{(2)}$: they do not describe the same physical system as they do not take the field into account in the same way.

Table 6.1. Work done on a dielectric

Field produced by	Interaction with field: included	Interaction with field: not included
fixed charges	$-(\boldsymbol{P} \cdot \delta\boldsymbol{E}_0)$	$(\boldsymbol{E}_0 \cdot \delta\boldsymbol{P})$
fixed potentials	$(\boldsymbol{E}_1 \cdot \delta\boldsymbol{P})$	$-(\boldsymbol{P} \cdot \delta\boldsymbol{E}_1)$

The various possible definitions that we have thus obtained for the *energy "of the dielectric"* share this energy out differently among the various parts of the complete system formed by the external conductors, the field, and the dielectric proper. In fact, only the global expression (6.89) which includes the field sources has an intrinsic significance. All other expressions, defined through differences, must be regarded as convenient calculational tools suited to some physical situation or others. The *external sources* for the field are *eliminated* in all cases, but this elimination leads to different results, depending on the nature of the sources, which are *fixed charges* in the case of (6.91) to (6.96), and *fixed potentials* in the case of (6.97) to (6.101). Moreover, in each of those two situations one can *either include or not* in the physical system the *interaction of the dielectric with the field.* These possibilities lead to introducing the four different definitions (6.91), (6.96), (6.97), and (6.101) of the work received by unit volume of the dielectric, which are summarized in Table 6.1. In all cases the *conjugate variables* are the *polarization* $\boldsymbol{P}$ and the so-called *"applied" field* $\boldsymbol{E}_0$ or $\boldsymbol{E}_1$ that would be produced by the fixed charges or potentials, if there were no dielectric. However, in the first and the last case, the *field* plays the rôle of *"position"* and the *polarization* that of *"force"*, whereas the opposite situation occurs in the second and third cases. It is important to distinguish those cases clearly in order correctly to identify the natural variables of the thermodynamic potentials which always are the "position" variables ξ_α. The latter may differ from the extensive variables A_i of § 6.1.2 in the case of long-range forces.

It is also important to note that the *first point of view* is the best suited for the *microscopic study of dielectrics.* The applied field $\boldsymbol{E}_0$, which does not include the field produced by the polarization of the dielectric, then occurs in the Hamiltonian through (6.94). It is coupled to the elementary charge observables, the average dipole moment of which per unit volume can be identified with the polarization.

We have restricted ourselves so far to an uncharged dielectric, assuming that the only free charges were the q_α carried by the external conductors. If the system studied not only can be polarized, but also can carry a density $\varrho(\boldsymbol{r})$ of *free charges*, the total work (6.88) can be expressed as

$$\delta \mathcal{W}_{\text{tot}} = \int \left(\boldsymbol{E} \cdot \delta \boldsymbol{D}\right) d^3\boldsymbol{r} - \int \Phi\, \delta\varrho\, d^3\boldsymbol{r},$$

where the second integral does not include the external charges. Hence we find an additional contribution to $\delta\mathcal{W}^{(0)}$, equal to $\int \varrho\, \delta\Phi_0\, d^3\boldsymbol{r}$. Similarly, in the four cases of Table 6.1, the contribution of the free charges to the work results from the replacement of $\boldsymbol{P}$ by $-\varrho$ and of $\boldsymbol{E}_0$, or $\boldsymbol{E}_1$, by Φ_0, or Φ_1.

The same considerations hold for *magnetic substances* in thermal and magnetostatic equilibrium. In general, the electric and magnetic fields are assumed to be generated by charges and currents carried by external conductors. To change the fields we must apply on those charges external forces, which during a time δt provide the *total* work

$$\delta \mathcal{W}_{\text{tot}} = -\,\delta t \int (\boldsymbol{j} \cdot \boldsymbol{E})\, d^3\boldsymbol{r} = \int (\boldsymbol{j} \cdot \delta \boldsymbol{A})\, d^3\boldsymbol{r} + \int \Phi\, \delta\varrho\, d^3\boldsymbol{r}, \tag{6.102}$$

where we have used relations (6.87) for the field $\boldsymbol{E}$ in terms of the potentials. Equation (6.102) extends to the regions where there are free charges and where *currents* flow; by using the Maxwell equations (6.87) we can change it into an expression in terms of the *fields*:

$$\delta \mathcal{W}_{\text{tot}} = \int (\boldsymbol{H} \cdot \delta \boldsymbol{B})\, d^3\boldsymbol{r} + \int (\boldsymbol{E} \cdot \delta \boldsymbol{D})\, d^3\boldsymbol{r}. \tag{6.103}$$

We are thus, by including in the system the sources of the fields, led to consider the inductions $\boldsymbol{B}$ and $\boldsymbol{D}$ as "position" variables and the fields $\boldsymbol{H}$ and $\boldsymbol{E}$ as "forces". From now on, we focus on the first term of (6.103), as we have already discussed the second term.

(i) Let us define $\boldsymbol{B}_0$ as the magnetic induction that would occur if we removed the magnetizable substance *without acting upon the external currents*. We can, for instance, imagine that these currents flow in superconducting spools, to avoid dissipation into heat, without them being connected to generators. Under those conditions we can subtract from the magnetic contribution to (6.103) the part which survives when there is no magnetic substance; this defines a work

$$\delta \mathcal{W}^{(0)} = \int (\boldsymbol{H} \cdot \delta \boldsymbol{B})\, d^3\boldsymbol{r} - \frac{1}{\mu_0} \int (\boldsymbol{B}_0 \cdot \delta \boldsymbol{B}_0)\, d^3\boldsymbol{r}.$$

Evaluating it explicitly gives

$$\begin{aligned} \delta \mathcal{W}^{(0)} &= \int (\boldsymbol{H} \cdot [\delta \boldsymbol{B} - \delta \boldsymbol{B}_0])\, d^3\boldsymbol{r} + \int ([\boldsymbol{B} - \boldsymbol{B}_0] \cdot \delta \boldsymbol{H}_0)\, d^3\boldsymbol{r} \\ &\quad - \int (\boldsymbol{M} \cdot \delta \boldsymbol{B}_0)\, d^3\boldsymbol{r}. \end{aligned}$$

The first term can be transformed as follows:

$$\begin{aligned}\int (\boldsymbol{H} \cdot [\delta \boldsymbol{B} - \delta \boldsymbol{B}_0]) \, d^3\boldsymbol{r} &= \int (\boldsymbol{H} \cdot \operatorname{curl}[\delta \boldsymbol{A} - \delta \boldsymbol{A}_0]) \, d^3\boldsymbol{r} \\ &= \int (\operatorname{curl} \boldsymbol{H} \cdot [\delta \boldsymbol{A} - \delta \boldsymbol{A}_0]) \, d^3\boldsymbol{r} \\ &= \int (\boldsymbol{j} \cdot [\delta \boldsymbol{A} - \delta \boldsymbol{A}_0]) \, d^3\boldsymbol{r},\end{aligned}$$

and it vanishes because we assume that no work is done on the circuits generating the field. The second term similarly vanishes, and we are left with

$$\delta \mathcal{W}^{(0)} = -\int (\boldsymbol{M} \cdot \delta \boldsymbol{B}_0) \, d^3\boldsymbol{r}. \tag{6.104}$$

This work can be assigned to the transformation of the magnetic substance, but it includes a contribution associated with its interaction with the field. The substance is brought into the field without other actions on the circuits; this *induces currents* and hence the currents $\boldsymbol{j}$ and $\boldsymbol{j}_0$ will be different. Part of $\delta \mathcal{W}^{(0)}$ is used to achieve this.

(ii) The physical situation of interest is the one where the *external currents are kept constant* while the magnetizable substance is being introduced. If the latter were not present, they would produce the induction $\boldsymbol{B}_1$. The work received by the substance, including its interactions with the field, for an arbitrary change in the parameters, is obtained by subtracting from (6.103) the value of the work when there is no substance present, that is,

$$\delta \mathcal{W}^{(1)} = \int (\boldsymbol{H} \cdot \delta \boldsymbol{B}) \, d^3\boldsymbol{r} - \frac{1}{\mu_0} \int (\boldsymbol{B}_1 \cdot \delta \boldsymbol{B}_1) \, d^3\boldsymbol{r}.$$

This expression can again be changed into

$$\begin{aligned}\delta \mathcal{W}^{(1)} &= \int \big([\boldsymbol{H} - \boldsymbol{H}_1] \cdot \delta \boldsymbol{B}\big) \, d^3\boldsymbol{r} + \int \big([\delta \boldsymbol{H} - \delta \boldsymbol{H}_1] \cdot \boldsymbol{B}_1\big) \, d^3\boldsymbol{r} \\ &\quad + \int \big(\boldsymbol{B}_1 \cdot \delta \boldsymbol{M}\big) \, d^3\boldsymbol{r},\end{aligned}$$

where the first two terms,

$$\int \big([\boldsymbol{H} - \boldsymbol{H}_1] \cdot \delta \boldsymbol{B}\big) \, d^3\boldsymbol{r} = \int \big([\boldsymbol{j} - \boldsymbol{j}_1] \cdot \delta \boldsymbol{A}\big) \, d^3\boldsymbol{r}$$

and

$$\int \big([\delta \boldsymbol{H} - \delta \boldsymbol{H}_1] \cdot \boldsymbol{B}_1\big) \, d^3\boldsymbol{r} = \int \big(\big[\delta \boldsymbol{j} - \delta \boldsymbol{j}_1\big] \cdot \boldsymbol{A}_1\big) \, d^3\boldsymbol{r},$$

vanish because the currents are kept constant. We have thus

$$\delta \mathcal{W}^{(1)} = \int (\boldsymbol{B}_1 \cdot \delta \boldsymbol{M}) \, d^3\boldsymbol{r}. \tag{6.105}$$

Using this we get by integration the total work released *by the generators to maintain the external currents* constant and *by an observer to bring the substance* from infinity. This work appears as a change in the free energy $F^{(1)}$, defined by

$$F^{(1)}(T,\boldsymbol{M}) - F^{(1)}(T,0) = \int_0^1 d\lambda \int \left(\boldsymbol{B}_1(T,\lambda\boldsymbol{M})\cdot\boldsymbol{M}\right) d^3\boldsymbol{r}. \tag{6.106}$$

(iii) If we are solely interested in the work, $\delta\mathcal{W}^{(2)}$, done *on the substance itself with the external currents being kept constant*, and not in the electric work provided by the generators, we must subtract the latter from $\delta\mathcal{W}^{(1)}$. This electric work is equal to $\delta\int(\boldsymbol{A}\cdot\boldsymbol{j})\,d^3\boldsymbol{r}$ or to $\delta\int(\boldsymbol{A}_1\cdot\boldsymbol{j}_1)\,d^3\boldsymbol{r}$, depending on whether or not the magnetic substance is present. We get thus

$$\begin{aligned}
\delta\mathcal{W}^{(2)} - \delta\mathcal{W}^{(1)} &= -\delta\int\{(\boldsymbol{A}\cdot\boldsymbol{j}) - (\boldsymbol{A}_1\cdot\boldsymbol{j}_1)\}\,d^3\boldsymbol{r} \\
&= -\delta\int\{(\boldsymbol{A}\cdot\boldsymbol{j}_1) - (\boldsymbol{A}_1\cdot\boldsymbol{j})\}\,d^3\boldsymbol{r} \\
&= -\delta\int\{(\boldsymbol{A}\cdot\mathrm{curl}\boldsymbol{H}_1) - (\boldsymbol{A}_1\cdot\mathrm{curl}\boldsymbol{H})\}\;d^3\boldsymbol{r} \\
&= -\delta\int\{(\boldsymbol{B}\cdot\boldsymbol{H}_1) - (\boldsymbol{B}_1\cdot\boldsymbol{H})\}\;d^3\boldsymbol{r} \\
&= -\,\delta\int(\boldsymbol{B}_1\cdot\boldsymbol{M})\;d^3\boldsymbol{r},
\end{aligned}$$

or

$$\delta\mathcal{W}^{(2)} = -\int(\boldsymbol{M}\cdot\delta\boldsymbol{B}_1)\;d^3\boldsymbol{r}. \tag{6.107}$$

This is the point of view which is best suited for a *microscopic study of magnetic substances.* In fact, in the Hamiltonian (2.65) of an arbitrary system of spinless charged particles with charges e_i we can choose the vector potential in the form $\boldsymbol{A}(\boldsymbol{r}) = \frac{1}{2}[\boldsymbol{B}_1\times\boldsymbol{r}]$, if $\boldsymbol{B}_1$ is a uniform field produced by external currents assumed to be fixed. This *applied field* occurs in $\widehat{H}$ as a parameter and the corresponding "force" observable, defined by (5.13), or,

$$\begin{aligned}
\frac{\partial\widehat{H}}{\partial\boldsymbol{B}_1} &= -\sum_i e_i\left[\widehat{\boldsymbol{r}}_i\times\{\widehat{\boldsymbol{p}}_i - e_i\widehat{\boldsymbol{A}}_i\}\right] \\
&= -\sum_i e_i\left[\widehat{\boldsymbol{r}}_i\times\widehat{\boldsymbol{v}}_i\right] \;\equiv\; -\,\widehat{\boldsymbol{M}},
\end{aligned} \tag{6.108}$$

can, apart from a sign, be identified as the *total magnetic moment observable*, in agreement with expression (6.107) for the macroscopic work. One can, by integrating (6.107) and (6.108), also identify the mean value of the Hamiltonian with the macroscopic internal energy. These results can be extended to the case of particles with spin, with magnetic moments $\widehat{\boldsymbol{m}}_i$ equal to $-\mu_{\mathrm{B}}\widehat{\boldsymbol{\sigma}}_i$ for electrons, as the applied magnetic field $\boldsymbol{B}_1$ then occurs in the Hamiltonian through the term $-\sum_i(\boldsymbol{B}_1\cdot\widehat{\boldsymbol{m}}_i)$. The terms in the Hamiltonian describing the interactions between currents and magnetic moments are responsible for the total field $\boldsymbol{B}$ occurring in the macroscopic Maxwell equations (6.87). This analysis justifies expression (6.107) for the magnetic work which we assumed without proof in § 1.3.3.

Altogether, for *magnetic substances* we have again the same ambiguities in the definition of work, of energy, and of conjugated variables as for dielectrics. The polarization $\boldsymbol{P}$ is simply replaced by the magnetization $\boldsymbol{M}$, and the "applied" electric fields $\boldsymbol{E}_0$ (for fixed external charges) or $\boldsymbol{E}_1$ (for fixed external potentials) by the applied magnetic inductions $\boldsymbol{B}_0$ (for the case when one does not act upon the external circuits) or $\boldsymbol{B}_1$ (for fixed external currents). The last case (6.107) is now the best suited for a microscopic analysis, as the applied field $\boldsymbol{B}_1$ produced by fixed external currents is coupled to the magnetic moment observable.

The extreme variety of substances in existence, dielectrics, piezoelectrics, diamagnetics, paramagnetics, ferromagnetics, ..., is associated with a large variety of forms for the electromagnetic thermodynamic potentials. As usual, thermodynamics provides us with general relations, while statistical mechanics allows us to determine the particular form of the thermodynamic potentials according to the circumstances. We have seen an example in Chapter 1 which enabled us, in particular, to study a magneto-caloric effect – cooling through adiabatic demagnetization – where electromagnetism and thermodynamics are intimately connected.

Summary

The macroscopic equilibrium states of an isolated system are determined by the maximum entropy principle, a consequence of the maximum statistical entropy principle which itself follows from probability and information theory. The thermodynamic entropy can be interpreted as the microscopic disorder (6.3) associated with given values of the macroscopic variables. The intensive variables, which are the partial derivatives of the entropy with respect to the extensive variables, become balanced at equilibrium.

The thermodynamic potentials are useful to determine the equilibrium state of a system in contact with sources of heat, work, matter, ... , and to evaluate their mutual exchanges. Methods involving differential calculus, Legendre transforms, and Jacobians provide general relations between response coefficients. The latter also satisfy inequalities which can be derived from the fact that the entropy is concave.

We have reviewed various applications: the stability of equilibria, critical points, phase separation, thermal engines, as well as osmotic, chemical, elastic, and electromagnetic equilibria. In particular, we established relation (6.78) between chemical potentials which governs chemical and similar equilibria, and we have discussed in detail the concept of work for dielectric and magnetic substances.

Exercises

6a Equation of State, Specific Heats, and Chemical Potential

Assume that the equation of state $\mathcal{P}(v,T)$ of a fluid, where v is the molar volume, has been determined experimentally. What can one say about the specific heats? about the chemical potential? Applications: the van der Waals equation $(\mathcal{P}+a/v^2)(v-b) = RT$; a perfect monatomic gas, in which case one can use dimensional analysis.

Solution:

All equilibrium properties will be known once we have determined a thermodynamic potential. As the natural variables involved here are v and T we must calculate the Massieu function $\Psi_{\rm C}$ or the free energy $F = -T\Psi_{\rm C}$, starting from

$$\frac{\partial F(T,N,\Omega)}{\partial \Omega} = -\mathcal{P}\left(\frac{\Omega N_{\rm A}}{N}, T\right).$$

Integrating from some fixed reference volume v_0 we find

$$F(T,N,\Omega) = -\frac{N}{N_{\rm A}}\int_{v_0}^{\Omega N_{\rm A}/N} \mathcal{P}(v,T)\,dv + N\varphi(T),$$

where the form of the integration constant takes the extensivity of F into account; the unknown function $\varphi(T)$ is independent of the density. The entropy, chemical potential, and specific heats per mole can now be derived:

$$S = -\frac{\partial F(T,N_{\rm A},\Omega)}{\partial T} = \int_{v_0}^{v} \frac{\partial \mathcal{P}(v,T)}{\partial T}\,dv - N_{\rm A}\varphi'(T),$$

$$\mu = \frac{\partial F}{\partial N} = \frac{F}{N} + \frac{\Omega}{N}\,\mathcal{P}\left(\frac{\Omega N_{\rm A}}{N}, T\right),$$

$$C_{\rm v} = -T\,\frac{\partial^2 F}{\partial T^2} = \int_{v_0}^{v} T\,\frac{\partial^2 \mathcal{P}}{\partial T^2}\,dv - N_{\rm A}T\varphi''(T),$$

$$C_{\rm p} = C_{\rm v} + \frac{Tv\alpha^2}{\kappa_T} = C_{\rm v} - T\left(\frac{\partial \mathcal{P}}{\partial T}\right)^2 \Big/ \frac{\partial \mathcal{P}}{\partial v}.$$

Extra information follows from the fact that the entropy is concave, which is expressed through (6.50) and (6.51), that is, $C_{\rm v} > 0$ and $\partial\mathcal{P}/\partial v < 0$. For all T the function φ must therefore satisfy the condition

$$\varphi''(T) < \min_{v}\frac{1}{N_{\rm A}}\int_{v_0}^{v}\frac{\partial^2\mathcal{P}}{\partial T^2}\,dv.$$

As φ depends only on T, measuring a specific heat at low density, where the gas is perfect, is sufficient to determine φ'', and thus F apart from a linear function of T.

For the van der Waals fluid we find for one mole

$$
\begin{aligned}
F(T,N_{\rm A},v) &= -\frac{a}{v} - RT\,\ln(v-b) + N_{\rm A}\varphi(T),\\
S &= R\ln(v-b) - N_{\rm A}\varphi'(T),\\
\mu &= -kT\ln(v-b) + \frac{kTv}{v-b} - \frac{2a}{vN_{\rm A}} + \varphi(T),\\
C_{\rm v} &= -N_{\rm A}T\varphi''(T),\\
C_{\rm p} &= -N_{\rm A}T\varphi''(T) + \frac{R^2T}{RT-2a(v-b)^2/v^3},
\end{aligned}
$$

with $\varphi'' < 0$. The specific heat at constant volume is independent of the density.

The results for the perfect gas correspond to $a = b = 0$. Dimensional analysis then achieves almost completely the determination of all equilibrium properties of a perfect monatomic gas. In fact, the expression for F shows that the combination $\ln v - \varphi(T)/kT$ must be dimensionless. Hence, $\varphi(T)/3kT$ must behave as the logarithm of a quantity which has the dimension of a length and which is constructed uniquely from k, $\hbar$, T, and the atomic mass m. This leads to

$$
F(T,N,\Omega) = NkT\,\ln\left[\frac{N}{\Omega}\,c\left(\frac{\hbar}{mkT}\right)^{3/2}\right].
$$

There remains a single unknown, the dimensionless constant c which cannot be determined except through a statistical physics calculation. In Chap.7 we shall see that c equals $(2\pi)^{3/2}/\mathrm{e}$. This constant only occurs, additively, in S/Nk and μ/kT. The molar specific heats $C_{\rm v} = \frac{3}{2}R$ and $C_{\rm p} = \frac{5}{2}R$ are completely determined from the equation of state $\mathcal{P}v = RT$ and dimensional analysis.

Note. Using (6.21) shows that the convexity conditions of $U(S,N,\Omega)$,

$$
\frac{\partial^2 U}{\partial S^2} \geq 0, \qquad \frac{\partial^2 U}{\partial S^2}\,\frac{\partial^2 U}{\partial \Omega^2} - \left(\frac{\partial^2 U}{\partial S\partial\Omega}\right)^2 \geq 0,
$$

are equivalent, after a Legendre transformation, to

$$
\frac{\partial^2 F}{\partial T^2} \leq 0, \qquad \frac{\partial^2 F}{\partial \Omega^2} \geq 0, \qquad \forall\ T \text{ and } \Omega.
$$

For one mole of a van der Waals fluid and $T < T_{\rm c} = 8a/27Rb$, the quantity

$$
\frac{\partial^2 F(T,v)}{\partial v^2} = -\frac{2a}{v^3} + \frac{RT}{(v-b)^2}
$$

is negative between two v values which describe the limits of metastability of the liquid and gas phases. Just as we did for S in § 6.4.6, we are led to replace, for given T, the $F(v)$ curve determined above by its convex envelope. This contains, between the values $v_{\rm l}$ and $v_{\rm g}$, corresponding to the liquid and the gas in stable coexistence, a straight line segment which is tangential to $F(v)$ at both ends. Its slope $-\mathcal{P}_{\rm s}(T)$ determines the saturated vapour pressure at the temperature considered, $T < T_{\rm c}$, that is, the height of the plateau in the isotherms between $v_{\rm l}$ and $v_{\rm g}$. From the expression for F in terms of the equation of state we find, for fixed T,

$$F(v_l) - F(v_g) = (v_g - v_l)\mathcal{P}_s = \int_{v_l}^{v_g} \mathcal{P}(v)\, dv.$$

This equation determines the position of the liquefaction plateau in the $(v, \mathcal{P})$ plane and reflects analytically the Maxwell construction (§ 9.3.3).

6b Isotherms, Adiabats, and Absolute Temperature

Knowing the network of isotherms of a mole of fluid in the $(v, \mathcal{P})$ plane we can determine a relative temperature scale. Let us assume that we also know the adiabat network. Are those two networks arbitrary? Can one use them to derive the absolute temperature and the entropy of each state, characterized by v and $\mathcal{P}$? Application: constant $\mathcal{P}v$ isotherms, constant $\mathcal{P}v^\gamma$ adiabats.

Hints:

Condition (6.40) which connects $v, \mathcal{P}, S$, and T is necesssary and sufficient for the existence of a thermodynamic potential such as $U(S, v)$, as it is equivalent to the identity of the mixed derivatives,

$$\frac{\partial}{\partial S}\frac{\partial U}{\partial v} = -\left(\frac{\partial \mathcal{P}}{\partial S}\right)_v = \frac{d(\mathcal{P}, v)}{d(v, S)}, \qquad \frac{\partial}{\partial v}\frac{\partial U}{\partial S} = \left(\frac{\partial T}{\partial v}\right)_S = \frac{d(T, S)}{d(v, S)}.$$

It expresses the fact that the *work provided* by the fluid during an infinitesimal Carnot cycle equals the *heat it receives* because the *Jacobian* $d(\mathcal{P}, v)/d(T, S)$ equals the *ratio of the areas* $\oint \mathcal{P}\, dv$ and $\oint T\, dS$.

Let us characterize the isotherms by a relative temperature θ, a function $\theta(v, \mathcal{P})$ which has a constant value along each isotherm. We try to reparametrize these isotherms by the absolute temperature which is an unknown function, $T(\theta)$, of θ. Similarly, knowing the adiabat network gives us a function $\sigma(v, \mathcal{P})$ and we want to determine the entropy $S(\sigma)$ in terms of the parameter σ. Condition (6.40) gives

$$J \equiv \frac{d(\mathcal{P}, v)}{d(\theta, \sigma)} = \frac{dT}{d\theta}\frac{dS}{d\sigma}.$$

This relation implies that the isotherm and adiabat networks cannot be arbitrary: they must be such that the Jacobian J, known as function of θ and σ, *can be factorized* as $J = f(\theta)g(\sigma)$, which is expressed as $\partial^2 \ln J/\partial\theta\partial\sigma = 0$. In fact, knowing the isotherm network and *two* adiabatic curves only, we can reconstruct the whole adiabat network by solving that equation, which is invariant under a reparametrization of the curves. If it is satisfied, we find T and S, respectively, as the primitives of $af(\theta)$ and $g(\sigma)/a$. The arbitrary multiplicative factor a corresponds to the choice of the unit of temperature; the additive integration constants of T and S are not determined by the isotherm and adiabat networks – we could have expected this for S but not for T.

In the example $\mathcal{P}v = \theta$, $\mathcal{P}v^\gamma = \sigma$, we have

$$J^{-1} = \begin{pmatrix} v & \mathcal{P} \\ v^\gamma & \gamma v^{\gamma-1}\mathcal{P} \end{pmatrix} = (\gamma - 1)\sigma,$$

which can be factorized with $f(\theta) = 1$, $g(\sigma) = 1/(\gamma - 1)\sigma$. We find

$$T = a\theta + b = a\mathcal{P}v + b,$$
$$S = \frac{1}{a(\gamma-1)} \ln \sigma + c = \frac{1}{a(\gamma-1)} \ln \mathcal{P}v^{\gamma} + c.$$

For $b = 0$ we find the properties of a perfect gas ($a = 1/R$), but fluids with a $\mathcal{P}v = R(T - b)$ equation of state are allowed.

6c Interfaces and Capillarity

We indicated in § 6.1.2 that the extension of the methods of thermodynamics to inhomogeneous systems often requires the introduction of non-local terms, either in the entropy or in the energy. Even if we are not dealing with forces with a macroscopic range, we are thus led to violate the extensivity in those regions of space where the properties change rapidly, for instance, in the vicinity of an interface between a liquid and its saturated vapour. We want to study the variation with the height z, across the interface, of the density $n(z)$, from the value $n_{\rm l}$ of the liquid as $z \to -\infty$ to the value $n_{\rm g}$ of the gas as $z \to +\infty$, in a fluid kept at a constant temperature. We must look for the minimum of the free energy $F\{T, n\}$ considered as a functional of the unknown density $n(\boldsymbol{r})$. We take for F the semi-empirical form

$$F\{T,n\} = \int d^3\boldsymbol{r} \, \left\{ f\left[T, n(\boldsymbol{r})\right] + \tfrac{1}{2} a(T) \left[\nabla n(\boldsymbol{r})\right]^2 + mgz\, n(\boldsymbol{r}) \right\},$$

where the integration is over a fixed volume and where the total number of particles $N = \int n \, d^3\boldsymbol{r}$ is also fixed. The first term, calculated as if the energy and the entropy were additive, involves the free energy density $f[T, n]$ of a fluid of density n, if it were homogeneous. The second term, with $a > 0$, describes a *short-range non-local effect which tends to make the fluid homogeneous* and which comes either from the microscopic entropy or from the microscopic energy; in Exerc.9e we shall derive it in a statistical mechanics model. The last term is the gravitational energy which tends to separate the phases.

1. Write down the equations which determine $n(\boldsymbol{r})$. Introduce a Lagrangian multiplier μ to take into account the constraint on N.

2. According to *Landau's model for phase transitions* the behaviour of f near the critical point $T_{\rm c}$, $n_{\rm c}$, $\mu_{\rm c}$ is dominated by the terms

$$f[T,n] = \tfrac{1}{2} b(T - T_{\rm c})(n - n_{\rm c})^2 + \tfrac{1}{4!} c(n - n_{\rm c})^4 + \mu_{\rm c} n,$$

apart from a possible additive constant, where b and c, as well as a, are positive constants. Neglect the gravitational term and show that there exists only one homogeneous phase for $T > T_{\rm c}$, and also for $T < T_{\rm c}$, $\mu \neq \mu_{\rm c}$ in which case the density takes on different values $n_{\rm l}$ for $\mu = \mu_{\rm c} + 0$ and $n_{\rm g}$ for $\mu = \mu_{\rm c} - 0$.

3. When $n_g < N/\Omega < n_l$ for $T < T_c$ we must take $\mu = \mu_c$. The system is then necessarily heterogeneous. Calculate the density $n(z)$ as function of the height z, taking into account that the gravitational term is small compared to the others. How does the thickness of the liquid-gas interface change as $T \to T_c$?

4. Show that the grand potential of the liquid in equilibrium with its saturated vapour contains not only a term proportional to the volume – taking the same value per unit volume in the two phases – and the gravitational energy, but also a term proportional to the *area of the interface.* Evaluate the latter term. Show that there is a *surface tension*, also called *capillary force*, which tends to reduce the area of the interface. What becomes of this force as $T \to T_c$?

Answers:

1. The minimum of the free energy for fixed N is equivalent to that of the grand potential:

$$\frac{\delta}{\delta n(\boldsymbol{r})}\left[F - \mu \int d^3\boldsymbol{r}\, n(\boldsymbol{r})\right] \equiv \frac{\partial f[T, n(\boldsymbol{r})]}{\partial n} - a(T)\nabla^2 n + mgz - \mu = 0,$$

$$\int d^3\boldsymbol{r}\, n(\boldsymbol{r}) = N.$$

2. The solution of

$$b(T - T_c)(n - n_c) + \frac{c}{6}(n - n_c)^3 + \mu_c - \mu = 0,$$

found graphically, is unique for $T > T_c$. Among the 3 solutions which exist for $T < T_c$ the two extreme ones are local minima; the lowest minimum is reached for the largest value of n if $\mu > \mu_c$, and for the smallest value of n if $\mu < \mu_c$, with

$$n \to n_l = n_c + \sqrt{\frac{6b(T_c - T)}{c}}, \qquad \text{if} \quad \mu \to \mu_c + 0,$$

$$n \to n_g = n_c - \sqrt{\frac{6b(T_c - T)}{c}}, \qquad \text{if} \quad \mu \to \mu_c - 0.$$

3. From the equation

$$\frac{d}{dn}\left(f - \mu_c n\right) - a\frac{d^2 n}{dz^2} = 0$$

it follows that

$$\frac{d}{dz}\left[f - \mu_c n - \frac{1}{2}a\left(\frac{dn}{dz}\right)^2\right] = 0.$$

The integration constant ε must be such that $dn/dz \to 0$ as $z \to \pm\infty$ in the two, gas and liquid, phases. By construction, $f - \mu_c n$ has the same minimum ε in these two phases, whence we find that $\varepsilon = -3b^2(T_c - T)^2/2c$. The final result is

$$dz = -\sqrt{\frac{a}{2}} \int \frac{dn}{\sqrt{f - \mu_c n - \varepsilon}} = -2\sqrt{\frac{3a}{c}} \int \frac{dn}{6b(T_c - T)/c - (n - n_c)^2},$$

or

$$n = n_c - \sqrt{\frac{6b(T_c - T)}{c}} \tanh\left[\sqrt{\frac{b(T_c - T)}{2a}}\,(z - z_0)\right].$$

The rôle of the field of gravity is to fix the orientation of the interface $z = z_0$ and the side $z < z_0$ on which one finds the liquid. The density falls from n_l to n_g over a thickness of the order of $\left[2a/(T_c - T)b\right]^{1/2}$. The thickness of the interface increases and becomes macroscopic close to the critical point, where it diverges, while $n_l - n_g$ tends to zero.

4. The grand potential density, which depends on $\boldsymbol{r}$ through n,

$$f[T, n] - \mu n + \tfrac{1}{2}\, a(\nabla n)^2 + mgzn,$$

at $\mu = \mu_c$ equals

$$f - \mu_c n + \frac{1}{2}\, a \left(\frac{dn}{dz}\right)^2 = \varepsilon + a \left(\frac{dn}{dz}\right)^2,$$

where we have dropped the gravitational term. It gives, apart from the volume term $\varepsilon\Omega$, a positive contribution in the interface region, proportional to the area σ of this interface, with a coefficient

$$\begin{aligned}\varphi &= a \int_{-\infty}^{+\infty} dz \left(\frac{dn}{dz}\right)^2 = a \int_{n_g}^{n_l} dn \left|\frac{dz}{dn}\right|^{-1} \\ &= \sqrt{\frac{ac}{12}} \int_{n_g}^{n_l} dn \left[\frac{6b(T_c - T)}{c} - (n - n_c)^2\right] = \frac{8}{c}\sqrt{\frac{ab^3}{2}}\,(T_c - T)^{3/2}.\end{aligned}$$

When the geometry changes in such a way that the interface area increases by $d\sigma$, while the total volume, T and $\mu = \mu_c$ stay constant, an increase in the grand potential can be identified as an amount of work done, equal to $\varphi\, d\sigma$. The coefficient φ can thus be interpreted as a force, the surface tension per unit length of the border of the interface. It vanishes as $T \to T_c$ while the difference between the phases disappears.

6d Critical Behaviour in the Landau Model

Many phase transitions have the following distinguishing feature which involves a qualitative change with temperature: an extensive variable, the "order parameter", vanishes in the high-temperature, disordered phase and becomes non-vanishing in the low-temperature, more ordered phase. For instance, in an Ising ferromagnet (Exerc.9a) the order parameter is the magnetization M in the z direction. When there is no field, it vanishes above the Curie point while below it, it can spontaneously take on one of the two values $\pm M_s$. In the presence of a field $B(\boldsymbol{r})$ in the z direction and at a temperature T, the order parameter $M(\boldsymbol{r})$ at equilibrium is a function which has to be determined. To do this we use the method of § 6.3.3 and look for the minimum of the free energy as functional of $M(\boldsymbol{r})$. Landau has suggested to take for the free energy in the vicinity of the critical point T_c, which here is the Curie point, the empirical form

$$F\{T,M\} = \int d^3\boldsymbol{r} \left\{f[T,M(\boldsymbol{r})] + \tfrac{1}{2}a[\nabla M]^2 - B(\boldsymbol{r})M(\boldsymbol{r})\right\}, \quad (6.109)$$

$$f[T,M] = \tfrac{1}{2}b(T-T_c)M^2 + \tfrac{1}{4!}cM^4, \quad (6.110)$$

where f describes the extensive part, the only one which exists for a homogeneous system; the last term of (6.109) is the coupling between the order parameter and the external field, and the term with ∇M represents the tendency for the material to become homogeneous. The same model has been used in Exerc.6c for describing the liquid-vapour transition near the critical point, replacing $M(\boldsymbol{r})$ by $n(\boldsymbol{r}) - n_c$, $B(\boldsymbol{r})$ by $-mgz + \mu - \mu_c$, and the free energy by the grand potential. Landau's model can also be extended to study other transitions through including cubic terms, describing, for instance, asymmetry in the liquid-vapour transition, or terms of a degree higher than 4, for tricritical points. One can also introduce several order parameters, such as the components of the magnetization for a ferromagnet which can become magnetized in arbitrary directions or the concentrations for a multi-phase mixture.

1. Determine for a homogeneous equilibrium situation with $B = 0$ the behaviour as function of $|T - T_c|$ of the spontaneous magnetization M_s below T_c, of the magnetic susceptibility $\chi = \partial M/\partial B$, and of the specific heat C at constant volume, on both sides of T_c. How does $M(B)$ behave at $T = T_c$ in the case of a homogeneous field B which tends to zero?

2. If the field $B(\boldsymbol{r})$ is inhomogeneous and weak, the magnetization behaves like $\delta M(\boldsymbol{r}) \equiv M(\boldsymbol{r}) - M_s \sim \int d^3\boldsymbol{r}\, \chi(\boldsymbol{r} - \boldsymbol{r}')B(\boldsymbol{r}')$. Calculate $\chi(\boldsymbol{r} - \boldsymbol{r}')$ for $T > T_c$, $T < T_c$, and $T = T_c$, and give the value of its range λ.

3. Recalling Exerc.4a, show that in canonical equilibrium the fluctuation ΔM of the magnetic moment in unit volume when there is no field diverges as $T \to T_c$. Give an expression for the correlation $\langle \delta M(\boldsymbol{r})\delta M(\boldsymbol{r}')\rangle$ in terms of the non-local susceptibility $\chi(\boldsymbol{r} - \boldsymbol{r}')$ for $B = 0$, and show that the correlation length diverges as $|T - T_c| \to 0$.

Results:

1. We find

$$M_s = \left[\frac{6b(T_c - T)}{c}\right]^{1/2}, \qquad \text{vanishes as } (T_c - T)^{1/2}.$$

$$\chi = \frac{1}{b(T - T_c)}, \quad T > T_c, \qquad \text{diverges at } T_c.$$

$$\chi = \frac{1}{2b(T_c - T)}, \quad T < T_c, \qquad \text{diverges at } T_c.$$

$$C = \frac{3b^2 \Omega T}{c}\, \theta(T_c - T), \qquad \text{discontinuous at } T_c.$$

$$M(B) \sim \left(\frac{6B}{c}\right)^{1/3}, \quad T = T_c, \qquad \text{singular as } B^{1/3}.$$

2. Expanding to first order in $\delta M(\boldsymbol{r})$ and Fourier transforming the equation

$$b(T - T_c)M + \frac{c}{6}M^3 - a\nabla^2 M - B(\boldsymbol{r}) = 0,$$

we find

$$\chi(\boldsymbol{r} - \boldsymbol{r}') = \frac{e^{-|\boldsymbol{r}-\boldsymbol{r}'|/\lambda}}{4\pi a|\boldsymbol{r} - \boldsymbol{r}'|},$$

with

$$\lambda = \left[\frac{a}{b(T - T_c)}\right]^{1/2}, \qquad T > T_c,$$

$$\lambda = \left[\frac{a}{2b(T_c - T)}\right]^{1/2}, \qquad T < T_c,$$

$$\lambda = \infty, \qquad T = T_c.$$

3. We have

$$\Delta M = (kT\chi)^{1/2} \propto |T - T_c|^{-1/2}.$$

$$\langle \delta M(\boldsymbol{r})\delta M(\boldsymbol{r}')\rangle = \frac{1}{\beta^2}\frac{\delta^2 \ln Z}{\delta B(\boldsymbol{r})\delta B(\boldsymbol{r}')} = \frac{1}{\beta}\frac{\delta M(\boldsymbol{r}')}{\delta B(\boldsymbol{r})} = kT\chi(\boldsymbol{r} - \boldsymbol{r}');$$

the correlation length λ diverges as $|T - T_c|^{-1/2}$.

Note. Long-range correlations occur in the vicinity of a critical point, even though the microscopic interactions may have a range of the order of Å. For instance, for the liquid-vapour transition near criticality, λ may become μm or larger. Light, with wavelengths of 0.4 to 0.6 μm, is then scattered strongly by the fluctuations of the material which looks milky: "*critical opalescence*". Strong fluctuations and long range correlations indicate pathological behaviour at the critical point. They are neglected in Landau's theory and must be taken into account by a correct theory of critical phenomena.

Wilson's theory responds to this objective. Along the lines of the remarks made in §§ 5.7.3 and 6.1.3 one imagines that the canonical partition function Z is calculated in two stages, by first taking the trace over the microscopic variables with *exact constraints* on the values of the order parameter $M(\boldsymbol{r})$ in each volume element. Those are free to vary in canonical equilibrium, and Z will subsequently be obtained as a weighted integral over their possible values. They play the rôle of the λ variables of (5.88) so that the probability distribution of $M(\boldsymbol{r})$ is proportional to

$$p\{M\} \propto \mathrm{e}^{-\beta F\{T,M\}}, \tag{6.111}$$

where the two terms $U-TS$ of $F\{T,M\}$ have the same origin as in § 5.7.3. This free energy under constraints, $F\{T,M\}$, is taken to be *Landau's free energy* (6.109). The required partition function, which is the normalization constant of (6.111),

$$Z = \mathrm{e}^{-\beta F} = \int dM(\boldsymbol{r})\, \mathrm{e}^{-\beta F\{T,M\}}, \tag{6.112}$$

is found in the second stage of the calculation through functional integration over all values of $M(\boldsymbol{r})$ at each point. Sufficiently far from the critical point the minimum of $F\{T,M\}$ is well pronounced so that the free energy F is just that minimum, in accordance with the saddle-point equation (5.92). We see thus that Landau's theory is justified in that region. However, in the immediate vicinity of the critical point, where the coefficient of the term in M^2 in (6.110) vanishes, the statistical fluctuations of $M(\boldsymbol{r})$, given by (6.111), become important. As a result, the corrections to the saddle-point method (Exerc.5b) make sufficiently large contributions to (6.112) to invalidate the critical behaviour found above in Landau's theory. Wilson and his followers showed that the form (6.110) was sufficient to calculate the correct critical behaviour and they actually did this, starting from (6.112), thus obtaining results in agreement with experiments for the various kinds of observed critical points.

6e Equilibrium of Self-Gravitating Objects

The size and the structure of objects such as planets or stars of various types and ages are governed by an interplay between the thermodynamic properties of matter and the gravitational forces within the object. At each point, the entropy density $s(\boldsymbol{r})$ is a function of the internal energy density $u(\boldsymbol{r})$ and of the densities of the constitutive particles. The total energy $E = U + E_{\rm G}$ is the sum of the internal energy $U = \int u(\boldsymbol{r}) d^3\boldsymbol{r}$ and of the self-gravitational energy

$$E_{\mathrm{G}} = -\frac{G}{2}\int d^3\boldsymbol{r}\, d^3\boldsymbol{r}'\,\frac{\varrho(\boldsymbol{r})\varrho(\boldsymbol{r}')}{|\boldsymbol{r}-\boldsymbol{r}'|}\,,$$

where $\varrho(\boldsymbol{r})$ is the mass density. Forgetting about the radiation emitted by the star and about the nuclear reactions which may take place in its core, we assume that equilibrium is reached.

1. As a simple model, consider a homogeneous spherical object of radius R and volume Ω. Show that $E_{\mathrm{G}} = -3GM^2/5R$, and that the equilibrium radius is determined from the equation of state by the condition $3\mathcal{P}\Omega = -E_{\mathrm{G}}$. Show that the object is unstable if the adiabatic compressibility is larger than $3/4\mathcal{P}$.

2. In the general case of a non-uniform density, the gravitational potential, related to $\varrho(\boldsymbol{r})$ by $\nabla^2 V = 4\pi G\varrho$, is given by $V(\boldsymbol{r}) = -G\int d^3\boldsymbol{r}'\varrho(\boldsymbol{r}')/|\boldsymbol{r}-\boldsymbol{r}'|$. Show that the equation of hydrostatics $\nabla\mathcal{P} = -\varrho\nabla V$ is a consequence of the fact that the entropy is a maximum at equilibrium. Prove also the so-called *virial theorem* $3\int d^3\boldsymbol{r}\mathcal{P} = -E_{\mathrm{G}}$.

3. Many stellar objects can be approximately described as sets of non-interacting and non-relativistic particles, for instance, the Sun can be modelled as a perfect ionized gas of protons, electrons and He nuclei, or a neutron star as a Fermi gas of neutrons (Exerc.10e, Prob.9). The internal energy is then related to the pressure by $U = \frac{3}{2}\mathcal{P}\Omega$ (Exerc.13a). By using either dimensional arguments or the techniques of §6.3.5, write down the general form of the thermodynamic functions. Show that the internal energy and the gravitational energy are related by $U = -\frac{1}{2}E_{\mathrm{G}}$. The object loses energy by radiation; assuming that it remains at equilibrium, determine the resulting changes in its size, internal energy, entropy and temperature.

Hints:

1. For a fixed value of the total energy E, the entropy must be stationary under an infinitesimal change δR around equilibrium, that is,

$$\delta S = \frac{1}{T}\delta U + \frac{\mathcal{P}}{T}\delta\Omega = -\frac{1}{T}\delta E_{\mathrm{G}} + \frac{3\mathcal{P}\Omega}{T}\frac{\delta R}{R} = 0\,.$$

Its second variation for $\delta S = 0$, $T\delta^2 S = -\delta^2 E_{\mathrm{G}} + \delta\mathcal{P}\,\delta\Omega$, must be negative, which yields

$$\left(\frac{\partial\mathcal{P}}{\partial\Omega}\right)_S < \frac{\partial^2 E_{\mathrm{G}}}{\partial\Omega^2} = \frac{4}{9}\frac{E_{\mathrm{G}}}{\Omega^2} = -\frac{4}{3}\frac{\mathcal{P}}{\Omega}\,.$$

2. In an infinitesimal deformation which brings each point $\boldsymbol{r}$ to $\boldsymbol{r}+\delta\boldsymbol{r}$ where $\delta\boldsymbol{r}$ is a function of $\boldsymbol{r}$, the volume element $d^3\boldsymbol{r}$ becomes $d^3\boldsymbol{r}(1+\operatorname{div}\delta\boldsymbol{r})$ and the gravitational energy becomes $E_{\mathrm{G}} + \int d^3\boldsymbol{r}\,\varrho(\nabla V\cdot\delta\boldsymbol{r})$. The stationarity of $S - k\beta E$ with respect to $u(\boldsymbol{r})$ implies that the temperature is uniform and equal to $1/k\beta$; its stationarity under deformations yields

$$\int d^3\boldsymbol{r}\left(\frac{\mathcal{P}}{T}\operatorname{div}\delta\boldsymbol{r} - k\beta\varrho(\nabla V\cdot\delta\boldsymbol{r})\right) = 0$$

for any $\delta \boldsymbol{r}$ and hence $\nabla \mathcal{P} = -\varrho \nabla V$ at any point. The virial theorem is obtained by taking $\delta \boldsymbol{r} = \varepsilon \boldsymbol{r}$, in which case $\delta E_G = -\varepsilon E_G$.

We can alternatively start from the uniformity at equilibrium of the temperature and of the chemical potentials μ_j for each particle species. As in electrostatic equilibrium (§11.3.3), neither the density $n_j(\boldsymbol{r})$ nor the reduced chemical potential $\mu'_j(\boldsymbol{r})$ are uniform; the latter quantity, $\mu'_j(\boldsymbol{r}) = \mu_j - V(\boldsymbol{r})m_j$ is defined locally by (14.93) through a shift in the origin of the single-particle energies which eliminates the gravitational potential at $\boldsymbol{r}$. The Gibbs-Duhem relation, which expresses extensivity when there is no field, then has the form $sdT = d\mathcal{P} - \sum_j n_j d\mu'_j$, and the interpretation of differentials as changes in space provides $\nabla \mathcal{P} + \sum_j n_j m_j \nabla V = 0$.

3. The Massieu potential $-F/T$ depends on Ω and T only through the combination $T^3\Omega^2$, or through the dimensionless variable $\hbar^2/mkT\Omega^{2/3}$. From $u(\boldsymbol{r}) = \frac{3}{2}\mathcal{P}(\boldsymbol{r})$, we get $U = \frac{3}{2}\int \mathcal{P}(\boldsymbol{r})d^3r = -\frac{1}{2}E_G$. A decrease in the total energy $E = U + E_G = \frac{1}{2}E_G = -U$ produces a contraction $\delta R = R\delta E/U$, a rise $\delta U = -\delta E$ in the internal energy, and a decrease $\delta S = -\delta U/T$ in the entropy. The change in temperature, given by

$$\frac{C_V \delta T}{\delta U} = 4 - \frac{3}{\mathcal{P}\kappa_T} = \frac{2TC_V}{U} - 1,$$

is negative in the Fermi gas limit but positive in the classical limit which holds for a star in formation. Such paradoxical features arise because a decrease occurs in the gravitational energy, which is twice the radiated energy. When a star is being formed, it contracts and heats while radiating. If its mass is larger than $0.1\,M_\odot$, the temperature of its core eventually reaches values of the order to 10^7 K necessary for initiating thermonuclear fusion of hydrogen. A stationary regime is then reached, in which the emission of radiation from the surface of the star is balanced by the production of heat in its core by fusion.

7. The Perfect Gas

"«C'est pour le gaz!» hurlait un employé dans la porte qu'un enfant lui avait ouverte."

A. Camus, L'Exil et le Royaume

"La pureté de l'air entre pour beaucoup dans l'innocence des mœurs."

Balzac, Le Médecin de Campagne

"Front éternel paume parfaite
Puits en plein air essieu de vent"

Paul Éluard, Médieuses

"Nobody is perfect."

Billy Wilder, Some like it Hot (reply of Joe Brown to Jack Lemmon)

In the second part of the book which covers Chaps.7 to 13 we study the properties of various simple substances using the concepts and methods which we have developed earlier. We start with rarefied monatomic gases, which we describe in the perfect gas model as a set of non-interacting mass points following the laws of classical mechanics. We start by justifying this model (§ 7.1), showing, in particular, that the translational degrees of freedom of the molecules in a gas or liquid can to a very good approximation be treated by classical statistical mechanics, in contrast to the internal degrees of freedom of the atoms, associated with the motion of the electrons, or of the molecules, associated with the rotations and vibrations of the constituents relative to one another. The partition function technique then enables us to explain the macroscopic thermodynamic properties of those gases at equilibrium, which have been known experimentally for a long time (§§ 7.2 and 7.3).

We finish the chapter with an elementary introduction to kinetic gas theory (§ 7.4), which is based on two ingredients: (i) The trajectories of the atoms, which include brief collisions with one another, are treated using the laws of classical dynamics. (ii) We use statistics to deal with the velocity of each of the atoms, which are hardly correlated – whereas in the Boltzmann-Gibbs treatment the statistics deals with the complete *system*. Although kinetic theory applies only to *classical gases*, it has the advantage that it explains simply and in a mechanistic way not only their thermostatic properties, but also a large number of non-equilibrium phenomena, such as effusion, viscosity, heat conduction, or diffusion. It played an essential historical rôle in the development of statistical mechanics as it was the form in which the latter first appeared. Already foreseen by Bernoulli, who showed how the pressure of a gas could be calculated from the kinetic energy of the constituent molecules, kinetic theory was especially the achievement of Clausius, Maxwell, and Boltzmann in the second half of the nineteenth century. We have already described (Introduction and § 3.4) how difficult it was to get kinetic theory accepted by the scientific community. We shall discuss its most elaborated form, the Boltzmann equation, in Chap.15.

7.1 The Perfect Gas Model

As in most theories, we shall start by idealizing the physical system we are studying and we shall represent it by a model. This model is based upon three ideas: the structure of the constituent particles does not play any rôle, their interactions can be neglected, and they can be treated by classical mechanics.

7.1.1 Structureless Particles

In contrast to a solid where the interatomic distances are of the same order as the characteristic sizes of the molecules, which are a few Å, a gas is a substance where the constituent molecules are sufficiently far apart that they can be differentiated. Under "*normal conditions*", that is, room temperature and atmospheric pressure, one mole occupies a volume $\Omega = 22.4$ l, so that the typical intermolecular distances, $d = n^{-1/3} = (\Omega/N_{\text{A}})^{1/3} \simeq 3 \times 10^{-9}$ m, are an order of magnitude larger than their sizes.

We replace here each of the N *molecules*, which are assumed to be identical, by a *point particle* situated at its centre of mass. In other words, we *neglect the structure* of the molecules. We shall get rid of this hypothesis in Chap.8 where we study the rôle played by the internal structure of the constituent molecules of the gas and where we discuss the validity of the present model. Without going into the details of that discussion we note that we can restrict ourselves solely to the translational degrees of freedom of the molecules provided they are *practically all in their ground state* at the temperatures considered.

The neglect of the internal structure of the molecules requires, in particular, that their ground state be neither degenerate nor quasi-degenerate. Morover, the probability p_m for a molecule to be in an excited state m must be very small as compared to the probability p_0 for it to be in its ground state. If E_0 and E_1 are the energies of the lowest two states of the molecules at rest, the ratio p_1/p_0 for their occupation probabilities, which is given by the Boltzmann-Gibbs distribution, equals $\exp(-\beta\Delta E)$ where $\Delta E = E_1 - E_0$

is the *excitation energy.* This ratio is negligibly small, as well as all other ratios p_m/p_0, if

$$\Delta E \gg kT. \tag{7.1}$$

This condition is never satisfied at room temperature for diatomic or polyatomic molecules, as the excitation of their lowest levels, which describe rotations, needs only energies ΔE of the order of 10^{-3} eV, which is small compared to typical kinetic energies of the order of $kT \simeq \frac{1}{40}$ eV. This condition is satisfied (Exerc.8b) for a large number of monatomic gases, in particular, for the *inert gases* (or "*rare gases*") such as argon, neon, helium, or krypton where ΔE is of the order of ten or twenty eV; in the present chapter we shall restrict ourselves to those gases. To simplify the notation we choose the energy origin for each atom such that $E_0 = 0$.

Condition (7.1) implies that $p_m \ll p_0$ for *each one* of the excited states; however, it is not sufficient to ensure the condition $p_0 \simeq 1$, since we need, in fact, that after summation over *all* excited states

$$\sum_{m\neq 0} p_m = 1 - p_0 \ll p_0. \tag{7.2}$$

Ionization of an atom gives rise to a very large number of energy levels which are very closely spaced, even tending to become a continuum when the size of the container in which the atom is placed becomes large. It is therefore not at all clear, not even for an inert gas with an ionization energy ΔE_{i} of several tens of eV, that condition (7.2) can be satisfied when we take the summation over all ionized states into account. In fact, Exerc.8c shows that the degree of ionization is weak, provided

$$\frac{1}{n}\left(\frac{m_{\mathrm{e}}kT}{2\pi\hbar^2}\right)^{3/2} \mathrm{e}^{-\Delta E_{\mathrm{i}}/kT} \ll 1, \tag{7.3}$$

where m_{e} is the electron mass. Normally, condition (7.3) is amply satisfied and it is legitimate to use the perfect gas model. However, a *highly rarefied or very hot gas is always ionized* and cannot be described as a system of structureless, non-interacting atoms. We are then dealing with an ionized gas, or *plasma*, in which the long-range Coulomb interactions between the ions or the electrons cannot be neglected under any circumstances whatsoever.

Let us also note that *quantum mechanics* plays an essential, though hidden, rôle to guarantee the existence of perfect gases. In fact, in classical physics an atom or molecule with a structure would not have a ground state separated from excited states by a finite gap ΔE, and (7.1) could not be satisfied. Morover, (7.3) would also be violated in the limit as $\hbar \to 0$. The fact that one can treat the atoms of the rare gases as point particles is based upon a quantum phenomenon, the *freezing-in of the internal degrees of freedom* of each atom which has a nearly unit probability of remaining in its ground state, even at room temperatures (§ 8.3.1). We have already seen how this phenomenon works in § 1.4.4: if the temperature is low as compared to a characteristic temperature Θ which is such that $k\Theta$ is of the order of magnitude of the excitation energy of the first excited state, the system practically remains in its ground state and the excited states play no rôle. For the rare gases Θ is very high, of the order of 10^5 K, and the atoms thus behave as structureless entities, frozen in their ground state when $T \ll \Theta$.

7.1.2 Non-interacting Particles

The constituent atoms of a rare gas, or, more generally, the molecules in any gas, interact with one another through forces with a range of the order of a few Å, which is short compared to the distances apart. In the perfect gas model we *neglect these interactions* at thermal equilibrium. This assumption is justified because the gas density is low, even though the strength of the forces, in particular, the short-range repulsion, becomes very large below 1 or 2 Å. In fact, assuming to a first approximation that the molecules are distributed uniformly, the average volume occupied by each of them is $n^{-1} = \Omega/N$, whereas the volume inside which the interactions with the other molecules play a rôle is of order δ^3 where δ is the range of the interactions. The probability that a molecule feels the interactions with its neighbours is thus $n\delta^3$ so that the interaction can be neglected provided

$$n\delta^3 \ll 1, \tag{7.4}$$

that is, provided *the range of the forces is short compared to the intermolecular distances.* Under normal conditions, for $\delta \simeq 3$ Å, we have $n\delta^3 \simeq 10^{-3}$ and we are justified to replace the Hamiltonian in the Boltzmann-Gibbs equilibrium distribution by an approximate Hamiltonian without interactions. The model Hamiltonian then includes only the kinetic energies of the molecules, while their structure is neglected, and the potential confining the molecules to the container in which the gas is enclosed. We shall see in Chap.9 that the interactions between the molecules are responsible for corrections to the perfect gas laws which become important when the density increases, and that they are essential to explain the origin and the properties of the liquid state where (7.4) does not hold.

Nevertheless, even in a rarefied gas the interactions play an important rôle in non-equilibrium situations. Because of their short range and due to the low density, the interactions give rise to brief binary collisions. In each collision the total momentum and the total energy of the pair of molecules are conserved; for a rare gas with an excitation energy ΔE which is large compared to typical kinetic energies, of the order of $kT \simeq \frac{1}{40}$ eV, an atom remains in its ground state during a collision with another atom so that the collisions are *elastic* and the total *kinetic* energy is conserved. Even though they are *rare*, these collisions explain properties such as the viscosity or the thermal conductivity of the gas (§§ 7.4.5 and 15.3.3). They are also responsible for the fact that a gas which initially is prepared in an arbitrary state, which may, for instance, be a heterogeneous state, tends to an equilibrium macrostate. Like the interactions between the spins in Chap.1 they, in fact, make it possible for the available energy to be distributed between the molecules in the most random manner *during the processes which lead to equilibrium.* The use of canonical distributions is thus justified by the existence of interactions between the molecules, even though these interactions hardly contribute at all to the energy and can be neglected in the equilibrium formalism.

As a model for the container in which the molecules are enclosed we take a *"box" potential* $V(\boldsymbol{r})$ which is zero inside, and infinite outside the container. A molecule hitting a wall is thus elastically reflected. In reality these collisions are inelastic as the solid walls are made up of molecules which can vibrate more or less strongly according to the temperature and which can thus absorb or give off energy. Collisions are the mechanism through which the gas exchanges work with a moving wall (Exerc.7f). On the other hand, they contribute during the establishment of thermal equilibrium to the equalization of the temperatures of the gas and of the wall through heat energy exchanges.

7.1.3 Classical Particles

The perfect gas model contains a last simplification: we assume that the motion of the *centre of mass* of the molecules, or of the atoms of the rare gases which we are considering here, obeys the laws of *classical mechanics.* The formalism which we must use to describe the macro-state is thus that of *classical statistical mechanics* (§§ 2.3, 3.3, and 4.3.4). This assumption is always satisfied in the gas phase, but we shall see in Chaps.10 to 13 that there are many other substances the constituents of which must be treated quantum mechanically. We shall also in § 10.3.4 recover the results of the present chapter as the low density limits of the general properties of non-interacting quantum gases.

Let us determine the *validity domain* of the classical approximation. For this a dimensional analysis will suffice. In canonical equilibrium the state of the gas is characterized by the density n, the temperature T, and the mass m of the constituent particles. If the latter behave like point particles and do not interact, there are no other parameters. Combining these quantities with one another and with the fundamental constants $\hbar$ and k we can construct only one dimensionless variable, and the particles can therefore be treated classically provided

$$n\lambda_T^3 \ll 1. \tag{7.5}$$

We have defined the *"thermal length"* λ_T at a temperature T by

$$\boxed{\lambda_T^2 \equiv \frac{2\pi\hbar^2}{mkT}}, \tag{7.6}$$

where the factor 2π has been introduced to simplify later formulae. The thermal length λ_T is the only quantity with the dimension of a length which can be constructed starting from the temperature. It is of the same order of magnitude as the de Broglie wavelength h/p of a particle with a kinetic energy $p^2/2m$ of the order of kT. Later we shall see that at a temperature T the mean kinetic energy of the particles is just equal to $\frac{3}{2}kT$. The condition (7.5) means therefore that at equilibrium the particles have *de Broglie wavelengths which are small as compared to their distances apart.*

Another interpretation of condition (7.5) refers to the Heisenberg inequality $\Delta x \, \Delta p > \frac{1}{2}\hbar$ for the statistical fluctuations in a position variable and in its conjugate momentum. The Ehrenfest equations (2.29) for the average position and momentum of a particle reduce to the classical Hamiltonian equations (2.64) if the fluctuations can be neglected, that is, if the extent of the wavepacket in phase space is small compared to the characteristic variables for the motion of the centre of the packet. The classical approximation will thus be valid, if the fluctuations Δx in the particle coordinates are small compared to the characteristic interparticle distance $d = n^{-1/3}$, and if the fluctuations Δp in the momentum are small compared to the characteristic momentum, equal to h/λ_T at a temperature T, as we have just seen. In agreement with what we said at the end of § 2.3.4 these conditions are compatible only provided $d\,h/\lambda_T \gg h$, the same inequality as (7.5).

We can write down yet another condition equivalent to (7.5) by noting that in the grand canonical ensemble the variable n is replaced by the *chemical potential* μ. The only dimensionless quantity is then $\alpha = \mu/kT$ which can take on any value between $-\infty$ and $+\infty$, and which changes in the same direction as n. We thus expect that the condition (7.5) is equivalent to $-\mu \gg kT$, that is, to $e^\alpha \ll 1$. Dimensional analysis and the equivalence between ensembles also show that $n\lambda_T^3$ can only be a function of $\alpha = \mu/kT$. In fact, we shall see that $n\lambda_T^3 = e^\alpha$ in the classical limit, so that the condition for the validity of the classical approximation can be written in the form

$$n\,\lambda_T^3 \;=\; e^\alpha \;\ll\; 1. \tag{7.7}$$

For oxygen under normal conditions, $d = n^{-1/3}$ is equal to 30 Å, whereas λ_T equals 0.2 Å; as a result the left-hand side of (7.7) is as small as 2×10^{-7} and the classical treatment is well justified. The chemical potential μ equals -0.37 eV which in absolute magnitude is much larger than $kT = \frac{1}{40}$ eV. Comparison with these values shows us that (7.7) remains valid, (i) even if the temperature is lowered to 1 K in which case one loses a factor 5×10^3, (ii) even for lighter molecules such as H_2 for which one loses a factor 60, and (iii) even if the density increases by a factor 10^3. The use of *classsical* statistical mechanics is thus legitimate for *all gases* and even for *almost all liquids* down to solidification, as the typical density of a liquid is of the order of 10^3 times that of a gas under normal conditions. The only exception is liquid helium in the form of its two isotopic forms ^{4}He and ^{3}He, at temperatures $T < 10$ K; in that case, the high density, the low mass, and the low temperatures conspire to make (7.7) invalid (Chap.12). It is true that hydrogen is lighter, but it is solid and not liquid at these temperatures. Condition (7.7) is often violated for the atoms in solids and always for the electrons in metals, whose mass is only 1/1800 times that of the proton (Chap.11), and for photons which have zero mass (Chap.13).

One should note that even in a gas where the classical perfect gas model is justified at equilibrium, it is not always legitimate to treat the collisions classically. Their possible quantum features are taken into account by the formalism used in Chap.15 where they are studied.

Altogether, the rare gases, where the interatomic distances are larger both than the range of the interatomic forces and than the thermal length, and where the excitation energy is much larger than kT, can be satisfactorily described by the perfect gas model – except under extreme conditions, realized in astrophysics, where the temperature is so high or the density so low that the gas is ionized. For other gases one must almost always take the structure of the molecules into account (Chap.8). For compressed gases and especially for liquids, the intermolecular interactions play an essential rôle even though one can treat the motion of their centre of mass classically (Chap.9).

7.2 The Maxwell Distribution

The simplicity of the perfect gas model enables us to study it in any of the ensembles. We shall here work mainly in the canonical ensemble, but we shall also recover the same results in the grand canonical and microcanonical formalisms.

7.2.1 The Canonical Phase Density

According to our earlier discussions a perfect gas of N structureless molecules is governed by the classical Hamiltonian which is a function of the position coordinates and their conjugate momenta,

$$H_N\big(\boldsymbol{r}_1, \boldsymbol{p}_1, \ldots, \boldsymbol{r}_N, \boldsymbol{p}_N\big) \;=\; \sum_{i=1}^{N} \left(\frac{\boldsymbol{p}_i^2}{2m} + V(\boldsymbol{r}_i) \right). \tag{7.8}$$

The potential $V(\boldsymbol{r}_i)$ represents the box in which the molecules are enclosed: it is *zero inside* the volume Ω and *infinite outside* it. It may also include an external potential applied to the molecules such as the potential of the gravitational field (§ 7.3.3).

In classical statistical physics a macro-state of the gas is described by a phase density D_N, that is, a probability distribution for the $6N$ coordinates $\boldsymbol{r}_1$, $\boldsymbol{p}_1$, ..., $\boldsymbol{r}_N$, $\boldsymbol{p}_N$ of the molecules in phase space (§ 2.3.2). In canonical thermal equilibrium D_N is the *classical Boltzmann-Gibbs distribution* (4.46):

$$D_N \;=\; \frac{1}{Z_N}\, \mathrm{e}^{-\beta H_N}. \tag{7.9}$$

In evaluating the normalization constant, that is, the *canonical partition function*,

$$Z_N = \mathrm{Tr}\, e^{-\beta H_N},$$

the trace must be replaced in the classical limit by an integration over phase space with measure (2.55). Using expression (7.8) for the Hamiltonian we find

$$\begin{aligned} Z_N &= \int d\tau_N\, e^{-\beta H_N} \\ &= \frac{1}{N!} \int \prod_{i=1}^{N} \left(\frac{d^3\boldsymbol{r}_i\, d^3\boldsymbol{p}_i}{h^3} \exp\left\{ -\beta \left[\frac{\boldsymbol{p}_i^2}{2m} + V(\boldsymbol{r}_i) \right] \right\} \right). \end{aligned} \tag{7.10}$$

The contributions from each of the N particles factor out (§ 4.2.5) so that Z_N can be written as

$$Z_N = \frac{(Z_1)^N}{N!} \tag{7.11}$$

in terms of the single-molecule partition function,

$$Z_1 = \int \frac{d^3\boldsymbol{r}\, d^3\boldsymbol{p}}{h^3} \exp\left[-\beta \left\{ \frac{\boldsymbol{p}^2}{2m} + V(\boldsymbol{r}) \right\} \right]. \tag{7.12}$$

The integration over $\boldsymbol{r}$ gives a factor Ω for a box potential and the integrations over the components of $\boldsymbol{p}$ can again be factorized. Using the formula (see the table of formulae at the end of the book)

$$\int_{-\infty}^{+\infty} e^{-ay^2}\, dy = \sqrt{\frac{\pi}{a}} \tag{7.13}$$

we find

$$Z_1 = \frac{\Omega}{h^3} \left(\frac{2m\pi}{\beta} \right)^{3/2}. \tag{7.14}$$

Using Stirling's formula, where we keep only the extensive contributions to $\ln Z_N$ and introducing the notation (7.6), we find finally

$$Z_N(\beta) = \frac{\Omega^N}{N!\lambda_T^{3N}} \sim \left(\frac{\Omega e}{N\lambda_T^3} \right)^N. \tag{7.15}$$

7.2.2 The Momentum Probability Distribution

We now want to study the equilibrium distribution of the momenta $\boldsymbol{p}_i$ of the gas molecules. According to the definitions in § 2.3, $D_N \, d\tau_N$ represents the *probability that the gas consists of N molecules and that its representative point in the 6N-dimensional phase space lies within $d\tau_N$*. This information is too detailed: in order to obtain the probability distribution $g(\boldsymbol{p}) \, d^3\boldsymbol{p}$ of the momentum $\boldsymbol{p}$ of one of the particles we must examine how D_N depends on the three components of $\boldsymbol{p}$ and get rid of the $6N - 3$ other coordinates through integration. Because of the symmetry of D_N in the set of all particles and its factorized exponential form (7.8), (7.9), we find without calculations

$$\boxed{g(\boldsymbol{p}) \, d^3\boldsymbol{p} \; \propto \; \mathrm{e}^{-p^2/2mkT} \, d^3\boldsymbol{p}} \; . \tag{7.16}$$

The normalization constant follows from (7.13) and we thus get for the probability that a molecule has a momentum within a volume element $d^3\boldsymbol{p}$ around $\boldsymbol{p}$ in momentum space:

$$g(\boldsymbol{p}) \, d^3\boldsymbol{p} \;=\; \left(2\pi mkT\right)^{-3/2} \mathrm{e}^{-p^2/2mkT} \, d^3\boldsymbol{p}. \tag{7.17}$$

The *distribution of the velocities* $\boldsymbol{p}/m$ of the molecules in thermal equilibrium is therefore proportional to the *exponential of the kinetic energy divided by kT*. This is *Maxwell's law* (1860), which one can consider as the reduction to a single molecule of the canonical Boltzmann-Gibbs distribution.

In fact, the original proof by Maxwell was much earlier than the work by Boltzmann and Gibbs (§ 4.2.1). It was based upon the isotropy of $g(\boldsymbol{p})$ and a hypothesis, which was not justified *a priori*, that the three components of $\boldsymbol{p}$ were statistically independent. As an exercise one may show that these two properties are sufficient to find the form (7.16) for $g(\boldsymbol{p})$. For Maxwell $g(\boldsymbol{p}) \, d^3\boldsymbol{p}$ simply represented the *fraction* of molecules which had a momentum within $d^3\boldsymbol{p}$. Boltzmann introduced in 1877 the concept of an *ensemble* of systems, emphasizing the probability law for the $6N$ particle coordinates in phase space. That enabled him to derive (7.16) from the microcanonical distribution (§ 7.2.5) without needing Maxwell's factorization hypothesis.

Let us generalize this result and evaluate the *reduced single-particle density* $f(\boldsymbol{r}, \boldsymbol{p})$ defined in § 2.3.5. We remind ourselves that $f(\boldsymbol{r}, \boldsymbol{p}) \, d^3\boldsymbol{r} \, d^3\boldsymbol{p}$ is the *average number of molecules in the one-particle phase space volume element* $d^3\boldsymbol{r} \, d^3\boldsymbol{p}$ and that it follows from the density in phase through (2.81), or

$$\begin{aligned} f(\boldsymbol{r}, \boldsymbol{p}) \, d^3\boldsymbol{r} \, d^3\boldsymbol{p} \;&=\; \frac{d^3\boldsymbol{r} \, d^3\boldsymbol{p}}{h^3} \, N \int \frac{1}{N!} \prod_{i=2}^{N} \left(\frac{d^3\boldsymbol{r}_i \, d^3\boldsymbol{p}_i}{h^3} \right) \\ &\quad \times D_N\big(\boldsymbol{r}, \boldsymbol{p}, \boldsymbol{r}_2, \boldsymbol{p}_2, \ldots, \boldsymbol{r}_N, \boldsymbol{p}_N\big); \end{aligned} \tag{7.18}$$

the measure $d\tau_N$ associated with D_N in the N-particle phase space is multiplied by a factor N since (7.18) represents the sum of the probabilities that each of the particles 1, 2, ..., N successively occupies the volume element $d^3\boldsymbol{r}\, d^3\boldsymbol{p}$ while the positions and momenta of all the other particles are arbitrary. Using in (7.18) the expression (7.8), (7.9), (7.11), and (7.12) for the perfect gas density in phase, we find

$$f(\boldsymbol{r},\boldsymbol{p}) = \frac{N}{Z_1 h^3}\, \mathrm{e}^{-\beta[p^2/2m+V(\boldsymbol{r})]}. \tag{7.19}$$

This reduced density can be factorized:

$$f(\boldsymbol{r},\boldsymbol{p}) = n(\boldsymbol{r})\, g(\boldsymbol{p}), \tag{7.20}$$

where g is the Maxwell distribution (7.17). The factor

$$n(\boldsymbol{r}) = N \left[\int d^3\boldsymbol{r}\; \mathrm{e}^{-\beta V(\boldsymbol{r})}\right]^{-1} \mathrm{e}^{-\beta V(\boldsymbol{r})}, \tag{7.21}$$

equal to the constant N/Ω for a gas in a box if there is no external field, is nothing but the *density of molecules at the point* $\boldsymbol{r}$ since $d^3\boldsymbol{r} \int d^3\boldsymbol{p}\, f(\boldsymbol{r},\boldsymbol{p})$ is the average number of molecules, with arbitrary momenta, within the volume element $d^3\boldsymbol{r}$. Expression (7.20) reflects the absence of correlations between the positions and the momenta of the molecules in the classical limit. This property may not hold for quantum gases.

7.2.3 Applications of the Maxwell Distribution

To use the Maxwell distribution for practical applications we must remember that, if we change variables, *the probability* $g(\boldsymbol{p})\, d^3\boldsymbol{p}$ *remains invariant* and the probability measure $g(\boldsymbol{p})$ must be multiplied by the Jacobian. As an example, let us calculate the average number of molecules per unit volume, $dn(v)$, with an absolute magnitude of the velocity lying between v and $v+dv$. We must integrate $g(\boldsymbol{p})\, d^3\boldsymbol{p}$ between spheres with radii mv and $m(v+dv)$, which leads to

$$\begin{aligned} dn(v) &= n \int_{mv<p<m(v+dv)} g(\boldsymbol{p})\, d^3\boldsymbol{p} = 4\pi n\, g(\boldsymbol{p})\, p^2\, dp \\ &= 4\pi n\, m^3 g(mv)\, v^2\, dv. \end{aligned} \tag{7.22}$$

The curves of Fig.7.1 give examples of *molecular velocity distributions*. Typical velocities are of the order of the sound velocity in the gas considered, which is 300 ms^{-1} in air, as the average molecular velocity is

$$\frac{1}{n}\int v\, dn(v) = \sqrt{\frac{8}{\pi}\frac{kT}{m}},$$

while the sound velocity is

$$\sqrt{\gamma\,\frac{\mathcal{P}}{nm}} = \sqrt{\gamma\,\frac{kT}{m}}. \tag{7.23}$$

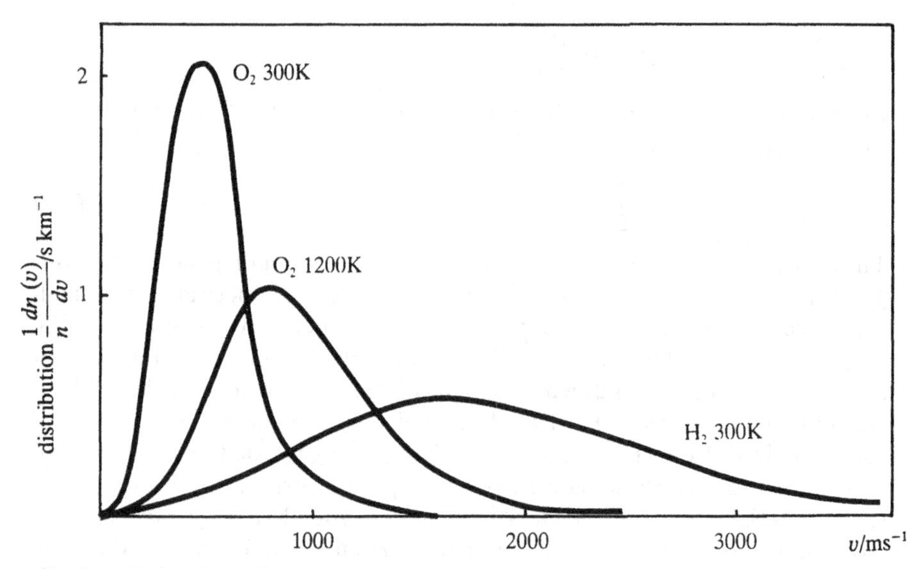

Fig. 7.1. Molecular velocity distributions

Here γ is the specific heat ratio $C_\mathrm{p}/C_\mathrm{v}$, and we shall see in Chap. 8 that $\gamma = \frac{5}{3}$ for the perfect gas and $\gamma = \frac{7}{5}$ for diatomic gases.

The molecular velocity distribution has an effect upon the *shape of the spectral lines* emitted or absorbed by a gas, because of the Doppler shift. This effect (Michelson 1892) is important in astrophysics for the determination of the temperatures of stellar objects (Exerc. 7d).

The *composition of planetary atmospheres* is directly affected by the molecular velocity distribution. In fact, the escape velocity from a star, $\sqrt{2GM/R}$, where G is the gravitational constant, depends on the mass, M, and the radius, R, of the star. Molecules with larger velocities can escape from the gravitational field of the star. The curves of Fig. 7.1 show that this escape is easier for light molecules and that it is helped by high temperatures. This enables us to understand why there is no hydrogen in the Earth's atmosphere, even though that element is so very abundant in the Universe. If the temperature at the top of the atmosphere is sufficiently high and if the molecules are sufficiently light, we have in equilibrium an appreciable fraction of molecules with velocities above the escape velocity, which for the Earth is 11 000 ms^{-1}. Those molecules can escape and at equilibrium will be replaced by others which, in their turn, escape. This has happened for hydrogen. For oxygen and nitrogen the molecular masses are sufficiently large that the losses are insignificant, even at the rather high temperatures prevailing when the Earth was formed. On the Moon the escape velocity is so low that the whole atmosphere has had time to escape.

The Maxwell distribution enables us to express the average kinetic energy of the molecules as a function of the temperature. We have

$$\left\langle \frac{p^2}{2m} \right\rangle = \int d^3\boldsymbol{p}\; g(\boldsymbol{p})\; \frac{p^2}{2m},$$

which can be calculated, using Cartesian coordinates, (7.13), and its derivative with respect to a:

$$\left\langle \frac{p^2}{2m} \right\rangle = \frac{3}{2\beta} = \frac{3}{2}kT. \tag{7.24}$$

The average kinetic energy of the molecules is thus proportional to the temperature and is *independent of the nature of the gas.* Numerically, it is small compared to 1 eV, as kT equals $\frac{1}{40}$ eV at room temperatures. The value of the proportionality coefficient in (7.24) is a particular case of the equipartition theorem for energy (§ 8.4.2) which assigns at equilibrium, to each classical degree of freedom on which the Hamiltonian depends quadratically, an average energy of $\frac{1}{2}kT$. Here the degrees of freedom are the translations in the three space directions which are governed by the $\boldsymbol{p}^2/2m$ term in the Hamiltonian. Note that the *dispersion of the kinetic energy* round that average value is large, as the relative statistical fluctuation given by (8.51) equals $\sqrt{2/3}$. If initially all molecules had the same kinetic energy (7.24), their collisions with one another would change these individual energies and they would end up with a distribution in agreement with Maxwell's law.

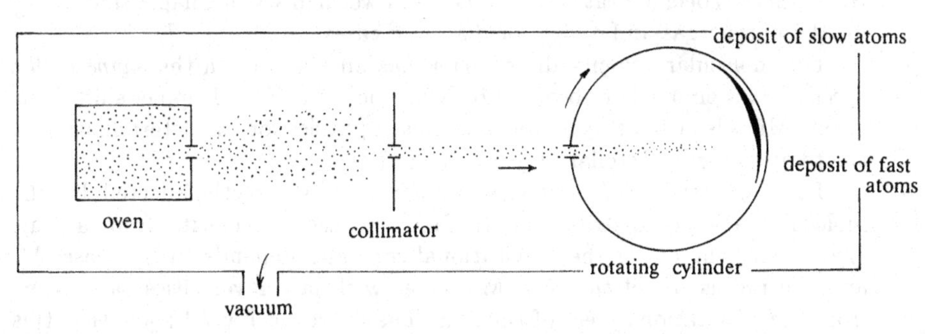

Fig. 7.2. Experimental determination of the Maxwell distribution

Let us finally note that one can *directly measure the molecular velocity distribution* of a gas and thus experimentally check Maxwell's law at the microscopic level. To do this, consider a vessel containing a gas and being maintained at a temperature T which we can control. The wall of the vessel is pierced by a very small hole which connects it with an empty enclosure and through which the gas can escape through *effusion* (§ 7.4.1)) with a small flux. The characteristics of the molecular jet which is thus produced allow us to reconstruct the velocity distribution in the vessel: after being collimated the jet consists of molecules with momenta all in the same direction but with different absolute magnitudes; as an exercise one can show, using Eq.(7.50), that the probability distribution of the velocities in the jet is proportional to $v^3 g(mv)\, dv$, which makes it possible to use it to find the distribution $g(\boldsymbol{p})\, d^3\boldsymbol{p}$ in the vessel (Exerc.7d). Figure 7.2 shows a method for measuring the velocities in the jet: a molecular jet of vaporized silver enters a rotating cylinder and the molecules are deposited on different parts of the cylinder, depending on their velocity. The measure of the thickness of the deposit gives us $g(\boldsymbol{p})$. The first experiment of this kind for directly measuring the Maxwell distribution was performed by Otto Stern (1920).

7.2.4 Grand Canonical Equilibrium of a Gas

Including the number of molecules N as one of the state variables and using a *canonical* ensemble is appropriate for the study of a *closed system* which cannot exchange matter with the outside world. If the volume of the gas considered can exchange molecules with its environment, which may consist either of an identical gas surrounding it, or of solid walls which *adsorb* molecules (Exerc.4b), or of *another*, liquid or solid, *phase* with which it is in equilibrium (§ 9.3 and Prob.8), or if the gas can undergo *chemical reactions* (§ 8.2.2), it is better to use the chemical potential μ as state variable instead of N and to work with a *classical grand canonical* ensemble. The density in phase D is then a *set of functions* D_N which at equilibrium have the form

$$D_N = \frac{1}{Z_G}\, e^{-\beta H_N + \alpha N}. \tag{7.25}$$

The formalism resembles that of § 7.2.1, but now includes summations over N. The *grand partition function*,

$$Z_G = \sum_{n=0}^{\infty} Z_N\, e^{\alpha N}, \tag{7.26}$$

can be calculated using (7.11) and we find

$$\ln\, Z_G(\alpha, \beta) = Z_1(\beta)\, e^{\alpha}, \tag{7.27}$$

where Z_1 is given by (7.12). If there is no external field, (7.14) gives

$$\ln\, Z_G(\alpha, \beta) = \Omega\, \lambda_T^{-3}(\beta)\, e^{\alpha}. \tag{7.28}$$

The thermodynamic properties can be derived either from Z_N, given by (7.15), or from Z_G, given by (7.28), using the formalisms of §§ 4.2.6 or 5.6.6. In § 7.3 we shall use the canonical ensemble. To see, without explicit calculations, that the same results are obtained when we start from the grand canonical ensemble it is sufficient to prove that the two *ensembles are equivalent* (§ 5.5.3) by checking that $\ln\, Z_G$ and $\ln\, Z_N$ derive one from the other through a Legendre transformation (§ 6.3). To do that, we start from (7.28) and use the relation $\partial \ln\, Z_G / \partial\alpha = N$ to connect N and α, whence follows that

$$e^{\alpha} = \frac{N\lambda_T^3}{\Omega}. \tag{7.29}$$

Eliminating α between (7.28) and (7.29) then gives

$$\ln Z_G - \alpha \frac{\partial \ln Z_G}{\partial \alpha} = N - N \ln \frac{N}{Z_1} \sim \ln Z_N, \tag{7.30}$$

which is the same equation as (7.11) when $N \gg 1$. The equivalence of the canonical and the grand canonical ensembles is here a consequence of Stirling's formula. As an exercise one can also check this equivalence, as in § 5.5.3, by evaluating the sum (7.26) over N, or the integral (5.52) over α, using the saddle-point method.

The evaluation of the *reduced density* $f(\boldsymbol{r}, \boldsymbol{p})$ in the grand canonical ensemble starts from (7.18), except that D_N is here given by (7.25) and that we must sum over N. That summation is easy; it gives the result

$$f(\boldsymbol{r}, \boldsymbol{p}) = \frac{1}{h^3} e^{\alpha - \beta H_1(\boldsymbol{p}, \boldsymbol{r})}, \tag{7.31}$$

which is the same as (7.20), as the *molecular density* can now be expressed in terms of α through

$$n(\boldsymbol{r}) = \lambda_T^{-3} e^{\alpha - \beta V(\boldsymbol{r})}. \tag{7.32}$$

Using (7.29) we can check the equivalence of (7.21) and (7.32).

In Chap.10 we shall find that the perfect gas is the classical limit of a non-interacting *quantum gas* when condition (7.7) is satisfied. We shall then use the grand canonical ensemble, the only one which enables us to make simple calculations. Expressions (7.27) for Z_G and (7.32) for the density will appear as the limits of the more general expressions (10.52) and (10.53) when $e^{\alpha} \ll 1$. We shall also study *non-equilibrium gases* (Chap.15) in the grand canonical formalism, as we shall need to take into account exchanges of particles between neighbouring volume elements; inside each of them a quasi-equilibrium will be realized so that the reduced density $f(\boldsymbol{r}, \boldsymbol{p})$ will have the form (7.31) where α and β can vary from point to point in space.

Although the reduced single-particle density $f(\boldsymbol{r}, \boldsymbol{p})$ is the same for the canonical and for the grand canonical ensembles, there occurs a small difference in the calculation of the reduced two-particle density $f_2(\boldsymbol{r}, \boldsymbol{p}, \boldsymbol{r}', \boldsymbol{p}')$, defined by (2.82), with a possible summation over N. As an exercise, one can obtain in the canonical ensemble:

$$f_2(\boldsymbol{r}, \boldsymbol{p}, \boldsymbol{r}', \boldsymbol{p}') - f(\boldsymbol{r}, \boldsymbol{p}) f(\boldsymbol{r}', \boldsymbol{p}') = -\frac{1}{N} f(\boldsymbol{r}, \boldsymbol{p}) f(\boldsymbol{r}', \boldsymbol{p}'), \tag{7.33a}$$

and in the grand canonical ensemble:

$$f_2(\boldsymbol{r}, \boldsymbol{p}, \boldsymbol{r}', \boldsymbol{p}') - f(\boldsymbol{r}, \boldsymbol{p}) f(\boldsymbol{r}', \boldsymbol{p}') = 0. \tag{7.33b}$$

As a result, there is *no correlation* between the molecules of the perfect gas in the grand canonical equilibrium. There is a weak correlation between them in canonical equilibrium, produced by the fact that the total number of molecules is fixed; this correlation clearly vanishes in the thermodynamic limit.

7.2.5 Microcanonical Equilibrium

The phase density D_N of the classical microcanonical equilibrium (§4.3.4) is constant in the region of phase space where

$$U \;<\; H_N(\boldsymbol{r}_1,\boldsymbol{p}_1,\ldots,\boldsymbol{r}_N,\boldsymbol{p}_N) \;<\; U+\Delta U, \tag{7.34}$$

and vanishes elsewhere. If ΔU is sufficiently small, we can replace D_N by

$$D_N \;\propto\; \delta(H_N - U), \tag{7.35}$$

except when we evaluate the entropy (3.36) which involves the volume W of the region (7.34) with measure $d\tau_N$.

It is inconvenient to use (7.34) or (7.35), as the contributions from the various molecules do not factorize, even though there are no interactions. The calculations involving the microcanonical ensemble are therefore in general carried out using techniques which reduce it to other ensembles. For instance, Fourier transforming (7.35),

$$D_N \;\propto\; \frac{1}{2\pi \mathrm{i}} \int_{\beta_0-\mathrm{i}\infty}^{\beta_0+\mathrm{i}\infty} d\beta \; \mathrm{e}^{-\beta H_N+\beta U}, \tag{7.36}$$

we see that D_N can be written as a simple integral over canonical distributions that we can easily factorize. Starting from (7.36) we can prove the equivalence of the microcanonical and canonical ensembles using the saddle-point method (§5.5.3 and Exerc.5b).

As a further exercise about the equivalence of ensembles, let us start directly from (7.35) and find again the Maxwell distribution (7.16). We must integrate (7.35) over the momenta of all particles, bar one, which leads to

$$g(\boldsymbol{p}) \;\propto\; \int \prod_{i=2}^{N} d^3\boldsymbol{p}_i \; \delta\left(\frac{p^2}{2m} + \sum_{i=2}^{N} \frac{p_i^2}{2m} - U\right).$$

In the $3(N-1)$-dimensional space we take as variables the length

$$P \;=\; \sqrt{\sum_{i=2}^{N} \boldsymbol{p}_i^2}$$

and the $3N-4$ angles fixing the direction of the vector $\boldsymbol{P}$. Apart from a constant factor which is equal to the surface area of the hypersphere in $3(N-1)$ dimensions, we get

$$g(\boldsymbol{p}) \;\propto\; \int_0^{\infty} P^{3N-4}\, dP \; \delta\left(\frac{p^2}{2m} + \frac{P^2}{2m} - U\right).$$

Let us remind ourselves that the δ-distribution has the following property (see the formulae at the end of the book): if the x_i are the zeroes, which we are assuming to be simple, of $f(x)$, we have

$$\delta[f(x)] = \sum_i \frac{1}{|f'(x_i)|}\,\delta(x - x_i).$$

This can easily be proved by taking as variable $f(x)$ instead of x in the vicinity of each x_i, and introducing the appropriate Jacobian. As a result we have

$$\delta\left(\frac{p^2}{2m} + \frac{P^2}{2m} - U\right) = \frac{m}{P}\,\delta\left(P - \sqrt{2mU - p^2}\right),$$

and hence, apart from a new constant factor,

$$g(\boldsymbol{p}) \propto \left(1 - \frac{p^2}{2mU}\right)^{(3N-5)/2}.$$

In the thermodynamic limit, where $N \to \infty$ with U/N finite, this expression tends to (7.16), provided we put $kT = 2U/3N$. We have thus again found the Maxwell distribution and also the relation (7.24) between the average kinetic energy of the molecules and the temperature.

7.3 Thermostatics of the Perfect Gas

7.3.1 Thermostatic Pressure

The *free energy* which follows from (7.15) is

$$F(T, N, \Omega) = NkT\left(\ln\frac{N\lambda_T^3}{\Omega} - 1\right). \tag{7.37}$$

As we are dealing with a thermodynamic potential, we can derive all equilibrium properties of the perfect gas from it.

In particular, according to (5.75), the derivative of (7.37) with respect to Ω gives the *equation of state*

$$\boxed{\mathcal{P}\Omega = NkT}. \tag{7.38}$$

The theory has thus enabled us to prove the *perfect gas law* which was established experimentally in the form

$$\mathcal{P}\Omega = \frac{N}{N_{\mathrm{A}}}RT \tag{7.39}$$

between the seventeenth and the nineteenth centuries by studying gases in the low density limit (Boyle 1661, Mariotte 1676, Charles and Gay-Lussac 1802). The so-called "perfect gas temperature" scale, defined empirically by

(7.39), can thus be identified with the absolute temperature of statistical physics and of the Second Law of thermodynamics. The gas constant, $R = 8.3$ J K^{-1} mol^{-1}, is then identified with kN_A, where $N_A = 6 \times 10^{23}$ is Avogadro's number and where $k = 1.38 \times 10^{-23}$ J K^{-1} is Boltzmann's constant, introduced in Chap.3 to adjust the units of entropy and of temperature to the values used in macroscopic experimental physics.

To find the same result in the grand canonical ensemble we start from (7.28), write down the *grand potential*

$$A(T, \mu, \Omega) = -kT\Omega\lambda_T^{-3}\,e^{\mu/kT}, \tag{7.40a}$$

and then we eliminate μ from (5.76) and (5.78) to write

$$NkT = -kT\,\frac{\partial A}{\partial \mu} = -A = \mathcal{P}\Omega. \tag{7.40b}$$

From Eqs.(7.40a and b) we find the *chemical potential* μ as a function of the density and the temperature. In Chap.8 we shall study some physical and chemical properties which follow directly from this expression.

7.3.2 Thermal Properties

Using (5.65) we find from (7.37) the thermal properties of the perfect gas by taking the derivative with respect to the temperature T. The *entropy* $S = -\partial F/\partial T$ can thus be expressed as a function of the temperature through the *Sackur-Tetrode formula* (1911–13)

$$\boxed{S = Nk\left(\ln\frac{\Omega}{N\lambda_T^3} + \frac{5}{2}\right)}, \tag{7.41}$$

where λ_T is given by (7.6). A consequence of (7.41) is the relation between volume and temperature for an *adiabatic expansion* of the perfect gas,

$$\Omega T^{3/2} = \text{constant}, \tag{7.42}$$

which, if we use (7.38), can also be written in the form

$$\mathcal{P}\,\Omega^{5/3} = \text{constant}. \tag{7.43}$$

This behaviour, which has been amply verified experimentally for rarefied mon-atomic gases, does not hold for gases where the molecules have a less simple structure (Chap.8).

Expression (7.41) becomes *negative* at low temperatures when λ_T becomes of the order of the intermolecular distance d. This, in agreement with the statement made at the beginning of § 3.3.2, is a sign of the failure of the classical approximation used here. In fact, as we saw in § 7.1.3, the condition for the validity of this approximation is just $\Omega/N\lambda_T^3 \gg 1$, that is, $S/Nk \gg 1$.

This implies that the entropy (7.41) per molecule is always not only positive, but also large compared to k, whenever the perfect gas model is justified.

The *extensivity* of the entropy of the perfect gas, obvious from (7.41), would not have been satisfied without the presence of $N!$ in the denominator of (7.10), which gives rise to the term $-Nk \ln N$ in (7.41). We have stressed in §§ 2.3.2 and 2.3.4 that this factor $1/N!$ which is connected with the indistinguishability of the molecules has a *quantum origin*. Even though it is based upon classical statistical mechanics, the theory of a perfect gas contains this ingredient which cannot be justified strictly in the framework of classical physics and which is necessary to ensure the extensivity and to resolve the Gibbs paradox (§ 8.2.1).

The *internal energy* which follows from (5.64) and (7.41) equals

$$\boxed{U = \tfrac{3}{2}NkT} \quad . \tag{7.44}$$

As the Hamiltonian in our model reduces to the translational kinetic energy of the molecules, the expression for the internal energy per molecule is the same as the average value (7.24) of the kinetic energy. We see thus that *the internal energy of the perfect gas depends solely on the temperature.* This property, which we shall generalize in Chap.8 to real gases at low density, is *Joule's law*: Joule established it experimentally by showing that a *sudden expansion* of a gas, without exchanging mechanical or thermal energy with the outside, also was *isothermal* (Gay-Lussac 1807, made more precise by Joule in 1845).

A microscopic study, based upon kinetic theory, of an *adiabatic expansion* (Exerc.7f) moreover enables us to understand, using merely conservation of mechanical energy, how molecules, by shedding some of their kinetic energy to the piston in the form of *work*, change their momentum distribution $g(\boldsymbol{p})$, which thus decreases the value of $\langle p^2 \rangle$ and with it the temperature according to (7.42).

We find the *specific heat at constant volume* by taking the derivative of (7.44) with respect to the temperature, whence

$$C_{\rm v} = \frac{3}{2}Nk = \frac{3}{2}R\,\frac{N}{N_{\rm A}}. \tag{7.45}$$

To find the specific heat *at constant pressure* we must add to the change in internal energy the work, $\mathcal{P}\,d\Omega$, done on the outside during the expansion, in accordance with the general result (6.41). We then get *Mayer's relation*

$$C_{\rm p} - C_{\rm v} = Nk. \tag{7.46}$$

The ratio of the specific heats is in this case

$$\gamma \equiv \frac{C_{\rm p}}{C_{\rm v}} = \frac{\kappa_T}{\kappa_S} = \frac{5}{3}. \tag{7.47}$$

According to (6.44) it is equal to the ratio of the isothermal and adiabatic compressibilities, and hence we have $\kappa_S = 1/\gamma \mathcal{P}$. This explains the presence of the index γ in Eq.(7.43) for the adiabatic expansion, and as a consequence, its appearance in expression (7.23) for the *sound speed* in the gas.

These results can easily be checked by experiments which give us the specific heats through calorimetry and γ either by measuring the sound speed or by studying adiabatic expansions. Mayer's relation is well satisfied for all gases at low densities. On the other hand, (7.45) and (7.47) are only found for monatomic gases. We shall see in Chap.8 how the structure of the molecules affects the various macroscopic properties of gases.

7.3.3 The Rôle of an Applied Field

In §§ 7.3.1 and 7.3.2 we have restricted ourselves to a perfect gas for which the molecules were not subject to any applied external field. Including an applied field does not present any difficulties, as the potential $V(\boldsymbol{r})$ occurring in (7.12) can be arbitrary. We have seen that the kinetic and the spatial properties are independent of one another, as is, especially, seen in the factorization (7.20) of the reduced density. The density $n(\boldsymbol{r})$ given by (7.21) or (7.32) now changes from one point to another. The internal energy can be split into the sum of a contribution associated with the field and a kinetic contribution which is the same as its value when there is no field.

For instance, in a *field of gravity* (Exerc.7a) we have

$$V(\boldsymbol{r}) = mgz,$$

and the density is proportional to

$$n(\boldsymbol{r}) \propto e^{-mgz/kT}. \tag{7.48}$$

It decreases exponentially with height, the effect being more pronounced when the molecular mass is larger. Equation (7.48) is *Laplace's barometer formula* which has been checked in the laboratory (Jean Perrin 1908) for dilute solutions, where the particles of the solute behave like the molecules of a perfect gas. In the atmosphere the situation is less simple as the gas is not in isothermal equilibrium: exchange of heat between layers of air at different altitudes is difficult, there are convection currents, and radiation plays an important rôle.

The situation is similar in the case of a gas enclosed in a *vessel rotating* with an angular velocity ω and in thermal equilibrium with the vessel (Exerc.7b). If $r_\perp$ is the distance from the axis of rotation we find that the density of the gas in this relative equilibrium is proportional to

$$n(\boldsymbol{r}) \propto -e^{m\omega^2 r_\perp^2/2kT}. \tag{7.49}$$

The centrifugal effects tend to increase the density at the periphery, the more strongly, the larger the mass of the molecules involved. This property is the basis of an *isotope separation* process, which might compete with gas diffusion but which is much less used in industry, called *ultracentrifuging*: in a centrifuge rotating at high velocities the concentration of the heavy isotope is larger at the periphery than at the centre.

7.4 Elements of Kinetic Theory

In this section we shall again prove some of the earlier results using a more direct and older method than the general formalism of equilibrium statistical physics used so far, and which is particularly useful for studying gases – kinetic theory, due to Maxwell and Boltzmann. This method will allow us, moreover, to deal with problems where the gas is not in thermal equilibrium. We shall limit ourselves here to some elementary discussions, postponing until Chap.15 a more quantitative theory.

7.4.1 Effusion

Kinetic theory is based upon the analysis of the elementary mechanical processes, *free motion* of the molecules between two successive collisions, *collisions* between those molecules, and, furthermore, upon calculating *averages* over the system of gas molecules and striking the *balance*. Its simplest application is *effusion*. A gas, at thermal equilibrium, is enclosed in a vessel with a wall which is pierced by a very small hole through which molecules can escape so that the density will decrease (Exerc.7e). In practice this phenomenon may also concern the passage of a gas through porous walls (Exerc.7g) where each pore is modelled as a hole through which molecules can pass. The only rôle played by collisions is to maintain equilibrium conditions up to the immediate vicinity of each hole. The free path of molecules between successive collisions (§ 7.4.5) is assumed to be large compared to the hole size so that we can neglect the interactions between the molecules during the time that they enter and pass through the hole. As a result the dynamics of effusion, which is a *non-equilibrium* process, is governed mainly by the flux of molecules which arrive at the appropriate angle of incidence at the opening ΔS of each hole.

We wish thus to evaluate the *number of molecules* ΔN with momentum $\boldsymbol{p}$ (within $d^3\boldsymbol{p}$) which *impinge upon the element of area* ΔS at an angle of incidence θ between times t and $t+\Delta t$. Let the x-direction be at right angles to ΔS so that $\cos\theta = p_x/p$ (Fig.7.3). Following the motion of molecules backwards in time and neglecting collisions we see that at time $t - \Delta t$ those which will hit the area ΔS during the time interval Δt are necessarily situated in a slanted cylinder of base ΔS and length $v\Delta t = p\Delta t/m$, that is, of height $p_x\Delta t/m$. The volume of that cylinder is $\Delta S\, p_x \Delta t/m$, and the total number of molecules contained in it is therefore

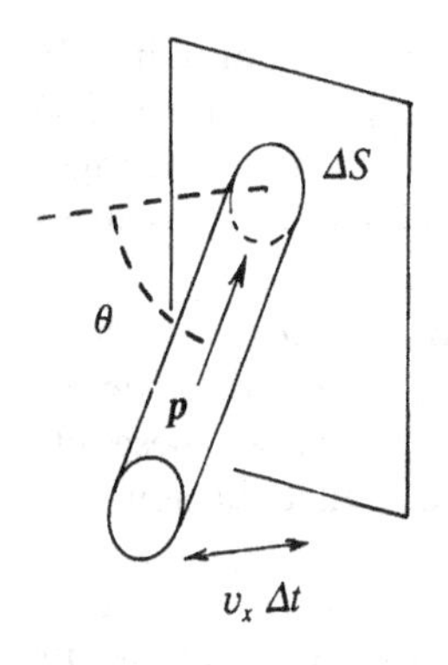

Fig. 7.3. Calculating the flux of molecules through a hole

$$n\,\Delta S\,\frac{p_x}{m}\,\Delta t,$$

where n is the density of molecules in the region and at the time considered. Among all these molecules we want to distinguish the ones which are aimed towards the area ΔS with a momentum $\boldsymbol{p}$ within $d^3\boldsymbol{p}$. The hole is sufficiently small that the equilibrium in its vicinity is not perturbed – except, of course, for the molecules which pass through it – and the momenta of the molecules in the cylinder are therefore distributed according to Maxwell's law (7.17), so that the number we look for is given by

$$\boxed{\Delta N \;=\; n\,\Delta S\,\frac{p_x}{m}\,\Delta t\; g(\boldsymbol{p})\, d^3\boldsymbol{p}}\;. \tag{7.50}$$

Assuming that the porous wall separates the gas from vacuum we get the *rate of effusion* by integrating (7.50) over those values of $\boldsymbol{p}$ which are in the appropriate directions determined by the geometry of the hole – for instance, over $p_x > 0$, if all molecules which hit the hole at any angle of incidence emerge on the other side. This gives us for the number of molecules escaping per unit time, apart from a geometric factor,

$$\frac{dN}{dt} \;\propto\; -nS\,\sqrt{\frac{kT}{m}}, \tag{7.51}$$

where S is the total area of the pores. If there is gas on both sides, we must strike the *balance of passages in both directions* (Exerc.7g); this gives, if the temperature is uniform, a rate of effusion which is proportional to the difference in the densities n, that is, to the *difference in pressure* at the two sides.

Formula (7.51) shows that effusion is the faster, the *smaller the mass* of the molecules. This property poses a problem for the construction of balloons, as they must be filled by a light gas (helium) which thus rather easily passes through the cover. It also implies that effusion of a gas mixture through a large number of very small holes in a porous wall selects the molecules as function of their mass, thus producing *isotope separation*; this process is, in practice, the only one used on an industrial scale to enrich uranium in its 235 isotope. The difficulty remains that the difference in mass between the 235 and 238 isotopes is so small, which means that one must have many passages through porous walls before obtaining an appreciable change in the relative concentration (Exerc.14a).

Effusion through a hole followed by collimation also provides us with an efficient laboratory process, often used to produce molecular jets with average velocities which depend on the initial thermal motions. Of course, these jets are not in equilibrium and have a non-Maxwellian velocity distribution.

As the faster particles escape preferentially, effusion produces a *cooling* of the residual gas which reaches new equilibria while the molecules escape. As an exercise one can show, by integrating (7.50) after weighting it with $p^2/2m$, that the kinetic energy lost with the escaping molecules produces a loss in internal energy according to

$$\frac{dU}{dt} = 2kT\,\frac{dN}{dt}. \tag{7.52}$$

It then follows from (7.44) that the temperature decreases as $N^{1/3}$ for a monatomic gas.

7.4.2 Kinetic Pressure

In kinetic theory the pressure is interpreted as resulting from the collisions of the molecules with the wall of the vessel in which the gas is enclosed. Each collision imparts to the wall a small momentum. During a time interval Δt, a macroscopic surface area ΔS of the wall undergoes a large number of collisions and the average over all these collisions gives rise to a pressure force which is proportional to ΔS. On the macroscopic scale one thus observes a *collective effect*, with temporal and spatial fluctuations which can be neglected (Bernoulli 1738).

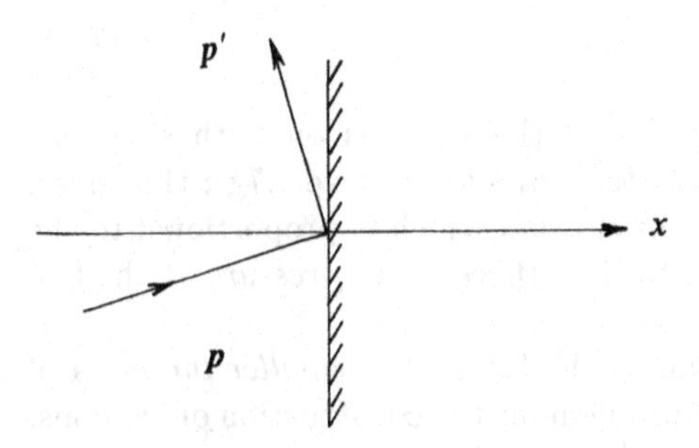

Fig. 7.4. Scattering of a molecule by the wall

In order to calculate this kinetic pressure we consider a surface element ΔS of the wall which between the time t and the time $t+\Delta t$ scatters a certain number of molecules. We choose the x-axis at right angles to the wall. Let us consider one collision during which the momentum of the molecule changes from $\boldsymbol{p}$ to $\boldsymbol{p}'$ (Fig.7.4). The wall has a complex microscopic structure so that the scattering is not a specular reflection, that is, we have not necessarily $p_x = -p'_x$, $p_y = p'_y$, $p_z = p'_z$, and the details of the process are complicated. However, as the duration of the collision is very short, typically 10^{-11} s, as compared to Δt, we can apply the *shot theorem* in order to evaluate the integral of the force $\boldsymbol{F}(t)$ exerted by the molecule on the wall over the time interval Δt:

$$\int_t^{t+\Delta t} dt' \, \boldsymbol{F}(t') \; = \; \boldsymbol{p} - \boldsymbol{p}'. \tag{7.53}$$

This is a simple consequence of the equation of motion, $d\boldsymbol{p}/dt = \boldsymbol{\varphi}$, where $\boldsymbol{\varphi}$ is the force applied to the molecule at time t'.

The total impulse received by the wall during the interval Δt is the sum of (7.53) over all collisions taking place during that interval. In (7.50) we have evaluated the number of molecules of momentum $\boldsymbol{p}$, within $d^3\boldsymbol{p}$, reaching the surface area ΔS during the time interval Δt. We similarly need the number of molecules of momentum $\boldsymbol{p}'$, within $d^3\boldsymbol{p}'$, which leave the surface area ΔS during the interval Δt. In a stationary regime the distribution of the molecules before and after their collisions with the wall must be the same so that the latter number is equal to

$$n \, \Delta S \, \frac{|p'_x|}{m} \, \Delta t \; g(\boldsymbol{p}') \, d^3\boldsymbol{p}'. \tag{7.54}$$

Summing (7.53) over all molecules which during the interval Δt reach ΔS and leave it, we get the integral of the total force $\boldsymbol{F}_{\text{tot}}(t)$ that they exert on this surface area element:

$$\begin{aligned}\int_t^{t+\Delta t} dt' \, \boldsymbol{F}_{\text{tot}}(t') \; &= \; \int_{p_x>0} n \, \Delta S \, \frac{p_x}{m} \, \Delta t \, g(\boldsymbol{p}) \, d^3\boldsymbol{p} \, \boldsymbol{p} \\ &\quad - \int_{p'_x<0} n \, \Delta S \, \frac{|p'_x|}{m} \, \Delta t \, g(\boldsymbol{p}') \, d^3\boldsymbol{p}' \, \boldsymbol{p}' \\ &= \; \Delta S \; \Delta t \, \frac{n}{m} \int d^3\boldsymbol{p} \; g(\boldsymbol{p}) \; p_x \boldsymbol{p}.\end{aligned} \tag{7.55}$$

Because of the inertia of the wall one observes macroscopically the time average of this force. As expected, it is at right angles to the wall, because $g(\boldsymbol{p})$ is isotropic, and is proportional to its surface area. We have thus *proved the existence of a pressure* $\mathcal{P}$ which is uniform and constant in time, and we have derived an expression for it:

$$\mathcal{P} \; = \; \frac{n}{m} \int d^3\boldsymbol{p} \; g(\boldsymbol{p}) \; p_x^2 \; = \; \frac{n}{m} \; \langle p_x^2 \rangle \, . \tag{7.56}$$

Using the isotropy of $g(\boldsymbol{p})$ and the homogeneity of the gas we obtain

$$\mathcal{P} = \frac{N}{\Omega}\frac{2}{3}\left\langle\frac{p^2}{2m}\right\rangle = \frac{2U}{3\Omega}. \tag{7.57}$$

We find thus that the pressure is *proportional to the density and to the average kinetic energy* of the molecules. It has not been necessary to use the explicit form of the distribution $g(\boldsymbol{p})$; we needed only its isotropy (Exerc.13a). Expression (7.57) would thus be valid, even if equilibrium had not been established.

We took as a model of the wall a simple plane; this may appear to be too rough an approximation as we work at the atomic level. Nevertheless, the above result remains unaltered, if one makes up the balance of the momenta of the molecules which in one direction or another traverse a plane situated away from the actual solid wall itself. The molecules which carry the momenta $\boldsymbol{p}$ and $\boldsymbol{p}'$ in (7.55) are then not the same, but this does not make any difference.

We note that the principle of this calculation of the pressure by kinetic theory differs significantly from the one which follows from the general formalism of partition functions and thermodynamic potentials. We have here defined the pressure as the resultant of the forces exerted by the gas molecules on the wall whereas the thermodynamic pressure was more globally defined starting from the work done on the gas through a quasistatic compression. It is true that we have found again the same result, but it is not surprising that the calculations were so different, as the two definitions of pressure were not *a priori* equivalent on the microscopic scale.

7.4.3 Kinetic Interpretation of the Temperature

Comparing (7.57) with the empirical equation of state (7.39) of rarefied gases enables us to identify, apart from a multiplying constant, the *average kinetic energy* of the gas molecules with the absolute temperature:

$$\left\langle\frac{p^2}{2m}\right\rangle = \frac{3}{2}\frac{R}{N_{\mathrm{A}}}T = \frac{3}{2}kT. \tag{7.58}$$

We thus find here again the relation (7.24) between the energy and the temperature and a microscopic explanation of the properties of the Joule expansion.

This microscopic interpretation of the temperature has historically played an important rôle in showing that, at least in a perfect gas, a purely thermodynamic concept such as the absolute temperature can be understood by using only mechanistic ideas. This was a first step towards the unification, using microscopic and statistical physics, of theories which on the macroscopic scale looked very different – mechanics and thermodynamics.

Nevertheless, even though kT has the dimension of an energy, this quantity is directly related to energetic properties of a particle only for a perfect gas. In fact, as we have seen in Chaps.5 and 6, β rather than kT is the fundamental quantity; it is conjugated to the energy of the system through its relation with the entropy and it has the dimensions of an inverse energy.

7.4.4 Transport in the Knudsen Regime

The main interest of kinetic theory is not so much the calculation of equilibrium properties as the study of processes where this equilibrium has been disturbed by external perturbations. In fact, the kinetic approach is rather more complicated than that of § 7.3, but more general and more powerful. In § 7.4.1 effusion has provided us with a simple example which demonstrates the effectiveness of the kinetic theory for non-equilibrium situations; in that case the presence of a hole prevents the establishment of equilibrium.

That example illustrates, moreover, a situation where the collisions between the molecules, which can be neglected when equilibrium has been established, can still be neglected notwithstanding the fact that the system is evolving. In effusion it is legitimate to forget about collisions when the dimensions of the holes are sufficiently small so that there are no collisions in them. The same holds whenever the gas system which we are considering has either very small dimensions or a very low density. A situation of this kind is called a *Knudsen regime* or a *ballistic regime* and the phenomena are ultimately governed essentially by the collisions of the molecules with the walls. Under such conditions transfer of heat from a hot to a cold wall, for instance, is due to the fact that the collisions are inelastic: they make the kinetic energy of the molecules increase or decrease on average, depending on whether the wall is hot or cold. In § 11.3.4 we shall see that a semiconductor behaves like a gas; the charge carriers, negative electrons and positive holes, play the rôle of non-interacting classical particles; in the processes leading to equilibrium their collisions with one another are replaced by the collisions with impurities in the semiconductor. The ballistic regime is then relevant for some thin parts of semiconductor devices and for the very small samples involved in the technologies presently being developed in microelectronics.

The ballistic regime is reached when the *dimensions of the system* are small compared to the *mean free path*, that is, the average distance traversed by the molecules between successive collisions. Below we shall give its order of magnitude.

7.4.5 Local Equilibrium and Mean Free Path

Nevertheless, transport phenomena occur most often not in the ballistic regime, but in the Boltzmann regime, or *local equilibrium regime*, or *hydrodynamic regime*, in which the prime rôle is played by collisions inside the substance. Generally speaking, one will be dealing with a *transport phenomenon* whenever one forces a quantity which should be uniform in equilibrium to vary from point to point. Let us, for instance, assume that initially the temperature is not uniform. Thermostatics indicates what will be the final state of the isolated system (§§ 5.1.2 and 6.1), but does not tell us anything about the evolution in time of *energy transport* processes associated with the equalization of the temperatures. Kinetic theory gives an intuitive picture of this effect of *approach to equilibrium* (§ 4.1.5): the molecules situated in the hotter regions are faster; those which move from the hotter to the cooler regions therefore carry with them a larger kinetic energy than those which, in equal numbers, move in the opposite direction. This transported energy can macroscopically be interpreted as heat and the phenomenon which we have just described is none other than heat conduction in a gas.

We have disregarded here convection, a transfer mechanism which is very important in gases and which is connected with *transport of matter.* On the microscopic level the mass flux vanishes on average in pure conduction processes, whereas the particle fluxes moving in the two opposite directions through a surface element are not balanced when there is convection. This global transfer contributes significantly to the equalization of the temperatures.

If one counteracts the tendency towards equilibrium, for instance, if a gas is put in contact at opposite ends with two thermostats at different temperatures, which perturb it permanently and prevent its temperature from becoming uniform, one obtains after a very short time a *stationary non-equilibrium* state (§ 4.1.4): the temperatures of the thermostats hardly change notwithstanding the transfer of heat from one to the other, as they are very large; there is a permanent energy flux passing through the gas, but its macro-state does not change with time. This heat conduction is on the macroscopic scale characterized by a *transport coefficient*, the *heat conductivity*, λ, which is the ratio of the heat current density, that is, the energy flux per unit area, to the temperature gradient.

We shall in Chaps.14 and 15 return to this analysis of transport phenomena by generalizing it and making it more precise. Here we are satisfied with a qualitative approach based upon the balance (7.50) of the displacements of the molecules in the gas and upon the idea that the collisions enable them to exchange energy or momentum. We have indicated (§ 7.1.2) how these collisions are responsible for the approach to equilibrium of the gas; the mechanism is the same for transport phenomena, which are akin to the approach to equilibrium even though the regime is stationary. To simplify the discussion, we restrict ourselves here, as in Chap.15, to a gas with molecules which are practically frozen in their ground state (§ 7.1.1) so that each collision is *elastic.* The strongly repulsive nature of the forces at short distances makes it possible to treat the collisions as *impacts* which last very briefly and to model the molecules as impenetrable *spheres* of radius $\frac{1}{2}\delta$, where the distance of closest approach δ is of the order of 1 to 4 Å.

The properties of the collisions are then characterized by the *mean free path* l which is the average distance traversed by a molecule before it hits another. To estimate the order of magnitude of l we neglect the motion of all molecules but one, forget for the time being about the interactions of the latter molecule, and assume that it has travelled, rectilinearly, a distance L. A sphere of radius δ centred on the molecule considered sweeps through a cylinder of length L and volume $\pi\delta^2 L$. The number of collisions that it undergoes when one restores the interactions is then given by the number of other molecules with centres inside this cylinder; it equals $\pi\delta^2 Ln$, if the density is n. As a result, the mean free path between two successive collisions is of the order of

$$l \sim \frac{1}{\pi n \delta^2}. \tag{7.59}$$

Under normal density conditions we find, assuming δ to lie between 1 and 3 Å, a mean free path of 1 to 0.1 μm which is much larger than the mean distances between the molecules (30 Å). In rarefied gases, the mean free path may become large as it reaches the order of km at a pressure of 10^{-6} to 10^{-7} Torr.

The characteristic time associated with collisions is the *period between successive collisions* of a molecule. It follows from (7.59) and the mean speed of the molecules (§7.2.3). Under normal conditions its order of magnitude is 10^{-9} s.

In the local equilibrium regime the mean free path is much shorter than the distances over which macroscopic quantities such as the temperature change appreciably. Similarly, the period between collisions of one molecule is much shorter than the times over which the external parameters change. On the other hand, each collision suffered by a molecule makes its momentum and its energy change in a random manner, governed by the distribution of the momenta of the other particles in the region considered. The molecule loses the memory of its preceding motion – it even loses its identity when it collides with another molecule of the same kind – and after a few collisions we may assume that the probability for its momentum components is the same as that of the medium. We can thus assume that the gas consists of volume elements which are large as compared to the mean free path and therefore practically at equilibrium, but sufficiently small that we can consider parameters such as the temperature and the chemical potential to be uniform inside them. In this *local equilibrium regime* transport phenomena, and especially heat transfer from one end of the sample to another, are the result of a very large number of collisions between the molecules: the excess kinetic energy in the hot region is transferred gradually, from point to point *over short distances* through the intermediary of the collisions. In contrast, in the Knudsen regime where l is large as compared to the dimensions of the vessel distant regions can have direct exchanges with one another through molecular motions and equilibrium is not established locally.

7.4.6 Heat Conductivity and Viscosity

In order to exemplify the methods used by kinetic theory to study various transport phenomena we shall evaluate by a rather rough, but semi-quantitative, method the *heat conductivity*. Assuming that the gas is in local equilibrium with a fixed temperature gradient, we must determine the kinetic energy flux of the molecules, taking their collisions into account. The essential idea which we shall use systematically in this book whenever we are dealing with a non-equilibrium phenomenon, is to consider the *detailed balance of exchanges.* Here we are dealing with kinetic energy exchanges across a surface area at right angles to the temperature gradient, which are caused because molecules cross that area in one direction or the other.

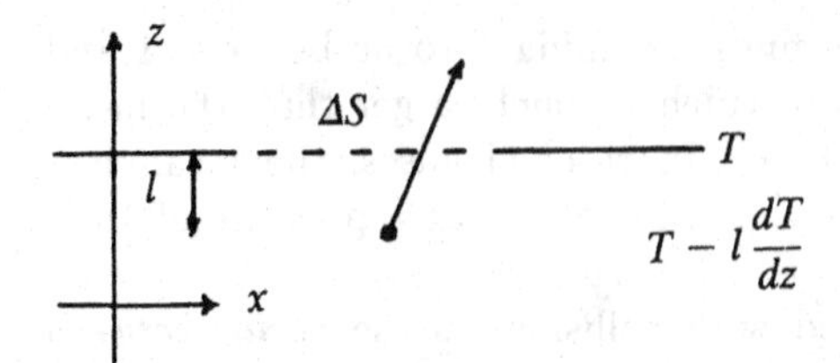

Fig. 7.5. Transport across an area

Let us assume that the temperature gradient is directed vertically and let us consider the molecules which cross, from below to above, the horizontal surface area ΔS during a time interval Δt (Fig.7.5). According to (7.51) their number is proportional to

$$n\,\Delta S\,\sqrt{\frac{kT}{m}}\,\Delta t, \tag{7.60}$$

where T is the temperature at the height considered; we drop numerical factors which do not affect the results in any crucial manner. It is reasonable to assume that the molecules were *thermalized at their last collisions*, which occurred below the area ΔS on average at a distance l of the order of the mean free path. The kinetic energy transported on average by each of the molecules is therefore equal to

$$\frac{3}{2}kT(z-l) \sim \frac{3}{2}k\left(T - l\,\frac{dT}{dz}\right),$$

which is the kinetic energy (7.58) associated with the temperature $T(z-l)$ prevailing at the altitude $z-l$ whence the molecules come. The kinetic energy transported from below to above is thus, apart from a numerical factor, equal to

$$n\,\Delta S\,\sqrt{\frac{kT}{m}}\,\Delta t\,k\left(T - l\frac{dT}{dz}\right). \tag{7.61}$$

From above to below the flux of molecules is still (7.60) as the heat conduction occurs in a permanent regime, without convection, that is, without mass transfer on average; however, the average kinetic energy transported per molecule is now

$$\frac{3}{2}k\left(T + l\,\frac{dT}{dz}\right).$$

Altogether, the heat flux crossing ΔS from below to above is thus – apart from a numerical factor – equal to

$$\Delta Q \propto -n\,\Delta S\,\sqrt{\frac{kT}{m}}\,kl\,\frac{dT}{dz},$$

and the corresponding energy current density equals

$$\boldsymbol{J}_E \propto -n\,\sqrt{\frac{kT}{m}}\,kl\,\nabla T. \tag{7.62}$$

Using (7.59) we thus find for the heat conductivity

$$\lambda = -\frac{J_E}{|\nabla T|} \propto n\,\sqrt{\frac{kT}{m}}\,kl \propto \sqrt{\frac{kT}{m}}\,\frac{k}{\delta^2}. \tag{7.63}$$

This expression, obtained by Clausius (1862) and Maxwell (1866), shows two properties: the conductivity is *independent of the density* and it is *proportional to* $\sqrt{T}$. Experiments have made it possible to check these properties for all densities where the mean free path is large as compared to the interatomic distances d, in order that the perfect gas model may be used, but sufficiently short that we are not in the Knudsen regime. The quantitative theory of Chap.15 will confirm the result (7.63) by giving us expressions for the transport coefficients in terms of the characteristics of the collisions between the molecules.

Other transport phenomena can be approached in the same way. In all cases one imposes a gradient of a quantity which characterizes local equilibrium; there then appears a current for the conjugate quantity. The transport coefficient characterizes the *response* of the system, that is, the current which it creates to react on the imposed departure from equilibrium.

In this way *viscosity* is associated with a *transport of momentum.* A gas which moves with a uniform velocity u in the horizontal x-direction is in a state of *equilibrum* (Exerc.4e). In this case we must include the total momentum amongst the constants of motion in the Boltzmann-Gibbs distribution; we shall see in § 14.2.1 that the intensive quantity, conjugated to the momentum with respect to the entropy, is $-\boldsymbol{u}/T$. The momentum distribution of the molecules still obeys Maxwell's law, but centred on $\boldsymbol{p} = m\boldsymbol{u}$ rather than on $\boldsymbol{p} = 0$. If we create a departure from equilibrium by letting the velocity $u(z)$ in the x-direction depend on the height z, the molecules have in *local equilibrium* an average momentum $m\boldsymbol{u}(z)$ corresponding to that height. The transfer of molecules from below to above and from above to below, due to thermal motions, then produces momentum exchanges which are not in balance. A total momentum $\Delta\boldsymbol{P}$ is in this way imparted during the time interval Δt through the surface area ΔS by the gas layer below it to the gas layer above it. The latter thus receives a momentum $\Delta\boldsymbol{P} \equiv \boldsymbol{F}\,\Delta t$, and hence, a force $\boldsymbol{F}$ is exerted by the layers below on the layers above each surface element. This *force*, due to a *transfer of momentum* from the region where the motion is faster to the region where it is slower, can macroscopically be interpreted as the Newtonian *viscosity* associated with the difference in velocity between layers. The evaluation of $\Delta \boldsymbol{P}$ is again based upon the number of molecules (7.60) crossing ΔS either from below to above or from above to below. The first have on average a momentum

$$mu(z-l) \sim mu(z) - ml\,\frac{du}{dz}$$

in the x-direction, as they were thermalized at a height $z - l$ according to an appropriately decentred Maxwell distribution $g(\boldsymbol{p})$. The average momentum of the molecules crossing ΔS from above to below is similarly $mu(z) + ml\,du/dz$ so that we have, in the x-direction,

$$\Delta P \propto -n\,\Delta S\,\sqrt{\frac{kT}{m}}\,\Delta t\,ml\,\frac{du}{dz}. \tag{7.64}$$

The viscosity η is defined as the ratio between the horizontal force per unit surface area and the velocity gradient. It is thus equal to

$$\eta \equiv \frac{\Delta P}{\Delta S \Delta t}\Big/\frac{du}{dz} \propto n\,\sqrt{\frac{kT}{m}}ml \propto \frac{\sqrt{mkT}}{\delta^2}. \tag{7.65}$$

The viscosity of a gas is thus *independent of the density* and *increases with the temperature* as $\sqrt{T}$. This result is physically reasonable as the vertical transport of momentum is favoured by an increase in the average velocity of the molecules. Moreover, equations (7.63) and (7.65) show that, apart from a numerical factor, the *ratio of the viscosity to the heat conductivity* is proportional to the molecular mass of the gas,

$$\frac{\eta}{\lambda} \propto \frac{m}{k} = \frac{mN_{\mathrm{A}}}{R}, \tag{7.66}$$

and is independent of the size δ of the molecules. These remarkable results, predicted by Maxwell in 1860, were regarded to be surprising at that time: everybody knows, in fact, that the viscosity of liquids, such as lubricating oils or honey, *decreases* with increasing temperature. Moreover, unless one remembers to associate the viscosity force with transfer of momentum, it is not obvious that the viscosity of a gas is independent of the density and that it is as large as is expressed by (7.65). The unexpected nature of the results predicted by Maxwell using the present theory led him to check them by experiments which lasted several years. Their complete success contributed to a confirmation of the young kinetic theory. The difficulties in measuring the viscosity and thermal conductivity in gases are connected with the fact that one must avoid turbulent and convective motions. The experiments are easier to perform in liquids but it has not been easy to explain those, starting from microscopic phenomena and using statistical physics.

Summary

The perfect gas model, which is justified for a theory of inert gases at low enough densities, is based upon the description of molecules as non-interacting structureless point particles in the classical limit. In one or other of the canonical ensembles it gives us the equilibrium properties of those gases, such as the Maxwell distribution (7.16) for the molecular velocities, the equation of state (7.38), the entropy (7.41), and the specific heat (7.45).

Kinetic theory enables us to explain mechanistically pressure and transport phenomena. By using the idea of a mean free path for the molecules and the balance equation (7.50), we derive the properties of effusion rates, thermal conductivity, and viscosity of gases.

Exercises

7a Barometric Equation

The molecules of the atmosphere are in the Earth's gravitational field. Idealize the atmosphere as an isothermal perfect gas at rest and at equilibrium.

1. Evaluate the variation of the atmospheric pressure with height.
2. The air contains impurities – rare gases, large molecules, dust particles. Calculate the distribution of these impurities with height.

Solution:

1. From (7.48) we get $d\mathcal{P}/dz = -\mathcal{P}mg/kT$ in agreement with the macroscopic hydrostatic equilibrium equation $d\mathcal{P} = \varrho g\, dz$. The pressure decreases by 12 % over 1000 m for a density $\varrho = 1.3$ kg m^{-3}.

2. Equation (7.48) is valid for each of the components, which are thus distributed with height as $\exp(-z/z_0)$ over an average height $z_0 = kT/mg$ which depends on m. It follows that, for 300 K, $z_0 = 9000$ m for N_2; this is the correct order of magnitude, even though the hypothesis of isothermal equilibrium is incorrect at those heights; $z_0 = 64\ 000$ m for He; $z_0 = 800$ m for tetraethyl lead, which is an additive for motor fuels and the dilution of which therefore remains practically unchanged in the air at the top of the Eiffel tower; $z_0 = 1$ m for macromolecules with a molar mass of 2×10^5 u; $z_0 = 1$ μm for small dust particles of mass 0.3×10^{-12} g.

Application: Measurement of Avogadro's Number. In 1909 Jean Perrin measured z_0 for granules suspended in a liquid. The determination of the mass of the granules then gave him an experimental value for the Boltzmann constant, and hence for Avogadro's number R/k. To observe the distribution of the granules with height he employed a small cell with a depth of 0.1 mm in which he deposited a droplet which he flattened using a cover-glass. By changing the focus of a microscope one can observe successive horizontal layers; the optical field of the microscope is restricted so that one can count each time only a small number of granules. The mass of the grains follows from their density and their diameter. The density was measured in different ways: for example, by comparing the mass of pure water and that of the suspension which occupies the same volume, or by measuring the residual mass after desiccation at 110°C. One can estimate the diameter of the granules by following their fall under the action of the gravitational field when one distributes the emulsion in a sufficiently tall column in a capillary tube: the grains fall in a uniform motion to produce ultimately the exponential distribution which one is studying; the acting force is equal to the apparent weight of the grain in the liquid; the friction force is given by Stokes's law ($6\pi\eta va$) which is a function of the viscosity η, the velocity v, and the radius a. One can also obtain the mass of the grains directly by producing, for instance in a centrifuge, a precipitation on the walls of the cell, which makes it possible to count them; as one knows the strength of the solution and its total mass, one finds the mass of the grains.

7b Isotope Separation by Ultracentrifuging

1. Consider a perfect gas at equilibrium in a cylinder of radius R and volume Ω rotating with a high angular velocity ω around the z-axis. Write down the corresponding classical Boltzmann-Gibbs distribution. Note (§ 4.3.3) that the constants of motion now are the energy and the component J_z of the total angular momentum $\boldsymbol{J} = \sum_i [\boldsymbol{r}_i \times \boldsymbol{p}_i]$. We must thus introduce not only the multiplier β, but also a multiplier associated with J_z; identify this multiplier with $-\beta\omega$ by evaluating the mean velocity at each point.

2. Find the same result also by changing the frame of reference: determine first the Lagrangian and then the Hamiltonian in the rotating frame (§ 2.3.3) and write down the canonical equilibrium in that frame; the collisions with the walls force the gas to follow the rotation of the cylinder.

3. Use the distribution thus obtained to find, through integration over the momenta, an expression for the molecular density as a function of the distance from the cylinder axis. What happens in the case of a gas mixture? When using ultracentrifuges for isotope separation one injects hexafluoride UF_6, which is a gaseous uranium compound, into a rotating cylinder. The UF_6 consists of a mixture containing the two isotopes ^{238}U and ^{235}U, and one wants to enrich it in the rarer fissile 235 isotope which is fuel for nuclear power stations. How does the ratio of the concentrations change as a function of the distance to the axis for a cylinder of radius $R = 20$ cm, rotating with 10 000 revolutions per minute at $T = 100$°C?

Solution:

1. The fluid carried along by the walls of the rotating vessel acquires a non-vanishing average angular momentum $\langle J_z \rangle$ around the axis of rotation. This angular momentum is a constant of the motion. In order to be able to assign to it a definite value one associates with it a Lagrangian multiplier λ, in exactly the same way as one associates the multiplier β with the energy in canonical equilibrium. The average $\langle J_z \rangle$ will be a function of λ. Let us give N not on average, but exactly – one

could also work with a multiplier α for $\langle N\rangle$. The density in phase which represents the rotating gas is thus of the form

$$D = \frac{1}{Z}\,\mathrm{e}^{-\beta H-\lambda J_z} = \frac{1}{Z}\,\exp\left\{-\sum_i\left[\frac{\beta p_i^2}{2m}+\lambda\left(x_i p_{yi}-y_i p_{xi}\right)\right]\right\},$$

with the usual measure $d\tau_N$. The energy and the average angular momentum are given by

$$U = -\frac{\partial}{\partial\beta}\ln Z, \qquad \langle J_z\rangle = -\frac{\partial}{\partial\lambda}\ln Z.$$

We must still find the physical meaning of the multipliers β and λ, the values of which are fixed by energy and angular momentum reservoirs connected with the gas, that is, by the walls of the vessel. The multiplier β associated with energy exchanges can be identified with $1/kT$ using the usual thermodynamic reasoning. On the other hand, λ is mechanical in nature and to identify it we need to compare our microscopic description with the macroscopic description of fluid mechanics. To do this we write down the single-particle reduced density

$$f(\boldsymbol{r},\boldsymbol{p}) \propto \exp\left\{-\frac{\beta p^2}{2m}-\lambda\left(xp_y-yp_x\right)\right\}$$

$$= \exp\left\{-\frac{\beta}{2m}\left(\boldsymbol{p}+\frac{m}{\beta}[\boldsymbol{\lambda}\times\boldsymbol{r}]\right)^2+\frac{m\lambda^2}{2\beta}\left(x^2+y^2\right)\right\},$$

whence we find the velocity distribution at a point $\boldsymbol{r}$ to be proportional to

$$\exp\left\{-\frac{m}{2kT}\left(\boldsymbol{v}+\frac{1}{\beta}[\boldsymbol{\lambda}\times\boldsymbol{r}]\right)^2\right\}.$$

The mean velocity of the fluid at the point $\boldsymbol{r}$ is thus equal to $-[\boldsymbol{\lambda}\times\boldsymbol{r}]/\beta$. It can be identified with the velocity $[\boldsymbol{\omega}\times\boldsymbol{r}]$ in a uniform rotation with angular velocity ω, provided we put $\omega = -\lambda/\beta$. On the macroscopic scale the gas rotates as a rigid body; on the microscopic scale the molecular velocities keep fluctuating, as in a gas at rest, and they are larger than the global rotation speed. The angular momentum is imparted to the gas when the molecules collide with the rotating walls, which changes the Maxwell distribution at every point, shifting its origin. The walls play the rôle of an angular momentum reservoir; their motion is characterized by a certain angular velocity, and the angular velocities ω of the fluid and of the walls become equal at equilibrium, exactly like the equalization of the temperature through energy exchanges.

2. The Lagrangian can be taken as remaining invariant under any change of reference frame, because the stationary action principle is independent of the frame. On the other hand, the Hamiltonian is changed, and we need the latter to write down the equilibrium conditions as it is a constant of the motion. For a single particle with position $\boldsymbol{r}'$ and velocity $\boldsymbol{v}'$ in the rotating frame, the Lagrangian in the two frames is given by

$$L_1 = \tfrac{1}{2}mv^2 = \tfrac{1}{2}m(\boldsymbol{v}'+[\boldsymbol{\omega}\times\boldsymbol{r}'])^2.$$

Hence we get the conjugate momentum of $\boldsymbol{r}'$:

$$\boldsymbol{p}' = \frac{\partial L_1}{\partial \boldsymbol{v}'} = m(\boldsymbol{v}' + [\boldsymbol{\omega} \times \boldsymbol{r}']),$$

and the Hamiltonian of the particle in the rotating frame is

$$H_1' = (\boldsymbol{p}' \cdot \boldsymbol{v}') - L_1 = \frac{\boldsymbol{p}'^2}{2m} - (\boldsymbol{\omega} \cdot [\boldsymbol{r}' \times \boldsymbol{p}']).$$

The equilibrium in the rotating frame is given by the canonical density in phase,

$$D = \frac{1}{Z} \mathrm{e}^{-\beta H'},$$

where H' is the sum of H_1' over the N particles. To switch back to the original coordinates, we note that $\boldsymbol{p}'$ and $[\boldsymbol{r}' \times \boldsymbol{p}']$ can be derived from $\boldsymbol{p}$ and $[\boldsymbol{r} \times \boldsymbol{p}]$, respectively, by means of the same change of coordinates that leads from $\boldsymbol{r}$ to $\boldsymbol{r}'$. Hence we get $H' = H - (\boldsymbol{\omega} \cdot \boldsymbol{J})$, and D can be identified with the earlier expression, provided $\boldsymbol{\lambda} = -\beta\boldsymbol{\omega}$.

3. Integrating the reduced density $f(\boldsymbol{r}, \boldsymbol{p})$ over $\boldsymbol{p}$ gives us, after normalization, the molecular density as function of the distance $r_\perp$ from the axis:

$$n(\boldsymbol{r}) = \frac{N}{2\Omega kT} \frac{m\omega^2 R^2}{\exp(m\omega^2 R^2/2kT) - 1} \exp\left(\frac{m\omega^2 r_\perp^2}{2kT}\right).$$

The increase of n with $r_\perp$ reflects the effect of the centrifugal force on the molecules, which is the more efficient, the larger their mass and the lower the temperature. For a mixture the two species are independent of each other and the ratio of the concentrations varies as

$$\frac{n_1(\boldsymbol{r})}{n_2(\boldsymbol{r})} = \frac{N_1 m_1}{N_2 m_2} \frac{\exp(m_2\omega^2 R^2/2kT) - 1}{\exp(m_1\omega^2 R^2/2kT) - 1} \exp\left[\frac{(m_1 - m_2)\omega^2 r_\perp^2}{2kT}\right].$$

The relative enrichment of ^{235}U of the gaseous UF_6 in the centre and its depletion at the walls are given by

$$\frac{n_1(0)}{n_2(0)} \bigg/ \frac{n_1(R)}{n_2(R)} = \exp\left[\frac{(m_2 - m_1)\omega^2 R^2}{2kT}\right] = 1.02.$$

The gas extracted from the periphery is depleted and that remaining at the centre enriched, as compared to the mixture which was injected originally. Starting from a natural isotopic mixture which contains 0.7 % of ^{235}U, this process can produce, after a few tens of stages, percentages of the order of the 3 % used in nuclear power stations, and it is thus more efficient than gas diffusion (Exerc.7g). However, the technology is difficult to master – fast rotation, leak-free shaft-seals, lubrication, corrosion – so that gas diffusion remains the main industrial process.

Comments. As in the case of equilibrium of a gas in a gravitational field (Exerc.7a) we could have obtained the result by a *macroscopic* calculation from thermodynamics and fluid mechanics, using locally the perfect gas laws and the balance between the forces, here centrifugal forces and pressure gradients. The above considerations must be regarded a microscopic justification of such a calculation.

Usually, when one writes down the condition for a system to be in canonical equilibrium one does not inquire about the Lagrangian multipliers for dynamical constants of motion such as the angular or the linear momentum. To be rigorous, one should do so; afterwards one should fix the value of these multipliers by requiring that on average the angular and linear momenta vanish: the system is at rest. In fact, for symmetry reasons these quantities vanish at the same time as the corresponding multipliers. For instance, we have

$$\begin{aligned}\langle J_z \rangle &= -\frac{\partial \ln Z}{\partial \lambda} = Nm\omega R^2 \left[\frac{1}{1-\exp(-m\omega^2 R^2/2kT)} - \frac{2kT}{m\omega^2 R^2} \right] \\ &\underset{\omega \to 0}{\sim} \tfrac{1}{2}\omega\, NmR^2.\end{aligned}$$

Therefore, if one omits a constant of the motion such as J_z by not introducing the appropriate multiplier, it amounts to *introducing a zero multiplier*, and hence to requiring, as we wanted to do, that on average the angular and the linear momenta vanish.

The method used under 1 to identify the hydrodynamic velocity on the microscopic scale can also be applied to non-equilibrium situations (§ 14.2.1). This velocity, which is much lower than the individual molecular velocities, appears as a *local* statistical property. One should note that two different methods were used to identify macroscopic quantities. The comparison with thermodynamics is suitable for the temperature, the chemical potential, and the global pressure of a homogeneous system, writing the work as $-\mathcal{P}\, d\Omega$. For quantities which vary from one point to another, such as the local velocity or the local pressure in a non-uniform system, we must compare the local microscopic results with macroscopic hydrodynamics.

In the change of frame under 2 the velocity, which is equal to $\boldsymbol{p}/m$ in the fixed frame, becomes $\boldsymbol{v}' = \boldsymbol{p}'/m - [\boldsymbol{\omega} \times \boldsymbol{r}']$ in the rotating frame so that the linear momentum $m\boldsymbol{v}'$ is no longer equal to the momentum $\boldsymbol{p}'$. The situation resembles that of a particle of charge q in a magnetic field, which has a velocity $(\boldsymbol{p} - q\boldsymbol{A})/m$. In fact, whereas positions and velocities are physical quantities, *momenta* have a certain amount of *arbitrariness* which is connected with the fact that we can change the Lagrangian by adding to it a time derivative without changing the equations of motion. For instance, in a Galilean transformation with velocity $\boldsymbol{u}$, the procedure of 2 where the Lagrangian is assumed to be invariant gives $\boldsymbol{p}_i' = \boldsymbol{p}_i$, whereas $\boldsymbol{v}_i' = \boldsymbol{v}_i - \boldsymbol{u}$; the Hamiltonian becomes $H' = H - (\boldsymbol{u} \cdot \boldsymbol{P})$, where $\boldsymbol{P}$ is the total momentum. However, there is also another procedure which better exhibits the Galilean invariance; it consists in adding to the Lagrangian the ineffective term

$$-\sum_i m_i \left((\boldsymbol{v}_i' \cdot \boldsymbol{u}) + \frac{1}{2}\boldsymbol{u}^2 \right) = \frac{d}{dt} \sum_i m_i \left(\frac{1}{2}\boldsymbol{u}^2 t - (\boldsymbol{r} \cdot \boldsymbol{u}) \right),$$

at the same time as changing from the $\boldsymbol{r}_i, \boldsymbol{v}_i$ to the $\boldsymbol{r}_i', \boldsymbol{v}_i'$ coordinates; in that case the momentum which is conjugate to $\boldsymbol{r}_i'$ is $\boldsymbol{p}_i'' = \boldsymbol{p}_i - m_i\boldsymbol{u} = m_i\boldsymbol{v}_i'$, and not $\boldsymbol{p}_i' = \boldsymbol{p}_i$, and the Hamiltonian $H'' = H - (\boldsymbol{u} \cdot \boldsymbol{P}) + \frac{1}{2}M\boldsymbol{u}^2$ has in terms of the $\boldsymbol{p}_i''$ exactly the same form as H in terms of the $\boldsymbol{p}_i$.

The expression for the angular momentum $\langle J_z \rangle$ is to lowest order in ω the same as for the rotation of a cylinder with uniform density, which has a moment of inertia equal to $\frac{1}{2}NmR^2$. However, the larger density at the periphery of the cylinder makes the moment of inertia increase with ω in an important manner, as $m\omega^2R^2/2kT$ is equal to 2.5 for the numerical values of the problem, which multiplies the moment of inertia by a factor 5.

The energy

$$U = -\frac{\partial \ln Z}{\partial \beta} = \frac{3}{2}NkT + \frac{1}{2}\omega \langle J_z \rangle$$

contains a contribution due to the motion, in agreement with the macroscopic dynamics. One can check that the entropy (4.14) also depends on the rotational velocity, but only to order ω^4. It decreases with ω, as the rotation produces changes in density which *increase the spatial order.*

7c Relativistic Gas

Evaluate the grand partition function of a hypothetical non-interacting very hot classical gas with "relativistic" energies:

$$\varepsilon(\boldsymbol{p}) = \sqrt{c^2p^2 + m^2c^4} \sim cp.$$

Hence find the equation of state and the internal energy. Compare the results with those for the non-relativistic perfect gas and for the photon gas (Chap.13).

Solution. We get from (7.27)

$$A = -kT\,\frac{\Omega}{h^3}\,\mathrm{e}^{\alpha} \int d^3\boldsymbol{p}\;\mathrm{e}^{-\beta\varepsilon(\boldsymbol{p})}$$

$$= -\frac{\Omega}{h^3}\,\frac{8\pi(kT)^4}{c^3}\,\mathrm{e}^{\mu/kT} = -\mathcal{P}\Omega = -NkT,$$

$$U = 3NkT.$$

The equation of state is the same as for the non-relativistic perfect gas – it is independent of the functional relation for $\varepsilon(\boldsymbol{p})$ – but the internal energy is twice as large.

For a photon gas, the number of particles is not conserved, and $\mu = 0$; moreover, quantum effects are important. In this case the average number of photons at equilibrium and the pressure are equal to (Chap.13)

$$N = \Omega\left(\frac{kT}{\hbar c}\right)^3 \frac{2}{\pi^2}\,\zeta(3),$$

$$\mathcal{P} = \frac{U}{3\Omega} = \frac{\pi^2(kT)^4}{45(\hbar c)^3} = \frac{\pi^4}{90}\,\frac{NkT}{\Omega}.$$

The relation between $\mathcal{P}$ and U is the same as for the state has been changed by a factor which comes partly from the non-conservation of the number of photons and partly from their, quantum, indistinguishability.

7d Doppler Profile of a Spectral Line

1. Find from the Maxwell law the probability distributions for the absolute magnitude of the velocity, for the kinetic energy, and for the components $v_{\parallel}$ and $v_{\perp}$ of the velocity along and at right angles to a particular direction.

2. Let λ_0 be the wavelength of an emission line corresponding to a transition between two energy levels of a molecule or an atom; we assume the line to be infinitesimally thin – "natural width" of the order of $\Delta\lambda/\lambda_0 \sim 10^{-9}$. If the molecule moves, the observed wavelength is shifted because of the Doppler effect. Under the usual experimental conditions where the molecules are in the gas phase, thermal motion produces a broadening of the line which should be calculated, together with the strength of the line, as function of λ. Numerical application: hydrogen lines at room temperature.

3. Evaluate the thickness of the deposit as function of the angle θ over which the cylinder has rotated in the experiment described at the end of § 7.2.3 for a cylinder with radius R and angular velocity ω.

Hints:

1. Do not forget that, if one changes variables, the probability density is not invariant, but its product with the volume element remains unchanged.

2. From the Doppler shift

$$\frac{\lambda - \lambda_0}{\lambda_0} = \frac{v}{c},$$

we get

$$I(\lambda)\, d\lambda \propto \exp\left[-\frac{(\lambda - \lambda_0)^2 mc^2}{2\lambda_0^2 kT}\right],$$

$$\sqrt{\frac{kT}{mc^2}} = 0.5 \times 10^{-5}.$$

The Doppler broadening is small, but not negligible as compared to fine structure splittings, which are a few Å for Na, that is, of relative order 10^{-4}. It is of the order of the hyperfine splitting: the existence of the 21 cm line of H gives rise in the visible spectrum of the hydrogen atom to a splitting of relative order $0.6 \times 10^{-6}/21\times10^{-2} = 0.3\times10^{-5}$. This is the reason for the interest in recent "recoilless" spectroscopy techniques.

3. The thickness is proportional to

$$\frac{1}{\theta^5} \exp\left(-\frac{2R^2\omega^2 m}{kT\theta^2}\right).$$

7e Liquid Nitrogen Trap

An efficient way to produce a good vacuum in a vessel is to condense the residual gases – in particular, the oil vapours coming from the pumps – on a cold surface. This is the "liquid nitrogen trap" which connects the vessel through a small hole with a tube plunged into liquid nitrogen.

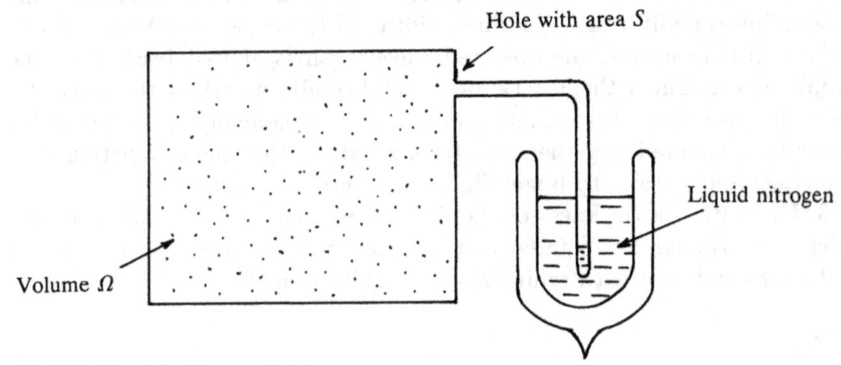

Fig. 7.6. Liquid nitrogen trap

1. What is the number of particles passing per unit time through the hole S? Assume that the diameter of the hole is small as compared to the mean free path and count all particles reaching its aperture from above.

2. Assume that each molecule entering the hole is "trapped", that is, condenses in the tube. Assume also that the temperature of the gas in the volume Ω is maintained constant through collisions with the walls. How does the pressure change with time? Starting from $\mathcal{P}_0 = 1$ Torr (mm Hg), how long will it take to reach $\mathcal{P} = 10^{-6}$ Torr? Take for oil vapour at room temperature $\langle v \rangle \simeq 100$ m/s; $\Omega = 1$ litre; $S = 1$ mm^2.

3. How are the above results changed if we take into consideration the molecules which pass the hole in the opposite direction? Assume that on the cold side the gas is in equilibrium at the temperature T_s of liquid nitrogen and at its corresponding saturated vapour pressure $\mathcal{P}_s$.

4. What happens, if the vessel is thermally insulated?

Answers:

1. $$\frac{dN}{dt} = -\frac{1}{4} nS\langle v \rangle = -nS\sqrt{\frac{kT}{2\pi m}}.$$

2. $$\mathcal{P} = \mathcal{P}_0 \exp\left(-\frac{S}{\Omega}\sqrt{\frac{kT}{2\pi m}}\, t\right).$$

$\mathcal{P}/\mathcal{P}_0 = 10^{-6}$ in 10 minutes. The mean free path does, actually, not exceed a value of 1 mm until the pressure has dropped to 10^{-3} Torr, so that our calculation is not realistic at the beginning of the process.

3.
$$\frac{dN}{dt} = -nS\sqrt{\frac{kT}{2\pi m}} + \frac{S\mathcal{P}_s}{\sqrt{2\pi mkT_s}}.$$

$$\mathcal{P} = \mathcal{P}_s\sqrt{\frac{T}{T_s}} + \left(\mathcal{P}_0 - \mathcal{P}_s\sqrt{\frac{T}{T_s}}\right)\exp\left[-\frac{S}{\Omega}\sqrt{\frac{kT}{2\pi m}}\,t\right].$$

The pressure does not tend to zero, as $t \to \infty$.

4. The temperature of the gas decreases with its density, following (7.52) for structureless molecules. In that case integration leads to

$$T^{-1/2} = T_0^{-1/2} + \frac{1}{6}\frac{S}{\Omega}\sqrt{\frac{k}{2\pi m}}\,t,$$

$$\mathcal{P} \propto \left\{t + 6\frac{\Omega}{S}\sqrt{\frac{2\pi m}{kT_0}}\right\}^{-8};$$

the decrease is slightly faster than in the isothermal case to begin with, but much slower later on.

Note. The efficiency of this process is primarily limited by adsorption of molecules on the walls where they remain stuck. To improve this, one degasses these walls by heating them, which reduces the adsorption (Exerc.4b).

7f Adiabatic Expansion

Consider a cylinder of length L and cross-section σ containing a monatomic perfect gas and closed by a moving piston which is assumed to move with a uniform velocity w (Fig.7.7). Assume that the collisions with the walls are elastic, and especially those with the piston in the frame fixed to the piston.

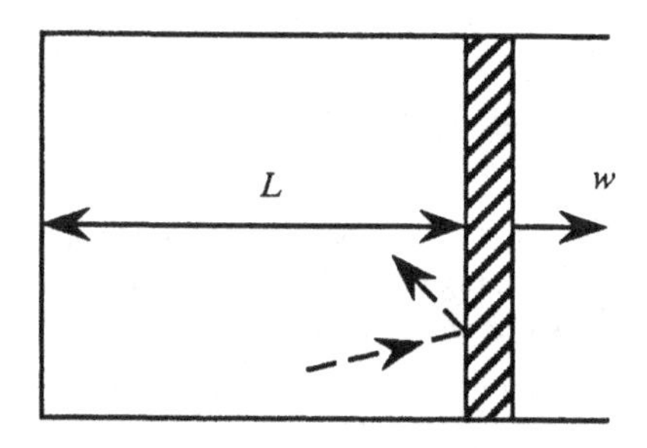

Fig. 7.7. Adiabatic expansion

1. What changes must one make to Eq.(7.50)?

2. Write down an expression for the energy given by the gas to the piston during a time interval dt by analysing the collision mechanism, assuming that initially the gas is in equilibrium at a temperature T.

3. Evaluate this energy to first order in w. If w is small as compared to the molecular velocities and if the mean free path is short compared to the

dimensions of the cylinder, one may assume that the gas is able to thermalize, that is, that its energy at all times is distributed amongst the molecules according to the Maxwell distribution. For that case write down the relation between the volume and the temperature during the expansion. Compare the result with (7.42).

4. How does the entropy change to the next order in w? What happens when w becomes very large and positive?

Solution:

1. By considering the frame fixed to the piston we see that the height of the infinitesimal cylinder considered in Fig.7.3 is changed to $(v_x - w)\Delta t$ so that (7.50) is changed to

$$n\,\Delta S\,\left(\frac{p_x}{m} - w\right)\,\Delta t\,g(\mathbf{p})\,d^3\mathbf{p}.$$

2. Analysing a collision in the moving frame, we see that it changes p_x, p_y, p_z into $-2p_x + 2mw$, p_y, p_z and that only molecules with $p_x > mw$ reach the piston. Hence, the energy given to the piston is

$$\begin{aligned} -\frac{dU}{dt} &= \frac{2n\sigma w}{m}\int_{p_x>mw} g(\mathbf{p})\,d^3\mathbf{p}\,(p_x - mw)^2 \\ &= \frac{4n\sigma wkT}{\sqrt{\pi}}\int_0^\infty dx\,x^2\,\exp\left\{-\left[x + w\sqrt{\frac{m}{2kT}}\right]^2\right\}. \end{aligned}$$

3. If $\frac{1}{2}mw^2 \ll kT$ the above expression reduces to

$$\frac{dU}{dt} \sim -n\sigma wkT = -\frac{N}{\Omega}\frac{d\Omega}{dt}kT.$$

Hence, using (7.44), we find

$$\frac{3\,dT}{2T} = -\frac{d\Omega}{\Omega}.$$

We have found again the equation (7.42) for the adiabatic expansion of a monatomic gas. Kinetic theory has enabled us to prove that a change in volume which is *slow* as compared to the molecular velocities *conserves the entropy.*

4. The next term in w gives

$$\frac{3}{2}\frac{dT}{dt} = -\frac{1}{\Omega}\frac{d\Omega}{dt}\,T\left(1 - w\sqrt{\frac{8m}{\pi kT}}\right),$$

$$\frac{dS}{dt} = Nk\left(\frac{3}{2T}\frac{dT}{dt} + \frac{1}{\Omega}\frac{d\Omega}{dt}\right) = \frac{N}{L}w^2\sqrt{\frac{8km}{\pi T}}.$$

The entropy increases, independent of the direction of the piston motion.

If $\frac{1}{2}mw^2 \gg kT$ most molecules are too slow to hit the piston. The gas does not give off any energy to the piston and we are back at a Joule expansion.

Comments. The above calculation of the dissipation is, in fact, too simplistic: it does not take properly into account the collisions, which play a rôle in the non-adiabatic processes, as can be seen from § 7.4.6. A correct theory must distinguish more clearly the Knudsen and the local equilibrium regimes.

One could compare this theory of the adiabatic expansion of a perfect gas with the behaviour of photons during the expansion of the Universe (§ 13.2.2): in both cases a dynamic process which is slow as compared to the particle velocities – here the motion of the piston, there the expansion of the Universe changing the photon momenta through the Doppler effect – is equivalent to an expansion at constant entropy.

7g Isotope Separation by Gas Diffusion

Two isotopic gas mixtures of uranium 238 and 235 hexafluoride, $^{238}UF_6$ and $^{235}UF_6$, at different pressures and with different concentrations, are separated by a porous wall. The pressures to the left and to the right are maintained at fixed values $\mathcal{P}_l$ and $\mathcal{P}_r$, where $\mathcal{P}_l \gg \mathcal{P}_r$. Write down the equations governing the evolution, through effusion across the pores of the wall, of the numbers of molecules of both types on the two sides of the wall. Assume that the concentration of the light isotope on the right stabilizes around a stationary value and evaluate an upper limit for that concentration as a function of the concentration on the left. How many diffusion stages does one need at least to change from natural uranium with 0.7 % of ^{235}U to enriched uranium with 3 %, which is used in most industrial reactors?

Effusion, called "gas diffusion" in this context, is the main process for uranium treatment: it covers 98 % of the production. One uses barriers with about 10^{10} pores per cm^2. In Exerc.14a we shall indicate how the diffusion stages are organized in the Eurodif factory in the Rhône valley. In Exerc.8a we shall also give entropy and energy estimates for this process.

Solution. Let the subscript 1 and 2 indicate, respectively, $^{235}UF_6$ and $^{238}UF_6$, and l and r the two sides of the wall. Integrating (7.50) over $p_x > 0$ we get

$$\frac{dN_{1r}}{dt} = \frac{\Delta S}{\sqrt{2\pi kT}} \frac{1}{\sqrt{m_1}} \left(\mathcal{P}_l C_{1l} - \mathcal{P}_r C_{1r}\right) = -\frac{dN_{1l}}{dt}.$$

We have assumed that each molecule which enters a pore from one side emerges on the other side and we have denoted the molecular concentrations by C ($C_1 + C_2 = 1$). If the concentrations on the right are stabilized, we have

$$\frac{dN_{1r}}{N_{1r}} = \frac{dN_{2r}}{N_{2r}},$$

that is,

$$\frac{1}{\sqrt{m_1}} \left(\frac{\mathcal{P}_l C_{1l}}{\mathcal{P}_r C_{1r}} - 1\right) = \frac{1}{\sqrt{m_2}} \left(\frac{\mathcal{P}_l \left(1 - C_{1l}\right)}{\mathcal{P}_r \left(1 - C_{1r}\right)} - 1\right);$$

the pressures and the concentrations on the left may go on changing. At best, with $\mathcal{P}_l \gg \mathcal{P}_r$ and with a small C_{1l}, the concentration C_{1r} on the right reaches a value

$C_{11}\sqrt{m_2/m_1} = C_{11}\sqrt{352/349} = 1.0043C_{11}$. The increase in concentration after a single operation is thus very small. To change from natural uranium (0.7 %) to uranium enriched to 3 % which is used in nuclear power stations needs therefore a very large number of successive effusions: even with unit efficiency and a very large pressure difference one would need 340 barrier passages. In practice the Eurodif factory has 1400 diffusion stages which operate in a stationary regime.

8. Molecular Properties of Gases

"On peut concevoir les parties intégrantes de l'air comme de petits filaments contournés en forme de spires flexibles et élastiques, et leur assemblage à peu-près comme un paquet de coton ou de laine cardée que l'on peut réduire en un plus petit volume lorsqu'on le presse, mais qui tend toujours à se remettre dans son premier état. Cette idée n'est qu'une esquisse bien grossière de la nature de l'air; et j'avoue qu'il y a peut-être cent contre un à parier, que les parties de cet élément n'ont point la figure que je leur attribue; parce que pour les supposer telles, je n'ai d'autre raison que leur flexibilité et leur ressort, et qu'elles peuvent être élastiques avec cent figures différentes d'un fil spiral: aussi lorsque j'adopte cette hypothèse avec la plupart des Physiciens, je ne prétends point dire ce qu'elles sont, mais seulement ce qu'elles peuvent être."

Abbé Nollet, Leçons de Physique Expérimentale, 1775

"L'atmosphère était saturée de gaz sulfureux, d'hydrogène, d'acide carbonique, mêlés à des vapeurs aqueuses .... Bientôt, avec ce bruit, les combinaisons chimiques se trahirent par une vive odeur, et les vapeurs sulfureuses saisirent à la gorge l'ingénieur et son compagnon. Voilà ce que craignait le capitaine Nemo! murmura Cyrus Smith, dont la figure pâlit légèrement."

Jules Verne, L'Ile Mystérieuse

In the following chapters we shall successively drop the three simplifying hypotheses which defined the perfect gas model (§ 7.1). We start by taking into account the *structure of the gas molecules*, which we had taken schematically to be point particles. We shall see that this structure plays an essential rôle in the determination of the macroscopic properties of most gases at equilibrium at sufficiently low densities. In particular, we shall use molecular spectroscopy (§ 8.4.1) to find the *thermal properties* of gases (§ 8.1), the properties of *mixtures* (§ 8.2), and their *chemical equilibrium* laws (§ 8.2.2). We shall study the quantum phenomenon of the *freezing in* of the internal molecular degrees of freedom which occur when the temperature is sufficiently low (§ 8.3.1), and the simplifications brought in by the *classical limit*, which is valid at sufficiently high temperatures (§ 8.4.2). We shall also in §§ 8.3 and 8.4 discuss some characteristic examples showing how the structure of the molecules in a gas is reflected in macroscopic effects.

8.1 General Properties

8.1.1 The Internal Partition Function

Let us consider one of the gas molecules, with centre of mass position $\boldsymbol{r}$, total momentum $\boldsymbol{p}$, and total mass m. Let $\boldsymbol{\varrho}_\alpha$ be the coordinates of the various constituents of the molecule, the atomic nuclei and electrons, with respect to its centre of mass, μ_α their masses, and $\boldsymbol{\pi}_\alpha$ their momenta in the centre of mass frame. We have $\sum \mu_\alpha \boldsymbol{\varrho}_\alpha \equiv 0$ and $\sum \boldsymbol{\pi}_\alpha \equiv 0$. The total kinetic energy of the constituents of the molecule can be written in the form

$$\frac{\boldsymbol{p}^2}{2m} + \sum_\alpha \frac{\boldsymbol{\pi}_\alpha^2}{2\mu_\alpha}.$$

The interaction potentials depend only on the relative positions, the relative momenta, and the spins $\boldsymbol{s}_\alpha$ of the nuclei and the electrons, so that the motion of the centre of mass can be split off. We can thus write the *Hamiltonian of the molecule* in the form

$$\frac{\widehat{\boldsymbol{p}}^2}{2m} + \widehat{h}, \tag{8.1}$$

where $\widehat{h}$ depends only on the internal observables $\widehat{\boldsymbol{\varrho}}_\alpha$, $\widehat{\boldsymbol{\pi}}_\alpha$, $\widehat{\boldsymbol{s}}_\alpha$. The energy levels ε_q of the molecule, which are associated with its ground state and its excited states, are the discrete eigenvalues of $\widehat{h}$. As an exercise, one could show, using ideas from §§ 2.1.5, 2.3.3, and 14.3.1, that the specific form (8.1) is just a consequence of *invariance* under translations, rotations, and Galilean transformations.

In the present chapter we neglect the interactions between the molecules, which is legitimate as long as the gas is sufficiently rarefied. The *Hamiltonian of the gas* is then the sum of the Hamiltonians (8.1) of the N molecules, which for the moment we assume to be all identical:

$$\widehat{H}_N = \sum_{i=1}^N \left(\frac{\widehat{\boldsymbol{p}}_i^2}{2m} + V(\widehat{\boldsymbol{r}}_i) \right) + \sum_{i=1}^N \widehat{h}_i. \tag{8.2}$$

We have denoted by $V(\boldsymbol{r})$ the potential which confines the molecules within the volume Ω of the box, as in (7.8). The two terms in (8.2) commute and can thus be treated separately. The first one acts in the Hilbert space of the centre of mass variables, where the observables are the positions $\boldsymbol{r}_i$ and the momenta $\boldsymbol{p}_i$ of the molecules; the second one acts in the Hilbert space of the internal variables, itself the direct product of the individual Hilbert spaces for each molecule.

The *kinetic part* is exactly the same as if the molecules had been replaced by structureless and non-interacting point particles. As in the perfect gas idealization we can treat that part in the framework of classical statistical mechanics whatever gas we are considering. In fact, we have seen (§ 7.1.3) that the condition (7.5) for the validity of the classical approximation is satisfied for all gases. (When one takes the structure of the molecules into account, condition (7.7) must be modified into $\zeta e^{\alpha} \ll 1$ by subtracting from $\alpha = \mu/kT$ the contribution (8.13) from the internal variables.)

On the other hand, the *internal degrees of freedom*, described by the Hamiltonians $\widehat{h}_i$ which are all similar, must be treated quantum mechanically. Indeed, the discrete nature of the energy levels ε_q, that is, the eigenvalues of $\widehat{h}$, cannot be accounted for by classical mechanics. As they refer to the internal variables of different molecules, the various Hamiltonians $\widehat{h}_i$ commute with one another and the energy levels of $\widehat{H}_N$ are, apart from the kinetic part, obtained simply by taking the sum of the energies ε_{q_i} of each molecule.

The evaluation of the *canonical partition function* is most easily carried out by taking advantage of the *factorization* of the kinetic contribution, on the one hand, which has already been evaluated in § 7.2.1, and of the contributions from the various molecules, on the other hand. We are, in fact, dealing with situations where the conditions for the applicability of the rules of § 4.2.5 hold. Denoting by Tr_N the trace over the Hilbert space of the internal variables of the N molecules and by tr the trace over the internal space of a single molecule, and using the classical limit (2.69) for the space of the external variables, we find

$$\begin{aligned} Z_N(\beta) &= \int d\tau_N \, \exp\left\{-\beta \sum_i \frac{\boldsymbol{p}_i^2}{2m}\right\} \, \text{Tr}_N \, \exp\left\{-\beta \sum_i \widehat{h}_i\right\} \\ &= \frac{1}{N!}\left[\frac{\Omega}{h^3}\int d^3\boldsymbol{p} \; e^{-\beta p^2/2m}\right]^N \left[\text{tr}\; e^{-\beta \widehat{h}}\right]^N \\ &= \frac{1}{N!}\left(\frac{\Omega}{\lambda_T^3}\right)^N \zeta^N, \end{aligned} \tag{8.3}$$

where λ_T again denotes the function of the temperature defined by (7.6). The internal degrees of freedom give rise to N identical factors ζ, each of which defines the *internal partition function* of a single molecule:

$$\boxed{\zeta(T) = \text{tr}\; e^{-\beta \widehat{h}} = \sum_q e^{-\varepsilon_q/kT}} \; . \tag{8.4}$$

The evaluation of ζ involves only the energy levels of a single, isolated molecule at rest. In the evaluation of the sum (8.4) we must bear in mind that if an energy level ε_q is d-fold degenerate, it must be counted d times as we are summing over states q and not over energy levels.

Even though the various results of the present chapter could be obtained in the canonical ensemble by starting from (8.3), we shall in what follows work with the grand canonical ensemble. By summing (8.3) over N, as in (7.26), we obtain the *grand potential*

$$A(T, \mu, \Omega) = -kT \ln Z_G = -kT \ln \sum_N Z_N \, e^{\alpha N}$$
$$= -\frac{\Omega kT}{\lambda_T^3} \zeta(T) \, e^{\mu/kT}, \tag{8.5}$$

which we shall use to find the thermodynamic properties of a real gas at equilibrium and at sufficiently low density.

The macroscopic rôle of the structure of the molecules at equilibrium is completely reflected in the internal partition function ζ, a function of T which occurs multiplicatively in the grand potential. In the free energy which follows from (8.3) $-NkT \ln \zeta$ occurs additively. The theory of the thermodynamic and chemical properties of gases therefore requires the explicit evaluation of (8.4); this will be our task in §§ 8.3 and 8.4.

The fact that the molecules are all made up from elementary constituents of the same kind poses a number of questions which we have avoided when we went from (8.1) to (8.2). First of all, if we consider, for instance, a gas of N oxygen molecules, each oxygen nucleus acts in principle with the same potentials on the $2N - 1$ other oxygen nuclei and on the $16N$ electrons; however, the approximate Hamiltonian (8.2) contains only its interactions with the other nucleus and with the 16 electrons of the molecule to which it belongs. The fact that we have neglected the other interactions, thus breaking the symmetry between the elementary constituents of the molecules, is justified by the short range of the forces and the low density of the gas. By associating in (8.2) each nucleus and each electron with a definitely chosen molecule we have eliminated *ipso facto* all configurations where an electron, or more generally a group of particles, switches from one molecule to another. The approximation is thus well suited for our aim of describing a gas consisting of *untouchable* molecules of a well defined kind without any possibilities for ionization or chemical reactions. Just as the Hamiltonian (7.8) of the perfect gas did not produce any energy exchanges, (8.2) cannot account for transfers of constituents between different molecules. We shall, nevertheless, see (§ 8.2.2) that the present approach is sufficient to describe chemical equilibria in gases where such transfers are allowed. It is true that during the establishing of equilibrium the interactions between the constituents of different molecules play an essential rôle in each collision: they are responsible for a redistribution of these constituents in the most random way possible which characterizes chemical equilibrium – just as for a perfect gas the collisions produce energy exchanges which govern the approach to equilibrium (§ 7.1.2). However, in both cases it is legitimate to treat the molecules as being independent of one another *once equilibrium has been established*, as long as the range of the forces is short as compared to the distances between the molecules, or the mean free path long as compared to these distances.

Another important aspect of the indistinguishability of the constituent particles, the electrons and nuclei, is the Pauli principle (Chap.10) according to which the Hilbert space only contains wavefunctions which are symmetric under the exchange of the coordinates of two particles, if we interchange atomic nuclei containing an even number of nucleons, or antisymmetric, if we interchange odd nuclei or electrons. Here the Pauli principle manifests itself in two ways. First of all, we must take it into account for *each molecule* when constructing the eigenstates of

$\widehat{h}$ which are characterized by the index q; in the calculation of the trace (8.4) we must thus watch out and retain only those states which have the *proper symmetry character* (§ 8.4.5). Next, we should apply the same restriction when we go over to the *whole system* and interchange molecules. Nevertheless, we have earlier constructed the micro-states of the system as tensor products characterized by the set of quantum numbers q_i, without bothering explicitly about the Pauli principle. In fact, as at low densities bound states of the various molecules do not overlap one another, the symmetrization or anti-symmetrization of the global wavefunction of each micro-state does not change its energy $\sum_i \varepsilon_{q_i}$, the only quantity which occurs in the calculation of (8.3). Hence, the only trace of the quantum indistinguishability which remains is the following: when we integrate over the centre of mass coordinates and the momenta of the molecules in the classical limit, the measure $d\tau_N$ in phase space *contains a factor* $1/N!$ which we have carefully included in (8.3). This factor is in the classical limit the remainder of the symmetry or antisymmetry of the wavefunctions under a simultaneous exchange of the coordinates of the elementary particles which make up two different molecules (§ 2.3.4 and Chap.10).

Let us finally note that expression (8.4) for the internal partition function does not make sense literally, if we want to be precise. In fact, if we include in the spectrum of $\widehat{h}$ the *continuum* which describes the ionized states of the molecule – or, more generally, the excited states when it is split up – the contribution from this continuum to (8.4) is infinite. This is a real difficulty for a *very dilute* or a *very hot* gas for which condition (7.3) is violated; we need a different approach for that case. It is then appropriate to treat the gas as a reactive mixture (§ 8.2.2) consisting (i) partly of neutral atoms, the internal partition function of which will be calculated by retaining in (8.4) only the bound states, (ii) partly of free electrons, and (iii) partly, finally, of ionized atoms having their own internal partition function (Exerc.8c). For the *not too rarefied* gases which we study in the present chapter there are practically no free electrons or ionized atoms, and it is legitimate to retain only the bound states q in (8.4) which then becomes a discrete sum. There is still a divergence, since there are an infinite number of bound states describing an electron moving at large distances around the ion which is the remainder of the molecule; they all have energies practically equal to the ionization energy. However, amongst those states we can forget about the ones with dimensions larger than the intermolecular distances: they are comparable to the ionized states and can be dropped under the same circumstances. As a result, when condition (7.3) is satisfied, it is legitimate to omit from the series (8.4) the higher-order terms for which the excitation energies $\varepsilon_q - \varepsilon_0$ get close to the ionization or dissociation energy of the molecule. We can disregard the divergence of this series, which is thus *dominated by its first terms* for which the $\varepsilon_q - \varepsilon_0$ are not large as compared to kT.

8.1.2 The Equation of State

As indicated in §5.6.4, we can find from (8.5) all properties of the gas in thermal equilibrium by taking derivatives. The *pressure* is given by (5.78),

$$\mathcal{P} = -\frac{\partial A}{\partial \Omega} = -\frac{A}{\Omega} \tag{8.6}$$

whereas the *density* follows from (5.76):

$$\frac{N}{\Omega} = -\frac{1}{\Omega}\frac{\partial A}{\partial \mu} = -\frac{A}{\Omega kT}. \tag{8.7}$$

The equation of state which we obtain by eliminating μ from (8.6) and (8.7) is

$$\mathcal{P}\Omega = NkT. \tag{8.8}$$

We see thus that for low density gases *the internal structure of the molecules does not affect the equation of state at all*, which remains the same as for a perfect gas, in agreement with experiments.

8.1.3 Thermal Properties

The *entropy* which follows from (5.72) and (8.5), where $\lambda_T \propto T^{-1/2}$, is equal to

$$S = -\frac{\partial A}{\partial T} = -\frac{5A}{2T} + \frac{\mu A}{kT^2} - A\frac{d}{dT}\ln\zeta,$$

or, if we use (8.7)

$$\frac{S}{N} = \frac{5}{2}k - \frac{\mu}{T} + kT\frac{d}{dT}\ln\zeta. \tag{8.9}$$

We can from this equation and Eq.(5.71) find the *internal energy* per molecule of the gas:

$$\frac{U}{N} = \frac{A}{N} + T\frac{S}{N} + \mu = \frac{3}{2}kT + kT^2\frac{d}{dT}\ln\zeta. \tag{8.10}$$

This expression differs from expression (7.44) for the perfect gas by the extra term coming from the internal degrees of freedom of the molecules. It is, nevertheless, remarkable that the right-hand side of (8.10) still *depends only on the temperature*. This property also follows from the equation of state (8.8), using the thermodynamic arguments of Exerc.6a. As a result, for all low density gases we get the *Joule expansion* property: a sudden irreversible expansion of a gas into a vacuum does not change its temperature, if there is no exchange of heat or work with the exterior.

Similarly, in a reversible, isothermal expansion the gas does not change its internal energy U. On the other hand, the work received during heating at constant pressure equals $-\mathcal{P}\,d\Omega = -Nk\,dT$. Hence, by striking an energy balance, we get *Mayer's relation* between the specific heats at constant pressure and constant volume, which has been well supported experimentally for low density gases, and which also follows from (6.41) and (8.8):

$$C_\text{p} - C_\text{v} = Nk. \tag{8.11}$$

We find the *specific heat at constant volume* from (8.10) by taking a derivative:

$$C_{\mathrm{v}} = \frac{3}{2}Nk + \frac{d}{dT}\left(T^2 \frac{d}{dT} \ln \zeta\right) Nk. \tag{8.12}$$

To the perfect gas term is added a second term which is associated with the *thermal excitation of the internal degrees of freedom*, and which can be obtained from the energy levels of the molecule, if we use (8.4). These levels are known experimentally with a high precision thanks to spectroscopic data, as the emitted or absorbed wavelengths λ are inversely proportional to the energy differences between the levels, $\Delta\varepsilon = hc/\lambda$. Remember that 2 eV corresponds to visible light, namely, $\lambda = 6000$ Å, so that room temperature ($\frac{1}{40}$ eV) corresponds to the infrared. The specific heats of gases can thus be derived easily from data on the infrared spectra of the molecules in the gas, and experiments have checked this *relation between thermal and spectroscopic properties*.

We shall proceed more theoretically in §§ 8.3 and 8.4 by calculating directly, though approximately, the form of ζ for various kinds of molecules. Our starting point will be the Hamiltonian $\widehat{h}$ which describes the nuclei and electrons, interacting through Coulomb forces. This will enable us to derive the specific heat of a gas from the structure of its molecules.

8.1.4 The Chemical Potential

The definition, the physical interpretation, the properties, and the importance of the chemical potential μ, which characterizes the tendency of a gas to give up molecules, have been discussed in § 5.6.3. As function of the density and the temperature it follows from (8.5) and (8.7) that

$$\begin{aligned} \frac{\mu}{kT} &= \ln\left(\frac{N}{\Omega}\lambda_T^3\right) - \ln \zeta(T) \\ &= \ln \frac{N}{\Omega} - \frac{3}{2} \ln \frac{mkT}{2\pi\hbar^2} - \ln \zeta(T). \end{aligned} \tag{8.13}$$

The dependence of μ on variables other than the temperature is solely through the first term; for given T it logarithmically *increases with the density*, or, what amounts to the same, with the pressure: the more it is compressed, the more easily the gas gives up molecules.

If we choose as energy zero for a molecule the energy of its ground state, we have $\zeta > 1$. The condition $N\lambda_T^3/\Omega \ll 1$ for the validity of the classical approximation for the translational degrees of freedom then implies that the *chemical potential of the molecules of a gas is negative.* We have estimated (§ 7.1.3) the kinetic contribution to μ for oxygen under normal conditions; if one adds to it the contribution from the rotation of the molecules (§ 8.4.3) one gets an order of magnitude of $\mu \simeq -0.5$ eV, to be compared with $kT \simeq \frac{1}{40}$ eV. This sign of μ implies that a gas loses free energy when it gains molecules at constant volume and temperature. Just as its volume tends to increase when it is surrounded by a vacuum because its pressure is positive, the gas would tend to absorb molecules which would hypothetically be available at zero chemical potential, that is, quantum molecules at low temperature and low density.

Finally, μ decreases with increasing temperature either at constant density or at constant pressure: a cold gas tends to give up molecules to a hotter gas at the same density or the same pressure.

8.1.5 The Entropy

We get from (8.9) and (8.13) an expression for the entropy as function of the temperature and the number of particles,

$$\frac{S}{Nk} = \ln \frac{\Omega}{N\lambda_T^3} + \frac{5}{2} + \frac{d}{dT}\left[T \ln \zeta(T)\right], \tag{8.14}$$

which generalizes the Sackur-Tetrode formula (7.41). Using (8.4) we can check that the last term in (8.14), which is the entropy associated with the internal quantum degrees of freedom of a molecule, is positive. The preceding terms are also positive, if we bear in mind (7.5), so that the entropy of the gas is certainly positive under the conditions where (8.14) is valid.

The equation for an adiabatic expansion follows immediately from (8.14) combined with the equation of state (Exerc.8d).

When the temperature decreases or the density increases, the entropy (8.14) decreases and one ends up by reaching a regime where the present approximations become incorrect. On the one hand, *interactions* between molecules start to play a part: dense gases or liquids (Chap.9); on the other hand, *quantum* effects begin to dominate: solidification (Chap.11). The way expression (8.14) for the entropy of a gas goes over into those for the low-temperature condensed, liquid or crystalline, phases plays an important rôle in the *experimental check of the Third Law.* In fact, (8.14) *is defined without an additive constant* and must go over into an entropy which vanishes as $T \to 0$ (§ 5.4.2). Let us, for instance, assume that there is only one solid phase, in which the specific heat at constant pressure is $C_{\rm p}^{\rm s}(T)$, and let L be the latent sublimation heat. The change in entropy under a transformation at constant pressure from the temperature $T_{\rm s}$ where the gas solidifies to the absolute zero is then equal to

$$S_{T_{\rm s}} - S_{T=0} = \frac{L}{T_{\rm s}} + \int_0^{T_{\rm s}} \frac{C_{\rm p}^{\rm s}(T)}{T}\, dT. \tag{8.15}$$

The check on whether the entropy actually vanishes at the absolute zero thus reduces to comparing the right-hand side of (8.15), which can be determined experimentally by calorimetric measurements, with that of (8.14), which one evaluates for $T = T_{\rm s}$ in the gas phase. This check has been carried out for a large number of substances; it shows that the entropy, normalized so that it vanishes at the absolute zero, contains, indeed, for a gas the additive constant occurring in (8.14). Note that this expression depends not only on

Boltzmann's constant k, but also on Dirac's constant $\hbar$ which can thus be measured indirectly by calorimetry.

8.2 Gas Mixtures

8.2.1 Non-reacting Mixtures

The above formalism can easily be generalized to a gas containing several kinds of molecules, X_1, X_2, ... When the gas is chemically inert, that is, when all numbers N_1, N_2, ... are *conserved*, we must introduce in the grand canonical equilibrium, apart from the Lagrangian multiplier β associated with the energy, a multiplier α_j for each kind of molecule X_j, and define the grand canonical ensemble by summing over N_1, N_2, ... In this way the grand potential A for a mixture of gases now depends on *a chemical potential for each kind of molecule.* In the classical limit the volume element (2.59) in phase space contains here the symmetry factor $1/N_1!N_2!\ldots$. The canonical partition function can then be factorized into a product of contributions like (8.3). The same is true for the grand partition function so that for two kinds of molecules, for example, the grand potential is a *sum of terms* like (8.5):

$$A(T, \Omega, \mu_1, \mu_2) = A_1(T, \Omega, \mu_1) + A_2(T, \Omega, \mu_2); \tag{8.16}$$

A_1 and A_2 differ in the values of the masses m_1, m_2 which occur in λ_T and in the expressions for the internal partition functions $\zeta_1(T)$ and $\zeta_2(T)$. Everything behaves as if we had two *independent systems occupying the same volume.*

The partial densities of each species in the gas are given by

$$\frac{N_j}{\Omega} = -\frac{A_j(T, \Omega, \mu_j)}{\Omega kT}; \tag{8.17}$$

these expressions should in practice be inverted into (8.13) when one gives the relative concentrations of the molecules rather than μ_1 and μ_2. The pressure, internal energy, and specific heat are sums of contributions from the various molecules. In particular, the *pressure* satisfies, because of (5.77) and (8.17), *Dalton's law* (1801),

$$\mathcal{P} = -\frac{A}{\Omega} = \frac{kT}{\Omega} \sum_j N_j, \tag{8.18}$$

which expresses that the different kinds of molecules in the gas contribute independently to the total pressure.

The entropy is also additive, in the sense that it is a sum of contributions (8.14), calculated as if *each* of the sets of molecules *occupied the whole volume* Ω. Let us, nevertheless, assume that we start from an initial situation where the molecules are *separated*, the N_1 molecules of kind X_1 being placed in a vessel of volume $\Omega_1 \equiv N_1\Omega/(N_1+N_2)$ and the N_2 molecules of kind X_2 in a vessel of volume $\Omega_2 \equiv N_2\Omega/(N_1+N_2)$. The two gases are at the same temperature T and thus at the same pressure. If we connect the two vessels, there occurs an *irreversible* transformation during which the molecules of the two species *mix*. The fact that the internal energy (8.10), which is conserved during the mixing, is independent of the volume has as a consequence that the temperature remains unchanged; the pressure, given by (8.18) in the final state, remains equally unchanged. The process is, nevertheless, distinguished by the existence of a *mixing entropy*. In fact, everything happens, for each set of molecules, as if the volume increased from Ω_1 or Ω_2 to $\Omega = \Omega_1 + \Omega_2$ at constant temperature; this makes the entropy increase – and the chemical potentials decrease – by an amount coming from the first term in (8.14),

$$\begin{aligned} \Delta S &= kN_1 \ln \frac{\Omega}{\Omega_1} + kN_2 \ln \frac{\Omega}{\Omega_2} \\ &= kN_1 \ln \frac{N_1+N_2}{N_1} + kN_2 \ln \frac{N_1+N_2}{N_2} \\ &\sim k \ln \frac{(N_1+N_2)!}{N_1!N_2!}. \end{aligned} \tag{8.19}$$

The mixing entropy measures the *increase in disorder* associated with the mixing of the molecules of the two species. It does *not* correspond to *any thermal phenomenon.* In its last form, (8.19) involves the number, $(N_1+N_2)!/N_1!N_2!$, of different ways to arrange the N_1 and the N_2 molecules, which can be distinguished between the two groups, but not within a group; we recognize therefore in ΔS the *information lost* when we randomly distribute all the molecules, starting from a situation where they were separated according to their type. If, conversely, we would wish to *separate the* X_1 *and* X_2 *molecules*, at given temperature and pressure, we should decrease the entropy of the gas. This is, according to the Second Law, impossible, unless we supply at least an amount of work equal to $T\Delta S$; the energy of the gas remains constant so that the work done is downgraded into heat given to the outside (Exerc.8a).

The *Gibbs paradox* (§ 3.4.3) originates from the existence of the mixing entropy. Before the end of the nineteenth century it seemed conceivable that one could change the nature of the molecules of a gas continuously. Under such an operation the mixing entropy (8.19) would remain invariant, except at the exact moment where the molecules X_1 and X_2 became identical; in fact, in that case, mixing them would not change anything. This discontinuity is one way of presenting the Gibbs paradox. In fact, the problem can no longer be put in those terms as we know that the structure of atoms and nuclei does not allow continuous changes which could be imagined to happen in classical physics.

In the form given in § 3.4.3 the Gibbs paradox shows up the relative and anthropocentric nature of the mixing entropy. As long as one does not know how to distinguish the X_1 and X_2 molecules, one cannot separate them, so that the fact whether or not we include the mixing entropy has no consequences – except as far as the Third Law is concerned. On the contrary, ΔS becomes relevant when one is dealing with the separation of the X_1 and X_2 molecules.

Let us finally note that the *quantum indistinguishability* also plays an important rôle in the existence of the mixing entropy. It is the reason for the appearance of the factor $1/N!$ in (8.3). If that factor were not there, the canonical entropy would contain an additional term $k \ln N!$ and it would *not be extensive*. This would lead to a paradox much worse than the Gibbs paradox: the mixing of two samples *of the same gas* containing N_1 and N_2 completely identical molecules, at the same temperature and pressure, would, in fact, make the entropy increase by the term (8.19); this, of course, is completely inadmissible.

8.2.2 Chemical Equilibria in the Gas Phase

Whether or not a gas is pure, inelastic collisions between molecules can produce transitions between the energy eigenlevels of the molecules; this ensures that the internal motions are brought in thermal equilibrium with the translations, as we have assumed them to do until now. In a gas mixture where chemical reactions may take place, collisions can, moreover, change the make-up of the molecules. In that case the numbers N_1, N_2, ... of the molecular types X_1, X_2, ... are no longer conserved, because of those collisions. The disappearance of some conservation laws implies that the chemical potentials μ_1, μ_2, ... are no longer independent variables, as they were in the non-reacting mixtures studied above; the numbers N_1, N_2, ... adjust themselves to values which make the disorder a maximum and which correspond to equilibrium. Our aim is to find their values.

As a preliminary stage, it is useful to consider situations which are not in chemical equilibrium and where the numbers of molecules of each species are assumed to be *frozen in*. These situations can be observed over rather long periods, if the reaction processes between the molecules are not very efficient; in any case, considering them is a useful preliminary for the construction of chemical equilibrium states. They are just described by the formalism of § 8.2.1. In particular, the interactions between the molecules are sufficiently weak so that they contribute only negligibly to the thermodynamic functions. In order afterwards to go over to chemical equilibrium situations we shall use the concept of chemical potential along the lines sketched in §§ 5.1.4 and 5.6.3 and worked out in detail in § 6.6.3. Before recalling the results of that study we shall discuss a simple example.

We shall consider the dissociation equilibium, $H_2 \leftrightarrows 2H$, of gaseous hydrogen at high temperatures and low densities. Neglecting the interactions between molecules, the Hamiltonian $\widehat{H}$ can be split into components,

$$\widehat{H}_{N_1 N_2} = \sum_{i=1}^{N_1} \left(\frac{\widehat{\boldsymbol{p}}_i^2}{2m_1} + \widehat{h}_i^{(1)} \right) + \sum_{i'=1}^{N_2} \left(\frac{\widehat{\boldsymbol{p}}_{i'}^2}{2m_2} + \widehat{h}_{i'}^{(2)} \right), \tag{8.20}$$

describing the dynamics of N_1 monatomic H molecules, of mass m_1, and N_2 diatomic H_2 molecules, of mass $m_2 = 2m_1$. Collisions produce transitions between eigenstates of $\widehat{H}$ with the same energy and with the same value of the total number of hydrogen atoms in the gas,

$$N = N_1 + 2N_2. \tag{8.21}$$

Let us, to begin with, assume that the transitions which change N_1 and N_2 are prohibited; we can then give separately the expectation values $\langle N_1 \rangle$ and $\langle N_2 \rangle$. The resulting grand canonical equilibrium is represented by the density operator

$$\left.\begin{aligned} \widehat{D} &= \frac{1}{Z_G}\, \mathrm{e}^{-\beta\widehat{H}+\alpha_1\widehat{N}_1+\alpha_2\widehat{N}_2}, \\ Z_G &= \sum_{N_1 N_2} \mathrm{e}^{\alpha_1 N_1+\alpha_2 N_2}\, \mathrm{Tr}\, \mathrm{e}^{-\beta\widehat{H}_{N_1 N_2}}, \end{aligned}\right\} \tag{8.22}$$

and the grand potential is given by (8.16). To describe the chemical equilibrium we now allow the $H_2 \leftrightarrows 2H$ transitions which may change N_1 and N_2 under the constraint (8.21). Apart from the energy there is only one other constant of motion, N, instead of N_1 and N_2, which we can take care of through a single Lagrangian multiplier α. The Hamiltonian is still (8.20), apart from negligible interaction terms, so that the grand canonical density operator can now be written in the form

$$\widehat{D} = \frac{1}{Z_G}\, \mathrm{e}^{-\beta\widehat{H}+\alpha(\widehat{N}_1+2\widehat{N}_2)}, \tag{8.23}$$

instead of (8.22). In the grand canonical formalism it is thus sufficient, for going from a *non-reacting mixture* to a gas in *chemical equilibrium*, to write down the *relations between the chemical potentials.* Here these relations have the form

$$\alpha_2 = 2\alpha_1 = 2\alpha, \tag{8.24}$$

and the grand potential of the H, H_2 mixture in chemical equilibrium follows simply from (8.16) through the replacement

$$A(T,\Omega,\mu) = A_1(T,\Omega,\mu) + A_2(T,\Omega,2\mu). \tag{8.25}$$

The total number N of atoms is equal to $-\partial A/\partial\mu$ whereas the number N_1 of monatomic molecules at equilibrium equals $-\partial A_1/\partial\mu$ and the number N_2 of diatomic molecules $-\partial A_2/\partial(2\mu)$.

Extending these results to the equilibrium of arbitrary reacting gases was the aim of § 6.6.3. One starts by listing the *possible reactions* k which transform the various kinds of molecules X_j of the gas into one another, writing them in the form

$$\sum_j \nu_j^{(k)} X_j \leftrightarrows 0. \tag{8.26}$$

In the above example there is only a single reaction $2\mathrm{H} - \mathrm{H}_2 \leftrightarrows 0$, and we have $\nu_1 = 2$, $\nu_2 = -1$. One then introduces a chemical potential μ_j for each kind of molecule and determines the relations between the thermodynamic quantities *as if the mixture were not reacting.* One obtains in this way the chemical equilibrium by *constraining the molecular chemical potentials* by the relations

$$\sum_j \nu_j^{(k)} \mu_j = 0. \tag{8.27}$$

In § 6.6.3 we justified the chemical equilibrium conditions (8.27) in different ways: (i) by looking for the *maximum of the entropy in two stages*, as indicated in § 6.1.3, the first stage leading to the thermal equilibrium of the non-reacting mixture, and the second one to the chemical equilibrium; (ii) by treating *each population* of molecules of one given kind as a *subsystem* of the gas, and studying the exchanges of particles between these subsystems, *superimposed in space*; (iii) by proceeding as above in the *grand canonical* ensemble. We also refer to § 6.6.3 for the interpretation of the relations (8.27).

In practice one wants to *determine the concentrations* in the gas of the molecules of the various kinds in chemical equilibrium. To calculate them it suffices to start from Eqs.(8.17) or (8.13) which are valid in thermal equilibrium, whether or not chemical equilibrium has been established, and to connect each number of molecules with the appropriate chemical potential; after that one can use the conditions for chemical equilibrium (8.27) to eliminate the chemical potentials. This gives us the required relations between the various numbers of molecules: one obtains as many equations as are necessary to evaluate the various concentrations as functions of the temperature, the volume – or the pressure – and the initial composition of the gas. The simplest case is the one where we have a single chemical reaction (8.26). Eliminating the chemical potentials from condition (8.27) and expressions (8.13) for each of the chemical species X_j then gives us a *relation between the partial densities* N_j/Ω,

$$\prod_j \left[\frac{N_j}{\Omega} \left(\frac{2\pi\hbar^2}{m_j kT} \right)^{3/2} \frac{1}{\zeta_j(T)} \right]^{\nu_j} = 1, \tag{8.28}$$

which determines them when Ω is given. When the chemical equilibrium is established at a *fixed total pressure*, elimination of Ω in (8.28) by using Dalton's Law (8.18) gives us a relation between the *relative concentrations*,

$$\mathcal{P}^{\sum_j \nu_j} \prod_j \left(\frac{N_j}{N_{\text{tot}}} \right)^{\nu_j} = K(T), \tag{8.29}$$

where $N_{\text{tot}} = \sum_j N_j$ is the total number of molecules in the gas. The quantity on the right-hand side, the so-called *equilibrium constant*,

$$K(T) \equiv \prod_j \left[\left(\frac{m_j}{2\pi\hbar^2} \right)^{3/2} (kT)^{5/2} \zeta_j(T) \right]^{\nu_j} , \tag{8.30}$$

depends for a given reaction only on the temperature.

Relation (8.28) or (8.29) is called the *mass action law*. Established in 1867 by Guldberg and Waage in the framework of kinetic theory, it was rediscovered a little later by Gibbs using purely thermodynamic arguments. Statistical mechanics enables us to use (8.4) to determine the equilibrium constant (8.30) starting from the energy levels of the various species of molecules, which themselves can be calculated theoretically or supplied experimentally by spectroscopy; indirect information about the $\zeta_j(T)$, and thus about $K(T)$, is also obtained by calorimetry (§ 8.1.3). In our example of the dissociation of hydrogen molecules the mass action law can be written as

$$\mathcal{P} \frac{N_1^2}{(N_1+N_2)N_2} = K(T) \equiv \left(\frac{m_1}{4\pi\hbar^2} \right)^{3/2} (kT)^{5/2} \frac{[\zeta_1(T)]^2}{\zeta_2(T)} , \tag{8.30'}$$

which determines N_1/N_2 as function of the temperature and the pressure, once we know the lowest levels of the hydrogen atom and the hydrogen molecule. If there are *several reactions* which can occur between the various kinds of molecules in the gas, each of them leads to an equation like (8.28) or (8.29).

If we know the thermodynamic potentials (8.16) of the mixtures, we are also able to determine theoretically the *thermal* or *mechanical* effects which accompany the chemical reactions in the gas. In particular, in § 6.6.3 we defined the reaction heat, at given temperature and pressure, and gave an expression (6.84) for it as the change in the enthalpy $H = U + \mathcal{P}\Omega$. From the energy (8.10) and the pressure (8.18) we find the *enthalpy of the mixture* of gases, whether or not in chemical equilibrium,

$$H = \sum_j N_j \left(\frac{5}{2}kT + kT^2 \frac{d}{dT} \ln \zeta_j \right) . \tag{8.31}$$

When expressed thus in terms of T, $\mathcal{P}$, and the N_j it has the features of depending *linearly* on the N_j and being *independent of* $\mathcal{P}$. For a reaction characterized by the ν_j coefficients the *reaction heat per mole* is the decrease in H following from changes $N_j = \nu_j N_\mathrm{A}$ in the numbers of molecules. It therefore is equal to

$$\mathcal{Q} = -N_\mathrm{A} k \sum_j \nu_j \left(\frac{5}{2}T + T^2 \frac{d}{dT} \ln \zeta_j \right) ,$$

or, if we use the definition (8.30),

$$\mathcal{Q} = -RT^2 \frac{d}{dT} \ln K(T) . \tag{8.32}$$

The reaction heats in gases therefore *depend solely on the temperature* and are directly *connected with the equilibrium constants* $K(T)$. In our example (8.30′) and (8.32) give us the dissociation heat of one mole of H_2 at a temperature T. The relation (8.32) between the equilibrium constant (8.30) and the reaction heat is *van 't Hoff's Law* (1884) which can also be proved by purely thermodynamic arguments for any reaction in the gas phase. That law has been well verified experimentally.

8.3 Monatomic Gases

In this section and the next we shall evaluate the internal partition function $\zeta(T)$ and the thermodynamic properties which follow from it for various kinds of gases.

8.3.1 Rare Gases; Frozen-In Degrees of Freedom

The molecules of inert, or rare, gases are monatomic and the energy levels ε_q of $\widehat{h}$ are those of the electron cloud in the potential of the nucleus, which is situated at the centre of mass. The ground state corresponds to a set of *filled electron shells* and the excitation energies are of the order of magnitude of the distance to the next shell, which is at least of the order of tens of eV, about 20 eV for helium. In the calculation of the internal partition function ζ, from which we derive the various physical quantities, we split off the contribution from the ground state. Its *multiplicity* is $g = 2s_n + 1$, where s_n is the spin of the nucleus, as the electron shells are filled and the ${}^1S_0$ electron state is non-degenerate, both with regard to the spin and with regard to the orbital motion. We thus have

$$\ln \zeta = -\frac{\varepsilon_0}{kT} + \ln \left[g + \sum_q{}' \mathrm{e}^{-(\varepsilon_q - \varepsilon_0)/kT} \right], \tag{8.33}$$

where the sum is over the excited states of the electrons in the atom.

It is convenient to express the excitation energies in temperature units and to introduce, as in § 1.4.4, the concept of a characteristic temperature associated with the excitation energies. We know that room temperature corresponds to $\frac{1}{40}$ eV so that typical electron excitation energies of, say, twenty eV correspond to a value of $\Theta_e = (\varepsilon_q - \varepsilon_0)/k$ of the order of 200 000 K. We call Θ_e the *characteristic temperature associated with the electronic degrees of freedom* of the atom. It is much higher not only than normal temperatures but even than those on stellar surfaces, so that the exponentials $\exp(-\Theta_e/T)$ which appear in (8.33) are negligible and the internal partition function reduces to *just the contribution from the ground state*

$$\ln \zeta = -\frac{\varepsilon_0}{kT} + \ln g. \tag{8.34}$$

This form implies that we must simply add to the internal energy of the perfect gas the energy $N\varepsilon_0$ of the molecules in their ground state, and to the entropy of the perfect gas a term $Nk \ln g$ which corresponds to the g^N possible configurations of the spins of the atomic nuclei, all with the same energy. This additive constant does not have any thermodynamic consequences, but it plays a rôle in the experimental check of Nernst's principle.[1] Similarly, the chemical potential contains a contribution due to the structure of the molecules and equal to $\varepsilon_0 - kT \ln g$. The *specific heat* (8.12) *is not changed*, when we take the structure of the molecules into account, if ζ has the form (8.34), so that for an inert gas

$$C_{\mathrm{v}} = \tfrac{3}{2}Nk. \tag{8.35}$$

We have indicated in § 7.1.1 that the inert gases behave practically as the idealized model of a perfect gas of structureless point particles, even though their molecules consist of a nucleus surrounded by a complex electron cloud. One says that *the electronic degrees of freedom are "frozen in"*: at any *temperature well below the characteristic temperature* Θ_{e} the electrons cannot be excited thermally since the fraction $\exp(-\Theta_{\mathrm{e}}/T)$ of excited states remains always negligible at equilibrium at those temperatures. Everything behaves, as if the electronic degrees of freedom did not exist (Exerc.8b).

On the other hand, if we assume for a moment that we reach *temperatures* T *of the order of magnitude of the characteristic temperature* Θ_{e}, we must add an electronic contribution to the specific heat, as the internal partition function no longer has the simple form (8.34). The electronic degrees of freedom are freed and the excited states play a rôle. This can clearly not happen in the case of rare gases, as we should reach 200 000 K, but we shall see examples of similar effects for more complex gases.

8.3.2 Other Monatomic Gases

Let us turn to monatomic gases other than the rare gases. The differences in energy between the ground state and the first excited states are much smaller than for the rare gases, but they still often provide characteristic temperatures of the order of 10 000 K. For instance, for sodium vapour the first excitation energy is that of the yellow 5893 Å doublet line corresponding to 2.1 eV or 24 000 K. The excited states therefore remain frozen in; the internal partition function retains the form (8.34), but the multiplicity $g = (2s_{\mathrm{n}}+1)g_{\mathrm{e}}$ now includes, on top of the contribution from the nuclear spin, a factor g_{e} coming from the possible existence of several electronic wavefunctions with

[1] Because the interactions between the nuclear spins are so weak, the quasi-degeneracy g^N of the ground state persists after solidification. The spins become ordered (*nuclear antiferromagnetism*) only at extremely low temperatures, of the order of nanokelvin, where the entropy decreases below $kN \ln g$ and tends finally to zero.

the same minimum energy. For instance, for the alkalines the electron ground state is $^2S_{1/2}$ and we have $g_e = 2$ which for ^{23}Na gives $g = 4$.

Nevertheless, when the lowest levels are split by magnetic interactions connected with the spins their differences are much smaller. For instance, for the hydrogen atom the hyperfine interaction between the spins of the proton and the electron gives a non-degenerate ground state of energy ε_0 and an excited triplet with energy $\varepsilon_0 + \delta\varepsilon$, where $\delta\varepsilon$, which is associated with the 21 cm line, is 6×10^{-6} eV, or, in temperature units, 0.07 K. In the internal partition function,

$$\zeta = e^{-\varepsilon_0/kT}\left(1 + 3e^{-\delta\varepsilon/kT}\right),$$

the temperature kT is now *much larger than the excitation energy* $\delta\varepsilon$ of this first excited level. We can therefore completely *neglect the hyperfine splitting* in the calculations of the thermodynamic properties, and write

$$\zeta = 4\,e^{-\varepsilon_0/kT}.$$

Everything happens in this case, as if the ground state were 4-fold degenerate.

Fine-structure splittings, which are due to spin-orbit interactions, have larger characteristic temperatures and it turns out that they will affect the thermal properties. For instance, for monatomic Cl, the ground state $^2P_{3/2}$, of multiplicity 4, is separated from the $^2P_{1/2}$ level, of multiplicity 2, by an energy $\delta\varepsilon = 0.11$ eV, corresponding to an infrared line of wavenumber 881 cm^{-1}. The associated characteristic temperature, $\delta\varepsilon/k = 1270$ K, can certainly not be neglected, even at the high temperatures at which one must work in order that the gas be not completely composed of Cl_2. From the internal partition function associated with these electron states,

$$\ln\zeta = -\frac{\varepsilon_0}{kT} + \ln\left(4 + 2\,e^{-\delta\varepsilon/kT}\right), \tag{8.36}$$

we find an *electronic contribution to the specific heat* which must be added to (8.35):

$$C_V^{\text{el}} = \frac{2x^2\,e^x}{(2\,e^x + 1)^2}\,Nk, \qquad x \equiv \frac{\delta\varepsilon}{kT}. \tag{8.37}$$

This contribution, shown in Fig.8.1, is negligible both at temperatures low as compared to $\delta\varepsilon/k$, where only the four sublevels of the ground state are equally populated, and at high temperatures, where all the six sublevels of the doublet are equally populated. In both limits $kT \ll \delta\varepsilon$ and $kT \gg \delta\varepsilon$, the internal state of the molecules is frozen in. The specific heat (8.37) shows a characteristic peak, the *Schottky anomaly*, at a value of kT of the order of the distance between the ground state and the first excited state. This kind of contribution has actually not been observed in the specific heats of monatomic gases because of experimental difficulties, but this shape often occurs for other kinds of substances. We have already encountered an example

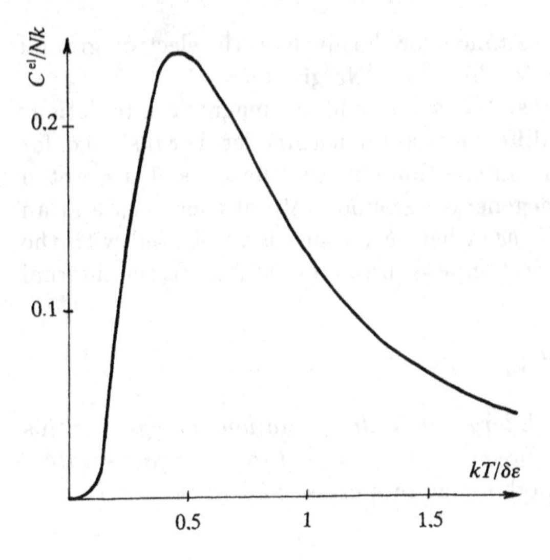

Fig. 8.1. Schottky anomaly in the specific heat

in § 1.4.4 for the specific heat of a paramagnetic substance placed in a fixed magnetic field.

8.4 Diatomic and Polyatomic Gases

8.4.1 The Born-Oppenheimer Approximation

In order to determine the internal partition function (8.4) and to find from it the thermal properties of the gas, we need to evaluate the eigenenergies ε_q of one molecule, for the ground state ε_0 and for those excited state which lie sufficiently close to the ground state that $\varepsilon_q - \varepsilon_0$ is *at most of the order of magnitude of* kT. To do this we shall use the so-called Born-Oppenheimer approximation (1927) which is of major interest not only in the theory of molecules, but also in the theory of solids (§ 11.1.1).

In the case of a molecule consisting of several atoms an analysis of the spectrum implies not only a study of the motion of the electrons, as in the case of a monatomic molecule, but above all a study of the relative motion of the nuclei. To carry that out we note that the atomic nuclei are several thousand times heavier than the electrons and therefore move much more slowly than the latter, if the kinetic and potential energies are of the same order of magnitude. The Born-Oppenheimer approximation now consists in solving the problem in *two stages.*

We start by studying the *motion of the electrons, neglecting that of the nuclei* which we assume to be fixed in some, arbitrary, positions. We must thus solve a Schrödinger equation describing the electrons, which interact through Coulomb forces, and are subject to a fixed external potential due to the nuclei and depending on the relative positions of the latter. For most molecules which have rather large binding energies this equation produces electron levels with spacings of the order of 1 eV, corresponding to characteristic temperatures of the order of 10 000 K. As a result, for most gases at room temperature the electron cloud is *frozen in into its lowest state.* The electronic degrees of freedom will therefore, as in the case of monatomic gases, lead only to supplying a multiplicity factor in ζ, if several electronic wavefunctions have the same, or almost the same, energy. In Chap.11 we shall see that the situation is not quite as simple when we apply the Born-Oppenheimer method to solids, since in that case the electron cloud is thermally excited at room temperatures. For the rather tightly bound and rather small molecules considered here we can, on the other hand, restrict ourselves to the single electronic eigenstate with minimum energy, for each, assumed given, arrangement of the nuclei.

To be more precise, the Hamiltonian of one molecule, in the model of § 8.1.1, has the form

$$\widehat{H} = \frac{\widehat{\boldsymbol{p}}^2}{2m} + \widehat{h} = \widehat{T}_{\mathrm{n}} + \widehat{T}_{\mathrm{e}} + \widehat{V}, \tag{8.38}$$

where $\widehat{T}_{\mathrm{n}}$ and $\widehat{T}_{\mathrm{e}}$ are the kinetic energies of the nuclei and the electrons and $\widehat{V}$ the total Coulomb interaction energy, which depends on both the electron coordinates $\widehat{\boldsymbol{r}}_{\mathrm{e}}$ and the nuclear coordinates $\widehat{\boldsymbol{R}}_{\mathrm{n}}$. The global translational kinetic energy of the molecule, $\widehat{\boldsymbol{p}}^2/2m$ is included in $\widehat{T}_{\mathrm{n}}$, as the mass of the electrons is small compared to the nuclear masses. The first stage of the Born-Oppenheimer method consists in looking for the ground state of the Schrödinger equation

$$\left[\widehat{T}_{\mathrm{e}} + \widehat{V}\left(\widehat{\boldsymbol{r}}_{\mathrm{e}}, \boldsymbol{R}_{\mathrm{n}}\right)\right] |\psi_{\mathrm{e}}\rangle = W(\boldsymbol{R}_{\mathrm{n}})\, |\psi_{\mathrm{e}}\rangle \tag{8.39}$$

in the Hilbert space of *only the electrons*, dropping $\widehat{T}_{\mathrm{n}}$ and regarding the $\boldsymbol{R}_{\mathrm{n}}$ not as operators, but as parameters. *The energy $W(\boldsymbol{R}_{\mathrm{n}})$ of the electronic ground state thus depends on the positions of the nuclei.*

In the second stage we study the *motion of the nuclei* for the lowest electron configuration which we have just determined. To do that we must reintroduce the term $\widehat{T}_{\mathrm{n}}$ from (8.38) that we omitted until now, and regard the $\widehat{\boldsymbol{R}}_{\mathrm{n}}$ again as operators which do not commute with $\widehat{T}_{\mathrm{n}}$ – a feature which did not occur in (8.39). The approximation made consists in assuming that the electron cloud, which is very mobile, adjusts itself instantly to the configuration of the nuclei which in this way feel the effect of the electrons indirectly. The Hamiltonian of the nuclei thus contains, on top of the kinetic energy $\widehat{T}_{\mathrm{n}}$, the energy $W(\widehat{\boldsymbol{R}}_{\mathrm{n}})$ which comes both from the Coulomb interaction between the nuclei and from their interaction with the electrons, after the electron coordinates have been eliminated as a result of their being frozen in into the lowest energy state of (8.39). The lowest energy levels of the molecule are thus finally obtained by looking for the eigenvalues of the Schrödinger equation

$$\left[\widehat{T}_{\rm n} + W(\widehat{\boldsymbol{R}}_{\rm n})\right] |\psi_{\rm n}\rangle = \varepsilon \, |\psi_{\rm n}\rangle \tag{8.40}$$

in the Hilbert space of *only the nuclei.*

We should note that the motion of the centre of mass of the molecule can be separated off in the Born-Oppenheimer approximation (8.40), as it could be done in the case of the exact Hamiltonian (8.1). The eigenenergies ε of (8.40) contain therefore a trivial contribution, the translational kinetic energy $p^2/2m$, which must be subtracted when we construct the required energies ε_q contributing to $\zeta(T)$.

For instance, for a *diatomic molecule* such as HCl the energy W depends only on the distance ϱ between the two nuclei. If for the moment we disregard the direct Coulomb interaction between the nuclei, the energy of the ground state of the 18 electrons is negative; it increases with ϱ from the binding energy of an atom of charge 18 – the combined charge of the Cl and H nuclei – for $\varrho = 0$, to the sum of the binding energies of the two separate, Cl and H, atoms for $\varrho = \infty$. To obtain W we must add to this function the repulsion between the H and Cl nuclei which becomes very large as $\varrho \to 0$. The result is the curve $W(\varrho)$ shown in Fig.8.2 where we dropped an additive constant. At small distances apart the direct repulsion dominates; at larger distances apart the binding energy of the electrons becomes dominant and $W(\varrho)$ shows a *pronounced minimum* near some value $\varrho = \overline{\varrho}$. The energy $W(\varrho)$ plays the rôle of an *effective interaction potential* for the nuclei in the Schrödinger equation (8.40), where the total energy of the system is equal to $W(\varrho)$ plus the kinetic energy $T_{\rm n}$ of the nuclei. The latter can be split into a sum of two terms,

$$\widehat{T}_{\rm n} = \frac{\widehat{\boldsymbol{p}}^2}{2m} + \frac{\widehat{\boldsymbol{\pi}}^2}{2\mu}, \tag{8.41}$$

where $\widehat{\boldsymbol{p}}$ is the momentum of the centre of mass of the diatomic molecule, m its total mass, $\widehat{\boldsymbol{\pi}}$ the relative momentum of the two nuclei, the masses of which are μ_1 and μ_2, and μ the reduced mass

$$\mu = \frac{\mu_1 \mu_2}{\mu_1 + \mu_2}. \tag{8.42}$$

Finally, if we drop, as in (8.1), the translational kinetic energy of the molecule, there remains for us the task to solve a Schrödinger equation (8.40), where the coordinates of the electrons and of the centre of mass have been eliminated, with an *effective internal Hamiltonian*

$$\widehat{h} \simeq \frac{\widehat{\boldsymbol{\pi}}^2}{2\mu} + W(\widehat{\varrho}), \tag{8.43}$$

which is the same as that of a single particle with coordinates $\widehat{\boldsymbol{\varrho}}$ in a central potential $W(\varrho)$.

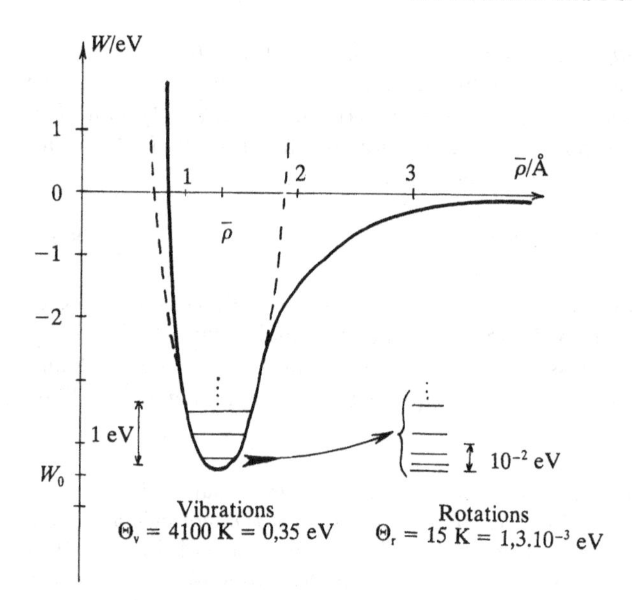

Fig. 8.2. Energy levels of the HCl molecule

The eigenvalues ε_q of (8.43) are found by *separating the angular and the radial variables* and they are characterized by the quantum numbers $q = l, m, n$ where l has a multiplicity $2l+1$ connected with the quantum number m. Moreover, we should include the quantum numbers of the nuclear spins which may give rise to additional degeneracies – or to quasi-degeneracies, as the magnetic interactions of those spins are negligibly small. In the rest of this chapter we shall study the *rotational* motion (associated with the quantum numbers l and m) and the *vibrational* motion (associated with the radial quantum number n) and their thermodynamic consequences.

A study of the quantum harmonic oscillator shows that the vibrational frequencies $\omega/2\pi$, and thus the spacing $\hbar\omega$ of the levels, are for a given potential inversely proportional to the square root of the mass of the oscillator. As the masses of the nuclei are much larger than the electron mass, one expects that the energy levels of $\widehat{h}$, associated with the relative motion of the nuclei, are much more closely spaced than the excited levels of the electron cloud – which we have justifiably assumed to be frozen in into its ground state. In fact, a numerical estimate of the inertia coefficients for the rotations and vibrations of diatomic molecules, defined by (8.63), shows that they are usually rather large. Hence the corresponding levels lie densely and the *characteristic rotation and vibration temperatures are much lower than the electronic characteristic temperatures.* For instance, for HCl the characteristic rotation temperature is $\Theta_r = 15$ K, and the vibration temperature $\Theta_v = 4100$ K, corresponding, respectively, to excitation energies of the order of 10^{-3} eV and 0.35 eV.

If the gas is at a *temperature well above these characteristic temperatures* quantization of the levels does not play any rôle as they lie densely on the scale kT; one can therefore treat the effective Hamiltonian (8.43) as a *classical Hamiltonian* and replace the calculation of the trace in (8.4) by an integration, as we saw in §2.3.2, which leads to

$$\zeta^{\rm cl} = \int \frac{d^3\boldsymbol{\pi} d^3\boldsymbol{\varrho}}{h^3} \, e^{-h(\boldsymbol{\pi},\boldsymbol{\varrho})/kT}. \tag{8.44}$$

As a first approximation, valid at temperatures which are high as compared to $\Theta_{\rm r}$ and $\Theta_{\rm v}$, we shall thus treat in § 8.4.3 the internal molecular rotational and vibrational variables by classical statistical mechanics using (8.44). Before doing this we shall prove the energy equipartition theorem which will be useful in that analysis.

Expression (8.44) must in actual cases be multiplied by a multiplicity factor g similar to the one in § 8.3.2. Moreover, if the two atoms of the molecule are *indistinguishable* we must introduce a factor $\frac{1}{2}$, which is a special case of the factor $1/S$ of (2.59), to compensate for the fact that a single configuration of the molecule is represented by two different points in phase space, $\boldsymbol{\pi}$, $\boldsymbol{\varrho}$ and $-\boldsymbol{\pi}$, $-\boldsymbol{\varrho}$. These constant factors do not affect the specific heats, but appear, for instance, in expression (8.14) for the entropy and in the mass action law through (8.30); they thus play an important rôle in chemical thermodynamics. Ehrenfest and Trkal recognized the importance of the symmetry factor S in this context in 1921.

8.4.2 The Energy Equipartition Theorem

One of the problems of statistical mechanics consists in determining *how the energy of a system is distributed* over its various degrees of freedom. We have seen that the general answer to this question is obtained by writing down that the temperatures associated with the independent degrees of freedom become equal. The result takes a particularly simple form for all problems *in classical statistical mechanics where the Hamiltonian is quadratic* in each of the phase space variables which occur in it. Let $x_1, \ldots, x_n$ be those variables; they can be either coordinates or momenta. The Hamiltonian is supposed to be a sum of n terms of the form

$$H = \sum_{j=1}^{n} h_j = \sum_{j=1}^{n} \frac{1}{2}\alpha_j x_j^2; \tag{8.45}$$

the α_j are arbitrary positive constants, which can be interpreted as elastic force coefficients if x_j is a position coordinate, and as inverse masses or inertia coefficients if x_j is a momentum. In thermal equilibrium the internal energy $U = \langle H \rangle$ is the sum of the average energies $\langle h_j \rangle$ associated with the n degrees of freedom.

The energy equipartition theorem states that under those conditions the *internal energy per degree of freedom* $\langle h_j \rangle$ *is equal to* $\frac{1}{2}kT$, whatever the value of the constants α_j:

$$\langle h_1 \rangle = \ldots = \langle h_n \rangle = \frac{1}{2} kT = \frac{U}{n}. \tag{8.46}$$

The total energy is *equally distributed* over all degrees of freedom and it is proportional to the temperature.

To prove this result we calculate the classical canonical partition function corresponding to (8.45), which has the form

$$\begin{aligned} \zeta^{\text{cl}} &= a \int_{-\infty}^{+\infty} dx_1 \ldots dx_n \exp\left[-\frac{1}{2}\beta \sum_j \alpha_j x_j^2\right] \\ &= a \prod_{j=1}^{n} \sqrt{\frac{2\pi}{\beta \alpha_j}}. \end{aligned} \tag{8.47}$$

The factor a comes from the coefficients in (2.55) or (2.59). We get the $\langle h_j \rangle$ through taking the appropriate derivatives:

$$\langle h_j \rangle = -\frac{1}{\beta} \alpha_j \frac{\partial}{\partial \alpha_j} \ln \zeta^{\text{cl}} = \frac{1}{2} kT. \qquad QED$$

The *variance* of the h_j similarly has a universal form since, according to (4.13), it is given by

$$\Delta h_j^2 = \frac{1}{\beta^2} \alpha_j^2 \frac{\partial^2}{\partial \alpha_j^2} \ln \zeta^{\text{cl}} = \frac{1}{2} (kT)^2 = 2 \langle h_j \rangle^2. \tag{8.48}$$

If we add all degrees of freedom, we get

$$\frac{\langle (\widehat{H} - U)^2 \rangle}{U^2} = \frac{2}{n}. \tag{8.49}$$

The equipartition theorem remains valid, if the system depends on other variables $y_1, \ldots, y_m$ which do not occur in the Hamiltonian and which have values in a finite domain. It can even be generalized to the case when the coefficients α_j *depend on these extra variables* $y_1, \ldots, y_m$, as one can see from (8.47) by first integrating over the $x_1, \ldots, x_n$ variables. One should, however, take care, in practical applications of the equipartition theorem, *not to include the y-kind variables among the n degrees of freedom.* To avoid any mistakes in the counting it is advisable to check that the Hamiltonian actually has the form (8.45).

For instance, in the case of a perfect gas, the $3N$ position variables $\boldsymbol{r}_1, \ldots, \boldsymbol{r}_N$ do not contribute, as we must freely integrate over them; there remain the $3N$ classical quadratic translational degrees of freedom corresponding to the momenta $\boldsymbol{p}_1, \ldots, \boldsymbol{p}_N$, and we find again the internal energy (7.44) and the average kinetic energy per molecule (7.24). Another example was given in § 5.7.3: the oscillating mirror is a classical one-dimensional harmonic oscillator with a Hamiltonian depending on the two conjugated variables θ and p_θ. Its average energy is kT in thermal equilibrium with its kinetic and potential energies being equal to $\frac{1}{2}kT$ in accordance with (5.91). In this case, θ does not vary from $-\infty$ to $+\infty$ like the x_j in (8.47). This has no consequences, since the probability law (5.90) is concentrated on θ-values not too large as compared to $(kT/C)^{1/2}$ and Exerc.5e has shown that this is small. It is thus legitimate to extend the integration range for θ from $-\infty$ to $+\infty$ in the calculation of the partition function.

We should, nevertheless, not forget the conditions for the applicability of the equipartition theorem which is restricted to *classical* statistical mechanics and to *quadratic* Hamiltonians. In particular, the resulting specific heats do not depend on the temperature, a property which is necessarily violated at low temperatures, as the corresponding entropy,

$$S = \int^T C \frac{dT}{T}, \tag{8.50}$$

would tend to $-\infty$ at zero temperature. We thus see again that it is necessary to use quantum mechanics at low temperatures in order to explain the fact that specific heats vanish, so that the integral (8.50) can converge at $T = 0$. During the last quarter of the nineteenth century the disagreement of (8.46) with specific heat measurements on polyatomic gases provided the opponents of kinetic theory with one of their main objections. In fact, we shall see below that the reason for this disagreement was not statistical mechanics, but its classical approximation: one needs to understand the quantization of the energy levels to explain why the vibrational degrees of freedom are frozen in – the main source of the violation of the equipartition theorem. The high-temperature behaviour of the specific heat of hydrogen (Figs.8.3 and 8.5) exemplifies another cause for violations: anharmonicity.

8.4.3 Classical Treatment of Polyatomic Gases

Let us return to expression (8.44) for the internal partition function of a *diatomic molecule* in the classical limit. The temperature is assumed to be high, but not high enough for the distance apart ϱ to differ appreciably from its equilibrium value $\overline{\varrho}$ where the potential W has a minimum. We can therefore replace the latter by its quadratic approximation, a harmonic well centred around $\overline{\varrho}$,

$$W(\varrho) \simeq W_0 + \tfrac{1}{2}\mu\omega^2(\varrho - \overline{\varrho})^2, \tag{8.51}$$

shown by the dashed line in Fig.8.2, and characterized by $W_0 = \varepsilon_0$, the minimum energy in classical mechanics, and by the coefficient ω which follows from the curvature of the potential. We thus obtain from (8.51) an approximation for the internal partition function:

$$\begin{aligned}
\zeta^{\mathrm{cl}}(T) &= \frac{4\pi\overline{\varrho}^2}{h^3} \int d^3\boldsymbol{\pi}\, d\varrho \, \exp\left[-\frac{\pi^2}{2\mu kT} - \frac{\mu\omega^2(\varrho-\overline{\varrho})^2}{2kT} - \frac{\varepsilon_0}{kT}\right] \\
&= \frac{T}{\Theta_{\mathrm{r}}} \frac{T}{\Theta_{\mathrm{v}}} \, \mathrm{e}^{-\varepsilon_0/kT}, \qquad (8.52) \\
\Theta_{\mathrm{r}} &\equiv \frac{\hbar^2}{2\mu\overline{\varrho}^2 k}, \qquad \Theta_{\mathrm{v}} \equiv \frac{\hbar\omega}{k}, \qquad \Theta_{\mathrm{r}} \ll \Theta_{\mathrm{v}} \ll T.
\end{aligned}$$

The integration over ϱ, which is concentrated around $\overline{\varrho}$, could be extended to $-\infty$. By substituting (8.52) into (8.10) we find the classical contribution from the rotations and vibrations of the molecules to the internal energy,

$$U^{\mathrm{cl}}_{\mathrm{r,v}} = NkT^2 \frac{d}{dT} \ln \zeta^{\mathrm{cl}} = N\varepsilon_0 + 2NkT, \qquad (8.53)$$

and thus to the specific heat, which is $2Nk$. As a result, for a *diatomic* gas at temperatures *above the rotation and vibration characteristic temperatures*, but below the electronic characteristic temperature, the total specific heat at constant volume is equal to

$$C_{\mathrm{v}} = \tfrac{7}{2}kT. \qquad (8.54)$$

One could have derived this result directly from the equipartition theorem. In fact, the internal partition function (8.52) depends on four variables, $\boldsymbol{\pi}$ and ϱ, which occur quadratically in the Hamiltonian and thus provide a contribution $2kT$ to the the internal energy per molecule, which must be added to the translational energy $\frac{3}{2}kT$. Note that the two angular variables θ and φ, which characterize the orientation of the molecule, do not occur in the Hamiltonian and therefore do not count when we apply the equipartition theorem.

However, for most diatomic molecules room temperature lies *between* the characteristic temperatures for rotation and for vibration, which we gave in § 8.4.1 for the case of HCl, and *one does not observe the value (8.54) for the specific heat.* As $T \gg \Theta_{\mathrm{r}}$ the rotational levels are still lying densely in the temperature region considered: in Fig.8.2 the rotational levels shown on top of the vibrational ground state level are drawn on a much expanded scale. Thus we can still treat the molecular rotations classically. On the other hand, as $T \ll \Theta_{\mathrm{v}}$ *the vibrational degrees of freedom remain frozen in*: if we consider the Schrödinger equation (8.59) for ϱ and its conjugate momentum, only the radial wavefunction corresponding to the lowest energy needs be considered, and ϱ is practically restricted to the value $\overline{\varrho}$, within a margin $\sqrt{\hbar/2\mu\omega} \ll \overline{\varrho}$. The effective Hamiltonian (8.43) now only depends on the angular variables θ and φ, characterizing the orientation of $\boldsymbol{\varrho}$, and on their conjugate momenta p_θ and p_φ; we can treat it in the framework of classical statistical mechanics, since $T \gg \Theta_{\mathrm{r}}$. The molecule can thus be idealized as a dumbbell, a *classical rigid linear rotator* consisting of the two nuclei at a fixed distance apart $\overline{\varrho}$ and being able to rotate around the centre of mass. The Hamiltonian (8.43) reduces to the classical rotational Hamiltonian,

$$h = \frac{1}{2\mathcal{I}} \left(p_\theta^2 + \frac{p_\varphi^2}{\sin^2\theta} \right) + \varepsilon_0, \tag{8.55}$$

where the moment of inertia $\mathcal{I}$ equals $\mu\overline{\varrho}^2$. Hence we find the approximate internal partition function

$$\begin{aligned} \zeta^{\rm cl}(T) &= \int \frac{d\theta\, d\varphi\, dp_\theta\, dp_\varphi}{h^2} \exp\left\{ - \left[p_\theta^2 + \frac{p_\varphi^2}{\sin^2\theta} \right] \Big/ 2\mathcal{I}kT - \frac{\varepsilon_0}{kT} \right\} \\ &= \frac{T}{\Theta_{\rm r}} \, {\rm e}^{-\varepsilon_0/kT}, \qquad \frac{\hbar^2}{2\mathcal{I}k} \equiv \Theta_{\rm r} \ll T \ll \Theta_{\rm v}; \end{aligned} \tag{8.56}$$

from it we get the rotational specific heat of the molecules, Nk, and hence the total specific heat,

$$C_{\rm v} = \tfrac{5}{2} Nk, \tag{8.57}$$

in agreement with most experiments carried out on *diatomic gases at room temperature.*

Here again, (8.57) follows directly from the equipartition theorem. At the temperatures considered, $\Theta_{\rm r} \ll T \ll \Theta_{\rm v}$, only the rotational degrees of freedom are unfrozen and they are treated classically. The internal partition function (8.56) per molecule involves two variables, p_θ and p_φ, quadratically; the variables θ and φ do not contribute, even though θ occurs in the coefficient of p_φ^2, since they are y-type variables, as defined in § 8.4.2. These two rotational degrees of freedom per molecule are added to the three translational degrees of freedom to give a total internal energy of $\frac{5}{2}NkT$ and the specific heat (8.57).

In the case of a *polyatomic gas*, consisting of molecules containing three or more atoms, there are more vibrational degrees of freedom, but they usually remain frozen in at room temperature. On the other hand, the characteristic rotational temperatures are very low, as the moments of inertia increase with the size of the molecule and with the mass of the constituents; they are of the order of 10 K for NH_3. Under those conditions, the molecule behaves as a classical rigid body which can rotate, and its orientation depends on the three Euler angles ψ, θ, φ. The Hamiltonian depends quadratically on the three conjugate momenta p_ψ, p_θ, p_φ, which are the three degrees of freedom contributing to the equipartition theorem. Hence we find a total specific heat equal to

$$C_{\rm v} = 3Nk. \tag{8.58}$$

8.4.4 Quantum Treatment of Diatomic Gases

We know how to calculate the thermal properties of gases in the two extreme cases, when the temperature is much lower than the characteristic temperatures of certain internal degrees of freedom of the molecules, which remain frozen in, or much higher, in which case these degrees of freedom give a constant contribution to the specific heat. In the intermediate regions, where the temperature is of the order of a particular characteristic temperature, we can find the behaviour of the specific heat only by evaluating the internal partition function $\zeta(T)$ in the framework of quantum mechanics.

A correct quantum treatment must take into account the *discrete nature of the energy levels.* This we have done in preceding sections for the electron motion, but we must still solve the Schrödinger equation with the effective Hamiltonian (8.43) of the nuclei, for temperature values where the semi-classical approximation of § 8.4.3 is no longer valid.

As in § 8.3.2, the magnetic coupling between the atomic and nuclear spins can be neglected and the effective interaction potential between the nuclei is practically independent of the spins. Nevertheless, we must, as in the case of the monatomic gases, take the nuclear spins into account for the calculation of the multiplicity of the energy levels; the latter can also include a factor due to the degeneracy of the ground state level of the electron cloud.

A last important quantum effect for molecules is the *Pauli principle* (Chap. 10). In § 8.1.1 we have stressed that the eigenenergies ε_q involved in the calculation of ζ are associated with wavefunctions which must have a well defined symmetry character with respect to the exchange of the coordinates of indistinguishable particles. For electrons, the antisymmetry of the eigenfunctions of (8.39) is essential in the calculation of the shape of the effective potential $W(\varrho)$ between the nuclei. The *indistinguishability of the nuclei* shows up more directly for a molecule containing identical atoms: the eigenfunctions of the effective Hamiltonian of the nuclei, (8.40), must be *symmetric* under an exchange of those nuclei, if they are *bosons*, that is, if the total number of protons and neutrons in each of them is even so that they have integer spin, and *antisymmetric* if they are *fermions* – nuclei with an odd number of nucleons and a half-odd-integral spin.

Let us first of all consider *diatomic molecules with distinguishable atoms*, such as HCl, with wavefunctions which are not subject to any symmetry restrictions. The Hamiltonian (8.43) of the relative motion of the nuclei is invariant under rotation; its eigenfunctions $\psi(\boldsymbol{\varrho})$, characterized by the three quantum numbers l, m, n, can therefore be split into a product of a spherical harmonic $Y_l^m(\theta,\varphi)$ which is an eigenfunction of the angular Laplacian with a radial function $R_{ln}(\varrho)$. We must still solve the *radial* eigenvalue equation

$$\left[-\frac{\hbar^2}{2\mu}\left(\frac{d^2}{d\varrho^2}+\frac{2}{\varrho}\frac{d}{d\varrho}-\frac{l(l+1)}{\varrho^2}\right)+W(\varrho)\right]R_{ln}(\varrho) \quad = \quad \varepsilon_{ln}R_{ln}. \qquad (8.59)$$

The eigenfunctions of the lowest states, the only ones involved in the evaluation of ζ, remain localized near $\varrho=\overline{\varrho}$. We can thus replace the effective interaction $W(\varrho)$ by (8.51) and the centrifugal potential $\hbar^2 l(l+1)/2\mu\varrho^2$ by its average value $\hbar^2 l(l+1)/2\mathcal{I}$, where $\mathcal{I}$ is the moment of inertia $\mu\overline{\varrho}^2$. Putting $x = \varrho - \overline{\varrho}$, a variable which we can extend to $-\infty$ as the wavefunction is practically equal to zero when $\varrho \ll \overline{\varrho}$, and introducing $\Phi_{ln}(x) = \varrho R_{ln}(\varrho)$, we can rewrite (8.59) in the form

$$\left[-\frac{\hbar^2}{2\mu}\frac{d^2}{dx^2} + \frac{1}{2}\mu\omega^2 x^2\right]\Phi_{ln}(x) = \left[\varepsilon_{ln} - W_0 - \frac{\hbar^2 l(l+1)}{2\mathcal{I}}\right]\Phi_{ln}(x). \tag{8.60}$$

On the left-hand side we see the Hamiltonian of a one-dimensional harmonic oscillator with eigenvalues $\hbar\omega(n+\frac{1}{2})$.

As a result the *energy levels* are given by

$$\varepsilon_{ln} = \varepsilon_0 + l(l+1)k\Theta_r + nk\Theta_v, \tag{8.61}$$

where we have introduced the ground state energy

$$\varepsilon_0 = W_0 + \tfrac{1}{2}\hbar\omega, \tag{8.62}$$

and the rotational and vibrational characteristic temperatures,

$$\Theta_r \equiv \frac{\hbar^2}{2\mathcal{I}k} = \frac{\hbar^2}{2\mu\overline{\varrho}^2 k}, \qquad \Theta_v \equiv \frac{\hbar\omega}{k}. \tag{8.63}$$

They are characterized by the quantum number $l = 0, 1, 2, \ldots$ of the orbital angular momentum, which describes the *rotation* of the diatomic molecule, and the radial *vibrational* quantum number $n = 0, 1, 2, \ldots$, which is associated with the oscillations about $\overline{\varrho}$ of the distance between the nuclei. The second angular quantum number m gives rise to a multiplicity $2l+1$ for each level (8.61). We have schematically shown these levels for HCl in Fig.8.2; as Θ_r is much lower than Θ_v, the levels are organized in rotational bands which are repeated regularly on top of each vibrational level with zero angular momentum.

The internal partition function following from (8.61) turns out to be a product of several factors:

$$\zeta(T) = g\,\zeta_r(T)\,\zeta_v(T)\,e^{-\varepsilon_0/kT}, \tag{8.64}$$

with

$$g = (2s_1+1)(2s_2+1)g_e, \tag{8.65}$$

$$\zeta_r(T) = \sum_{l=0}^{\infty}(2l+1)\,e^{-l(l+1)\Theta_r/T}, \tag{8.66}$$

$$\zeta_v(T) = \sum_{n=0}^{\infty} e^{-n\Theta_v/T} = \frac{1}{1-e^{-\Theta_v/T}}, \tag{8.67}$$

where g is the degeneracy due to the spins s_1 and s_1 of the nuclei and to the electronic state. These equations cover the various regimes studied in the earlier approximations. When $T \ll \Theta_v$, the vibrations are *frozen in* and $\zeta_v = 1$; when $T \ll \Theta_r$ the rotations are frozen in, and $\zeta_r = 1$. On the other hand, if $T \gg \Theta_r$, (8.66) tends to the integral

$$\begin{aligned} \zeta_r &\to \int_0^\infty dl\,(2l+1)\,\mathrm{e}^{-l(l+1)\Theta_r/T} \\ &= \int_0^\infty dx\,\mathrm{e}^{-x\Theta_r/T} = \frac{T}{\Theta_r}, \end{aligned} \tag{8.68}$$

as the sum is over a large number of levels; similarly, if $T \gg \Theta_v$, (8.67) tends to

$$\zeta_v \to \frac{T}{\Theta_v}. \tag{8.69}$$

Using the definitions (8.63), one checks easily that these two limits lead to the classical expressions (8.52) and (8.56).

The resulting specific heat(8.12) can be split into a sum of three terms, the *translational* specific heat, the same as for a perfect gas, the *rotational* specific heat,

$$C_{\text{rot}} = \frac{d}{dT}\left(T^2 \frac{d}{dT} \ln \zeta_r\right) Nk, \tag{8.70}$$

and the *vibrational* specific heat,

$$C_{\text{vib}} = \frac{d}{dT}\left(T^2 \frac{d}{dT} \ln \zeta_v\right) Nk = \left[\frac{\Theta_v/2T}{\sinh(\Theta_v/2T)}\right]^2 Nk. \tag{8.71}$$

The last two both tend to zero at low temperatures and to their semi-classical value Nk at high temperatures. Their form, though, is different; in particular, (8.71) increases monotonically, whereas (8.70) has a maximum. We show the results in Fig.8.3 for HD, where D is the deuteron, the hydrogen isotope consisting of a proton and a neutron. At very low temperatures, just above the liquefaction temperature, only the three translational degrees of freedom contribute. As T increases one observes the defreezing of the rotations, with C_v changing from $\frac{3}{2}Nk$ to its semi-classical value $\frac{5}{2}Nk$ at room temperatures ($\Theta_r \ll T \ll \Theta_v$). The vibrations in turn contribute at very high temperatures, where the experimental results start to differ from the simple theory given above: at those temperatures one should, in fact, take the anharmonicity of the potential $W(\varrho)$ into account.

For heavier diatomic gases we have always $\Theta_r \ll T$; the rotations can be treated classically and always give a contribution Nk. Similarly, the Θ_v temperatures are lower than for hydrogen (2200 K for O_2) and theory gives us the curve (8.71) when T becomes of the order of Θ_v. Figure 8.4 shows excellent agreement with experimental data.

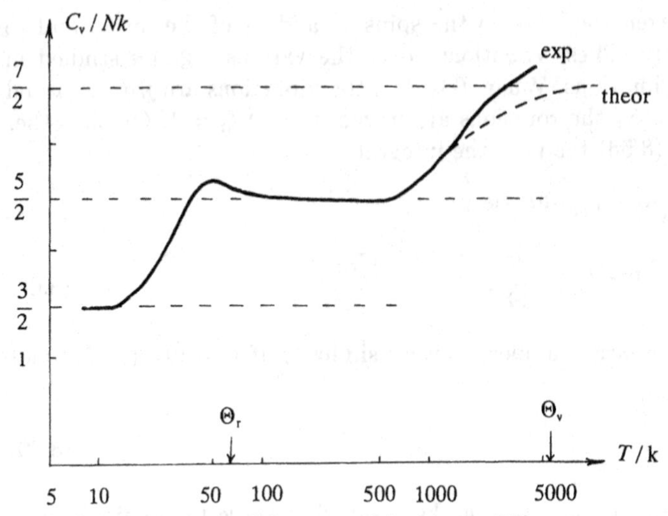

Fig. 8.3. Specific heat of deuterated hydrogen, HD

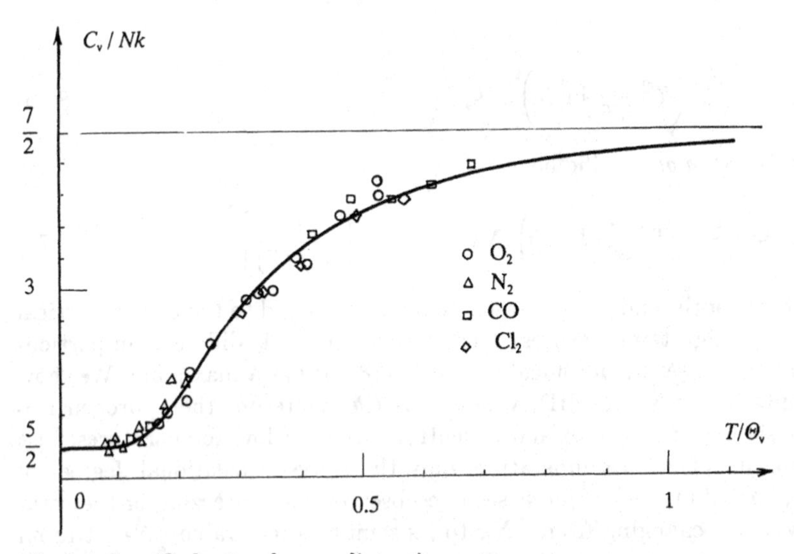

Fig. 8.4. Specific heats of some diatomic gases

Figure 8.4 includes the O_2, N_2, and Cl_2 molecules which have *indistinguishable atoms*. We shall see in § 8.4.5 that this changes the form of $\zeta_r(T)$. However, when $T \gg \Theta_r$, this change reduces to dividing (8.66) by a factor 2 which does not alter the theoretical curve in Fig.8.4.

8.4.5 The Case of Hydrogen

The effect of the *indistinguishability of the atoms* which make up each molecule of a gas shows up strikingly in the specific heats of the various hydrogen isotopes. Let us, for instance, compare HD and H_2. The interactions between electrons and nuclei are the same for the two molecules. The nuclear spins differ, but the resulting change in the hyperfine coupling can at the temperatures considered be completely neglected. The reduced masses μ differ by a factor $\frac{4}{3}$ which results in a slight shift of the characteristic temperatures: since W remains unchanged, Θ_r varies as μ^{-1}, and Θ_v as $\mu^{-1/2}$. However, these differences are insufficient to explain the large difference in the shape of the rotational specific heats of HD and H_2 which are shown in Figs.8.3 and 8.5. The important effect here is the fact that the two nuclei in H_2 are identical. These two nuclei, simple spin-$\frac{1}{2}$ protons, are fermions and the wavefunctions must therefore be antisymmetric under their exchange (Chap.10); this changes the counting of the eigenstates of (8.43).

The eigenfunctions of the molecular Hamiltonian are the product of a spin part and a space part, which itself is a function of the relative position $\boldsymbol{\varrho}$ of the nuclei, of the form $R_{ln}(\varrho)Y_l^m(\theta,\varphi)$. Let us, to start with, assume that the two nuclear spins are coupled into a singlet state; this function is antisymmetric under an exchange of the spin coordinates, and the space part of the wavefunction must therefore be symmetric under an exchange of the space coordinates. The latter changes $\boldsymbol{\varrho}$ to $-\boldsymbol{\varrho}$ in the wavefunction; under this space reflexion the spherical harmonics transform according to

$$Y_l^m(\pi-\theta,\varphi+\pi) \;=\; (-)^l\, Y_l^m(\theta,\varphi),$$

whereas R_{ln} remains unchanged. We must therefore retain only those wavefunctions for which l is *even* when the nuclear spins are in the *singlet* state. One usually calls all those states of the hydrogen molecule *parahydrogen* and they lead to the following contribution to the internal partition function:

$$\zeta_{\text{para}}(T) \;=\; \sum_{l=0,2,4,\ldots} (2l+1)\,\mathrm{e}^{-l(l+1)\Theta_r/T}, \tag{8.72}$$

while the vibrations remain frozen in at the temperatures considered. Similarly, when the nuclear spins are coupled into one of the three *triplet* states, the spin wavefunction is symmetric, and the spatial wavefunction must be antisymmetric and therefore have *odd* l values. This second family of molecular hydrogen states is called *orthohydrogen* and makes the following contribution to the internal partition function:

$$\zeta_{\text{ortho}}(T) \;=\; 3 \sum_{l=1,3,5,\ldots} (2l+1)\,\mathrm{e}^{-l(l+1)\Theta_r/T}, \tag{8.73}$$

where the factor 3 is due to the degeneracy of the spin states. The total internal rotational partition function,

$$\zeta(T) \;=\; \zeta_{\text{para}}(T) + \zeta_{\text{ortho}}(T), \tag{8.74}$$

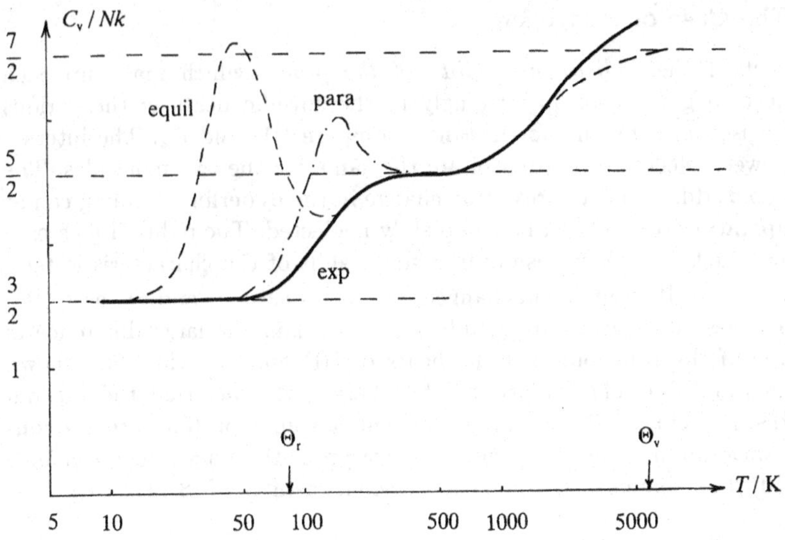

Fig. 8.5. Specific heat of hydrogen

differs from the HD one which equals $6\zeta_{\text{para}} + 2\zeta_{\text{ortho}}$, as the deuteron has spin 1. For D_2 we would find $6\zeta_{\text{para}} + \zeta_{\text{ortho}}$. From (8.74) it follows that for $T \ll \Theta_v$ the specific heat is equal to

$$\frac{C_v}{Nk} = \frac{3}{2} + \frac{d}{dT}\left[T^2 \frac{d}{dT} \ln\left(\zeta_{\text{para}} + \zeta_{\text{ortho}}\right)\right], \tag{8.75}$$

which is shown by the dashed line in Fig.8.5 and which exhibits a much more pronounced maximum than for HD. The agreement with the full drawn experimental curve is clearly extremely poor!

To resolve this puzzle, let us consider in more detail the mechanisms which bring the gas to equilibrium at temperatures where we want to measure the specific heat. Energy exchanges between molecules take place when they collide; they are the more difficult, the weaker the interactions between the molecules. In particular, whereas transitions between the various excited $l = 0, 2, 4, \ldots$ states of a parahydrogen molecule take place easily during a collision with another molecule, transitions from para- to orthohydrogen, or the other way round, are very rare: changing from a singlet to a triplet state needs, in fact, the flipping of one of the two nuclear spins, and we know that the interactions involving these spins are extremely weak. As a consequence, if hydrogen is cooled down without special precautions it will *not have time to reach equilibrium as far as the proportions of ortho- and parahydrogen are concerned*, and these proportions will remain fixed at their room temperature equilibrium values. Actually, it needs several days to attain equilibrium.

We can easily calculate these proportions. Each eigenstate of $\widehat{h}$, with energy ε_q, has a probability given by the Boltzmann-Gibbs distribution,

$$\frac{1}{\zeta}\, e^{-\varepsilon_q/kT}.$$

Hence, the numbers of molecules, N_{para} and N_{ortho}, of molecules in the states of para- or of orthohydrogen are, respectively,

$$N_{\text{para}} = N \frac{\zeta_{\text{para}}}{\zeta}, \qquad N_{\text{ortho}} = N \frac{\zeta_{\text{ortho}}}{\zeta}, \tag{8.76}$$

for a gas in *thermal equilibrium* at a temperature T. We could, of course, also have written down these relations directly by noting that the equilibrium for the reaction

$$H_2^{\text{para}} \leftrightarrows H_2^{\text{ortho}}$$

is a kind of chemical equilibrium with equal chemical potentials $\mu^{\text{para}} = \mu^{\text{ortho}}$. We can thus apply the mass action law (8.29), (8.30), which here reduces to

$$\frac{N_{\text{para}}}{N_{\text{ortho}}} = \frac{\zeta_{\text{para}}}{\zeta_{\text{ortho}}}.$$

In the case when $T \gg \Theta_{\text{r}}$ the two sums over l in (8.72) and (8.73) both tend to half the semi-classical integral (8.68). We see thus that the indistinguishability of the two nuclei introduces, in accordance with (2.59), a factor $1/\mathcal{S}$ with the symmetry number $\mathcal{S}$ being equal to 2. The factor 3 in (8.73) remains and describes the multiplicity of the nuclear spin states, so that (8.76) for $T \gg \Theta_{\text{r}}$ reduces to

$$\frac{N_{\text{para}}}{N} = \frac{1}{4}, \qquad \frac{N_{\text{ortho}}}{N} = \frac{3}{4}. \tag{8.77}$$

On the other hand, in the limit when $T \ll \Theta_{\text{r}}$, only the lowest state, with $l = 0$, contributes to ζ so that at very low temperatures the gas consists solely of parahydrogen, if it is at equilibrium. The equilibrium curve (8.75) takes these large changes in the proportions of para- and orthohydrogen into account. However, if one assumes that these proportions remain constant and equal to (8.77) when the temperature is lowered, one must compare the experimental full drawn curve with the specific heat of a *mixture consisting of the constant proportions* $\frac{1}{4}$ and $\frac{3}{4}$ of para- and orthohydrogen, that is, one must compare it with

$$\frac{C_{\text{V}}}{Nk} = \frac{3}{2} + \frac{d}{dT}\left[\frac{1}{4}T^2 \frac{d}{dT} \ln \zeta_{\text{para}} + \frac{3}{4}T^2 \frac{d}{dT} \ln \zeta_{\text{ortho}}\right], \tag{8.78}$$

instead of (8.75), and now the agreement between theory and experiment is excellent. While N_{para} and N_{ortho} remain frozen in into their values (8.77), as T decreases, their associated chemical potentials (8.13) move apart according to

$$\mu^{\text{ortho}} - \mu^{\text{para}} = kT \ln \frac{3\zeta_{\text{para}}}{\zeta_{\text{ortho}}} \underset{T \ll \Theta_{\text{r}}}{\longrightarrow} 2k\Theta_{\text{r}} - kT \ln 3;$$

such a discrepancy shows that when hydrogen is cooled down, it no longer remains in a true equilibrium state.

This explanation has been confirmed in several ways. First of all, one can repeat the measurements in the presence of a *catalyst* which accelerates the transitions between the parahydrogen and the orthohydrogen states; in that case one observes the equilibrium specific heat (8.75) shown by the dashed curve in Fig.8.5. One can also start at low temperatures where the equilibrium form is pure parahydrogen and *heat* the gas: in that case one observes the specific heat of pure parahydrogen,

$$\frac{C_V}{Nk} = \frac{3}{2} + \frac{d}{dT}\left(T^2 \frac{d}{dT} \ln \zeta_{\text{para}}\right), \tag{8.79}$$

which is shown by the dash-dot curve in Fig.8.5 and which differs quite considerably from both the equilibrium curve (8.75) and from the curve (8.78) that we find when we cool the gas down. Finally, spectroscopy enables us to determine the average occupations of the various quantum states of the molecules through measuring the line intensities and to check the predicted values of N_{para} and N_{ortho} in the different cases.

This study of the specific heat of hydrogen shows the importance of the *choice of variables* characterizing an equilibrium state – or rather a quasi-equilibrium state since the so-called equilibrium states usually are strictly speaking quasi-equilibria. In § 4.1 we have mentioned this important and sometimes subtle problem. Gaseous hydrogen is a typical example where the existence of a *quasi-constant of motion*, the ratio of para- to orthohydrogen, impedes the establishment of an ordinary grand canonical equilibrium when one lowers the temperature. An appropriate description needs the introduction of this *hidden variable*, with which we must associate a new Lagrangian multiplier: we must introduce two, rather than one, chemical potentials.

Summary

The effects of the internal structure of the molecules in a rarefied gas show up through the internal partition function (8.4), which is a factor of the grand potential. The equation of state of a perfect gas and the Joule law remain valid, but the specific heats are changed.

Mixing two different gases increases the entropy without involving heat or work. Writing down the relations between the chemical potentials enables us to establish the mass action law and the van't Hoff law for chemical equilibrium in a gas.

The specific heats of gases follow once we know the lowest energy levels of their molecules, which we can calculate using the Born-Oppenheimer approximation. One can associate with each, electronic, vibrational, rotational, degree of freedom a characteristic energy of the order of magnitude of the distances between the corresponding levels, or a characteristic temperature. Each degree of freedom is frozen in at temperatures which are low as compared to the characteristic temperature; at high temperatures one can treat them in the classical approximation, and often use the equipartition theorem.

Exercises

8a Entropy and Work in Isotope Separation

Starting from natural uranium, with a molecular concentration $C = 0.7\ \%$ of the ^{235}U isotope, one uses isotope separation of the UF_6 compound to produce enriched uranium, with $C' = 3.2\ \%$ of ^{235}U, which is used as fuel in nuclear power stations. UF_6 is a gas above 55 °C and one works at $T =$ 80 °C. Together with the enriched uranium one gets depleted uranium with $C'' = 0.2\ \%$ of ^{235}U. This latter value represents an economic compromise: having C'' too large produces wasted depleted uranium which still contains a significant fraction of ^{235}U; taking C'' too small involves too many cycles of the treatment needed to reject almost pure ^{238}U.

1. How much natural uranium must be treated in order to produce 1 kg of fuel? By how much does the entropy of the UF_6 which is involved increase for given temperature and pressure?

2. What is the ideal thermodynamic yield, characterized by the minimum work required in order to produce 1 kg of enriched uranium? Compare this with the electric energy actually consumed by the Eurodif factory, which is 9 MWh per kg.

3. In order to understand the origin of such a difference, consider a single porous barrier (Exerc.7g). The pressures $\mathcal{P}_l$ and $\mathcal{P}_r$ and the temperature T are kept at fixed values; in practice $\mathcal{P}_l/\mathcal{P}_r \simeq 5$ and $\mathcal{P}_l$ does not exceed atmospheric pressure for safety reasons. Write down the change in the entropy of the system of gases on both sides of the wall resulting from the effusion of UF_6 molecules during a time interval dt. What are the amounts of work and heat received altogether by these gases? In what limit would the process become reversible?

4. We assume that the effusion is preceded by an isothermal and reversible decompression and followed by an isothermal and reversible compression of the gas on the right-hand side, in order to compare states where all gases are at the reference pressure $\mathcal{P}_l$. What is the total amount of work provided in the totality of these processes? Why is it so much larger than the minimum work indicated by thermodynamics?

Solution:

1. Denoting the ^{235}U and ^{238}U isotopes by 1 and 2, and by N' and N'' the numbers of molecules in the enriched and depleted mixtures, conservation of the two kinds of molecules gives $N = N' + N''$, $CN = C'N' + C''N''$,whence we find that $N'' = N'(C' - C)/(C - C'')$. The numbers N' and N are therefore equal to

$$\begin{aligned} N' &= \frac{N_A}{0.238(1-C)+0.235C} = 2.5 \times 10^{24}, \\ N &= \frac{C'-C''}{C-C''}\,N' = 15 \times 10^{24}, \end{aligned}$$

and from this we get the mass to be treated, namely, 6 kg. The decrease in entropy when one separates this into an enriched and a depleted mixture is the difference between the associated mixing entropies (8.19), that is,

$$\begin{aligned} \frac{\Delta S}{k} &= N'C' \ln \frac{C'}{C} + N'(1-C') \ln \frac{1-C'}{1-C} \\ &+ N''C'' \ln \frac{C''}{C} + N''(1-C'') \ln \frac{1-C''}{1-C} \\ &\simeq N' \left[C' \ln \frac{C'}{C} - \frac{C'-C}{C-C''} C'' \ln \frac{C}{C''} + \frac{1}{2}(C'-C)(C'-C'') \right], \end{aligned}$$

which gives us $\Delta S \simeq 1.3$ J K^{-1}kg^{-1}.

2. The decrease in entropy means that we must provide at least an amount of work equal to $T\Delta S \simeq 0.5$ kJ kg^{-1}, for $T \simeq 350$ K. This amount is smaller by a factor 7×10^7 than the actual amount of work provided! Notwithstanding this deplorable theoretical yield, the operation remains economically profitable: the energy consumed in enriching the uranium represents only 3 % of the electrical energy produced afterwards in a reactor with the same uranium, which is 250 MWh per kg of fuel. In the apportioning of the cost of producing nuclear fuel, the enriching process hardly represents one quarter.

3. The numbers of molecules of the two UF_6 isotopes which pass through the wall during a time dt were evaluated in Exerc.7g. Denoting by σ the area of the pores, we found

$$\begin{aligned} dN_1 &= \frac{\sigma\, dt}{\sqrt{2\pi k T m_1}} \left(\mathcal{P}_l C_l - \mathcal{P}_r C_r \right), \\ dN_2 &= \frac{\sigma\, dt}{\sqrt{2\pi k T m_2}} \left[\mathcal{P}_l (1-C_l) - \mathcal{P}_r (1-C_r) \right]. \end{aligned}$$

In evaluating the change in (8.14) summed over the two sides and over the two isotopes we find

$$dS = dN_1 \ln \frac{\mathcal{P}_l C_l}{\mathcal{P}_r C_r} + dN_2 \ln \frac{\mathcal{P}_l (1-C_l)}{\mathcal{P}_r (1-C_r)}.$$

As the temperature is fixed the internal energy remains unchanged. The work received equals

$$\begin{aligned} -\mathcal{P}_l\, d\Omega_l - \mathcal{P}_r\, d\Omega_r &= -d\left(\mathcal{P}_l \Omega_l + \mathcal{P}_r \Omega_r\right) \\ &= -d\left[(N_{1l} + N_{2l} + N_{1r} + N_{2r})\, kT \right] = 0. \end{aligned}$$

and as a result the heat received also vanishes. The transformation would thus be reversible only if we had $dS = 0$. However, because of the form of dN_1 and dN_2 both terms in dS are positive, unless $\mathcal{P}_l C_l = \mathcal{P}_r C_r$ and $\mathcal{P}_l(1-C_l) = \mathcal{P}_r(1-C_r)$. The effusion will therefore only become reversible in the limit as the *pressures* and the *concentrations* are practically *equal* at both sides; that, however, implies that the process is extremely slow and inefficient.

4. In order to make N molecules pass from a pressure $\mathcal{P}_l$ to a pressure $\mathcal{P}_r$, at a given temperature, we must provide work equal to $kTN \ln \mathcal{P}_r/\mathcal{P}_l$. The totality of the operations therefore consumes an amount of work equal to

$$\delta\mathcal{W} \;=\; kT(dN_1 + dN_2)\ \ln\,\frac{\mathcal{P}_\mathrm{l}}{\mathcal{P}_\mathrm{r}},$$

and the system gives up to the outside altogether an amount of heat equal to $-\delta Q = \delta\mathcal{W}$. The total change in entropy is thus equal to

$$dS_0 \;=\; dS + \frac{\delta Q}{T} \;=\; -k\,dN_1\ \ln\,\frac{C_\mathrm{r}}{C_\mathrm{l}} + k\,dN_2\ \ln\,\frac{1-C_\mathrm{l}}{1-C_\mathrm{r}}.$$

One can check that $-dS_0$ is the decrease in entropy associated with a partal sorting out of the molecules through effusion, at given temperature and pressure, that is, the infinitesimal analogue of ΔS which was calculated in 1. The fact that dS is positive implies that

$$\delta\mathcal{W} \;>\; \delta\mathcal{W}_0 \;\equiv\; -T\,dS_0,$$

which shows the existence of a minimum work done in the separation, $\delta\mathcal{W}_0$, in agreement with the Second Law. In order that the effusion be efficient, we need a *large difference in pressure*; $\mathcal{P}_\mathrm{r}/\mathcal{P}_\mathrm{l}$ is of the order of 5 in the Eurodif factory. Morover, the *concentrations are very close to one another* on the two sides, as we saw in Exerc.7g; in practice $C_\mathrm{r}/C_\mathrm{l} - 1 \equiv \varepsilon \sim 2\times 10^{-3}$. These are two reasons which together make $\delta\mathcal{W}/\delta\mathcal{W}_0$ very large. To lowest order we find that

$$\begin{aligned}
\frac{\delta\mathcal{W}}{kT\,dN_1} &\sim \left(1+\frac{1}{C}\right)\ \ln\,\frac{\mathcal{P}_\mathrm{l}}{\mathcal{P}_\mathrm{r}} \;\simeq\; 230,\\
\frac{\delta\mathcal{W}_0}{kT\,dN_1} &\sim \varepsilon\ \left[1-\sqrt{\frac{m_1}{m_2}} - \frac{\varepsilon}{2(1-C)}\,\frac{\mathcal{P}_\mathrm{l}+\mathcal{P}_\mathrm{r}}{\mathcal{P}_\mathrm{l}-\mathcal{P}_\mathrm{r}}\right]\\
&\sim \frac{\varepsilon}{2}\ \left[\frac{m_2-m_1}{m_2} - \varepsilon\right] \;\simeq\; 6\times 10^{-6}.
\end{aligned}$$

Additional losses in efficiency are, for instance, caused by irreversibilities connected with the viscosity of the gas flow and by the large heat exchanges which are necessary to carry away the work done by the compressors and dissipated into heat. The above numbers, calculated for a single separation barrier, are compatible with the global value obtained in 2 for the 1400 stages in the factory: most of the consumed power is not used directly for the separation itself; it serves to perform the many, rather large compressions and decompressions which are necessary in order that the gas passes through the barriers with some degree of efficiency (Exerc.14a).

8b Excitation of Atoms by Heating

The ground state of helium is a $1s^2\ {}^1S_0$ state: two electrons in the $1s$ shell, with S indicating a zero orbital angular momentum, 1 the spin multiplicity, and 0 the total angular momentum. The excitations of the lowest excited states are characterized by the wavenumbers $1/\lambda = (\varepsilon_q - \varepsilon_0)/hc$ which are given in Table 8.1, in units of cm^{-1}. Table 8.1 also indicates the multiplicities of the levels. What is the fraction of atoms in each excited state in the Earth's atmosphere, where $T = 300$ K? and in the solar atmosphere, where $T = 6000$ K?

Table 8.1. Lowest excited states of the helium atom

$2s\ ^3S_1$	$2s\ ^1S_0$	$2p\ ^3P_2$	$2p\ ^3P_1$	$2p\ ^3P_0$	$2p\ ^1P_1$
159 850	166 272	169 081	169 081	169 082	171 129

Solution. The temperature is sufficiently low for us to have $Z \sim \exp(-\beta\varepsilon_0)$ so that the fractions are equal to $g_i \exp[-\beta(\varepsilon_i - \varepsilon_0)]$, where g_i is the multiplicity of the level. At room temperature, $\exp[\beta(\varepsilon_1 - \varepsilon_0)] = \mathrm{e}^{-767} \simeq 10^{-333}$ is completely negligible, even when it is multiplied by the total number of helium atoms in the Earth's atmosphere. In the solar atmosphere there are a fraction $3\exp[-\beta(\varepsilon_1 - \varepsilon_0)] \simeq 6.8 \times 10^{-17}$ of atoms in the ^3S state, 4.8×10^{-18} in the ^1S state, 2.2×10^{-17} in all three ^3P states, which are equally populated, and 4.5×10^{-18} in the ^1P state; these are small numbers, but significant when we remember how large Avogadro's number is.

8c Ionization of a Plasma: Saha Equation

Most of the matter in the Universe is so rarefied or so hot that it is ionized, while remaining globally neutral. It thus forms what has been called the "fourth state of matter", a plasma. Let us for the sake of simplicity consider a monatomic gas; we denote by ε_0 and g_0 the energy and the degree of degeneracy of the ground state of a molecule, and assume that it has only a finite number of excited bound states, with energies ε such that $\varepsilon - \varepsilon_0 \gg kT$. Let ΔE_{i} be the first ionization energy, the threshold above which an electron can be torn from the atom, leaving behind a positively charged ion in its ground state, with multiplicity g_+. We assume that this ion satisfies the same conditions for its excited states as the atom and that the temperature is sufficiently low that one can neglect multiple ionizations. Finally, we treat the electrons produced through thermal ionization as a classical perfect gas, neglecting the Coulomb interactions between these electrons and the ions. Write down the Saha equation, which gives us the ratio n_+/n_0 of the number of ionized atoms to the number of neutral atoms as function of the electron density n_{el}. Estimate the degree of ionization of hydrogen in a gas nebula where $T \simeq 10^4$ K, $n \simeq 10^{12}$ m^{-3}, and in the solar photosphere where $T \simeq 6000$ K, $n \simeq 10^{23}$ m^{-3}.

Answers. If we use (8.34), the mass action law (8.28) reduces to Saha's equation

$$\frac{n_+}{n_0} = \frac{1}{n_{\mathrm{el}}} \frac{2g_+}{g_0} \left(\frac{m_{\mathrm{e}}kT}{2\pi\hbar^2}\right)^{3/2} \mathrm{e}^{-\Delta E_{\mathrm{i}}/kT}.$$

The degree of ionization is found by writing $n_{\mathrm{el}} = n_+$, which expresses the electrical neutrality, and $n = n_0 + n_+$. For atomic hydrogen in the nebula, with $g_0 = 4 = 2g_+$ and $\Delta E_{\mathrm{i}} = 13.6$ eV, the ionization is nearly complete, $n_+ \sim n$, as

$$\frac{n_+^2}{n n_0} = 3.4 \times 10^8 \sim \frac{n}{n_0}.$$

In the photosphere, it is small but significant, as

$$\frac{n_+^2}{nn_0} = 4.3 \times 10^{-8}, \qquad \frac{n_+}{n} \sim 2 \times 10^{-4}.$$

8d Adiabatic Transformation and Specific Heats

Consider an arbitrary gas with specific heats C_p and C_v which may be functions of the temperature. Write down the relations between changes in pressure, volume, temperature, and energy, as functions of the specific heats, during an infinitesimal adiabatic transformation.

Answer. We have

$$\frac{d\mathcal{P}}{\mathcal{P}} = -\frac{C_\mathrm{p}}{C_\mathrm{v}}\frac{d\Omega}{\Omega} = \frac{C_\mathrm{p}}{Nk}\frac{dT}{T} = \frac{C_\mathrm{p}}{C_\mathrm{v}}\frac{dU}{NkT}.$$

8e Hemoglobin

A hemoglobin molecule (Hb) behaves like a trap able to capture and release one to four oxygen molecules (O_2). To simplify the discussion we assume, except in 5, that a *single* O_2 molecule can be fixed to each of the Hb molecules and that its energy is then equal to $-\varepsilon$, where $\varepsilon = 0.65$ eV; as energy origin we choose the energy of a free molecule at rest in air. When the blood is in contact with air in the lungs a chemical equilibrium is established between O_2 fixed in the blood in the form of oxyhemoglobin HbO_2 and O_2 in the air. Let n be the number of oxygen molecules per unit volume.

1. Write down the fraction f of oxyhemoglobin molecules as function of n and the temperature T. We treat here the oxygen in the air as a perfect gas of point particles. Calculate this fraction f_0 at body temperature, 37 °C, and normal atmospheric pressure; the air in the lungs contains 15% O_2 molecules.
2. The average distance between the two nuclei in the O_2 molecule is $\bar{\varrho} =$ 1.2 Å. In air the molecules can rotate freely; in HbO_2 the orientation of O_2 is fixed. Show that taking these facts into account does not change the results of the model of 1, provided we replace at the same time ε by a new value which must be determined. In what follows we shall for the sake of simplicity continue to treat the O_2 molecules as point particles with $\varepsilon = 0.65$ eV.
3. When carbon monoxide CO is present in the air, each Hb molecule can on a given site capture either an O_2 molecule, or a CO molecule, in which case Hb can no longer take part in the transfer of O_2 to the tissues. The binding energy of CO is $\varepsilon' = 0.78$ eV. Find the new value of f for the equilibrium which is now established, as function of n and of the CO molecule density n'. Starting from what fraction of CO in the air will the fraction f of oxyhemoglobin be divided by 2? What is then the fraction f' of hemoglobin with CO attached? What happens, if the air contains 5% of CO?

4. We assume that in the tissues irrigated by the blood the dissolved oxygen still behaves like a perfect gas with a partial pressure equal to 5 Torr. Calculate the fraction f_1 of oxyhemoglobin in the blood in equilibrium with these tissues. Is the result of the model treated here physiologically satisfactory?

5. In reality each hemoglobin molecule possesses four sites on which it can fix up to four O_2 molecules; the binding energy ε_1 of a first O_2 molecule is rather small, but the presence of this molecule changes the configuration of Hb in such a way that the binding energy ε of the next molecules on the other sites is larger. One makes a model of this behaviour by assuming for the sake of simplicity that each Hb can trap either *one* or *two* O_2 molecules, but *no more*, on two distinguishable sites. The binding energy of the *first* molecule is $\varepsilon_1 = 0.38$ eV, and that of the *second* one is $\varepsilon = 0.65$ eV. Express the average number f of O_2 molecules fixed at equilibrium to each Hb molecule. Calculate the numerical values of f_0 and f_1 for the conditions of 1 (lungs) and of 4 (tissues). Draw conclusions.

6. Hemoglobin can also bind carbon dioxide gas (CO_2), but on sites different from those for oxygen. For the sake of simplicity we again assume that there is just one site per Hb molecule to bind O_2 with an energy $\varepsilon = 0.65$ eV, and one other site which can bind CO_2 with an energy $\varepsilon'' = 0.65$ eV. However, the deformation of the oxyhemoglobin molecules mentioned in 5 has another consequence: it reduces the total binding energy of CO_2HbO_2 which is not equal to $\varepsilon + \varepsilon''$, but to $\varepsilon + \varepsilon'' - \delta$, with $\delta = 0.2$ eV. Write down the respective average numbers f and f'' of O_2 and CO_2 molecules fixed per Hb molecule. How does the existence of a reduction δ in the binding energy change f and f''? Discuss the physiological implications of this effect, on the one hand, for the tissues and, on the other hand, for the lungs, using numerical estimates. In particular, what is the consequence for the oxyhemoglobin in the tissues, where there is an excess of CO_2 characterized by a partial pressure of 80 Torr? What is the consequence of good oxygenation of the lungs, in which the partial pressure of CO_2 is 40 Torr?

Solution:

1. The same method as in Exerc.4b gives

$$f = \frac{1}{e^{-\alpha-\beta\varepsilon}+1}, \qquad e^{\alpha} = \left(\frac{2\pi\hbar^2}{mkT}\right)^{3/2} n.$$

In the lungs we find $e^{\alpha} \simeq 1.9 \times 10^{-8}$; the body temperature gives $\beta \simeq 37\ \text{eV}^{-1}$, whence $e^{-\beta\varepsilon} \simeq 2.7 \times 10^{-11}$ so that $f \simeq 1 - 1.4 \times 10^{-3}$ nearly equals 1: practically all hemoglobin has O_2 attached.

2. For O_2 in the air, the characteristic rotational and vibrational temperatures satisfy $\Theta_r \ll T \ll \Theta_v$. The internal partition function is thus given by half of (8.56), if we take into account the indistinguishability of the two atoms, or

$$\zeta = \frac{IkT}{\hbar^2} = \mu\bar{\varrho}^2 \frac{kT}{\hbar^2}$$
$$\simeq \frac{8 \times 10^{-3}}{6 \times 10^{23}} \cdot \left(1.2 \times 10^{-10}\right)^2 \cdot \frac{1.38 \times 10^{-23} \cdot 310}{\left(1.06 \times 10^{-34}\right)^2} \simeq 74.$$

For given n the rotation of the O_2 molecules in the air thus divides e^{α} by 74. As α only occurs in the combination $e^{\alpha+\beta\varepsilon}$ we may compensate for this effect by changing $\varepsilon = 0.65$ eV into $\varepsilon \simeq 0.77$ eV.

3. The new fraction of HbO_2 is

$$f = \frac{e^{\alpha+\beta\varepsilon}}{1 + e^{\alpha+\beta\varepsilon} + e^{\alpha'+\beta\varepsilon'}}.$$

If we use the notation $r = 0.15n'/n$ for the fraction of CO in the air, we find

$$e^{\alpha'} = e^{\alpha} \left(\frac{m}{m'}\right)^{3/2} \frac{r}{0.15} \simeq 1.6 \times 10^{-7}\, r,$$

so that

$$f^{-1} = f_0^{-1} + e^{\beta(\varepsilon-\varepsilon')+\alpha'-\alpha} \simeq 1 + 1000\, r.$$

As a result f is divided by 2 as soon as the air contains 0.1% of CO. Moreover, $f + f' \simeq 1$ and all the hemoglobin fixes either O_2 or CO. When $r = 5\%$, 98% of the Hb is unusable, in the form HbCO, and only 2% transports O_2. These numbers illustrate the poisonous character of CO.

4. As e^{α} is proportional to the partial pressure of O_2 we have here $e^{\alpha} \simeq 1.9 \times 10^{-8} \times 5/(760 \cdot 0.15) \simeq 8.4 \times 10^{-10}$, whence $f_1 = 0.97$. In this model the hemoglobin would only release 3% of its oxygen to the tissues, which would either make the feeding of them in oxygen very inefficient, or make it necessary for the blood to circulate much faster.

5. The average number of O_2 fixed per Hb is

$$f = \frac{2\left(e^{\alpha+\beta\varepsilon_1} + e^{2\alpha+\beta\varepsilon_1+\beta\varepsilon}\right)}{1 + 2e^{\alpha+\beta\varepsilon_1} + e^{2\alpha+\beta\varepsilon_1+\beta\varepsilon}}.$$

In the lungs, where $e^{\alpha_0} = 1.9 \times 10^{-8}$, we have

$$20 = e^{2\alpha_0+\beta\varepsilon_1+\beta\varepsilon} \gg 1 \gg e^{\alpha_0+\beta\varepsilon_1} = 0.03,$$

so that $f_0 \simeq 2$ (calculations give 1.9). The blood fixes the oxygen nearly as efficiently as in the first model. However, in the tissues, where $e^{\alpha_1} = 8.4 \times 10^{-10}$, we have

$$1 \gg e^{2\alpha_1+\beta\varepsilon_1+\beta\varepsilon} = 0.04 \gg e^{\alpha_1+\beta\varepsilon_1} = 1.3 \times 10^{-3},$$

so that $f_1 \simeq 0$ (calculations give 0.08). Practically all transported oxygen is released when it arrives. The adaptation of hemoglobin to its functions – f_0 takes its maximum possible value and f_1 the minimum value – is due to the difference between the two successive binding energies ε_1 and ε. This makes it possible for f to vary with e^{α}, that is, with the partial oxygen pressure, much faster than if $\varepsilon = \varepsilon_1$, in which case one finds again, apart from a factor 2, Langmuir's law as in 1.

In fact, this model with two sites applies not to hemoglobin but to myoglobin, the protein which takes care of oxygen exchanges within the muscles. In a more realistic model of Hb where the binding energy of a first molecule equals ε_1, those of a second or a third equal $\varepsilon_2 = \frac{1}{2}(\varepsilon_1 + \varepsilon)$, and that of the fourth equals ε, we find

$$f = \frac{4\left[z(1+z)^3 + bz^4\right]}{(1+z)^4 + b(1+z^4)}, \qquad z \equiv e^{\alpha+\beta\varepsilon_2}, \qquad b \equiv e^{\beta(\varepsilon-\varepsilon_1)/2} - 1,$$

which leads to an even better result: $f_0 \simeq 4(1-10^{-2})$ and $f_1 \simeq 10^{-2}$.

6. Each molecule has four possible configurations, Hb, HbO_2, CO_2Hb, and CO_2HbO_2. If α''/β is the chemical potential of CO_2, we find

$$f = \frac{1}{e^{-\alpha-\beta\varepsilon}\left(1+e^{\alpha''+\beta\varepsilon''}\right)\left(1+e^{\alpha''+\beta\varepsilon''-\beta\delta}\right)^{-1}+1},$$

$$f'' = \frac{1}{e^{-\alpha''-\beta\varepsilon''}\left(1+e^{\alpha+\beta\varepsilon}\right)\left(1+e^{\alpha+\beta\varepsilon-\beta\delta}\right)^{-1}+1}.$$

If $\delta = 0$, f is independent of the CO_2 pressure; similarly, f'' is independent of that of O_2. Denoting the value of f for $\delta = 0$, which was calculated in 1, by f_0, we get

$$f - f_0 = -f(1-f_0)\,\frac{e^{\beta\delta}-1}{e^{-\alpha''-\beta(\varepsilon''-\delta)}+1}.$$

The existence of δ thus leads to a *decrease* in the oxyhemoglobin fraction. This effect is the more marked, the larger is α'', that is, *the larger the partial pressure of* CO_2. It is negligible if $f_0 \simeq 1$, that is, if the fraction of HbO_2 when there is no CO_2 is large; if, however, the fraction of HbO_2 is small, it is *made even smaller* by the presence of CO_2. In particular, in the lungs f_0 is large, and f remains practically equal to 1. However, in the tissues, it is reduced so that O_2 is better released: the higher the production of CO_2, the larger the oxygenation of the tissues. One obtains the numerical value of f using $e^{\beta\delta} = 1.8 \times 10^3$, $e^{\beta\varepsilon} = e^{\beta\varepsilon''} = 3.7 \times 10^{10}$, and

$$e^{\alpha''} = 8.4 \times 10^{-10} \cdot \frac{80}{5}\left(\frac{32}{44}\right)^{3/2} = 8.3 \times 10^{-9}$$

for CO_2 at 80 Torr. This replaces $f_0 = 1$ found under 1 by a not much reduced $f = 0.83$, and $f_1 = 0.97$ found under 4 by a considerably diminished $f = 0.1$.

The situation is symmetric for CO_2 where

$$f'' - f_0'' = -f''(1-f_0'')\,\frac{e^{\beta\delta}-1}{e^{-\alpha-\beta(\varepsilon-\delta)}+1}.$$

In the lungs where the O_2 pressure is high, f'' is much smaller than f_0'', calculated for $\delta = 0$: the hemoglobin *gets rid more easily of the* CO_2 which it transports. In the tissues it attaches slightly less CO_2, but f'' luckily remains large. Numerically, we find for $\delta = 0$ that f_0'' would be almost equal to 1, both in the lungs – which would be annoying – and in the tissues; however, when $\delta = 0.2$ eV and when we take into account the partial pressures of O_2, we get $f'' = 0.2$ in the lungs and 0.9 in the tissues.

These interrelation effects carry the name of Christian Bohr (1904), the father of Niels and Harald Bohr.

9. Condensation of Gases and Phase Transitions

"Le chaleur de l'eau est indépendante de la violence de l'ébullition et de sa durée; l'eau moins comprimée par l'atmosphère bout plus tôt, et elle bout fort vite dans le vide."

d'Alembert, L'Encyclopédie

"La marmite de cuisson ou auto-cuiseur permet de réaliser une cuisson très rapide, puisque la température, à l'intérieur de la marmite, s'élève sous pression à 110 ou 120 °C."

Ginette Mathiot, Je sais cuisiner

"Les Arabes tirent le sel de l'eau par ébullition.

Chateaubriand, Itinéraire de Paris à Jérusalem

"Les alchimistes arabes désignèrent comme *al koh'l* toute poudre impalpable obtenue par sublimation, ainsi que tous principes volatils isolés par la distillation."

Marguerite Toussaint-Samat, Histoire Naturelle et Morale de la Nourriture, Bordas, 1987

In Chaps. 7 and 8 we restricted ourselves to the study of low density gases. When a gas is compressed, its molecules get closer to one another and the approximation which consists in neglecting their interactions is no longer valid. As the density increases there thus appear corrections to the perfect gas laws and we shall now give a theory for those. However, the occurrence of interaction terms in the Hamiltonian gives rise to a new mathematical difficulty: the partition function can no longer be factorized into the contributions of the single molecules. Calculating it involves more or less well controlled approximations, even when one introduces simplified models for the interactions (§§ 9.1 and 9.2).

The interactions between the molecules play a more interesting rôle, namely, they produce a phenomenon which is remarkable, although well known from experience: liquefaction. We know that if the temperature is not too high – more precisely, if it does not lie above the critical temperature – compressing a gas makes it *abruptly* go over into the liquid state, when one reaches the saturated vapour pressure. If we try to forget that this is a trivial fact and try to examine it critically, we find that it has surprising aspects: the thermodynamic quantities show a discontinuity or a singularity, and the fluid separates into two phases, the liquid and the vapour one. How can we explain that these two phases coexist and how can we explain their qualitatively different properties, if they consist of the same molecules which interact with the same forces? How can we derive from a single microscopic theory a discontinuous macroscopic behaviour? The same questions crop up for the numerous other instances of phase transitions which one observes on a macroscopic scale: solidification, the possibility to obtain various liquid phases for mixtures or various crystalline phases for solids, ferromagnetism, superconductivity, and so on. Their discontinuous nature is difficult to understand when we start from the microscopic structure of matter, and this has given rise to a long controversy: is there a sudden change in the forces between the elementary constituents? or is it possible to explain the existence of phases which are macroscopically so different starting from a single microscopic model? The second solution has won the day thanks to many researches which constitute one of the most spectacular contributions of statistical physics. We cite amongst the first important stages the elaboration of approximate microscopic theories of magnetism (P.Weiss, 1924; W.L. Bragg and E.J.Williams, 1934) and of liquefaction (J.Yvon, J.E. Mayer, 1937), which showed the appearance of a phase transition as a cooperative phenomenon – where, for instance, the magnetic moments orient themselves all in the same direction below the Curie temperature due to their mutual interactions. A decisive conceptual step was the first rigorous solution of a microscopic model which shows a phase transition, the two-dimensional Ising model (L.Onsager, 1944). More recently, the singular behaviour of the thermodynamic functions near a critical point, which had been known and studied experimentally for a long time, found its explanation thanks to the work by K.Wilson (1971). Below we shall restrict ourselves to an as simple as possible approximate theory for the gas-liquid transition (§ 9.3), while we shall study some other phase transitions in the form of exercises at the end of this chapter and of problems in the second volume.

9.1 Model and Formalism

In § 7.1 we saw that the perfect gas model, which is sufficient for describing rarefied monatomic gases, is based upon three approximations: the internal degrees of freedom of the molecules are frozen in, their centre of mass motion has a classical nature, and there are no interactions between the molecules. The first condition was abandoned in Chap.8 when we wanted to study rarefied di- and poly-atomic gases. On the other hand, we showed in § 7.1.3 that the second approximation – the use of classical statistical mechanics for the translational degrees of freedom – is valid not only for gases, but also for almost all liquids. However, we must take into account the interactions between the molecules, if we want to study the properties of a compressed gas or of a liquid.

9.1.1 Interactions Between Molecules

It is difficult to make a simple model of the interactions between di- and poly-atomic molecules as their collisions can produce transitions from one state to another in each of the molecules. The interactions thus affect the internal structure of the molecules. The situation is simpler for *monatomic fluids* and we shall restrict ourselves to those in the present chapter. In all cases the interactions between the molecules are due to the Coulomb forces between their constituents, the electrons and the nuclei. In the case of a monatomic gas, the electrons remain frozen in into their ground state at the temperatures of interest; when two atoms aproach one another, the deformation of the electron cloud produces an effective potential between them.

We sketched the theory of this interaction in § 8.4.1 where we were interested in the binding between two atoms produced by the effective forces which they exert upon one another. Here also the potential W is determined by the Born-Oppenheimer approximation as a function of the distance between a pair of atoms in the monatomic gas. It has the same shape as between Cl and H atoms: *attractive for long distances apart* and *repulsive for short distances apart.* On the other hand, it is much weaker than in § 8.4.1, as the two atoms considered do not have a bound state. For instance, for argon (Fig. 9.1) the minimum of W is of the order of 10^{-2} eV, rather than 1 eV for HCl. The position of the minimum, at a few Å, corresponds to the interatomic distances in the solid or liquid state: in fact, the potential energy tends to become as weak as possible at low temperatures while the kinetic energy becomes negligible, and the atoms then tend to arrange themselves so as to make W a minimum.

Of course, even though the elementary forces between the constituents, the nuclei and electrons, which are Coulomb forces, are two-body forces, the effective interactions between the atoms also contain three-, four-, ... body forces; however, at the usual gas densities and even at liquid densities, the probability for three or more atoms to be sufficiently close to one another so that the effects of these three-, four-, ... body forces can come into play is small. We shall therefore restrict ourselves mainly to a study of Hamiltonians which, apart from the kinetic energy, contain solely the binary potential which we have just described and which is a function of the relative positions of the molecular nuclei, taken in pairs.

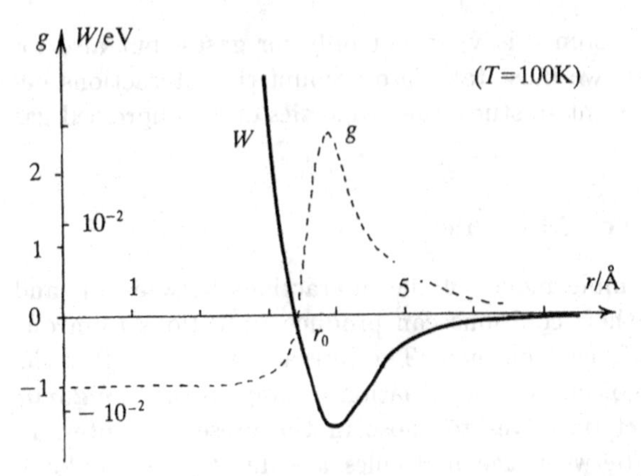

Fig. 9.1. The potential W between two argon atoms. The dashed curve shows $g \equiv e^{-\beta W} - 1$

Thus, we can write for the *effective Hamiltonian*

$$H_N = \sum_{i=1}^{N} \frac{p_i^2}{2m} + \sum_{i>j} W_{ij}, \tag{9.1}$$

where the potential $W_{ij} \equiv W(|\boldsymbol{r}_i - \boldsymbol{r}_j|)$ of the interaction between the molecules i and j depends only on their distance apart since they are monatomic, and has the shape shown in Fig.9.1. This potential W has a quantum mechanical origin as its calculation involves crucially the freezing-in of the electronic degrees of freedom into their ground state. Nevertheless, the effective Hamiltonian H_N, which results from eliminating the motion of the electrons and which depends on the positions and momenta of the *nuclei* of the atoms in the gas, can be treated *classically* for all monatomic fluids – bar helium at low temperatures – as we saw in § 7.1.3. The fluid is therefore finally satisfactorily described by a model of classical point particles with the Hamiltonian (9.1).

In order to take into account that the gas is confined to the volume Ω one must add to (9.1) a box potential $\sum_i V(\boldsymbol{r}_i)$ which vanishes inside Ω and is infinite outside it; in fact, as in Chaps.7 and 8, it will be sufficient to restrict the integrations over $\boldsymbol{r}_1, \ldots, \boldsymbol{r}_N$ to the domain Ω.

9.1.2 The Grand Potential of a Classical Fluid

The grand partition function is equal to

$$Z_G(\alpha, \beta) = \sum_{N=0}^{\infty} e^{\alpha N} \int d\tau_N \, e^{-\beta H_N}. \tag{9.2}$$

We can immediately integrate over the momenta, which leads to

$$Z_G = \sum_N \frac{e^{\alpha N}}{N!} \left[\int \frac{d^3 \boldsymbol{p}}{h^3} \, e^{-\beta p^2/2m} \right]^N \int_\Omega d^3 \boldsymbol{r}_1 \ldots d^3 \boldsymbol{r}_N \, e^{-\beta \sum_{i>j} W_{ij}}$$
$$= \sum_N \frac{e^{\alpha N}}{N!} \left(\frac{mkT}{2\pi \hbar^2} \right)^{\frac{3N}{2}} \int_\Omega d^3 \boldsymbol{r}_1 \ldots d^3 \boldsymbol{r}_N \, e^{-\beta \sum_{i>j} W_{ij}}. \qquad (9.3)$$

For the grand potential we get thus

$$A(T, \Omega, \mu)$$
$$= -kT \ln \left\{ \sum_N \frac{f^N}{N!} \int_\Omega d^3 \boldsymbol{r}_1 \ldots d^3 \boldsymbol{r}_N \, e^{-\beta \sum_{i>j} W_{ij}} \right\}, \qquad (9.4)$$

where we have written

$$f \equiv e^{\mu/kT} \lambda_T^{-3} \equiv e^{\mu/kT} \left(\frac{mkT}{2\pi \hbar^2} \right)^{3/2}. \qquad (9.5)$$

Once we know (9.4) we can derive from it the thermodynamic quantities through

$$dA = -\mathcal{P} \, d\Omega - S \, dT - N \, d\mu.$$

The presence of the interaction not only changes (through T) the thermodynamic or (through μ) the chemical quantities, but also the equation of state, which now differs from that for a perfect gas. Unfortunately, the canonical partition function which occurs in (9.4) for each N is a multiple integral over a very large number, $3N$, of variables, and its evaluation is impossible, even for the simplest models for the potential W. We must therefore in what follows appeal to various approximations without always being able to give a rigorous justification for them.

Let us bear in mind that (9.4) is based solely on two approximations: the interactions between the molecules are taken into account by a binary isotropic potential, which is realistic for a monatomic gas, and we use the classical limit. This expression is therefore justified not only for the *gas* phase, but also for the *liquid* phase of the substance studied. As we indicated at the start of this chapter, statistical mechanics thus provides a *single expression*, which contains a potential W, fixed once and for all, and which should describe the *liquid-vapour transition* as function of the two variables β and α, or T and μ. Nevertheless, nothing enables us to discern from (9.4) the presence of a line of singularities in the β, α-plane which would separate the two phases, liquid and vapour. The theory of the transition and of the critical point must start from (9.4), but we expect that it will be difficult mathematically. To give an idea of how complicated it is we note that even the proof itself of the extensivity, that is, of the existence of a limit for A/Ω as $\Omega \to \infty$ – which is such an essential property (§ 5.5.2) and which is the preliminary step for the study of A/Ω as function of β and α – is by no means easy. The proof requires that the potential W be repulsive at short distances apart and that its absolute magnitude decreases sufficiently fast at large distances apart.

Expression (9.4) can be extended to fluids containing several kinds of molecules and, in principle, can serve as the departure point, for instance, for a study of the *mixing* and *demixing* properties which give rise to phase transitions. A special, simple case is that of *dilute solutions* (Exerc.9f) where the particular form of (9.4) enables us to account for a large number of properties without having to elaborate the difficult theory of the liquid solvent itself.

The presence of interactions produces correlations between the particles, which are characterized by the two-body density f_2. The latter, defined by (2.82), depends here on the momenta only through a Maxwellian and on the positions only through the distance apart r of the two particles. The short-range repulsion between the latter leads to f_2 being very small for $r < r_0$; at large distances apart, the particles are no longer correlated and $f_2 \sim f_1 f_1$ tends to a constant value. In (2.83) we expressed the internal energy in terms of f_2. We shall see in (14.119) that the grand potential (9.4) can also be expressed in terms of pair correlations.

It is remarkable that (9.4) depends on the mass m of the molecules only through (9.5) and that a change in this mass can be absorbed in a change in the origin of $\alpha \equiv \mu/kT$. Besides, the Born-Oppenheimer method shows that two isotopes of the same element produce exactly the same potential W. As a result *two isotopes have the same equation of state and the same thermal properties*, both in the gas and in the liquid phase – except in the case of liquid helium at very low temperatures where the quantum effects become macroscopic as we shall see in Chap.12.

9.2 Deviations from the Perfect Gas Laws

9.2.1 The Maxwell Distribution

One property remains very simple, notwithstanding the presence of intermolecular interactions, namely, the velocity distribution. The density in phase,

$$D_N = \frac{1}{Z_G} e^{-\beta H_N + \alpha N}, \tag{9.6}$$

depends on the momenta only through the kinetic energy. As a result the number of molecules with momenta within a volume element $d^3\boldsymbol{p}$ around a value $\boldsymbol{p}$ equals

$$\left\langle \sum_i \delta^3(\boldsymbol{p}_i - \boldsymbol{p})\, d^3\boldsymbol{p} \right\rangle \propto e^{-\beta p^2/2m},$$

exactly as in the case of a system without interactions. It is remarkable that the *Maxwell distribution* is thus valid in the, gas or liquid, fluid phases, *even at higher densities.* The spatial correlations between molecules become very strong in the liquid, but their velocities remain statistically decoupled, at least as long as the classical approximation is valid.

9.2.2 Perturbation Methods

If the gas density remains sufficiently low, the molecules are practically always far from one another and only see the attractive tail of the potential W. The latter is not very strong and one expects that the lowest-order correction to the perfect gas laws can be obtained by expanding the grand potential in *powers of the interaction* W and only retaining the first-order terms. We find

$$\begin{aligned} A &= -kT \ln\Big\{\sum_N \frac{f^N}{N!} \int_\Omega d^3\boldsymbol{r}_1 \ldots d^3\boldsymbol{r}_N \\ &\quad \times \left[1 - \tfrac{1}{2}\beta N(N-1)W(|\boldsymbol{r}_1 - \boldsymbol{r}_2|) + \ldots\right]\Big\} \\ &= -kT \ln\left\{ e^{\Omega f} - \tfrac{1}{2}\beta f^2 e^{\Omega f} \int d^3\boldsymbol{r}_1\, d^3\boldsymbol{r}_2\, W_{12} + \ldots\right\} \\ &= -kT\Omega f - a\Omega f^2 + \ldots , \end{aligned} \tag{9.7}$$

where we have defined the constant a by

$$\begin{aligned} a &\equiv -\frac{1}{2\Omega} \int d^3\boldsymbol{r}_1\, d^3\boldsymbol{r}_2\, W_{12} \\ &= -\frac{1}{2} \int d^3\boldsymbol{r}\, W(r) \;=\; -2\pi \int_0^\infty r^2\, dr\, W(r). \end{aligned} \tag{9.8}$$

Note that the series expansion in (9.7) poses a convergence problem. The successive terms within the braces behave, in fact, as $a\Omega$, $a^2\Omega^2$, ... , and they become larger and larger, in the large volume limit. This could have been foreseen because A is extensive: Z_G is the exponential of a sum of terms which are proportional to the volume, and its expansion in powers of the interaction involves higher and higher powers of the volume. Actually, to establish (9.7), even when a is small, one should check that the terms in $a^n\Omega^n$ in Z_G have the proper coefficients needed to reconstitute an exponential. Put differently, one should expand A, rather than Z_G, directly in powers of W and check that each term is proportional to Ω.

As the potential appears in the combination βW, the expansion in powers of the potential is at the same time an expansion in powers of the inverse temperature and therefore it can provide a good approximation only at *high temperatures.*

The approximation used would, however, be valid only if the potential were everywhere weak. That is not the case at short distances, where W increases rapidly; the integral (9.8) is, moreover, dominated by just this interior region which must be eliminated, if one wants to obtain reasonable results. As we have seen, at *low densities* only the attractive tail of W should be involved and the repulsive core should not play any rôle. We should thus retain only the negative part of the potential and define a, rather than by (9.8), by

$$a = -2\pi \int_{r_0}^{\infty} r^2 \, dr \, W(r), \tag{9.9}$$

where r_0, which corresponds roughly to the point where W changes sign, is of the order of magnitude of atomic dimensions. More complicated methods are needed to take the repulsive core into account.

From the approximation (9.7) we get through differentiation the pressure,

$$\mathcal{P} = kTf + af^2,$$

and also the gas density,

$$\frac{N}{\Omega} = f + \frac{2a}{kT} f^2,$$

and the internal energy,

$$U = A + TS + \mu N = A - T\frac{\partial A}{\partial T} - \mu\frac{\partial A}{\partial \mu} = \frac{3}{2}\Omega f kT + 2a\Omega f^2.$$

Eliminating μ from those equations we find, to first order in the interaction potential,

$$\mathcal{P} = kT\frac{N}{\Omega} - a\left(\frac{N}{\Omega}\right)^2, \tag{9.10}$$

$$\frac{U}{N} = \frac{3}{2}kT - a\frac{N}{\Omega}. \tag{9.11}$$

9.2.3 The van der Waals Equation

The *equation of state* (9.10) contains a correction term as compared to the perfect gas law and it is a first step towards the establishment of the *van der Waals equation* (1873),

$$\left[\mathcal{P} + a\left(\frac{N}{\Omega}\right)^2\right]\left[\frac{\Omega}{N} - b\right] = kT, \tag{9.12}$$

which we know reproduces with rather a good accuracy the isotherms of simple gases and liquids. This empirical equation cannot receive a complete theoretical justification, but a qualitative argument allows us to understand the origin of the term b: whereas the term with a takes the long-range weak attractions into account, we still need to take into account the fact that the short-range (of order r_0) repulsions prevent the molecules to penetrate one another. Due to this the volume Ω available for each molecule is reduced to $\Omega - Nb$, where b is of the order of magnitude of the volume $4\pi r_0^3/3$ which is excluded due to the presence of each of the other molecules. If we replace Ω in the dominant term of (9.10) by $\Omega - Nb$, we find (9.12), and the numerical values (9.9) for a and r_0^3 for b are of the same order of magnitude as the empirical values.

A less cavalier justification for the replacement of Ω by $\Omega - Nb$ is provided by a model where the potential $W(r)$ is schematized as follows. For $r < r_0$, $W(r)$ is very large so that the particles cannot approach one another to distances less than r_0: everything takes place as if each particle were a *hard sphere* of radius $\frac{1}{2}r_0$. Outside the hard core, $r > r_0$, $W(r)$ is negative and small. Let us, to begin with, drop the regions $r > r_0$ of the potentials, and let us consider the term of order N in (9.4). The multiple integral is over the domain $r_{ij} > r_0$, $\forall\, i, j$, and the integrand is there equal to 1. Let us first of all integrate over $\boldsymbol{r}_N$; we must find the volume outside $N-1$ spheres, $r_{jN} > r_0$, $j < N$, of radius r_0 centred on the other particles. If those spheres do not overlap, we find $\Omega - 4\pi(N-1)r_0^3/3$. However, they may well overlap, as the distances between their centres are constrained in the integrations over the $\boldsymbol{r}_j$ to be larger than r_0, not larger than $2r_0$. Because of this we have underestimated the integral over $\boldsymbol{r}_N$, but the error is small if the density is sufficiently low so that $Nr_0^3 \ll \Omega$, in which case the molecules $j < N$ rarely approach one another closely in the integration. Iterating this process gives us for the multiple integral of (9.4)

$$\prod_{j=1}^{N} \left[\Omega - (j-1)\,\frac{4\pi r_0^3}{3} \right],$$

and hence

$$A \;=\; -kT\,\ln \left\{ \sum_{N=1}^{\infty} \frac{(-2bf)^N}{N!}\,\frac{\Gamma(N-\Omega/2b)}{\Gamma(-\Omega/2b)} \right\} \;=\; -\,\frac{kT\Omega}{2b}\,\ln\,(1+2bf).$$

We have introduced the coefficient

$$b \;=\; \tfrac{2}{3}\pi r_0^3, \tag{9.13}$$

which is *half the volume of the hard core*, and used the binomial formula (see end of the book). The resulting free energy, obtained through a Legendre transformation with respect to μ, is

$$\begin{aligned} F \;&=\; NkT \left[\ln\,\frac{N\lambda_T^3}{\Omega - 2Nb} + \frac{\Omega}{2Nb}\,\ln\left(1 - \frac{2Nb}{\Omega}\right) \right] \\ &\simeq\; NkT \left[\ln\,\frac{N\lambda_T^3}{\Omega - Nb} - 1 \right]. \end{aligned}$$

The approximation which we have just made is justified when Nb/Ω is sufficiently small, since the two expressions for F differ by less than 0.5×10^{-3} for $\Omega = 20Nb$, and by less than 3% for $\Omega = 4Nb$; moreover, their difference has the appropriate sign partly to correct the error we made above in the evaluation of A. Finally, we include the contribution from the $r > r_0$ part of the potential by treating it as a small perturbation, and we then find for the free energy the approximation

$$F = NkT\left(\ln\frac{N\lambda_T^3}{\Omega - Nb} - 1\right) + \frac{N^2}{\Omega}a, \tag{9.14}$$

where a is defined by (9.9)

Differentiating (9.14) with respect to Ω we get the van der Waals equation (9.12). Inversely, we could derive (9.14) from (9.12) using a thermodynamic argument (Exerc.6a) which is completed by knowing F at low densities, where it is given by (7.37).

The justification of replacing Ω by $\Omega - Nb$ is more rigorous *in one dimension* where b will be identified with the distance of shortest approach r_0, the one-dimensional equivalent of (9.13). To evaluate the integral (9.4) we order the particles in the box, of length L, according to

$$0 < r_1 < r_2 < \cdots < r_N < L,$$

which compensates for the factor $1/N!$. If we then choose as new variables

$$r_1' = r_1 - r_0, \quad r_2' = r_2 - 2r_0, \quad \ldots, \quad r_N' = r_N - Nr_0,$$

we can take the impenetrability condition, $r_{ij} > r_0$, exactly into account by reducing the integration domain to

$$0 < r_1' < r_2' < \cdots < r_N' < L - Nr_0.$$

The hard cores are thus eliminated by simply replacing L by $L' \equiv L - Nr_0$. Neglecting the $r_{ij} > r_0$ part of the potentials, we obtain *exactly*, even if Nr_0 is not small as compared to L,

$$F = NkT\left[\ln\frac{N\lambda_T}{L - Nr_0} - 1\right].$$

By using the $r_1', \ldots, r_N'$ as variables, one can then easily calculate the free energy to first order in the attractive $r > r_0$ part of $W(r)$, and the result is

$$F = NkT\left[\ln\frac{N\lambda_T}{L - Nr_0} - 1\right] - \frac{N^2}{L}a,$$

$$\begin{aligned} a \equiv & -\frac{L}{L'}\int_0^\infty dy\, e^{-yN/L'} \\ & \times\left[W(r_0 + y) + \frac{yN}{L'}W(2r_0 + y) + \frac{y^2N^2}{2L'^2}W(3r_0 + y) + \cdots\right] \\ \approx & -\int_{r_0}^\infty dr\, W(r) - \frac{N}{L}\int_{r_0}^{2r_0} dr\,(2r_0 - r)\,W(r) + \cdots. \end{aligned} \tag{9.15}$$

If the density is sufficiently low so that $Nr_0 \ll L$, expression (9.15) reduces to its first term, so that we have proved the one-dimensional equivalent of (9.9), (9.13), and (9.14).

The *internal energy* U, derived from (9.14), has the same form as (9.11) and is independent of b. In a *Joule expansion* the gas temperature therefore changes, in contrast to what happens in the case of a perfect gas. If the volume increases suddenly from Ω to $\Omega + \Delta\Omega$, the change in temperature,

$$\Delta T = \frac{2a}{3}\Delta\left(\frac{N}{\Omega}\right) = -\frac{2aN\Delta\Omega}{3k\Omega(\Omega+\Delta\Omega)}, \tag{9.16}$$

is always negative. The observed cooling at low densities and high temperatures is always very small in practice and in agreement with expression (9.16).

9.2.4 The Virial Series

Both in § 9.2.2 where, in (9.9), we changed the integration limit in the definition of a and in § 9.2.3 where we replaced the volume Ω by $\Omega - Nb$ we have used rather debatable heuristic arguments. We shall now construct more systematically an expansion of the thermodynamic quantities in *powers of the density.*

This expansion is called the virial expansion for historical reasons: one of the proofs uses the *Clausius virial theorem*

$$-\frac{1}{2}\,\langle(\boldsymbol{F}_i\cdot\boldsymbol{r}_i)\rangle = \left\langle\frac{p_i^2}{2m} - \frac{1}{2}\frac{d}{dt}(\boldsymbol{p}_i\cdot\boldsymbol{r}_i)\right\rangle = \frac{3}{2}kT, \tag{9.17}$$

where $\boldsymbol{F}_i$ is the total force exerted on particle i and where the virial is defined as the left-hand side of (9.17). It is also called the Ursell-Yvon-Mayer expansion after the authors who first constructed (1927–37) the general term of this expansion. Here we shall restrict ourselves to evaluating the lowest-order terms in the expansion. The contribution of the box potential to the virial is easily related to the pressure of the walls (§ 14.4.4).

Considering the general expression (9.4), (9.5) for the grand potential we note that f is none other than the density (7.32) which the gas would have had, if there were no interactions, for given temperature and chemical potential. It is thus natural to start with expanding A in powers of f. To do this, we introduce the function

$$g(r) = e^{-\beta W(r)} - 1, \tag{9.18}$$

which for the case of argon is shown by the dashed curve in Fig.9.1 for $T = 100$ K. Like the potential this function has a short range, but it is bounded by -1 in the region where W becomes infinite. If we expand the integrand of (9.4) in powers of g_{ij} up to third order, use the identical nature of the particles, and perform the integrations, we get

$$
\begin{aligned}
Z_{\mathrm{G}} &= \sum_N \frac{f^N}{N!} \int_\Omega d^3\boldsymbol{r}_1 \ldots d^3\boldsymbol{r}_N \prod_{i>j} (1+g_{ij}) \qquad (9.19)\\
&= \sum_N f^N \int_\Omega d^3\boldsymbol{r}_1 \ldots d^3\boldsymbol{r}_N \\
&\times \left[1 + \frac{1}{2(N-2)!} g_{12} + \frac{1}{2\cdot 2^2 (N-4)!} g_{12}g_{34} \right.\\
&+ \frac{1}{2(N-3)!} g_{12}g_{13} + \frac{1}{2^3\cdot 3!(N-6)!} g_{12}g_{34}g_{56}\\
&+ \frac{1}{2^2(N-5)!} g_{12}g_{13}g_{45} + \frac{1}{3!(N-4)!} g_{12}g_{13}g_{14}\\
&\left. + \frac{1}{2(N-4)!} g_{12}g_{23}g_{34} + \frac{1}{3!(N-3)!} g_{12}g_{23}g_{31} + \ldots \right]\\
&= \mathrm{e}^{\Omega f} \left[1 + \Omega f^2 B_1 + \frac{1}{2}\left(\Omega f^2 B_1\right)^2 + 2\Omega f^3 B_1^2 + \frac{1}{3!}\left(\Omega f^2 B_1\right)^3 \right.\\
&\left. + \left(2\Omega f^3 B_1^2\right)\left(\Omega f^2 B_1\right) + \frac{4}{3}\Omega f^4 B_1^3 + 4\Omega f^4 B_1^3 + \Omega f^3 B_2 + \ldots \right].
\end{aligned}
$$

Hence we find for the grand potential

$$
-\frac{A}{\Omega kT} = f + f^2 B_1 + f^3\left(2B_1^2 + B_2\right) + \frac{16}{3} f^4 B_1^3 + \ldots, \qquad (9.20)
$$

where the quantities B_1 and B_2 are functions of T defined by

$$
B_1 = \frac{1}{2}\int d^3\boldsymbol{r}\, g(r) = \frac{1}{2}\int d^3\boldsymbol{r} \left[\mathrm{e}^{-W(r)/kT} - 1\right], \qquad (9.21)
$$

$$
B_2 = \frac{1}{6}\int d^3\boldsymbol{r}_1\, d^3\boldsymbol{r}_2\, g(r_1) g(r_2) g(|\boldsymbol{r}_1 - \boldsymbol{r}_2|). \qquad (9.22)
$$

One sees easily that the power of f appearing in each term of (9.20) is equal to the number of particles which are involved in the integrations producing the coefficients B_1, B_1^2, B_2, ... We have thus constructed the start of the *expansion of A in powers of f*, that is, of $\mathrm{e}^{\mu/kT}$, and at the same time its expansion in g, keeeping all terms up to third order in either f or g.

A complete proof of (9.20) requires showing that the higher-order terms in (9.19), which contain higher and higher powers of Ω, together produce an exponential, thus generalizing what we saw at the start of the expansion. To carry this out it is convenient to represent each term in (9.19) by a diagram (Fig.9.2) formed of segments which connect, in all possible ways, points representing the particles concerned. Each segment provides a factor g_{ij} and each point a factor f, while the contribution from a diagram is obtained by integrating over the coordinates of the points; one can show that one must divide the quantity obtained in this way by the symmetry number of the diagram, that is, by the number of permutations of the labels of the particles which leave this diagram invariant. One can then prove that expression (9.19) is equal to the exponential of the contributions (9.20) of the *connected diagrams*, which are proportional to the volume Ω. This demonstrates the *extensivity* of the successive terms of the expansion of A and enables us to write these terms down systematically to each order in g_{ij} or in f.

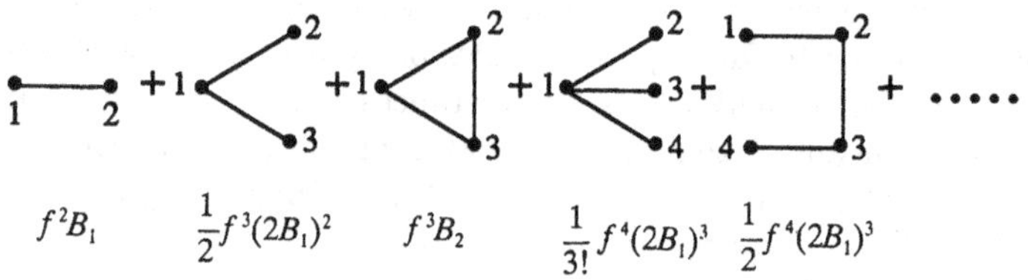

$f^2B_1 \qquad \frac{1}{2}f^3(2B_1)^2 \qquad f^3B_2 \qquad \frac{1}{3!}f^4(2B_1)^3 \qquad \frac{1}{2}f^4(2B_1)^3$

Fig. 9.2. The diagrams which contribute to (9.20)

The quantity f, which is proportional to $\exp(\mu/kT)$, tends to 0 with the density so that (9.20) produces the expansion of the thermodynamic quantities *in powers of the density.* To see this explicitly, let us find from (9.20) the density as function of f. To third order we find

$$\frac{N}{\Omega} = -\frac{1}{\Omega}\frac{\partial A}{\partial \mu} \approx f + 2f^2B_1 + 3f^3(2B_1^2 + B_2), \tag{9.23}$$

or, if we invert this,

$$f \approx \frac{N}{\Omega} - 2B_1\left(\frac{N}{\Omega}\right)^2 + (2B_1^2 - 3B_2)\left(\frac{N}{\Omega}\right)^3, \tag{9.24}$$

which, together with (9.5), gives us the expansion of the chemical potential in powers of the density:

$$\frac{\mu}{kT} \approx \ln\frac{N\lambda_T^3}{\Omega} - 2B_1\frac{N}{\Omega} - 3B_2\left(\frac{N}{\Omega}\right)^2. \tag{9.24$'$}$$

In order to express the other thermodynamic quantities as series in N/Ω we must eliminate μ from (9.20) and (9.24). It helps to carry out a Legendre transformation at the same time, that is, to change from the grand potential A to the thermodynamic potential associated with the variables T, Ω, N, that is, to the *free energy* F. Using (9.24′) we then find finally the expansion of F up to second order in powers of the density:

$$\begin{aligned} F(T,\Omega,N) &= A + \mu N \\ &= -NkT\left[\ln\frac{\Omega}{N\lambda_T^3} + 1 + B_1\frac{N}{\Omega} + B_2\left(\frac{N}{\Omega}\right)^2 + \cdots\right]. \end{aligned} \tag{9.25}$$

From the equivalent series (9.20) or (9.25) we obtain the expansion of all thermodynamic quantities in powers of e^{α} or of N/Ω, and, in particular, the *equation of state*

$$\mathcal{P} = -\frac{\partial F}{\partial \Omega} = \frac{N}{\Omega}kT - B_1\left(\frac{N}{\Omega}\right)^2 kT - 2B_2\left(\frac{N}{\Omega}\right)^3 kT + \cdots . \quad (9.26)$$

In contrast to the perturbation expansions of § 9.2.2, expressions (9.20), (9.25), or (9.26) enable us to face up to the pathology associated with the strong singularity of the potential at short distances, which was cured in § 9.2.3 only by rough means. Expanding in powers of the density leads, in fact, to expanding at the same time in powers of g, a function of r which remains bounded at short distances, in contrast to W.

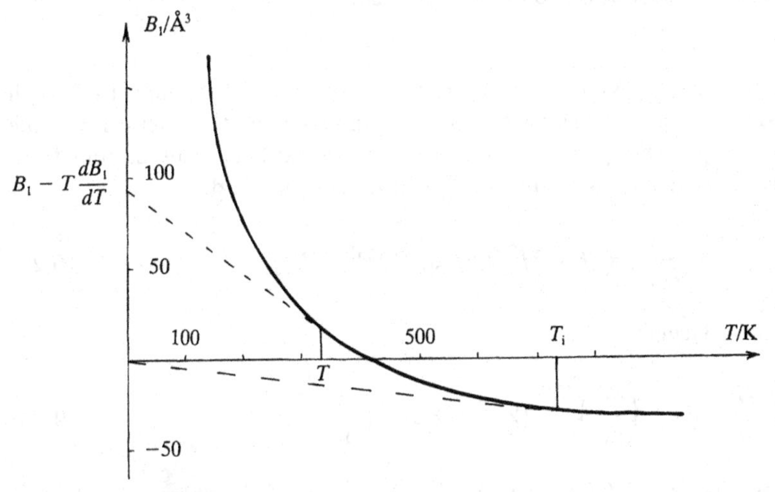

Fig. 9.3. The first virial coefficient B_1 for argon

In order to compare the expansions (9.25) and (9.26) to the results obtained earlier we shall restrict ourselves to first order in g and evaluate B_1, defined by (9.21), as function of the temperature. The behaviour of B_1 is shown in Fig.9.3 for the case of argon, the potential of which was shown in Fig.9.1. At high temperatures the repulsive part of the potential dominates (9.21) and B_1 is negative; on the other hand, when the temperature is lowered, the attractive region dominates and B_1 increases. One can model this behaviour by separating in B_1 the contribution from the two parts of the potential – weak attraction at distances larger than r_0 and strong repulsion at short distances. A reasonable approximation thus consists in replacing $g(r)$ by $-\beta W(r)$ for $r > r_0$ and by -1 for $r < r_0$, assuming that at the temperatures considered we have $-W(r) \ll kT$ for most of the region $r > r_0$ and $W(r) \gg kT$ for most of the region $r < r_0$. We thus treat again the attraction as a perturbation of first order in W and the repulsion as an infinitely hard core. This gives us

$$B_1 \simeq -2\pi\beta \int_{r_0}^{\infty} r^2\, dr\, W(r) + 2\pi \int_0^{r_0} r^2\, dr\, (-1)$$
$$\equiv \frac{a}{kT} - b. \tag{9.27}$$

This simple approximation for B_1 is clearly a rough one, but it reproduces qualitatively the characteristics of the $B_1(T)$ curve shown in Fig.9.3. From (9.26) and (9.27) we get at low densities for the pressure

$$\mathcal{P} = kT\frac{N}{\Omega}\left(1 + b\frac{N}{\Omega}\right) - a\left(\frac{N}{\Omega}\right)^2. \tag{9.28}$$

Expression (9.28) is the same as the expansion of the van der Waals equation up to second order in N/Ω, if a is defined by (9.9) and b by (9.13), that is, half the volume of the hard core. The virial expansion, which is more trustworthy than the van der Waals approximation, enables us to recover the latter, at least at rather low densities, and also to justify cutting off the integration domain as we did in § 9.2.2 when we defined a.

To higher orders the expansion of the van der Waals equation for $\mathcal{P}$ gives terms $kTb^{p-1}(N/\Omega)^p$. On the other hand, one can prove that successive terms in the virial expansion (9.25) of $-F/\Omega kT$ can be represented by diagrams similar to the ones of Fig.9.2, apart from two changes: (i) one must only take into account diagrams which remain connected when one takes away any one of the points; (ii) one must associate with each point a factor N/Ω, rather than f. The two terms written down in (9.25) represent the contributions from the first and the third diagram of Fig.9.2, the only ones which remain for the calculation of F. When T is sufficiently high, each diagram is dominated for each of its g_{ij} factors by the $r_{ij} < r_0$ region where g_{ij} is equal to -1. A diagram containing p points thus gives a contribution to F which is proportional to $kT\Omega r_0^{3(p-1)}(N/\Omega)^p$. For instance, B_2, defined by (9.22), can be calculated by taking as the integration variables the distances between the three particles r_1, r_2, r_3; the volume element d^3r_2 equals, when r_1 is fixed, $r_2 r_3\, dr_2\, dr_3\, d\varphi/r_1$ and we find (the integration domain is defined by $|r_2 - r_3| < r_1 < r_2 + r_3$, $r_1 < r_0$, $r_2 < r_0$, $r_3 < r_0$)

$$B_2 = -\frac{4\pi^2}{3}\int r_1 r_2 r_3\, dr_1\, dr_2\, dr_3 = -\frac{5\pi^2}{36} r_0^6. \tag{9.29}$$

The resulting $p = 3$ term in the expansion (9.26) is equal to $\frac{5}{8}kTb^2(N/\Omega)^3$. Its coefficient is not the same as that of the corresponding term $kTb^2(N/\Omega)^3$ in the expansion of the van der Waals equation. More generally, the expansion found here does not produce the geometric series expected from $(1 - bN/\Omega)^{-1}$, so that for higher than second order in N/Ω the van der Waals equation cannot be taken to be more than an empirical expression.

9.2.5 The Joule-Thomson Expansion

We shall use expression (9.25), taken only to first order in N/Ω, to explain the thermodynamic properties of the Joule-Thomson (or Joule-Kelvin) expansion which was studied in 1850 by James P. Joule and William Thomson – the later Lord Kelvin. We are dealing with the expansion of a gas from a vessel kept at a pressure $\mathcal{P}_1$ into a second vessel kept at a pressure $\mathcal{P}_2 < \mathcal{P}_1$ without exchange of heat with the surroundings. We assume that a stationary regime has been reached, where the temperatures of the two parts remain fixed at T_1 and T_2, the gas flowing in an *irreversible* manner at a rate controlled by a valve or a porous plug which connects the two vessels. The pressures $\mathcal{P}_1$, $\mathcal{P}_2$ and the initial temperature T_1 being given, we want to determine T_2 in this stationary regime.

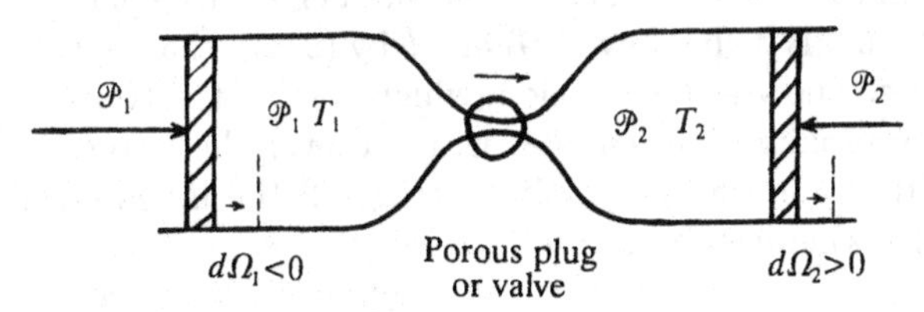

Fig. 9.4. Joule-Thomson expansion

During a time interval dt, a number $dN_2 = -dN_1$ of particles pass through the plug. In this process, the system does not exchange heat, but it exchanges work, with the outside, in contrast to the Joule expansion where the gas expands into an empty vessel. The energy balance provides

$$dU_1 + dU_2 = -\mathcal{P}_1\, d\Omega_1 - \mathcal{P}_2\, d\Omega_2,$$

which we can rewrite as

$$\frac{d(U_1 + \mathcal{P}_1\Omega_1)}{dN_1} = \frac{d(U_2 + \mathcal{P}_2\Omega_2)}{dN_2}. \tag{9.30}$$

Hence, if $\mathcal{P}_1$, $\mathcal{P}_2$, T_1, and T_2 do not vary with time, the *enthalpy* per particle, defined as function of $\mathcal{P}$ and T by

$$\frac{H}{N} \equiv \frac{U + \mathcal{P}\Omega}{N} = \frac{1}{N}\left[F - T\frac{\partial F}{\partial T} - \Omega\frac{\partial F}{\partial \Omega}\right], \tag{9.31}$$

is the same in the two vessels when the permanent regime is reached. To first order in the density or the pressure we find the enthalpy from (9.25), (9.31), (9.26):

$$\begin{aligned} \frac{H}{N} &= \frac{5}{2}kT + kT\frac{N}{\Omega}\left(T\frac{dB_1}{dT} - B_1\right) \\ &= \frac{5}{2}kT + \mathcal{P}\left(T\frac{dB_1}{dT} - B_1\right). \end{aligned} \tag{9.32}$$

If we write down the condition that (9.32) *remains constant during the expansion*, we determine in the $\mathcal{P}, T$-plane the lines along which the expansion may take place.

To have an idea about the shape of these expansion curves we replace B_1 in (9.32) by its approximate expression (9.27). The result is shown in Fig.9.5. We see that there exists an *inversion temperature*, which follows from the equation $B_1 - T\, dB_1/dT \simeq 2a/kT - b = 0$,

$$T_{\mathrm{i}} \simeq \frac{2a}{kb}, \tag{9.33}$$

above which the Joule-Thomson expansion *heats* the gas and *below* which it *cools* the gas.

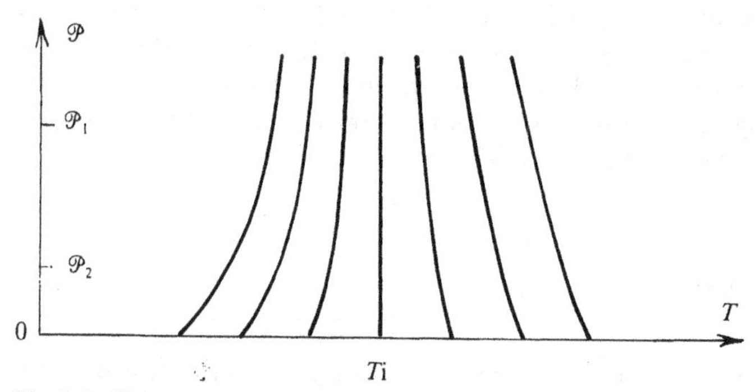

Fig. 9.5. Constant enthalpy curves

This is a general effect. In fact, it follows from the identity $dH = \Omega\, d\mathcal{P} + T\, dS + \mu\, dN$ that

$$\left(\frac{\partial H}{\partial T}\right)_{\mathcal{P}} = C_{\mathrm{p}},$$

so that at low densities, where expression (9.33) holds, the slope of a constant enthalpy curve is equal to

$$\frac{dT}{d\mathcal{P}} = -\left(\frac{\partial H}{\partial \mathcal{P}}\right)_T \Big/ \left(\frac{\partial H}{\partial T}\right)_{\mathcal{P}} = \frac{N}{C_{\mathrm{p}}}\left(B_1 - T\frac{dB_1}{dT}\right). \tag{9.34}$$

The bracket in (9.34) is in Fig.9.3 represented by the ordinate of the intersection of the tangent to the $B_1(T)$-curve with the vertical axis. We see that it changes sign at the temperature T_{i} where this tangent passes through the origin and we find again the inversion effect. For argon the inversion temperature T_{i} is 720 K, in qualitative agreement with (9.33).

As an exercise, one could prove, starting from (9.30), that the stationary regime is *stable* in a Joule-Thomson expansion: if, with $\mathcal{P}_1$, $\mathcal{P}_2$, and T_1 fixed, the temperature T_2' of the second vessel deviates from the value T_2 defined by $H_1/N_1 = H_2/N_2$, the difference $|T_2' - T_2|$ decreases with time, according to the relation $\left[T_2'(t) - T_2\right] N_2(t) = \text{constant}$.

The shape of the $B_1(T)$-curve and hence the inversion effect itself are consequences of the shape of the potential $W(r)$: at high temperatures the repulsion between the molecules dominates and the expansion produces heating; at lower temperatures the attraction dominates and that is the origin of the cooling. The effect has been widely observed experimentally; however, expression (9.34) is quantitatively exact only in the low pressure limit as it was derived using the truncated expansion (9.32). In fact, if the initial pressure is high, the inversion temperature depends on it and when we evaluate it we must take into account the next term in (9.32). The cooling through the Joule-Thomson expansion is used in practice for *obtaining low temperatures*, for instance, for liquefying gases; the lowering of the temperature, (9.34), is small – of the order of 0.1 K per atmosphere – but the process has the advantage of being able to function in a stationary regime and to be suitable for industrial applications (Linde, 1895).

9.3 Liquefaction

Expression (9.4) for the grand potential describes in principle not only the properties of gases, but also those of liquids and those of the liquid-vapour transition. However, in that region we do not have a small parameter at our disposal, so that approximation methods based upon series expansions are inappropriate. We shall appeal to a more roundabout approximation to determine the grand potential, namely, the *variational mean field or effective potential method* which will enable us to understand theoretically the properties of the liquid-vapour transition (§ 9.3.2, Prob.7). This kind of method, although rather rough, is efficient; it can be extended to other examples of phase transitions, such as sublimation and melting (Probs.8 and 10), ferromagnetism (Exerc.9a), ferroelectricity (Prob.5), or changes in crystalline phases (Probs.4 and 12).

9.3.1 The Effective Potential Method

We start from the following remark. Whereas it is practically impossible to evaluate the grand potential (9.4) for interacting particles, there exist simple probability distributions for which the calculation of the thermodynamic quantities and of mean values of observables does not present any difficulties. For the problem we are interested in here, we are dealing with densities in phase $\mathcal{D}$ which describe a gas of non-interacting molecules subject to an external potential $\mathcal{V}(\boldsymbol{r})$. The Hamiltonian in that case has the form

$$\mathcal{H}_N = \sum_{i=1}^{N} \left[\frac{p_i^2}{2m} + \mathcal{V}(\boldsymbol{r}_i) \right], \tag{9.35}$$

and the contributions from the different particles can be completely factorized. Of course, such distributions $\mathcal{D}$ differ from the Boltzmann-Gibbs distribution D associated with the Hamiltonian H_N with interactions, given by (9.1), which we wish to study; but we can try to choose $\mathcal{V}$ in such a way as to *simulate as well as possible* the properties of the real system by those of the model *without interactions* with an *adjustable potential* $\mathcal{V}$. We shall thus replace the two-body potential W_{ij} by an effective potential $\mathcal{V}_i$ *of independent particles* which will depend on the temperature and on the chemical potential and which *on average* describes the effect of all the other molecules on the molecule i. This procedure neglects the correlations which exist in the exact equilibrium state D as a result of the interactions W_{ij}, but they are partly simulated in the approximate state $\mathcal{D}$ through the mean field $\mathcal{V}(\boldsymbol{r})$.

The determination of the best possible approximation for $\mathcal{V}$ is based upon the *variational method* of § 4.2.2. Let $\widehat{D}$ be the grand canonical density operator which we are interested in and which is associated with the Hamiltonian (9.1), and let $\widehat{\mathcal{D}}$ be a *trial density operator* which is sufficiently simple to allow us to perform calculations and which we want to choose in the best possible way to replace $\widehat{D}$. For the case of a grand canonical equilibrium we can write inequality (4.10) in the form

$$\frac{1}{k} S(\widehat{\mathcal{D}}) - \beta \operatorname{Tr} \widehat{\mathcal{D}} \widehat{H} + \alpha \operatorname{Tr} \widehat{\mathcal{D}} \widehat{N} \; < \; \ln Z_{\mathrm{G}}, \tag{9.36}$$

or, if we define averages of the energy and of the number of particles with respect to the trial density operator $\widehat{\mathcal{D}}$,

$$\left.\begin{aligned} \mathcal{U} &= \operatorname{Tr} \widehat{\mathcal{D}} \widehat{H}, \\ \mathcal{N} &= \operatorname{Tr} \widehat{\mathcal{D}} \widehat{N}, \end{aligned}\right\} \tag{9.37}$$

in the form

$$A \; < \; \mathcal{A} \equiv \mathcal{U} - TS(\mathcal{D}) - \mu \mathcal{N}. \tag{9.38}$$

This inequality remains valid in the classical limit we are considering here. The best choice for $\mathcal{D}$ corresponds to an approximate grand potential $\mathcal{A}$ which is as close as possible to the exact grand potential A and is thus obtained by *looking for the minimum of* $\mathcal{A}$. This criterion gives us an approximate density in phase $\mathcal{D}$ which will enable us to calculate simply the averages of the various observables of interest and which we can reasonable hope to be sufficiently close to the exact, but impracticable, Boltzmann-Gibbs distribution D.

Let us apply this method to the gas with interactions and take as a trial density in phase $\mathcal{D}$ the one which the system would have in grand canonical equilibrium, if its Hamiltonian were simply the Hamiltonian (9.35) without interactions rather than (9.1):

$$\mathcal{D}_N = \frac{1}{\zeta}\,\mathrm{e}^{-\beta\mathcal{H}_N+\alpha N}. \tag{9.39}$$

We shall determine the adjustable arbitrary potential $\mathcal{V}(\boldsymbol{r})$ using the above variational method. The distribution $\mathcal{D}$ describes *uncorrelated* molecules with a *density at the point* $\boldsymbol{r}$ given by (7.32), that is, if we use the notation (9.5), by

$$n(\boldsymbol{r}) = \lambda_T^{-3}\,\mathrm{e}^{\mu/kT}\,\mathrm{e}^{-\mathcal{V}(\boldsymbol{r})/kT} = f\,\mathrm{e}^{-\mathcal{V}(\boldsymbol{r})/kT}. \tag{9.40}$$

The normalization constant of (9.39) is equal to

$$\zeta = \sum_N \frac{\mathrm{e}^{\alpha N}}{N!}\left[\int \frac{d^3\boldsymbol{p}\,d^3\boldsymbol{r}}{h^3}\,\mathrm{e}^{-(\beta p^2/2m)-\beta\mathcal{V}(\boldsymbol{r})}\right]^N = \mathrm{e}^{\int d^3\boldsymbol{r}\,n(\boldsymbol{r})}. \tag{9.41}$$

Let us for the moment restrict ourselves and take for $\mathcal{V}(\boldsymbol{r})$ a constant $\mathcal{V}$. Either $\mathcal{V}$ or n remains the only adjustable parameter, which must be determined variationally by looking for the minimum of (9.38). The approximate average number of particles is

$$\mathcal{N} = \int d^3\boldsymbol{r}\,n(\boldsymbol{r}) = \Omega n. \tag{9.42}$$

We also need the average energy $\mathcal{U}$. Note that this is the average (9.37) over the trial density in phase $\mathcal{D}$ of the *true Hamiltonian H*, and not of the trial Hamiltonian; in other words, the trial grand potential $\mathcal{A}$ which occurs in the inequality (9.38) is not equal to $-kT\ln\zeta$, in contrast to the exact quantity $A = U - TS - \mu N = -kT\ln Z_{\mathrm{G}}$. For the kinetic part, as in the case of a perfect gas, the equipartition theorem gives $\frac{3}{2}NkT$, where we must replace N by Ωn; this result is valid quite generally, even for the average $\mathrm{Tr}DH$ over the exact classical Boltzmann-Gibbs distribution. For the potential energy we get, using the fact that there are no correlations in the trial state,

$$\frac{1}{\zeta}\sum_N \mathrm{e}^{\alpha N}\int d\tau_N\,\mathrm{e}^{-\beta\mathcal{H}_N}\,\frac{N(N-1)}{2}\,W(|\boldsymbol{r}_1-\boldsymbol{r}_2|)$$
$$= \frac{1}{2}\int d^3\boldsymbol{r}_1\,d^3\boldsymbol{r}_2\,n(\boldsymbol{r}_1)n(\boldsymbol{r}_2)W(|\boldsymbol{r}_1-\boldsymbol{r}_2|).$$

As we might have expected, since the particles are uncorrelated, this expression for the potential energy is the same as that for a continuous medium of uniform density n which interacts with itself through the potential W. The exact value of the internal energy would be given by the general form (2.83) and would involve the reduced two-particle density f_2, a function of $|\boldsymbol{r}_1-\boldsymbol{r}_2|$ which is difficult to determine and which describes the correlations between the molecules. Altogether the approximate internal energy is therefore

$$\mathcal{U} = \tfrac{3}{2}kT\Omega n + \tfrac{1}{2}\Omega n^2\int d^3\boldsymbol{r}\,W(\boldsymbol{r}). \tag{9.43}$$

We still must evaluate the approximate entropy $S(\mathcal{D})$. To do this we need only note that (9.39) is nothing but the density in phase of a perfect gas with a chemical potential equal to $\mu - \mathcal{V}$ rather than to μ and with a density equal to n; expression (7.41) then gives

$$S(\mathcal{D}) \;=\; \Omega nk \left(\tfrac{5}{2} - \ln n\lambda_T^3\right). \tag{9.44}$$

Altogether, the *trial grand potential* $\mathcal{A}$ is equal to

$$\begin{aligned}\mathcal{A} &= \mathcal{U} - TS(\mathcal{D}) - \mu\mathcal{N} \\ &= kT\Omega\left[n \ln \frac{n}{f} - n\right] + \frac{1}{2}\Omega n^2 \int d^3\boldsymbol{r}\, W(\boldsymbol{r}),\end{aligned} \tag{9.45}$$

and we must look for its *minimum* with respect to the parameter $\mathcal{V}$, or, what is the same, with respect to n which is related to $\mathcal{V}$ through (9.40). We thus get the equation

$$kT \ln \frac{n}{f} + n \int d^3\boldsymbol{r}\, W(\boldsymbol{r}) \;=\; 0, \tag{9.46}$$

which can also be written as

$$\mathcal{V} \;=\; \int d^3\boldsymbol{r}\, W(\boldsymbol{r})\, n. \tag{9.47}$$

If, more generally, we look for the minimum of $\mathcal{A}$ with respect to a potential $\mathcal{V}(\boldsymbol{r})$ which varies arbitrarily in space, Eqs.(9.45) and (9.47) become

$$\begin{aligned}\mathcal{A} &= kT \int d^3\boldsymbol{r}\, n(\boldsymbol{r}) \left[\ln \frac{n(\boldsymbol{r})}{f} - 1\right] \\ &\quad + \frac{1}{2} \int d^3\boldsymbol{r}\, d^3\boldsymbol{r}'\, n(\boldsymbol{r}) n(\boldsymbol{r}') W(|\boldsymbol{r} - \boldsymbol{r}'|),\end{aligned} \tag{9.48}$$

$$\mathcal{V}(\boldsymbol{r}) \;=\; \int d^3\boldsymbol{r}'\, W(|\boldsymbol{r} - \boldsymbol{r}'|)\, n(\boldsymbol{r}'). \tag{9.49}$$

The latter condition, which expresses that the effective one-body potential $\mathcal{V}$ takes in the best possible manner the two-body potential W into account, has a simple intuitive meaning: the molecules i, which are distributed in space with a density n, *create at the point* $\boldsymbol{r}$ *an average potential*

$$\left\langle \sum_i W(|\boldsymbol{r} - \boldsymbol{r}_i|) \right\rangle \;=\; \mathcal{V}(\boldsymbol{r})$$

given by (9.49). We then assume that each molecule *moves independently of the others in this average potential* $\mathcal{V}(\boldsymbol{r})$ which it has helped to create; when we finally write down that thermal equilibrium is established, this implies that *the density* n *can be derived from* $\mathcal{V}$ through the equation (9.40) for independent particles. We must thus solve a pair of coupled equations, (9.40) and (9.49), which express approximately the density n of the molecules as function of the effective potential $\mathcal{V}$ to which they are subject, and *vice versa.* The procedure is the same as in the Hartree method (§ 11.2.1) which we shall use to describe the electrons in a solid: we shall replace in that case the Coulomb repulsion between electrons by a potential which is determined by successive approximations, assuming the density known and then deriving from it the potential as if we were dealing with a charged continuous fluid; after that we recalculate the density in thermal equilibrium for that potential, and so on, iterating the procedure until the coupled equations are satisfied. Such methods, based upon the use of a *self-consistent potential* which is determined from that potential itself, are currently employed in different fields of physics under various names: molecular field or Weiss field method in magnetism, Bragg-Williams method for alloys, self-consistent potential in nuclear physics, or "bootstrap" theory in elementary particle physics – the potential is assumed to pull itself up by its bootstraps.

If $\int d^3\boldsymbol{r}\, W(\boldsymbol{r})$ is *positive*, (9.46) has always one and only one solution which gives us the unique minimum of (9.45). The corresponding density n or the effective potential $\mathcal{V}$ depend on the temperature and the chemical potential. We then find from (9.46) the approximate grand potential of the fluid and hence its thermodynamic properties, with $\mathcal{A}$ depending on T and μ both explicitly and through n. To first order in W this expression is the same as the approximation (9.7), (9.8). However, it includes higher-order contributions in W and one may thus hope that it is a better approximation.

If $\int d^3\boldsymbol{r}\, W(\boldsymbol{r})$ is *negative*, (9.45) has no lower bound as $n \to \infty$. As a result, by virtue of (9.38) which is now an inequality, satisfied whatever the value of n, the grand partition function *diverges.* One cannot have a grand canonical equilibrium state in this case, where the ground state energy decreases as $-N^2$ as $N \to \infty$. Placed in a thermostat and a particle reservoir, the system would indefinitely capture particles, its density would become infinite, and all thermodynamic properties depending on the extensivity would be violated. In fact, the existence of a thermodynamic limit (§ 5.5.2) and the stability of matter require that the potential for the interaction between molecules is sufficiently *repulsive*, at least at short distances.

9.3.2 The Gas-Liquid Transition

The above method is only of academic interest if one tries to apply it directly to the kind of interactions which one finds in a real gas. In fact, the potential energy must be negative when the density is relatively low, and become positive at high densities, as is suggested by the shape of the two-body potential W. The variational method which we used does not enable us to take this property into account for the realistic two-body Hamiltonian (9.1), as the approximation (9.43) for the potential energy depends in that case solely on a single parameter, the integral of the potential, and is simply proportional to n^2. Nevertheless, a discussion of a *model Hamiltonian* which differs from (9.1), but retains its essential characteristics, will enable us to understand the microscopic origin of the gas-liquid transition without needing recourse to a more elaborate approximation technique than the method of § 9.3.1.

We therefore introduce a model system which we expect to have some resemblance to the fluid we are studying. Its properties are the following: the two-body potential is attractive and its integral, $-2a$, is negative; in order to avoid the catastrophe which then would make A divergent as we noted at the end of § 9.3.1, we add repulsive three-, four-, ... body potentials which are intended to prevent the density from becoming infinite. In the mean field approximation a three-body potential makes a contribution to $\mathcal{U}$ which is proportional to n^3. Denoting the value of that repulsive potential by c_1 and assuming that its range is of the order of the radius of a molecule, we have thus for the coefficient of the term in n^3 in $\mathcal{U}$

$$\frac{1}{6} \int d^3\boldsymbol{r}_1 \, d^3\boldsymbol{r}_2 \, d^3\boldsymbol{r}_3 \, W_3(\boldsymbol{r}_1, \boldsymbol{r}_2, \boldsymbol{r}_3) \equiv c_1 b^2 \Omega,$$

where b is of the order of magnitude of the volume (9.13) of a single molecule. Similarly, the four-, five-, ... body potentials give terms proportional to b^3n^4, b^4n^5, ... To simplify the discussion we shall choose their coefficients such that the *approximate internal energy* associated with the trial density (9.39) has the form (we have put $c = b^2c_1$)

$$\mathcal{U} = \frac{3}{2}kT\Omega n - a\Omega n^2 + \frac{c\Omega n^3}{1 - bn}, \tag{9.50}$$

instead of (9.43). The second term takes the attractive part of the two-body potential into account and the last term *simulates its repulsive part*, as the density n cannot exceed the value $1/b$ for which the energy $\mathcal{U}$ becomes infinite.

Another model, that of the *lattice gas*, also enables us to describe simply both the short-range repulsion and the long-range attraction. It consists in imagining that the molecules can only be situated on the sites of a lattice of mesh size r_0, rather than occupying arbitrary positions. Forbidding two molecules to occupy the same site imposes a minimum distance apart r_0 as if they had a hard core. Nothing prevents us then to introduce an attractive potential between molecules placed on different sites and after that to describe the liquid-gas transition by the effective potential method (Prob.7).

We must now, for variable n, look for the *minimum of the trial grand potential*

$$\mathcal{A}(n, T, \Omega, \mu) = kT\Omega \left(n \ln \frac{n}{f} - n \right) - a\Omega n^2 + \frac{c\Omega n^3}{1 - bn}, \tag{9.51}$$

the value of which will give us approximately the grand potential $A(T, \Omega, \mu)$. From the vanishing of the derivative of (9.51) with respect to n we then obtain the equation

$$kT \ln \frac{n}{f} - 2an + \frac{cn^2(3-2bn)}{(1-bn)^2} = 0,$$

or, if we use (9.5),

$$\mu = \mu(n) \equiv kT \ln n\lambda_T^3 - 2an + \frac{cn^2(3-2bn)}{(1-bn)^2}, \tag{9.52}$$

which *determines the density n as function of T and μ*. Eliminating μ from (9.51) and (9.52), and using a Legendre transformation as in (9.25), we find for the *free energy* the approximation

$$F = A + \mu N \simeq NkT \left(\ln \frac{N\lambda_T^3}{\Omega} - 1 \right) - \frac{aN^2}{\Omega} + \frac{cN^3}{\Omega(\Omega - Nb)}, \tag{9.53}$$

which is consistent with (9.44) and (9.50).

Hence we get the approximate *equation of state*

$$\mathcal{P} = -\frac{\mathcal{A}}{\Omega} = -\frac{\partial F}{\partial \Omega} = kTn - an^2 + \frac{cn^3(2-bn)}{(1-bn)^2}. \tag{9.54}$$

The isotherms of the equation of state (9.54) show the same behaviour as those of the van der Waals equation. In particular, there exists a critical temperature T_c below which (9.54) shows a minimum and a maximum (Fig.9.6), and $\mathcal{P}$ may even become negative. The equation of state (9.54) thus contradicts the general properties $\mathcal{P} > 0$, $\partial\mathcal{P}/\partial n < 0$ predicted by both thermostatics and equilibrium statistical mechanics.

However, this expression is not completely correct. In fact, Eq.(9.52) which we have used gives *all the maxima and minima* of the variational expression (9.51) whereas we should only have chosen the *absolute minimum* of $\mathcal{A}$. We must therefore discuss in more detail the solutions of Eq.(9.52). Let us, *for fixed Ω and T*, consider the function $\mu(n)$ which is the right-hand side of this equation, and let us write down its successive derivatives:

$$\begin{aligned} \frac{\partial \mu(n)}{\partial n} &= \frac{1}{\Omega}\frac{\partial^2 \mathcal{A}}{\partial n^2} = \frac{kT}{n} - 2a + \frac{2c}{b}\left[\frac{1}{(1-bn)^3} - 1\right], \\ \frac{\partial^2 \mu(n)}{\partial n^2} &= \frac{1}{\Omega}\frac{\partial^3 \mathcal{A}}{\partial n^3} = -\frac{kT}{n^2} + \frac{6c}{(1-bn)^4}, \\ \frac{\partial^3 \mu(n)}{\partial n^3} &> 0. \end{aligned} \tag{9.55}$$

We note that there exists a temperature T_c such that the function $\mu(n)$ has a point of inflection with a horizontal tangent; we find it by letting the first two equations (9.55) be equal to zero and this defines the *critical point* T_c, n_c, μ_c, $\mathcal{P}_c$ through the equations

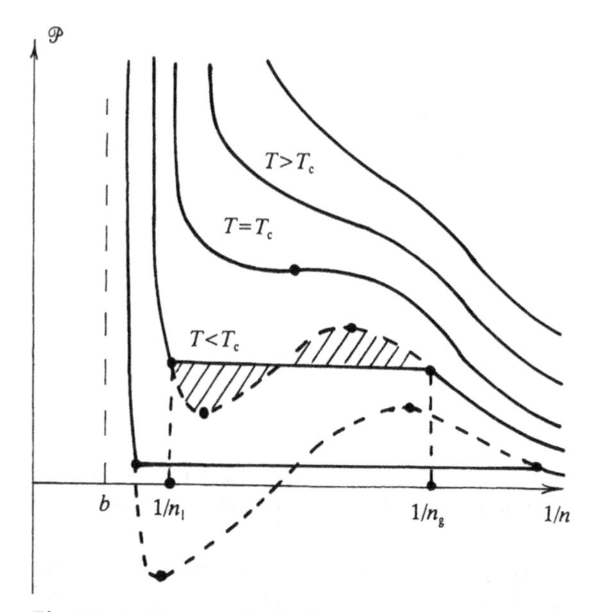

Fig. 9.6. Isotherms of a fluid with the internal energy (9.50)

$$\begin{aligned} kT_c &= \frac{6cn_c^2}{(1-bn_c)^4}, \\ \frac{1+2bn_c}{(1-bn_c)^4} &= 1+\frac{ab}{c}, \\ \mu_c &= \mu(n_c, T_c), \\ \mathcal{P}_c &= \mathcal{P}(n_c, T_c) \;=\; \frac{cn_c^3(2+bn_c)}{(1-bn_c)^4}. \end{aligned} \tag{9.56}$$

When the temperature lies above T_c, equations (9.55) show that $\mu(n)$ increases monotonically from $-\infty$ to $+\infty$ so that Eq.(9.52) has a single solution $n(\mu, T)$; therefore $\mathcal{A}$ has a single minimum which is equal to the approximate grand potential A (Fig.9.7). Expressions (9.53) and (9.54) for the free energy and the pressure, $\mathcal{P} = -A/\Omega$, are then valid whatever the value of n. In the $\mathcal{P}(1/n)$-diagram of Fig.9.6 the isotherms are monotonically decreasing curves, as, for given T,

$$\frac{\partial \mathcal{P}}{\partial(1/n)} = -\,n^3\,\frac{\partial \mu(n)}{\partial n}. \tag{9.57}$$

The situation is more complicated when $T < T_c$ where it follows from (9.55) and (9.56) that $\mu(n)$ has a maximum and a minimum. When μ is negative and large, the equation $\mu = \mu(n)$ has still only one solution and $\mathcal{A}(n)$ has therefore a single minimum A, reached at a low value of the density n which increases with μ (Fig.9.8). When μ increases, the equation $\mu = \mu(n)$

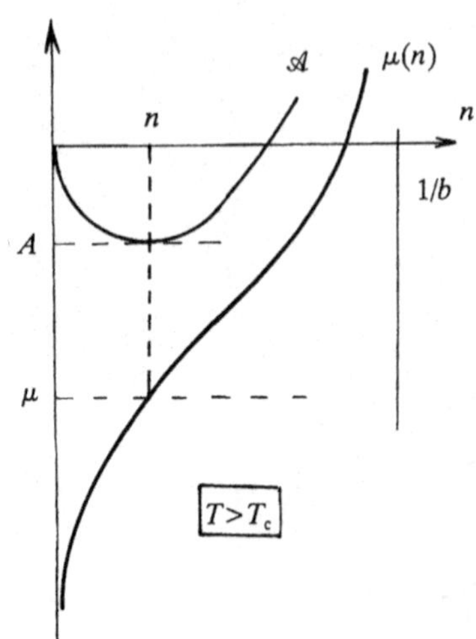

Fig. 9.7. The chemical potential $\mu(n)$ at a fixed temperature T above the critical temperature

changes from having one to having three solutions and a second minimum of $\mathcal{A}$ appears for a larger value of n; as μ increases, the two minima become deeper, the second one faster than the first one. As long as the latter remains the absolute minimum, it provides us with the required value of n and hence of $A = \mathcal{A}(n)$ and of the other physical quantities. The second solution of (9.52), which corresponds to a less deep minimum of $\mathcal{A}(n)$, and *a fortiori* the third one, reached for an intermediate value of n, which corresponds to a maximum, *must be rejected.* For a certain value $\mu_s(T)$ of μ the two minima have the same depth (Fig.9.8). There are thus *two solutions* $n_g(T)$ and $n_l(T)$ associated with that value of μ, which give the same value A_s for A. The latter, together with the values of n_g, n_l, and μ_s, are for given T determined by solving the equations

$$\begin{aligned} \mu_s &= \mu(n_g) = \mu(n_l), \\ A_s &\equiv -\mathcal{P}_s\Omega = \mathcal{A}(n_g) = \mathcal{A}(n_l). \end{aligned} \tag{9.58}$$

When μ goes on increasing beyond $\mu_s(T)$, the deepest minimum of $\mathcal{A}$ becomes the one on the right and n increases starting from the value n_l. From here on the solutions of (9.52) which must be discarded are the minimum on the left and the maximum. Finally, for even larger values of μ only the solution on the right remains and $\mathcal{A}(n)$ once again has a single minimum, this time for a high density which grows until it reaches $1/b$.

For a given T below T_c, we have thus found a *discontinuous* behaviour, which is characteristic of the so-called first-order transitions (§ 6.4.6). The density n, as function of μ, increases, but it jumps from $n_g(T)$, *the density of the saturated vapour*, to $n_l(T)$, *the minimum liquid density* at temperature T, when μ passes through $\mu_s(T)$. The fluid thus undergoes a sudden *change of state* at the *transition point* $\mu = \mu_s(T)$: it is a *gas* when $\mu < \mu_s(T)$ and in that case $n < n_g(T)$, and it is a *liquid* when $\mu > \mu_s(T)$ in which case

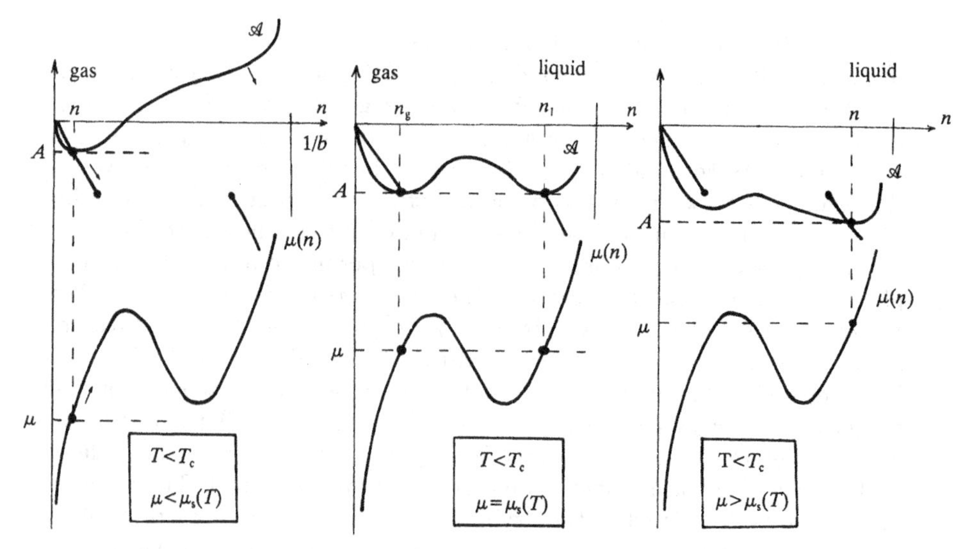

Fig. 9.8. The chemical potential $\mu(n)$ when $T < T_c$ and the solution of (9.52)

$n > n_l(T)$. In each of these two regions the approximate expressions (9.53) for the free energy and (9.54) for the pressure as functions of n are valid; however, we must exclude the parts with $n_g < n < n_l$ from the $F(n)$ and $\mathcal{P}(n)$ curves which, for constant T, are defined by (9.53) and (9.54). On the other hand, the $\mathcal{P}(\mu, T)$ *isotherms* are continuous as functions of μ with a discontinuous derivative. In fact, the Gibbs-Duhem identity (5.79), which can be written in the form

$$d\mathcal{P} = -d\left(\frac{A}{\Omega}\right) = \frac{S}{\Omega}\,dT + n\,d\mu, \tag{9.59}$$

gives a jump $n_l - n_g$ in the slope of the $\mathcal{P}(\mu)$ isotherm at $\mu = \mu_s$. As function of n an isotherm appears in Fig.9.8 as the locus of the absolute minima of $\mathcal{A}$, since we have $\mathcal{P} = -A/\Omega$. It therefore consists of *two sections* of the curve in Fig.9.6; the pressure increases with n according to (9.57), but it is in the present approximation not defined when $n_g(T) < n < n_l(T)$.

In the vicinity of $T = T_c$, the difference $n_l - n_g$ between the liquid and the saturated vapour densities can be evaluated by expanding Eqs.(9.58) near the critical point. We find

$$\frac{n_l - n_c}{n_c} \sim \frac{n_c - n_g}{n_c} \sim \left[\frac{3(T_c - T)(1 - bn_c)}{T_c(1 + bn_c)}\right]^{1/2}, \tag{9.60}$$

so that the densities of the two phases *tend towards one another* when $T \to T_c - 0$, whereas there exists *only a single fluid phase when* $T > T_c$.

Our model has thus enabled us to explain the well known properties of the gas-liquid transition: at a given temperature $T < T_c$ the *saturated vapour pressure* $\mathcal{P}_s(T)$ is a limit beyond which the gas cannot be compressed in thermal equilibrium without condensing as a liquid; the density $n_g(T)$ of the saturated vapour is the upper bound of the gas densities. Symmetrically, the liquid density cannot go lower than $n_l(T)$ and the liquid vaporizes if its pressure is lowered below $\mathcal{P}_s(T)$. The density has a discontinuity. However, the differences between the liquid and its saturated vapour diminish as T increases towards T_c and one can pass continuously from the one to the other beyond T_c. Coexistence of liquid and vapour is permitted provided these two phases have the same temperature, the *same chemical potential*, and the *same pressure* which are given by $\mu = \mu_s(T)$, $\mathcal{P} = \mathcal{P}_s(T)$. If the pressure is fixed and *less* than $\mathcal{P}_c$ the temperature which is defined by $\mathcal{P}_s(T) = \mathcal{P}$ is the *boiling temperature* of the liquid or the condensation temperature of the gas. It *increases with the pressure* according to the law (9.63) which will be given below. For instance, water already boils at 80 °C at an altitude of 5000 m where the atmospheric pressure has dropped by half. The increase of $\mathcal{P}_s(T)$ with T is used in the drying of dishes in dishwashers: by cooling the inside wall with cold water, one lowers the corresponding value of $\mathcal{P}_s(T)$, which makes water condense there, thus drying the air and the hot dishes. All these everyday facts to which we are accustomed, as well as the results of macroscopic experiments, have thus found a satisfactory microscopic theoretical foundation. Yet, we stressed at the start of this chapter how a phase change such as the liquid-gas transition looked a puzzling phenomenon from the microscopic point of view.

The discontinuity in the density between the liquid and the saturated vapour implies a discontinuity in the internal energy given by (9.50), the free energy given by (9.53), and the entropy given by (9.44). The latter, which in our approximation is equal to

$$S_g - S_l \;=\; Nk \,\ln \frac{n_l}{n_g}, \tag{9.61}$$

is a measure of the *sudden increase in disorder when the liquid evaporates.* It is associated with a remarkable thermal property of the transition, namely, the existence of a *vaporization latent heat* $L = T(S_g - S_l)$ which must be provided in order that, at given temperature T and pressure $\mathcal{P}_s(T)$, a liquid of N molecules can evaporate. Independent of the approximation used L is connected with the saturated vapour pressure through *Clapeyron's relation* (6.61). We find the latter also by using (9.59) to write down that the change in $\mathcal{P}_s(T)$ on the two sides of the $\mu_s(T)$ curve, in the two phases, is the same:

$$\frac{d\mathcal{P}_s}{dT} \;=\; n_g \left(\frac{S_g}{N} + \frac{d\mu_s}{dT}\right) \;=\; n_l \left(\frac{S_l}{N} + \frac{d\mu_s}{dT}\right). \tag{9.62}$$

From (9.62) we find the latent heat per mole,

$$L \;=\; T(S_g - S_l) \;=\; N_A T \frac{d\mathcal{P}_s}{dT} \left(\frac{1}{n_g} - \frac{1}{n_l}\right), \tag{9.63}$$

and also the slope of the saturation curve in the T, μ plane,

$$\frac{d\mu_s}{dT} = -\frac{1}{N}\frac{n_l S_l - n_g S_g}{n_l - n_g}. \tag{9.64}$$

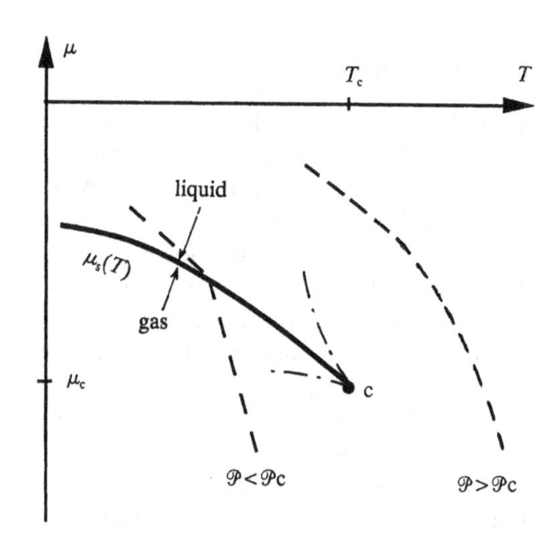

Fig. 9.9. The phase diagram in the T, μ plane

A convenient graphical representation of the transition consists in drawing the *phase diagram in the T, μ plane* (Fig.9.9). The *saturation curve* $\mu_s(T)$ separates the gas phase, which lies below it, from the liquid phase, which lies above it. It ends at the critical point around which the distinction between the two fluid phases loses its meaning. We have shown by dashed curves two *isobars* $\mathcal{P}$ = const; along these lines $A/\Omega = -\mathcal{P}$ is constant, so that the T, μ, $\mathcal{P}$ diagram is also a graphical representation of the grand potential as function of its natural variables. This enables us to consider it as an *abacus* providing us graphically with the various thermodynamic quantities. In fact, (9.59) shows that the horizontal and vertical distances between isobars are inversely proportional to the entropy per unit volume and to the density. The *slope of an isobar*,

$$\left(\frac{d\mu}{dT}\right)_{\mathcal{P}} = -\frac{S}{N}, \tag{9.65}$$

gives us directly the *entropy per particle*, and its curvature,

$$\left(\frac{d^2\mu}{dT^2}\right)_{\mathcal{P}} = -\frac{C_p}{NT}, \tag{9.66}$$

the specific heat at constant pressure. It follows from (9.65) and (9.66) that the isobars are decreasing and concave curves. Above the critical pressure, for which the isobar passes through the critical point, they are regular. Below it, they show a discontinuity in slope at the point where they cross the saturation curve. In fact, the pressure is continuous, but its partial derivatives (9.59) are not continuous for $\mu = \mu_s(T)$. The discontinuity in the slope of the isobar represents according to (9.65) the difference between the entropies per particle in the two phases, that is, L/NT.

9.3.3 Coexistence of Gas and Liquid Phases

The use of the grand canonical ensemble hides a difficulty. In practice, at the same time as the available volume Ω, the total number of particles N is given rather than the chemical potential. If, for a given $T < T_c$, N/Ω lies between the values $n_g(T)$ and $n_l(T)$, which we found above, our theory does not work as we do not find a solution which is admissible for Eq.(9.52). The reason for this is simple. We know that, if $n_g\Omega < N < n_l\Omega$, the fluid splits in space into two phases, one a liquid with density n_l and the other a gas with density n_g, which are in equilibrium with one another. However, our approximation in § 9.3.2 assumed a uniform density n. Let us therefore return to our variational treatment, allowing henceforth the effective potential $\mathcal{V}(\boldsymbol{r})$ and hence the *trial density* $n(\boldsymbol{r})$ to be *inhomogeneous*.

Let us, for instance, assume that we divide the available volume Ω into two parts, Ω_1 and $\Omega_2 = \Omega - \Omega_1$ in which the potential $\mathcal{V}$ takes on two distinct values $\mathcal{V}_1$ and $\mathcal{V}_2$. The interactions W, W_3, ... have a short range; we can therefore neglect the contributions to the trial internal energy $\mathcal{U}$ coming from the boundary between the two domains Ω_1 and Ω_2, so that the trial grand potential is equal to

$$\mathcal{A} = \mathcal{A}(n_1, T, \Omega_1, \mu_1) + \mathcal{A}(n_2, T, \Omega - \Omega_1, \mu), \tag{9.67}$$

where $\mathcal{A}(n, T, \Omega, \mu)$ is given by (9.51). We must look for the minimum of (9.67) with respect to the three adjustable parameters, n_1, n_2, Ω_1. When $\mu \neq \mu_s(T)$, this minimum is reached for $n_1 = n_2$ and does not differ from the one which we obtained above. However, along the transition line $\mu = \mu_s(T)$, $\mathcal{A}$ is a minimum not only in the two cases $\Omega_1 = \Omega$, $n_1 = n_g$ and $\Omega_1 = 0$, $n_2 = n_l$, that describe the uniform gas and liquid states which we studied earlier, but also for $n_1 = n_g$, $n_2 = n_l$, Ω_1 arbitrary. The grand potential per unit volume *thus remains constant when we let the fractions of the two phases vary* and also the shape of their boundary, when $\mu = \mu_s(T)$. In particular, we can choose Ω_1 such that $\Omega_1 n_g + \Omega_2 n_l = N$ so that the values $n_g < n < n_l$ which we had not found in § 9.3.2 are now obtained for a *separation of the fluid into two phases* with $\mu = \mu_s(T)$. Each point of the saturation curve $\mu_s(T)$ thus represents not only the two uniform phases with densities n_g and n_l, but also a *juxtaposition of these two phases in arbitrary ratios.*

We can easily generalize this argument to any number of domains. It then provides us for $\mu = \mu_s(T)$ with a whole family of solutions for which the density in each point takes either the value n_g or the value n_l and which describe a liquid in arbitrary regions of space and its saturated vapour elsewhere. The grand potential takes on the same value for all these solutions, so that *the pressure remains constant* when T is given while N/Ω varies between n_g and n_l. This describes the *evaporation plateau*, which is the part of each isotherm of Fig.9.6 between $1/n_g$ and $1/n_l$ and along which the system is made up of two separated phases.

Since the entropy (9.44) generalized to inhomogeneous states is an integral over the volume with an integrand which is a function of the density, for given T and Ω, the *entropy varies linearly with* N in the coexistence region $n_g\Omega < N < n_l\Omega$. (In fact, the volumes occupied by each of the two phases are themselves linear functions of N.) It also varies, for given T and N, linearly with Ω in this region. The same holds for the internal energy and for the free energy. This property implies that during the evaporation, for fixed temperature and pressure, *the heat supplied is proportional to the amount of evaporated liquid*, a well known experimental fact.

We have just found a large number of solutions which are characterized by the regions occupied by one or other of the two phases inside the vessel. In practice, *gravity* determines these regions. In order to see how the effective potential theory takes this fact into account we include the gravity term, which we have so far neglected, in the Hamiltonian. The variational energy (9.50) is replaced by

$$\mathcal{U} = \int d^3r \left[\frac{3}{2}kTn - an^2 + \frac{cn^3}{1-bn} + mgzn\right], \tag{9.68}$$

where z is the height and where $n = n(\boldsymbol{r})$ is the trial density which now depends on the position. At each point Eq.(9.52) is replaced by

$$\mu - mgz \equiv \mu_s - mg(z - z_s) = \mu(n), \tag{9.69}$$

where we have denoted by z_s the height for which $\mu - mgz_s$ equals $\mu_s(T)$. If the vessel is not too high so that the gravity term $mg|z - z_s|$ remains small, the solution of (9.69) is $n = n_g$ for $z > z_s$, $n = n_l$ for $z < z_s$. As a result, even though gravity hardly alters the thermodynamics, it plays an essential rôle in affecting the search for the absolute minimum of $\mathcal{A}$. Through (9.69) it requires that the density *decreases* with increasing height and it *chooses a particular solution*, where the liquid assembles at the bottom of the vessel, separated from its saturated vapour by a horizontal interface $z = z_s$.

The approximations made above give us a *sudden jump in the density at the interface.* A microscopic theory of the structure of the interface needs taking into account non-local contributions to the trial grand potential (Exerc.9e). This enables us to understand the origin of an interface free energy and of capillary forces. We thus justify the more macroscopic and semi-empirical approach to the same problem (Exerc.6c).

We have so far constructed only stable equilibrium states which are associated with the *absolute minimum* of the trial grand potential $\mathcal{A}$. In fact, a *local minimum* can describe approximately a *metastable phase* (§4.1.6, Exerc.9d, Prob.7). Let us, indeed, assume that, for reasons which we shall discuss later on, the density is constrained to stay within a given range during a period which is long according to our own time scale. The minimum of $\mathcal{A}$ *in this range* then describes in our model an equilibrium state which is metastable if elsewhere there exists a deeper minimum; the state of the system is described by the former minimum during a sizeable time, but eventually changes towards the latter if one waits sufficiently long. Let us consider a temperature much lower than the critical temperature; this enables us to distinguish two density ranges which are very different, that of the gas phase and that of the liquid. Let us, for instance, start from a stable gas state ($n < n_g(T)$) which in Fig.9.8 is described as the absolute minimum of $\mathcal{A}$, reached for the left-hand solution of the equation $\mu = \mu(n)$. Let us now make μ increase by raising the pressure at constant T, and let us assume that the density be constrained to remain within the range of gas densities. When μ passes $\mu_s(T)$ we get under that constraint for the minimum of $\mathcal{A}$ a *local* minimum, which is still associated with the left-hand solution of $\mu = \mu(n)$ and which describes a metastable state; the *absolute* minimum $n > n_l$ is not accessible. The system remains a gas, but its density increases continuously with μ beyond the density $n_g(T)$ of the saturated vapour. This phenomenon is often observed: *supersaturated vapour*. Symmetrically, if we start from a liquid phase $\mu > \mu_s(T)$, $n > n_l(T)$ and if we make μ decrease below $\mu_s(T)$ by lowering the pressure at fixed T, under conditions where the density is forced to remain in the liquid range, we find another metastable phase, a *superheated liquid*. In Fig.9.8 this phase is represented by a local minimum of $\mathcal{A}$ which is reached for a density just below $n_l(T)$; it continues the absolute minimum which for $\mu > \mu_s$ describes the stable liquid phase. Of course, neither of these two metastable phases can exist outside the T, μ region where the equation $\mu = \mu(n)$ has three solutions. This condition, for given T, gives an upper bound for μ and n for the supersaturated vapour and a lower bound for the superheated liquid. On the $\mathcal{P}(1/n)$ *isotherms* in Fig.9.6 the metastable phases are indicated by the descending parts of the dashed curves in the n_g, n_l interval. In the T, μ *phase diagram* of Fig.9.9, the region representing the supersaturated vapour continues the gas region by being superimposed upon part of the liquid region; it is bounded below by the saturation curve and above by another curve which ends at the critical point. Similarly, the region representing the superheated liquid, which is situated below the saturation curve, is superimposed upon a part of the gas region. The limits of metastability are sketched in Fig.9.9 as dash-dot curves; as an exercise one could determine their shape near the critical point, using (9.52), (9.55), and (9.56).

The physical relevance of such local minima of $\mathcal{A}$ is a question of *dynamics*. One can, for instance, only observe superheated liquid if the time it takes for this metastable phase to change to the true equilibrium state is very long. When T, N, and Ω are given this equilibrium state is a saturated vapour coexisting with liquid. The metastable superheated liquid can usually be reached, starting from the stable liquid for $\mu > \mu_s(T)$, either by slowly heating at fixed $\mathcal{P}$ or by lowering the pressure below $\mathcal{P}_s(T)$ at fixed T. The subsequent evolution of this metastable state towards the equilibrium state requires a partial boiling, which necessitates the formation of vapour *bubbles* with increasing sizes. However, the appearance of bubbles, or "*nucleation*", implies a sudden decrease in the density at some points, together with a slight increase in the liquid density; this dynamic phenomenon is governed by many factors. In particular, the probability for producing a bubble by statistical fluctuations is, according to (5.88), small if the free energy of the interface between the liquid and the bubble is large. The mechanism for the growing of the bubble depends on the viscosity and the heat conductivity of the fluid. Exterior factors also are involved in the nucleation, such as the surface conditions of the container. Another example is provided by *bubble chambers* which are used for the detection of elementary particles: liquid hydrogen is first suddenly decompressed and made metastable, and the perturbation produced by the passage of a charged particle produces a line of bubbles. If the probability for nucleation is small the fluid density needs an extremely long time to decrease to values $n \sim n_g$ which are necessary to establish the true equilibrium. Under those conditions a decompression or heating of the stable liquid, leading to the $\mu < \mu_s(T)$ region, has little chance of making it boil; one reaches a metastable equilibrium and the variational theory for this equilibrium is based upon looking for the minimum of the trial grand potential $\mathcal{A}$ *in the region of high densities* which are rather close to the initial density. In fact, this region is the only one which is accessible, if we are restrict ourselves to rather short times during which no bubbles are formed.

The densities $n_g(T)$ and $n_l(T)$ of the two stable phases which can coexist at the temperature T are analytically determined by Eqs.(9.58). The Maxwell construction gives a simple graphical solution of those equations, merely expressing that *at the two ends of the evaporation plateau* in Fig.9.6 the *chemical potential and the pressure take the same value.* The forbidden part of the isotherm is represented by the dashed part of the $\mathcal{P}(1/n)$ curve in Fig.9.6; along it $\mathcal{P}$ and μ, given by (9.54) and (9.52), change continuously from their initial values $\mathcal{P}_s(T)$, $\mu_s(T)$, which they take on for $n = n_g$ to the same values which they again reach when $n = n_l$. Between these points the derivatives of $\mathcal{P}(1/n)$ and $\mu(n)$ for constant Ω and T are related by (9.57). To write down the condition that μ takes the same values at the two ends of the evaporation plateau we integrate (9.57) by parts along the forbidden portion of the isotherm:

$$
\begin{aligned}
0 &= \int_{n_g}^{n_l} \frac{\partial \mu}{\partial n}\, dn \\
&= -\int_{1/n_l}^{1/n_g} \frac{\partial \mathcal{P}}{\partial(1/n)}\, \frac{1}{n}\, d\left(\frac{1}{n}\right) \\
&= \int_{1/n_l}^{1/n_g} \mathcal{P}\, d\left(\frac{1}{n}\right) - \mathcal{P}_s\left(\frac{1}{n_g} - \frac{1}{n_l}\right).
\end{aligned}
\tag{9.70}
$$

On the right-hand side we can identify the oriented area of the domain contained between the forbidden part of the isotherm and the plateau. The fact that it vanishes gives a microscopic theoretical justification of *Maxwell's rule* which determines the position of the evaporation plateau when the isotherm obtained from an approximate theory shows a minimum and a maximum.

In thermodynamics one usually gives the following argument for justifying the Maxwell construction: in an isothermal transformation along a closed circuit following the forbidden part of the isotherm and the plateau, the work received, which is equal to the oriented area enclosed, must be equal to zero because of the Second Law. This argument is not satisfactory, as it is based upon considering homogeneous, non-equilibrium states represented by the forbidden part of the isotherm: even though those which correspond to the descending parts of the isotherm and which are associated with relative minima of $\mathcal{A}$ can describe metastable states, the ascending part, associated with a *maximum* of $\mathcal{A}(n)$, has no physical meaning at all.

The variational method which we have used has the advantage of being applicable even at high densities, in contrast to the approximations of § 9.2 which only gave us the van der Waals equation as a debatable extrapolation from the low density region to the region where the fluid is a liquid. Nevertheless, one must wonder whether the singularities that we have found are perhaps a spurious phenomenon resulting from the approximation made. In fact, though the latter is partly controlled by the fact that the minimum of the trial function $\mathcal{A}$ is an upper bound to the exact value of the grand potential, nothing proves that a discontinuity in the slope of *that bound* indicates a true discontinuity in the slope of the corresponding physical quantity. Indeed, the outcome of the variational method may become inaccurate when the system shows large fluctuations. This happens in one dimension (Exerc.9c). This also happens in the *immediate vicinity of the critical point* where detailed experiments show significant discrepancies from the mean field theory. For instance, Eq.(9.60) predicts a decrease in $n_l - n_g \propto (T_c - T)^{1/2}$ as T increases and approaches T_c. Whereas this behaviour is almost correct when $T_c - T$ is of the order of one kelvin, precise experiments show that the exponent is not equal to $\frac{1}{2}$ when one approaches the critical point much more closely, but to 0.325 ± 0.002, whatever the substance. A correct theory of such exponents requires that one takes into account the *fluctuations* which are neglected here but which play an essential rôle at the critical point where the distinction between the liquid and the gas disappears (Exerc.6d). Nevertheless, even though the nature of the singularity at the critical point and the behaviour of the thermodynamic functions in that point are, in general, not given correctly by the effective potential approximation, the latter gives at least qualitative agreement with experiments and provides us with an excellent starting point for more sophisticated theories.

The existence of singularities in the thermodynamic functions is a consequence of the large size of the system. There are *no phase transitions in finite systems*. In fact, the partition function Z in that case is a sum of exponentials; each of them is a holomorphic fuction of its natural variables, such as β, and is positive when β is real. The thermodynamic potential associated with $\ln Z$ is thus also holomorphic in a strip along the real β-axis, where it cannot have any singularity. If one makes an analytical continuation to *complex* β-values, the only thing which can happen is the appearance of *zeroes in the partition function* giving rise to branch points of $\ln Z$ outside the real axis. A phase transition, at a real transition point β_0, can thus only appear *in the thermodynamic limit* if some zeros of $N^{-1} \ln Z$ in the complex β-plane pinch the real axis by tending on both sides to the point $\beta = \beta_0$ as $N \to \infty$ (Exerc.9b). Only in that case can the thermodynamic quantities show a behaviour which is different for $\beta < \beta_0$ and for $\beta > \beta_0$.

Let us also note that, although the liquid and gas phases are separated by a singularity when $T < T_c$, their properties can be derived from one another through an *analytical continuation* when one follows a contour in the T, μ-plane which encircles the critical point (Fig.9.9). In the variational approximation of § 9.3.2 this is clear: we started from a *single analytical* expression (9.53) and the singularity appeared because we truncated the domain for n. A similar situation occurs for other phase transitions such as *ferromagnetism* (Exerc.6d, 9a). In that case, above the critical Curie temperature and when there is no magnetic field $\boldsymbol{B}$ present, the magnetization spontaneously takes on a finite value and it is *oriented in an arbitrary direction.* Each of the possible orientations characterizes a phase, similar to each of the two liquid or gas phases. The singularity for $\boldsymbol{B} = 0$, $T < T_c$ in the thermodynamic quantities plays the rôle of the liquid-vapour coexistence curve in the T, μ-plane. Here also, we can go around the critical point by introducing a non-vanishing field, letting the orientation of $\boldsymbol{B}$ vary and finally letting $\boldsymbol{B}$ go to zero; we thus pass analytically from one direction of the spontaneous magnetization to another. Nevertheless, in a *fluid-solid* transition the two phases are *qualitatively different* and cannot be connected continuously; the phase diagram separates two regions *which are not in touch with one another* and there is no critical point around which one can go to connect them. In a variational theory of this transition (Prob.8) the approximations used to describe each of the – fluid and crystalline – phases are, moreover, different.

Let us finally stress that in the region where the two gas-liquid phases coexist *one cannot uniquely determine the state of the system* by looking for the minimum of the grand potential. We have, in fact, found below the critical temperature and for $\mu = \mu_s(T)$ two solutions n_g and n_l, not to mention the solutions which describe states where both phases coexist. For other phase transitions this same property shows up even more clearly. When a substance is cooled down below its Curie temperature in zero field, it becomes ferromagnetic and it acquires a spontaneous magnetization which is oriented in a *preferred direction*. However, if the substance had a single equilibrium macro-state, the latter would be isotropic as one expects *a priori* to observe rotational invariance; as a result, the magnetization should be zero. This is observed for $T > T_c$, but the *rotational invariance is spontaneously broken when* $T < T_c$. The magnetization is an *order parameter*, which must be introduced in the macroscopic description of the system (Exerc.6d). The increase in disorder, that is, in entropy, with temperature is reflected spectacularly in the behaviour of the order parameter M, as the latter is *zero* when $T > T_c$, and *non-zero* and increasing when T decreases below T_c (see Fig.9.13). Each low temperature phase, which is here characterized by the direction of $\boldsymbol{M}$, has *lost its rotational invariance property*; the latter, moreover, implies the very existence of several phases, as these can be derived from one another through rotation. Because of the equivalence of the phases, small perturbations are in practice sufficient to determine the direction of the magnetization, just as gravity separates the gas and the liquid phases. Other examples of transitions show similarly a spontaneous breaking of invariance, like the *crystallization of a fluid* (Probs.8, 10, and 13), *ferroelectricity* (Prob.5), or changes in *crystalline structure* (Probs.4, 12, and 19). In particular, in crystallization, the fluid, which is the only stable phase at high temperatures, is invariant under arbitrary translations or rotations; among this displacement group only certain operations, those which characterize the geometry of the crystal lattice, remain below the transition point.

The broken invariance is less obvious in the case of the liquid-vapour transition. One can discover it, though, in the *fluid on a lattice* model of Prob.7: whatever T, there is symmetry for $\mu = \mu_c$ between whether or not there is a particle on each of the sites, similar to the Ising model where the flipping of all spins leaves the Hamiltonian invariant for $B = 0$ (Exerc.9a). In fact, these two models show a complete formal similarity: one gets from one to the other by identifying σ_i with $2n_i - 1$, where $n_i = 0$ or 1 is the number of particles on site i and $\sigma_i = \pm 1$ is the spin on that site, and at the same time replacing $\mu - \mu_c$ by the magnetic field. Above T_c the invariance is not broken and the density $n = \langle n_i \rangle$ equals $\frac{1}{2}$ when $\mu = \mu_c$; below it we find two solutions n_g and n_l which are symmetric with respect to $\frac{1}{2}$. In the model of § 9.3.2 we recognize near the critical point the same symmetry by taking as order parameter $M = n - n_c + k(n - n_c)^2 + \ldots$ as is suggested by (9.60). An appropriate choice of k then enables us, by eliminating the term in M^3, to give exactly the same form to the expansion of the trial grand potential (9.51) around the critical point, as for the Landau expansion (6.109), (6.110) of the free energy of a ferromagnetic, through replacing $\mu - \mu_c$ by B. This formal similarity is the origin of the *universality of the critical phenomena*: the properties at the critical point are the same for seemingly different transitions – liquid-gas, Ising ferromagnetism in one direction, binary alloys, demixture of binary solutions – when one can establish a correspondence between their order parameters and their spontaneous invariance breaking.

Summary

When a gas is compressed, the interactions between its molecules give rise to corrections to the perfect gas laws which can be calculated by power expansions in the interaction or in the density. The equation of state is changed and an approximate theory enables us to justify the empirical van der Waals equation. A Joule expansion cools the gas; a Joule-Thomson expansion heats or cools it, according to whether the initial temperature lies above or below an inversion temperature which can be evaluated if one knows the intermolecular forces.

The liquefaction of a gas, taken as a prototype of a phase transition, can be theoretically studied by a variational approximation method applied to a model. In this way one can explain the various properties which are observed at the macroscopic scale: discontinuity in the density, evaporation plateau in the isotherms, evaporation heat, phase diagrams, separation of the fluid into two phases, critical point, broken invariance, metastable phases.

Exercises

9a Ferromagnetism

A simplified model of a ferromagnetic solid consists in considering a set of interacting spin-$\frac{1}{2}$ particles on a lattice with sites $i = 1, \ldots, N$. For substances which are weakly anisotropic we take for the interaction between two neighbouring spins the "*exchange force*" $-K(\widehat{\boldsymbol{S}}_i \cdot \widehat{\boldsymbol{S}}_j)$, where $\widehat{\boldsymbol{S}}_i$ denotes the three components of a spin operator (§ 10.1.4); this is the so-called *Heisenberg model.* We shall study a model with a simpler interaction, the *Ising model*, which is adequate for very anisotropic solids which can be magnetized along practically only one preferred axis, either in the $+z$- or in the $-z$-direction. Let $\widehat{\sigma}_i$ denote the operators $2\widehat{S}_i^z/\hbar$ which commute with one another and which have as eigenvalues $\sigma_i = \pm 1$, and let us assume that each spin has a magnetic moment $-2\mu_{\rm B} \boldsymbol{S}_i/\hbar$. The spin Hamiltonian in a magnetic field B applied along the $+z$-axis is then given by the expression

$$\widehat{H} = -\sum_{i>j} V_{ij} \widehat{\sigma}_i \widehat{\sigma}_j + \mu_{\rm B} B \sum_i \widehat{\sigma}_i,$$

where the V_{ij} depend only on the distance between the sites i and j. Even though the energy levels are known ($\sigma_i = \pm 1$) the model is still too complicated for a calculation of its equilibrium properties as the density operator $\widehat{D} = \exp(-\beta \widehat{H})/Z$ is too difficult to work with. We shall replace it by an approximate density operator $\widehat{\mathcal{D}}$ with which we can perform the calculations and which has the form

$$\widehat{\mathcal{D}} \propto \mathrm{e}^{-x \sum_i \widehat{\sigma}_i},$$

where x must be determined as well as possible to simulate $\widehat{D}$. To do that we use a variational method which is based on the fact that the free energy $F = -kT \ln Z = \mathrm{Tr} \widehat{D} \widehat{H} - TS(\widehat{D})$ is the minimum of the trial free energy $\mathcal{F} = \mathrm{Tr} \widehat{\mathcal{D}} \widehat{H} - TS(\widehat{\mathcal{D}})$ with arbitrarily varying $\widehat{\mathcal{D}}$ (§§ 4.2.2 and 9.3.1). In order to obtain the best possible approximation, we choose x such as to make $\mathcal{F}$ as small as possible, that is, as close to F as possible.

1. Write down and discuss the equation which determines x for the case when $B = 0$. Show the existence of a phase transition at a temperature $T_{\rm c}$ (Curie temperature) and determine it in the given approximation. In the low-temperature ferromagnetic phase each spin has a non-vanishing magnetic moment $-\mu_{\rm B} \langle \sigma_i \rangle$ and the solid has a *spontaneous magnetization* $M_{\rm s} = -\mu_{\rm B} \sum_i \langle \sigma_i \rangle / \Omega$; determine its behaviour near $T = T_{\rm c}$ and near $T = 0$.

2. Show that F and S are continuous at $T = T_{\rm c}$, when $B = 0$, but that the specific heat has a discontinuity. Explain qualitatively the experimental curve of Fig.9.10.

3. Write down and discuss the equation which determines x for the case when $B \neq 0$. Show that the system is *paramagnetic* for $T > T_{\rm c}$; evaluate its

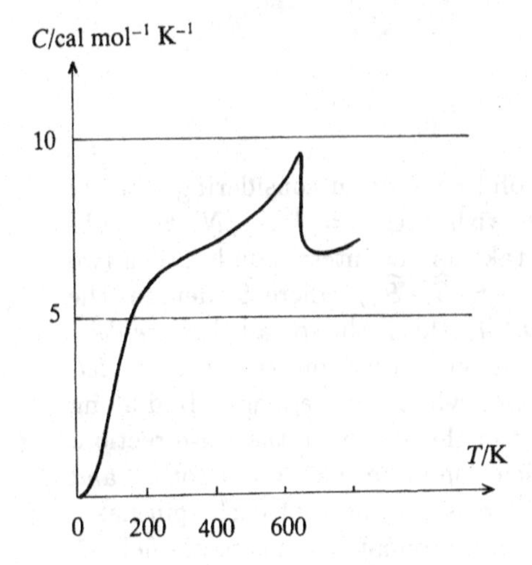

Fig. 9.10. Specific heat of nickel. This metal is ferromagnetic for $T < 631$ K

magnetic susceptibility for $T > T_c$ (*Curie–Weiss law*), for $T < T_c$, and as $T \to 0$. Hence deduce that the statistical fluctuation of the magnetic moment of a sample of volume Ω in zero field becomes infinite at the Curie point.

4. Compare the critical behaviour to that of the Landau theory (Exerc.6d) for the case of a uniform field.

One should note that, although the Ising model gives results in qualitative agreement with experiment, it is not justified for metals (Exerc.11f).

Solution:

1. The trial free energy can be written in the form

$$\mathcal{F} = \frac{\sum_{\{\sigma_i\}} e^{-x\sum_i \sigma_i} \left(- \sum_{j>k} V_{jk}\sigma_j\sigma_k \right)}{\sum_{\{\sigma_i\}} e^{-x\sum_i \sigma_i}} + kT \, \mathrm{Tr}\mathcal{D} \ln \mathcal{D}.$$

We evaluate it by noting that for the $\mathcal{D}$ distribution the σ_i are statistically independent and have an average value equal to

$$\mathrm{Tr}\,\mathcal{D}\sigma_i = -\tanh x,$$

and that the partition function associated with $\mathcal{D}$ is equal to (Exerc.4c)

$$\sum_{\{\sigma_i\}} e^{-x\sum_i \sigma_i} = (2\cosh x)^N.$$

We can thus find the trial entropy and, hence, putting $v = \sum_j V_{ij}$, the trial free energy,

$$\mathcal{F}(x, T, B) = -\tfrac{1}{2} N v \tanh^2 x + NkT(x \tanh x - \ln 2\cosh x). \tag{9.71}$$

We look for the minimum of $\mathcal{F}$ with respect to x amongst the solutions of $\partial\mathcal{F}/\partial x = 0$ which can be written in the form

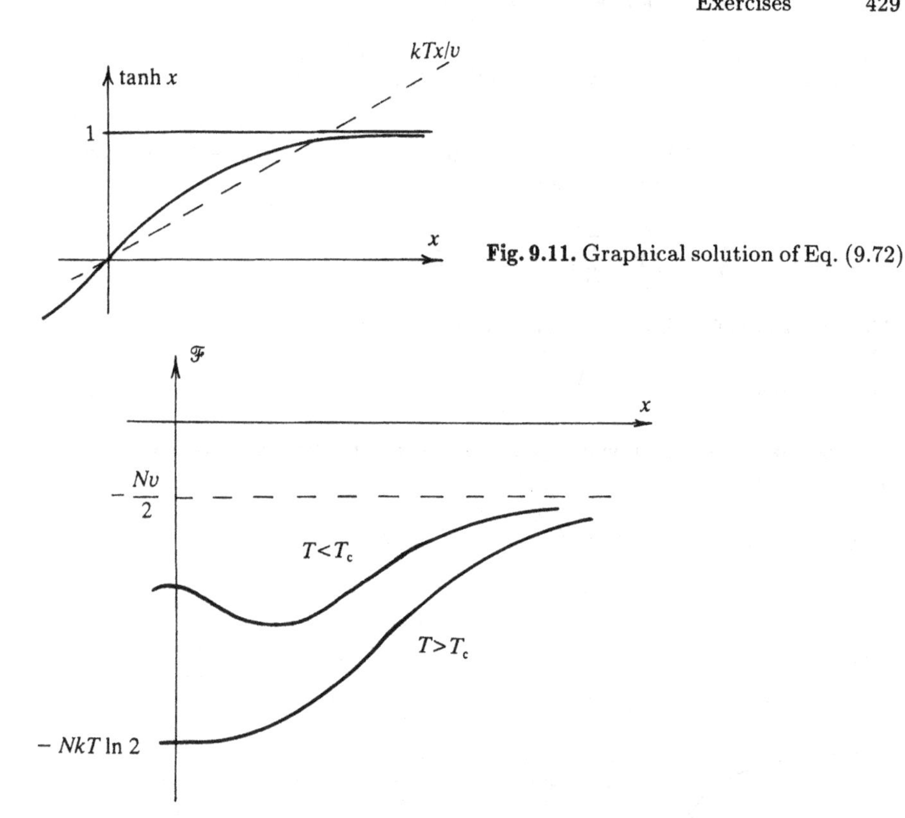

Fig. 9.11. Graphical solution of Eq. (9.72)

Fig. 9.12. The trial free energy $\mathcal{F}$ as function of x

$$\frac{kTx}{v} = \tanh x. \tag{9.72}$$

When $kT > v$ there is a single minimum of $\mathcal{F}$ at $x = 0$ which equals (Figs.9.11 and 9.12)

$$F = -NkT \ln 2.$$

When $kT < v$, the point $x = 0$ becomes a maximum and the minimum is reached for the two non-trivial solutions, $\pm x$ with $x > 0$, of Eq.(9.72). There is a phase transition at a temperature

$$T_c = \frac{v}{k}.$$

The invariance of the Hamiltonian and of $\widehat{D}$ under $\sigma_i \Rightarrow -\sigma_i$ when $B = 0$, which is associated with spin flips, or with a reversal of the z-axis, is spontaneously broken when $T < T_c$ as the approximation $\mathcal{D}$ for the state of the system does not have that symmetry when $x \neq 0$. We find two symmetric solutions for $T < T_c$, as for the fluids in § 9.3. The spontaneous magnetization is in the low-temperature $\pm x$ phases equal to

$$M_s = -\frac{N}{\Omega}\mu_B\langle\sigma_i\rangle = \pm\frac{N}{\Omega}\mu_B\tanh x = \pm\frac{N}{\Omega}\mu_B\frac{kT}{v}x.$$

We solve Eq.(9.72), as $T \to T_c$, by expanding $\tanh x$ as $x \to 0$; this gives us $x^2 \sim 3(1 - T/T_c)$ and hence (Fig.9.13)

$$M_s \sim \mu_B\frac{N}{\Omega}\sqrt{3}\sqrt{1-\frac{T}{T_c}}, \qquad T \to T_c - 0.$$

The expansion of $\tanh x$ as $x \to \infty$ gives, as $T \to 0$,

$$x = \beta v\left(1 - 2\,e^{-2\beta v}\,\ldots\right),$$

and the magnetization remains practically constant and equal to its maximum possible value,

$$M_s \approx \mu_B\frac{N}{\Omega}\left(1 - 2\,e^{-2v/kT}\right), \qquad T \to 0.$$

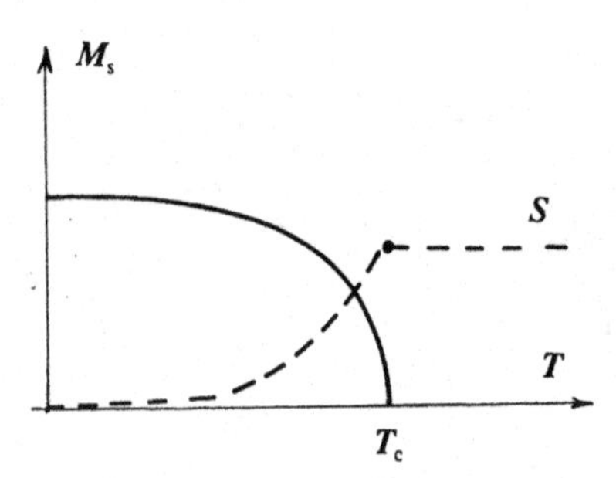

Fig. 9.13. The spontaneous magnetization and the entropy as functions of T

Note. One can find Eq.(9.72) in a simple way by assuming that the system of interacting spins behaves as a system of independent spins, each of which is subject to an effective magnetic field. The latter results from the interactions of that spin with all the other spins, themselves considered to be in thermal equilibrium in the effective field to which they are subjected. The effective Weiss field at i equals $\mathcal{B}_i = \sum_j V_{ij}\langle\sigma_i\rangle/\mu_B$ and the self-consistency equation $\langle\sigma_j\rangle = \tanh\beta\mu_B\mathcal{B}$ is the same as (9.72). This method is simpler than the variational method used above and it gives a physical meaning to $x = \mu_B\mathcal{B}/kT$, but it has the drawback of not giving a criterion for choosing from among the solutions of (9.72) or (9.74) the correct one, that is, the one which makes (9.71) a minimum.

2. The parameter x is continuous as $T \to T_c$ and this is therefore also true of F which is given by (9.71) and (9.72). To evaluate the entropy we use either $S = -k\,\mathrm{Tr}\mathcal{D}\ln\mathcal{D}$, or, using the fact that $\partial\mathcal{F}/\partial x = 0$,

$$\begin{aligned} S &= -\frac{dF}{dT} = -\frac{\partial\mathcal{F}}{\partial T} - \frac{\partial\mathcal{F}}{\partial x}\frac{dx}{dT} = -\frac{\partial\mathcal{F}}{\partial T} \\ &= kN[\ln(2\cosh x) - x\tanh x], \end{aligned}$$

and its continuity as $T \to T_c$, $x \to 0$ is obvious (Fig.9.13). The approximation is consistent with Nernst's principle, as

$$S \sim kN\,\mathrm{e}^{-2x}\,(2x+1) \sim \frac{2vN}{T}\,\mathrm{e}^{-2v/kT} \to 0,$$

as $T \to 0$. The specific heat, which vanishes when $T > T_c$, equals

$$C = T\,\frac{dS}{dT} = -\frac{kNx}{\cosh^2 x}\,T\,\frac{dx}{dT}$$

when $T < T_c$. Taking the derivative of (9.72) we find

$$\frac{k}{v} = \left[\frac{1}{x\cosh^2 x} - \frac{\tanh x}{x^2}\right]\frac{dx}{dT},$$

whence we get

$$C = Nk\,\frac{T}{T_c}\,\frac{2x^3}{\sinh 2x - 2x}.$$

The specific heat tends exponentially to zero as $T \to 0$: the system is frozen in into the state with maximum absolute magnitude of the magnetization. It increases with T up to the value $\frac{3}{2}Nk$, which is reached at $T = T_c - 0$, where $x \to 0$. Near T_c its behaviour,

$$C \approx \frac{3}{2}Nk\left(1 - \frac{3}{5}\,\frac{T_c - T}{T_c}\right),$$

is linear and it falls suddenly to zero when $T = T_c + 0$, remaining zero beyond that point. Its form is therefore that of a sawtooth. The specific heat of nickel can be explained by adding to the specific heat due to the lattice vibrations (§ 11.4.3), which increases from 0 to $3Nk$ – that is, to 6 calories per mole per degree – this spin specific heat which has an anomaly at the Curie point where it drops by $\frac{3}{2}Nk$.

3. If we change the Hamiltonian by adding the magnetic field term $\mu_B B\sum_i \sigma_i$, the trial free energy becomes

$$\mathcal{F} = -\tfrac{1}{2}Nv\tanh^2 x - N\mu_B B\tanh x + NkT(x\tanh x - \ln 2\cosh x), \qquad (9.73)$$

and its absolute minimum is, for $B > 0$, reached for the unique solution, or the largest of the three solutions, of the equation

$$kTx = v\tanh x + \mu_B B. \qquad (9.74)$$

This can be seen by noting that, when (9.74) is satisfied, the extrema of (9.73) are given by

$$F = -\tfrac{1}{2}N\mu_B B\tanh x - NkT\,\ln\left(2\cosh x - \tfrac{1}{2}x\tanh x\right),$$

which is a decreasing function of x. For fixed $B \neq 0$, x and F have no singularities as functions of T, in contrast to the case when $B = 0$. Since x and therefore M have the same sign as B, the solid is paramagnetic. As $B \to \infty$, (9.74) yields $\tanh x \to 1$ and we have saturation, as in ordinary paramagnetism where $v = 0$.

The magnetization is found by differentiation:

$$\Omega M = -\frac{dF}{dB} = -\frac{\partial \mathcal{F}}{\partial B} - \frac{\partial \mathcal{F}}{\partial x}\frac{dx}{dB} = \mu_B N \tanh x. \qquad (9.75)$$

When $T > T_c$ we have in the limit as $B \to 0$

$$x(kT - v) \sim \mu_B B,$$

or

$$\chi = \left.\frac{dM}{dB}\right|_{B=0} = \frac{N}{\Omega}\frac{\mu_B^2}{k(T-T_c)} \qquad \text{(Curie–Weiss Law)}.$$

At high temperatures ($T \gg T_c$) one finds again the Curie law (Exerc.4a). The susceptibility is positive (paramagnetism) and becomes infinite at the transition point as $T \to T_c - 0$. At $T = T_c$ the magnetization given by (9.74) and (9.75) varies as function of B according to

$$M \sim \mu_B \frac{N}{\Omega}\left(\frac{3\mu_B B}{kT_c}\right)^{1/3}, \qquad B \to 0.$$

When $T < T_c$ one gets, as expected, in the limit as $B \to \pm 0$,

$$M \to \pm M_s.$$

The magnetization tends to the spontaneous magnetization, and the spins align themselves due their interactions in a direction which is determined by the sign of B. The susceptibility is obtained by differentiating (9.74), (9.75) with respect to B:

$$\chi = \frac{dM}{dB} = \frac{N}{\Omega}\frac{\mu_B^2}{k(T\cosh^2 x - T_c)} \underset{B\to 0}{\longrightarrow} \frac{N}{\Omega}\frac{2\mu_B^2 x}{kT_c(\sinh 2x - 2x)}.$$

As $T \to T_c - 0$, we get, since $x^2 \sim 3(1 - T/T_c)$,

$$\chi \sim \frac{N}{\Omega}\frac{2\mu_B^2}{k(T_c - T)}, \qquad T \to T_c - 0.$$

The susceptibility becomes infinite, as when $T \to T_c + 0$, but with a coefficient which is twice as large. As $T \to 0$ we get, since $x \sim v/kT$,

$$\chi \sim \frac{N}{\Omega}\frac{4\mu_B^2\, e^{-2v/kT}}{kT} \to 0.$$

The system is frozen in into $\langle\sigma_i\rangle \simeq \pm 1$, and it no longer reacts to B.

The statistical fluctuation in the magnetic moment is given by $\Delta M^2 = \Omega kT\chi$ (Exerc.4a); it diverges thus as $|T - T_c|^{-1/2}$ when $T \to T_c \pm 0$.

4. The parameter x is small near $T = T_c$. Expanding (9.73) in powers of $m = M\Omega/N\mu_B = -\langle\sigma_i\rangle = \tanh x$ up to and including the fourth order terms gives

$$\frac{\mathcal{F}}{N} = -\frac{1}{2}vm^2 - \mu_B Bm + kT\left(-\ln 2 + \frac{1}{2}m^2 + \frac{1}{12}m^4\right).$$

The trial free energy $\mathcal{F}(M, T, B)$ has thus near the critical point $T_c = v/k$ the same form as the Landau expression (6.109), (6.110) for the macroscopic free energy of a homogeneous substance when the equilibrium is shifted by imposing a constraint upon $\langle\sigma_i\rangle$. It is therefore not surprising that the results obtained above in the vicinity of $T = T_c$ are the same as those of Exerc.6d.

9b Spins with Infinite-Range Interactions

Using the notation of the preceding exercise we assume that the interaction potential of the Ising model is a constant, $V_{ij} = v/N$, where the factor $1/N$ has been introduced to ensure that the interaction energy is extensive. This is not a realistic model, as two spins here interact in the same way whatever their distance apart, but it is interesting as it can be solved exactly in the limit of a macroscopic system ($N \to \infty$), even when there is a non-zero field present. Show, in fact, by using the saddle-point method (Exerc.5b) that the mean field approximation of Exerc.9a becomes exact as $N \to \infty$ for this potential.

Study the analytical properties of the canonical partition function Z in the vicinity of the critical point $B = 0$, $T_c = v/k$. Write down the saddle-point approximation for Z for large N, first when $B \neq 0$, then when $B = 0$, $T > T_c$, and finally when $B = 0$, $T < T_c$; in the last case expand to dominant order in $\tau = (T - T_c)/T_c$. Extend these results to complex values of τ for $B = 0$ and hence derive the positions of the *zeroes of the partition function* Z in the complex τ-plane. What are, for finite but large N, the singularities of the free energy, in zero field, as function of the temperature?

Solution:

We want to calculate

$$Z = \sum_{\{\sigma_i\}} \exp\left[\frac{\beta v}{2N}\left(\sum_i \sigma_i\right)^2 - \beta\mu_B B \sum_i \sigma_i\right].$$

A first method consists in summing over the σ_i for a given value of $S = \sum_i \sigma_i$ and then to sum over S which, using the saddle-point method, one can treat as a continuous variable in the limit as $N \to \infty$. In order to avoid combinatorial calculations, we shall use another method. It consists in getting for the summation over the σ_i to an expression which has a factorized form,

$$e^{-xS} \equiv \prod_i e^{-x\sigma_i}.$$

We use for this the identity

$$e^{kS^2} = \frac{1}{\sqrt{\pi k}} \int_{-\infty}^{+\infty} dy\, e^{-y^2/k - 2yS},$$

where $k \equiv \beta v/2N$, which leads to

$$Z = \frac{1}{\sqrt{\pi k}} \int_{-\infty}^{+\infty} dy\, e^{-y^2/k - xS},$$

where $x \equiv 2y + \beta\mu_B B$. Summing over the σ_i and taking x as integration variable, we get

$$Z = \sqrt{\frac{N}{2\pi\beta v}} \int_{-\infty}^{+\infty} dx \, \exp\left\{-N\left[\frac{(x-\beta\mu_B B)^2}{2\beta v} - \ln 2\cosh x\right]\right\}.$$

When N is large, (5.92) gives for the free energy

$$\begin{aligned} F &\sim NkT \min_x \left[\frac{(x-\beta\mu_B B)^2}{2\beta v} - \ln 2\cosh x\right] \\ &= \min_x \left[\mathcal{F}(x,T,B) + \frac{\cosh^4 x}{2Nv}\left(\frac{\partial\mathcal{F}}{\partial x}\right)^2\right], \end{aligned}$$

where $\mathcal{F}$ is the trial free energy (9.73) of the variational method. This expression has the same extrema as $\mathcal{F}$ itself and they are reached when x is given by (9.74).

The approximate results which were established in the preceding exercise for a general Ising model thus become exact as $N \to \infty$ in the particular case when $V_{ij} = v/N$. This is due to the fact that when the range of the forces becomes very large, each spin interacts with a very large number of other spins; it thus becomes legitimate to replace these spins by their mean equilibrium value – neglecting their fluctuations which cancel in relative magnitude as they are many – and thus to assume that each spin is subject to the effective mean field $\mathcal{B} = \sum_i v\langle\sigma_i\rangle/N\mu_B$.

When $B \neq 0$ only the highest saddle-point x, which corresponds to the absolute minimum of $\mathcal{F}$, contributes to Z and we get from (5.94)

$$Z \sim \left(1 - \frac{T_c}{T\cosh^2 x}\right)^{-1/2} e^{-F/kT},$$

where $\cosh^2 x - T_c/T$ vanishes only at the critical point. When $B = 0$, $T > T_c$, we find

$$Z \sim \left(1 - \frac{T_c}{T}\right)^{-1/2} e^{-\mathcal{F}(0,T,0)/kT} = \sqrt{\frac{1+\tau}{\tau}}\, 2^N.$$

When $B = 0$, $T < T_c$, there exist two saddle-points $\pm x$ which make the same contribution to Z; with small $\tau \equiv (T - T_c)/T_c$, $x^2 \sim -3\tau$, this gives

$$Z \sim \sqrt{\frac{2}{-\tau}}\, 2^N \exp\left(\frac{3}{4}N\tau^2\right).$$

When τ is complex, the above contributions from each saddle-point remain unchanged for small $|\tau|$, but only the highest saddle-point(s) which the integration contour passes through when it goes from $x = -\infty$ to $x = +\infty$ contributes. A study of

$$\mathrm{Re}\left[\frac{x^2}{2\beta v} - \ln 2\cosh x\right] \approx -\ln 2 + \mathrm{Re}\left(\frac{x^2\tau}{2} + \frac{x^4}{12}\right)$$

in the complex x-plane shows that, if $|\arg\tau| < \frac{3}{4}\pi$, the contour can be chosen in such a way that its highest point is $x = 0$. If $\frac{3}{4}\pi < |\arg\tau| \leq \pi$, only the two saddle-points $x = \pm\sqrt{-3\tau}$ contribute. In these two regions one finds again the results obtained for real τ, and Z does not have any zeroes there. On the other hand, when $\arg\tau = \pm\frac{3}{4}\pi$, the contour passes through the three saddle-points which have the same height and one finds

$$2^{-N}\,Z \sim \frac{1}{\sqrt{\tau}} + \sqrt{\frac{2}{-\tau}}\,\exp\left(\frac{3}{4}N\tau^2\right).$$

This expression is valid not only when $|\arg\tau| = \frac{3}{4}\pi$, but also in its vicinity where only one of its two terms dominates. All the same, we must have $|N\tau^2| \gg 1$, as otherwise the three saddle-points lie too close to one another for their contributions to be separable; very close to the critical point Z is analytical without zeroes.

By letting $\arg\tau = \pm\frac{3}{4}\pi \pm \alpha$ we obtain the zeroes of Z from

$$\exp\left(\mp\frac{3}{8}\pi\mathrm{i}\right) + \sqrt{2}\,\exp\left(\pm\frac{\pi\mathrm{i}}{8} + \frac{3}{4}N\tau^2\right) = 0,$$

which gives

$$|\tau|^2 \simeq \frac{4\pi}{3N}\left(2k - \frac{1}{2}\right), \qquad \alpha \simeq -\frac{\ln 2}{3N|\tau|^2},$$

where k is a positive integer. These zeroes are situated close to two semi-straights starting from the critical point. When N is large, they lie extremely densely. The free energy and the other thermodynamic quantities – which are all holomorphic along the real axis, even around $T = T_{\mathrm{c}}$ – thus have an infinity of branch points; these lie on two semi-straights which are symmetric with respect to the real axis. They become extremely densely packed towards T_{c} when N becomes large, and this separates in the thermodynamic limit the two regions $T > T_{\mathrm{c}}$ and $T < T_{\mathrm{c}}$ when $B = 0$. When $B \neq 0$, these regions remain connected near the real axis.

9c Linear Chain of Spins

The one-dimensional Ising model with interactions between nearest neighbours, which has as energy levels the expressions

$$-\frac{v}{2}\sum_{i=1}^{N-1}\sigma_i\sigma_{i+1} + \mu_{\mathrm{B}}B\sum_i \sigma_i, \qquad (\sigma_i = \pm 1)$$

can be solved exactly. Evaluate its free energy and its correlation function $\langle\sigma_i\sigma_j\rangle$ in zero field. In order to do this note that the summation over the σ_i can be replaced by a summation over the variables $\tau_i = \sigma_i\sigma_{i+1}$. Hence find the internal energy and the specific heat. Does the model have a phase transition? Make a comparison with the mean field approximation.

When the field is non-zero, the *transfer matrix* method consists in considering each factor

$$\exp\left(\tfrac{1}{2}\beta v\,\sigma_i\sigma_{i+1} - \beta\mu_{\mathrm{B}}B\sigma_i\right) \equiv \langle\sigma_i|T|\sigma_{i+1}\rangle,$$

which occurs in the calculation of Z, as a 2×2 matrix. The expression for Z is then found in terms of a product of $N-1$ identical matrices which can be evaluated by diagonalizing T. Use that method to evaluate the free energy in a non-zero field in the limit as $N \to \infty$.

Solution:

Taking σ_1 and τ_i as variables we have

$$Z = \sum_{\sigma_1,\{\tau_i\}} \exp\left(\frac{1}{2}\beta v \sum_i \tau_i\right) = 2\left(2\cosh\frac{1}{2}\beta v\right)^{N-1},$$

or, in the limit of large N,

$$F = -NkT \ln\left(2\cosh\frac{v}{2kT}\right).$$

The system does not have a phase transition, in contrast to what happens for a spin system in more dimensions.

The correlation function is given by

$$\langle\sigma_i\sigma_{i+m}\rangle = \langle\tau_i\tau_{i+1}\cdots\tau_{i+m-1}\rangle.$$

The τ_i variables are uncorrelated and $\langle\tau_i\rangle = \tanh\frac{1}{2}\beta v$. Hence, the correlation function decreases exponentially with distance as $\exp(-m/m_0)$; the correlation distance,

$$m_0 \equiv \left[-\ln\tanh\frac{v}{2kT}\right]^{-1},$$

increases from 0 to ∞ as T decreases from ∞ to 0.

To understand the absence of a phase transition, we note that in $F = U - TS$ the interaction energy is lowest when neighbouring spins are parallel, which tends to create correlations between all spins; however, statistical fluctuations raise the entropy, enough so that distant spins remain uncorrelated at all non-zero temperatures. In other words, for $1 \ll m \ll N$, σ_i and σ_{i+m} do not preferably take equal values, because the rather large number of configurations such that $\sigma_k\sigma_{k+1} = -1$ for $i < k < i+m$ prevents the order to set in between i and $i+m$. Only at zero temperature does the chain become ordered, with $\langle\sigma_i\sigma_{i+m}\rangle$ finite as $m \to \infty$, and all spins then point in the same direction, either $+1$ or -1; in more than one dimension, the limit of $\langle\sigma_i\sigma_{i+m}\rangle$ for $m \to \infty$ below T_c is M_s^2.

The internal energy is

$$U = F + TS = F - T\frac{\partial F}{\partial T} = -\frac{1}{2}Nv\tanh\frac{v}{2kT},$$

and the specific heat,

$$C = \frac{Nv^2}{4kT^2\cosh^2(v/2kT)},$$

tends to zero both at high and at low temperatures.

Note that the approximation of Exerc.9a is not correct in this case as its results are independent of the dimensionality. It is true that expression (9.71) is for all T an upper bound of the above exact solution, but the discontinuity in its second derivative at $T = v/k$ is factitious in one dimension. In fact, the Ising model has a phase transition in two or more dimensions and also in the case of very long range interactions (Exerc.9b). The mean field approximation is the better justified, the larger the dimensionality: it becomes exact when the dimensionality is infinite in which case each spin has an infinite number of neighbours, as in Exerc.9b.

The canonical partition function for $B \neq 0$ can be expressed in terms of the transfer matrix T as follows:

$$Z = \sum_{\{\sigma_i\}} \langle\sigma_1|T|\sigma_2\rangle\langle\sigma_2|T|\sigma_3\rangle \cdots \langle\sigma_{N-1}|T|\sigma_N\rangle \, \mathrm{e}^{-\beta\mu_\mathrm{B}B\sigma_N}$$
$$= \sum_{\sigma_1,\sigma_N} \langle\sigma_1|T^{N-1}|\sigma_N\rangle \, \mathrm{e}^{-\beta\mu_\mathrm{B}B\sigma_N}.$$

The matrix T has the form

$$T = \begin{pmatrix} \mathrm{e}^{\frac{1}{2}\beta v-\beta\mu_\mathrm{B}B} & \mathrm{e}^{-\frac{1}{2}\beta v-\beta\mu_\mathrm{B}B} \\ \mathrm{e}^{-\frac{1}{2}\beta v+\beta\mu_\mathrm{B}B} & \mathrm{e}^{\frac{1}{2}\beta v+\beta\mu_\mathrm{B}B} \end{pmatrix},$$

and its eigenvalues,

$$t_\pm = \mathrm{e}^{\frac{1}{2}\beta v} \cosh \beta\mu_\mathrm{B}B \pm \sqrt{\mathrm{e}^{\beta v} \sinh^2 \beta\mu_\mathrm{B}B + \mathrm{e}^{-\beta v}},$$

are positive. It can be diagonalized as

$$T = \sum_{\varepsilon=\pm 1} \begin{pmatrix} \mathrm{e}^{-b}A_{-\varepsilon} \\ \varepsilon\mathrm{e}^{b}A_\varepsilon \end{pmatrix} t_\varepsilon \, (\mathrm{e}^{b}A_\varepsilon \quad \varepsilon\mathrm{e}^{-b}A_{-\varepsilon}),$$

where $b \equiv \frac{1}{2}\beta\mu_\mathrm{B}B$ and

$$A_\pm \equiv \sqrt{\frac{1}{2} \pm \frac{\sinh 2b}{2\sqrt{\sinh^2 2b + \mathrm{e}^{-2\beta v}}}}.$$

Hence we find

$$Z = (1 \;\; 1)\, T^{N-1} \begin{pmatrix} \mathrm{e}^{-2b} \\ \mathrm{e}^{2b} \end{pmatrix} = \sum_\varepsilon t_\varepsilon^{N-1} \left(\mathrm{e}^{-b} A_{-\varepsilon} + \varepsilon\mathrm{e}^{b} A_\varepsilon\right)^2.$$

Only the largest eigenvalue $\varepsilon = +1$ contributes in the limit as $N \to \infty$ and this gives us the exact solution for the free energy,

$$F = -NkT \ln t_+.$$

Changing the boundary conditions does not affect this result.

9d Stable and Metastable Magnetic Phases

In order to describe the magnetism of substances with spin-1 magnetic atoms, we introduce the following schematic model. Let the N atoms, where N is large, be arranged at the sites, indicated by i, of a simple cubic lattice. We see that each of them has 6 nearest neighbours. We denote the z-component of each spin i, which can take the values $+1$, 0, and -1, by σ_i. We assume that the Hamiltonian has the form

$$H = J \sum_{\{i,j\}} (\sigma_i - \sigma_j)^2,$$

where J is a positive constant. The sum is over all *nearest neighbour* pairs $\{i, j\}$ of atoms in the lattice. The same kind of model is also used to describe alloys; the three values $\sigma_i = +1,\ 0$, and -1 then denote three different atoms which can occupy the site i.

1. Give expressions for the entropy S, the internal energy U, and the canonical partition function Z at zero temperature, and at high temperatures $(T \gg J/k)$.

2. The system is supposed to evolve according to the following mechanism. Time is made discrete and at each time only one of the sites i can change by altering the value of its σ_i; then another randomly chosen site changes, and so on. The average energy over a long period is fixed. At low temperatures the rate of this process is very small and the large value of N makes it impossible to reach true canonical equilibrium within a reasonable period. On an intermediate time scale we can, however, reach *quasi-equilibrium*. Let us, for instance, start from the pure state $\sigma_i = 0, \forall\, i$, and let us add a small amount of energy. The only accessible micro-states during the periods considered are those for which only a *small fraction of spins σ_i is different from zero*. Amongst that set of micro-states a Boltzmann-Gibbs distribution will be established; it describes a quasi-equilibrium macro-state, which represents the 0-*phase* of the system at low temperatures. We define similarly the +-*phase*; this is a quasi-equilibrium macro-state where the only accessible micro-states are those for which most σ_i are equal to $+1$, namely, only a number which is small as compared to N of σ_i differ from $+1$. Finally, in the −-*phase* most of the σ_i are equal to -1. We associate with each of these three phases a canonical partition function Z_0, Z_+, Z_-.

Evaluate for $T \ll J/k$ and N large the dominant behaviour of $\ln Z_0$, $\ln Z_+$, and $\ln Z_-$. One method consists in assuming, for instance for the 0-phase, that Z_0 can be factorized, when $T \ll J/K$, into contributions relating to each of the sites i; we then calculate the contribution from the site i by assuming that the σ_j on the sites which are the nearest neighbours of i remain equal to zero. Another possible method consists in evaluating to dominant order in N the eigenenergies and multiplicities of the accessible micro-states, neglecting configurations where two spins which are nearest neighbours are both changed. Find in this way the specific heat per site for the 0-phase at low temperatures.

3. We can, at the same low temperatures as above, reach true canonical equilibrium, either by waiting an extremely long time, or more simply by first bringing the system to a high temperature, where equilibrium can easily be established, and then progressively cooling it down to T (*annealing method*).

Use Z_0, Z_+, and Z_- to find the canonical partition function Z for this true equilibrium. What is the value of $\ln Z/N$ in the large N limit? Write down the fractions of each of the three, 0, +, −, phases in the true equilibrium state. What are the physical consequences of this?

Is the Nernst principle for the thermodynamic entropy satisfied in the three quasi-equilibrium phases? Is it satisfied in the true equilibrium state?

Solution:

1. The minimum energy, equal to zero, is reached when all the spins σ_i are equal to one another, which corresponds to the three micro-states $\sigma_i = +1$, $\sigma_i = 0$, $\sigma_i = -1$. We find $Z = 3$ and $S = k \ln 3$.

As $T \to \infty$ all 3^N micro-states are equiprobable, and hence $S = kN \ln 3$. The lattice contains three times as many bonds as sites. As a result, by taking an average over all the micro-states we find for the internal energy

$$U = \langle H \rangle = 3NJ \left\langle \left(\sigma_i - \sigma_j\right)^2 \right\rangle = 3NJ \, \frac{1}{3^2} \, (4 \times 1 + 2 \times 4) = 4NJ.$$

Hence we find $\ln Z = S/k - \beta U \approx N \ln 3 - 4N\beta J + \mathcal{O}(N\beta^2 J^2)$.

2. For a site i surrounded by neighbours $\sigma_j = 0$, the partial partition function, z_0, is the sum over the states $\sigma_i = 0, \pm 1$:

$$z_0 = 1 + 2\,\mathrm{e}^{-6\beta J},$$

as the energy is either 0 when $\sigma_i = 0$, or $6J$ when $\sigma_i = \pm 1$, in which case there are 6 nearest neighbours with $\sigma_j = 0$. Hence,

$$\ln Z_0 \simeq N \ln \left(1 + 2\,\mathrm{e}^{-6\beta J}\right) \simeq 2N\,\mathrm{e}^{-6\beta J}.$$

Similarly, in the +-phase, for a site i surrounded by 6 neighbours with $\sigma_j = +1$, the energy equals 0 when $\sigma_i = 1$, $6J$ when $\sigma_i = 0$, and $6 \times 2^2 J$ when $\sigma_i = -1$, so that

$$z_+ = 1 + \mathrm{e}^{-6\beta J} + \mathrm{e}^{-24\beta J},$$

and we find

$$\ln Z_+ \simeq N \ln \left(1 + \mathrm{e}^{-6\beta J} + \mathrm{e}^{-24\beta J}\right) \simeq N\,\mathrm{e}^{-6\beta J} \simeq \ln Z_-.$$

In the second method we note that the first excited states are obtained by changing one spin to +1 or −1 in the 0-phase. Their number is $2N$, and their energy $6J$. Changing two distant spins leads to $2^2 N(N-7)/2$ micro-states of energy $2 \times 6J$; in the case of two neighbouring spins we have a much smaller number, $2 \times 3N$, of micro-states with energy $10J$ and also $2 \times 3N$ micro-states of energy $14J$. Similarly, changing n spins gives about $(2N)^n/n!$ micro-states of energy $n \times 6J$, if we neglect configurations where neighbouring spins have been changed. Hence,

$$\ln Z_0 \simeq \ln \sum_{n=0}^{\infty} \frac{(2N)^n}{n!}\,\mathrm{e}^{-n6\beta J} = 2N\,\mathrm{e}^{-6\beta J}.$$

This method would enable us to find the next-order correction terms ($6N \exp(-10\beta J)$ $-14N \exp(-12\beta J) + \ldots$) in $\ln Z_0$. For the +- and −-phases, changing n spins from

$\sigma_i = +1$ (or -1) to 0 leads similarly, to highest order in N, to $N^n/n!$ micro-states with energy $n \times 6J$, whence

$$\ln Z_\pm \simeq \ln \sum_{n=0}^{\infty} \frac{N^n}{n!} \, \mathrm{e}^{-n6\beta J} = N \, \mathrm{e}^{-6\beta J}.$$

The internal energy in the three phases is equal to

$$U_0 = 2U_+ = 2U_- = -\frac{\partial}{\partial \beta} \ln Z_0 = 12NJ \, \mathrm{e}^{-6J/kT},$$

which gives us for the specific heat per site

$$c_0 = 2c_+ = 2c_- = 72 \, \frac{J^2}{kT^2} \, \mathrm{e}^{-6J/kT}.$$

3. As Z is a sum over all micro-states, without any restrictions, and as in the preceding subsection we have taken all the low-energy states into account separately in Z_0, in Z_+, and in Z_-, we have $Z = Z_0 + Z_+ + Z_-$. For large N, we have $\ln Z/N \simeq \ln Z_0/N$.

The probability for a micro-state of energy E is equal to $\mathrm{e}^{-\beta E}/Z$ so that the total probability for the micro-states of the 0-phase equals Z_0/Z. The relative fractions of the three phases are thus Z_0/Z, Z_+/Z, and Z_-/Z. When N is large, $Z_+/Z_0 = Z_-/Z \simeq 0$ so that at low temperatures, after annealing, one will always observe only the 0-phase, even though at $T = 0$ the three phases have the same probability. The low-teperature behaviour is governed not by the ground state, but by the low-lying excited states.

Nernst's principle is satisfied in all three cases, as it states that the entropy per unit volume, that is, S/N for large N, must tend to zero as $T \to 0$. In fact, we have

$$\frac{S_0}{N} \sim \frac{S}{N} \sim \frac{2S_+}{N} \sim \frac{2S_-}{N} \sim \frac{12J}{T} \, \mathrm{e}^{-6J/kT} \to 0.$$

9e Liquid-Vapour Interface

We assumed at the beginning of § 9.3.3 that the trial grand potential of an inhomogeneous fluid could be obtained by adding the contributions from the parts which have different densities. In actual fact, there are corrections coming from the regions where the density varies. Use the mean field approximation to write down expressions which replace (9.42), (9.43), and (9.44) when $n(\boldsymbol{r})$ is not constant. At the liquid-vapour interface the density $n(\boldsymbol{r})$ does not change suddenly, but gradually from n_l to n_g over a characteristic length which is rather large as compared to the range of the forces. We shall take the very-short-range repulsion into account by using the model of § 9.3.2 and by further assuming that the density is practically constant over this range. However, the attractive part of the potential – which gives rise to the second term in (9.50) when the density is uniform – is now sensitive to density variations. Write down to lowest order in $\nabla n(\boldsymbol{r})$ the new form of the trial grand potential, generalizing (9.51). Compare the result with the semi-empirical macroscopic theory of Exerc.6c.

Solution:

As the trial number of particles, the trial kinetic energy, and the trial entropy $\mathcal{S}(\mathcal{D})$ are integrals over the position in space, they can be written down simply by adding the contributions from each volume element $d^3\boldsymbol{r}$. However, the potential energy, which is an integral over *two* positions, changes its form because the range of the interaction is not equal to zero:

$$\mathcal{N} = \int d^3\boldsymbol{r}\, n(\boldsymbol{r}),$$

$$\mathcal{U} = \frac{3}{2}kT \int d^3\boldsymbol{r}\, n(\boldsymbol{r}) + \frac{1}{2}\int d^3\boldsymbol{r}\, d^3\boldsymbol{r}'\, n(\boldsymbol{r})\, n(\boldsymbol{r}+\boldsymbol{r}')\, W(\boldsymbol{r}'),$$

$$S(\mathcal{D}) = k \int d^3\boldsymbol{r}\, n(\boldsymbol{r}) \left\{\frac{5}{2} - \ln\left[n(\boldsymbol{r})\,\lambda_T^3\right]\right\}.$$

The only term in (9.51) which is changed in a non-trivial way is the two-particle attractive energy, $-a\Omega n^2$, which becomes

$$\frac{1}{2}\int d^3\boldsymbol{r}\, d^3\boldsymbol{r}'\; n(\boldsymbol{r})\, n(\boldsymbol{r}+\boldsymbol{r}')\, W(\boldsymbol{r}')$$

$$= \frac{1}{2}\int d^3\boldsymbol{r}\, d^3\boldsymbol{r}'\, n(\boldsymbol{r}) \left[n(\boldsymbol{r}) + \left(\boldsymbol{r}' \cdot \nabla n(\boldsymbol{r})\right) + \frac{1}{2}\sum_{\alpha,\beta} r'_\alpha r'_\beta\, \frac{\partial^2 n(\boldsymbol{r})}{\partial r_\alpha \partial r_\beta} + \cdots\right] W(\boldsymbol{r}').$$

The second term and the terms with $\alpha \neq \beta$ vanish because $W(\boldsymbol{r}')$ is isotropic. Integrating the last term by parts over $\boldsymbol{r}$ and defining

$$\int d^3\boldsymbol{r}'\, W(\boldsymbol{r}') = -2a, \qquad \int d^3\boldsymbol{r}'\, {r'_\alpha}^2\, W(\boldsymbol{r}') = -a',$$

we find

$$\mathcal{A}\{n(\boldsymbol{r}), T, \Omega, \mu\} = \int d^3\boldsymbol{r} \left[kT\left(n \ln\frac{n}{f} - n\right) - an^2 + \frac{cn^3}{1-bn}\right] + \frac{1}{2}a' \int d^3\boldsymbol{r}\, (\nabla n)^2.$$

This result is exactly the starting point of our considerations in Exerc.6c. The empirical coefficient $a(T)$ which we then introduced is the same as the coefficient a' which is here defined in terms of the attractive part of the intermolecular potential. The other terms also have the same form in the two cases. They are obtained by adding together the contributions from the various volume elements, evaluated for a homogeneous substance with a density which is constrained to have locally the value $n(\boldsymbol{r})$. The mean field theory enables us thus to justify microscopically the approach of Exerc.6c and to recover all its results, in particular, those about the shape of $n(\boldsymbol{r})$ across the interface and about the capillary energy. In actual fact, not only the interaction energy, but also the entropy, produces non-local terms, in $(\nabla n)^2$, with a temperature-dependent coefficient, if we take correlations between particles into account. One can see this from the virial expansion (9.25) where the first-order term becomes for the inhomogeneous case

$$-\frac{1}{2}kT \int d^3r\, d^3r'\, n(\boldsymbol{r})\, n(\boldsymbol{r}')g(\boldsymbol{r}-\boldsymbol{r}') \approx -kTB_1 \int d^3r\, n^2(\boldsymbol{r})$$
$$+\frac{1}{6}kT \int d^3r'\, r'^2 \left[e^{-W(r')/kT} - 1\right] \int d^3r\, (\nabla n)^2.$$

Note. The same idea can be applied to lattice models. In particular, if the field B in Exerc.9a is not uniform, or if for zero field one wants to study the wall between two ferromagnetic regions, where the magnetization points in different directions, one is led to assume that the variable x depends on the site. The trial free energy (9.73) becomes

$$\mathcal{F} = -\frac{1}{2} \sum_{ij} V_{ij} \tanh x_i \tanh x_j$$
$$+ \sum_i \left[-\mu_{\rm B} B_i \tanh x_i + kT\,(x_i \tanh x_i - \ln 2\cosh x_i)\right].$$

If x varies little on the scale of one cell, the first term of $\mathcal{F}$ produces, apart from $-\frac{1}{2}v\sum_i \tanh^2 x_i$, a contribution which is proportional to the square of the gradient of the order parameter $\langle\sigma_i\rangle = -\tanh x_i$, as expected. The surface separating the liquid and the gas in the lattice model of Prob.7 can be studied in the same way.

9f Dilute Solutions

1. Dilute solutions have properties which resemble those of perfect gases, notwithstanding the strong interactions of the solvent molecules with one another and with the molecules of the solute. In order to understand this fact we treat the solution as a classical fluid and neglect the interactions between the molecules of the solute. Start from the extension of (9.3) to a mixture of two kinds of molecules and show that the grand potential of a dilute solution has the form

$$A(T,\mu,\mu',\Omega) = A^0(T,\mu',\Omega) - \Omega kT\lambda_T^{-3}\zeta(T,\mu')e^{\mu/kT},$$

where A^0 refers to the pure liquid, and where ζ also depends only on the properties of the solvent. Compare this with (8.5).

2. This form of A will enable us to find many properties of dilute solutions (*Raoult's Laws*). Show that, if a solution contains several kinds of molecules which can react chemically, their densities obey the *mass action* law. If a semipermeable membrane lets the solvent through, but not the solute, it will be subject to *osmotic pressure* (§ 6.6.2); show that the latter can be calculated for a dilute solution as if the solute were a perfect gas, and the solvent were not there. Calculate the difference in the density of the solvent on the two sides of the membrane. Evaluate for a solution which is kept at constant pressure the density of the solvent as function of the concentration of the solute.

3. When a bottle contains water with CO_2 under pressure above it, the latter dissolves in the water. Assuming that the gas is pure CO_2, find the change with its pressure of the dissolved CO_2 concentration (*Henry's law*), neglecting ionization into CO_3H^- and H^+.

Consider the equilibrium between a solution (sugar in water) and its saturated vapour, which is assumed to consist purely of the solvent (water vapour). Calculate the *lowering of the saturated vapour pressure* and the *rise in the boiling temperature* due to the presence of the solute, in the limit of small concentrations. Find the same result by a hydrostatic argument involving the osmotic pressure (*van 't Hoff's law*). Calculate similarly the *lowering of the melting temperature* of a crystal of a solvent in equilibrium with the solution.

Note. This last effect is similar to what one observes when one adds salt to ice. There also the establishment of equilibrium leads to a lowering of the temperature. One goes from 0° to −20°C by adding one part of salt to four parts of crushed ice, which is an ancient process for making sorbet; at the same time the ice melts – this is applied to prevent icing in streets. Such thermal properties depend on the solvent and on the function $\zeta(T, \mu')$. However, the solution of NaCl is an *ionic* solution and the above theory must be modified to take into account the long-range Coulomb interactions between the dissolved ions, as in § 11.3.3.

Solution:

1. The general expression (9.3) for the grand partition function can be written as

$$Z_G = \sum_{NN'} \frac{e^{\alpha N + \alpha' N'}}{N!N'!} \lambda_T^{-3N} {\lambda'_T}^{-3N'} \int_\Omega d^3 r_1 \cdots d^3 r_N$$

$$\times\, d^3 r'_1 \cdots d^3 r'_{N'} \exp\left[-\beta \sum_{ik} W_{ik} - \beta \sum_{i>j} W'_{ij} \right],$$

where $W'_{ij} \equiv W'(r'_i - r'_j)$ indicates the interaction between molecules i and j of the solvent and $W_{ik} \equiv W(r'_i - r_k)$ the interaction between molecule i of the solvent and molecule k of the solute. The term with $N = 0$ is nothing but Z_G^0 for the pure solvent. To calculate the term with $N = 1$ we notice that before integration over r_1 the integrand is the same as the grand canonical phase density D^0 of the pure solvent; we can thus write this term in the form

$$Z_G^0\, e^\alpha \lambda_T^{-3} \int d^3 r_1 \left\langle e^{-\beta \sum_i W(r'_i - r_1)} \right\rangle_0 = Z_G^0\, e^\alpha \lambda_T^{-3} \Omega \zeta,$$

$$\zeta \equiv \left\langle e^{-\beta \sum_i W(r'_i)} \right\rangle_0,$$

where we have used the translational invariance of D^0 in calculating the average $\langle\ \rangle_0$ over D^0. Similarly, the term with $N = 2$ is equal to

$$Z_G^0 \frac{e^{2\alpha}}{2} \lambda_T^{-6} \int d^3r_1 \, d^3r_2 \left\langle e^{-\beta \sum_i W(r_i'-r_1)} e^{-\beta \sum_j W(r_j'-r_2)} \right\rangle_0 .$$

When the solution is dilute, the main contribution comes from points r_1 and r_2 which are at large distances apart, in which case the correlations between a solvent molecule i, close to r_1, and a solvent molecule j, close to r_2, are negligible – except at the critical point. We thus get approximately $Z_G^0 \left(e^\alpha \lambda_T^{-3} \Omega \zeta\right)^N /N!$ for the terms with $N \geq 2$ of Z_G. The summation over N produces the result which we wanted. The contribution to A from the solute has exactly the same form (8.5) as if it were a rarefied gas, except that the internal partition function of a molecule with a structure must be replaced by ζ. The latter is a function of the temperature and of μ', that is, of the density of the solvent, and it depends on the interactions between one molecule of the solute with the solvent.

2. The density of the dissolved molecules is

$$\frac{N}{\Omega} = -\frac{1}{\Omega}\frac{\partial A}{\partial \mu} = \lambda_T^{-3} \, \zeta(T,\mu') \, e^{\mu/kT},$$

which is connected with the chemical potential μ through the same relation (8.13) as in a gas, for a given chemical potential of the solvent. The formalism of § 8.2.2 which leads to (8.28) can thus be applied without changes to dilute solutions. However, the chemical properties which involve the pressure and the heat are changed because of the contribution A^0 from the solvent.

The pressure can be split into a sum of independent terms,

$$\mathcal{P}(T,\mu,\mu') = -\frac{A}{\Omega} = \mathcal{P}^0(T,\mu') + \frac{NkT}{\Omega},$$

and the contribution from the solute is the same as for a perfect gas. In osmosis μ' is the same on the two sides of the wall so that only the solute contributes to the osmotic pressure. The density of the solvent for given μ' depends on μ, and thus on N, according to

$$\frac{N'}{\Omega} = -\frac{1}{\Omega}\frac{\partial A}{\partial \mu'} = \frac{\partial \mathcal{P}(T,\mu,\mu')}{\partial \mu'} = \frac{N'^0(T,\mu')}{\Omega} + kT\,\frac{N}{\Omega}\frac{\partial}{\partial \mu'} \ln \zeta(T,\mu').$$

For given pressure and temperature the chemical potential μ' varies with N as follows

$$\frac{\partial \mathcal{P}^0(T,\mu')}{\partial \mu'} d\mu' + \frac{kT}{\Omega} dN = 0, \qquad \text{or} \qquad N'^0 \, d\mu' = -kT\, dN;$$

hence, the density of the solvent varies linearly as

$$d\left(\frac{N'}{\Omega}\right) = \frac{\partial}{\partial \mu'}\left(\frac{N'^0}{\Omega}\right) d\mu' + kT\,\frac{dN}{\Omega}\frac{\partial}{\partial \mu'} \ln \zeta$$

$$= \left(-\frac{N'^0}{\Omega}\kappa^0 + \frac{\partial}{\partial \mu'} \ln \zeta\right) kT\, d\left(\frac{N}{\Omega}\right),$$

where κ^0 is the isothermal compressibility of the pure solvent.

3. The chemical potential μ is the same in the gas and in the solution. Using (8.5) and (8.6) to eliminate it we find

$$\frac{N}{\Omega} = \zeta(T,\mu') \, \frac{\mathcal{P}}{kT\zeta_{\rm g}(T)},$$

where $\zeta_{\rm g}$ is the internal partition function of the CO_2 molecule in the gas. A sudden decompresion produces bubbles – hence the danger of surfacing after diving into deep water.

In the equilibrium between solution and water vapour the pressure $\mathcal{P}_{\rm s}(T,N/\Omega)$ and the chemical potential $\mu_{\rm s}'(T,N/\Omega)$, which have the same values in the two phases, depend on the density N/Ω of the solute in the liquid. Denoting by $\mathcal{P}_{\rm g}(T,\mu')$ the pressure in the gas, we have when $\mu' = \mu_{\rm s}'(T,N/\Omega)$

$$\mathcal{P}^0(T,\mu') + \frac{NkT}{\Omega} = \mathcal{P}_{\rm g}(T,\mu') = \mathcal{P}_{\rm s}\left(T,\frac{N}{\Omega}\right).$$

Differentiation with respect to $n \equiv N/\Omega$, directly and through μ', gives us

$$n_{\rm l}'^{0} \, \frac{\partial \mu_{\rm s}'}{\partial n} + kT = n_{\rm g}' \, \frac{\partial \mu_{\rm s}'}{\partial n} = \frac{\partial \mathcal{P}_{\rm s}}{\partial n},$$

where $n_{\rm l}'^{0}$ and $n_{\rm g}'$ denote the water densities in the pure liquid and in the gas, calculated for $\mu' = \mu_{\rm s}'(T,n)$. For small n we can replace $\mu_{\rm s}'(T,n)$ by $\mu_{\rm s}'(T,0)$ in $n_{\rm l}'^{0}$ and $n_{\rm g}'$, which are thus replaced by the densities of the two coexisting phases of the pure solvent; this leads to

$$\mu_{\rm s}'(T,n) = \mu_{\rm s}'(T,0) - \frac{kTn}{n_{\rm l}'^{0} - n_{\rm g}'},$$

$$\mathcal{P}_{\rm s}(T,n) = \mathcal{P}_{\rm s}(T,0) - \frac{kTn_{\rm g}'n}{n_{\rm l}'^{0} - n_{\rm g}'}.$$

This result does not hold near the critical point where the assumption that the vapour is made of pure solvent is no longer justified. It can be obtained directly by considering the equilibrium in the field of gravity of a solution in touch with the pure solvent across a semipermeable membrane, with both liquids having saturated vapour of the solvent above them. Because of the osmotic pressure nkT the solution-vapour interface is higher by z than the pure liquid-vapour interface. The pressures at the interfaces are, respectively, $\mathcal{P}_{\rm s}(T,n)$ and $\mathcal{P}_{\rm s}(T,0)$ and they differ by

$$\mathcal{P}_{\rm s}(T,n) - \mathcal{P}_{\rm s}(T,0) = \varrho_{\rm g} g z = nkT + \varrho_{\rm l} g z,$$

where $\varrho_{\rm g}$ and $\varrho_{\rm l}$ are the mass densities of the gas and of the solution; the latter is practically constant and equal to that of the pure liquid, whatever the height. This gives the same result as above. The change in the boiling temperature which is given by Clapeyron's formula (9.63) is, for given $\mathcal{P}$, equal to

$$T(n) = T(0) + \frac{RT^2}{L} \, \frac{n}{n_{\rm l}'^{0}}.$$

The theory is the same for the equilibrium between a solution and a crystal of the solvent, except that we must replace the gas density by the solid density. However, in Clapeyron's formula for crystallization the melting heat, L_{m}, is positive in the solid-liquid direction, whereas the evaporation heat L which we used above is positive in the liquid-gas direction; this alters the sign of the change in the transition temperature. The melting temperature is thus lowered as follows

$$T_{\text{m}}(n) = T_{\text{m}}(0) - \frac{RT^2}{L_{\text{m}}} \frac{n}{n_1'^0}.$$

9g Density of States of an Extensive System and Phase Separation

Consider a macroscopic system of volume Ω with energy levels which practically form a continuum; the number $W(E)$ of micro-states with energies between E and $E + \Delta E$ is proportional to ΔE provided ΔE is (i) sufficiently small so that $W(E)$ and $W(E + \Delta E)$ differ only slightly, and (ii) sufficiently large so that $W(E) \gg 1$. This enables us to define the density of states $\varrho(E, \Omega) = W(E)/\Delta E$. To simplify the discussion we disregard the variable N, the particle number, assuming, for instance, that one considers only configurations for which the density N/Ω has a value which is fixed once for all. The extension to a liquid-vapour transition, for instance, can easily be made by replacing in the discussion E/Ω by the two variables E/Ω and N/Ω. The *global* density of states ϱ refers to the micro-states of the whole, N-particle, system; it should not be confused with the *single-particle* density of states $\mathcal{D}(\varepsilon)$ which will be introduced in § 10.3.3 and which is proportional to Ω. By taking into account only those micro-states for which the system is homogeneous on the macroscopic scale we can define similarly a density of homogeneous states $\varrho_{\text{h}}(E, \Omega)$, assumed to be twice differentiable.

1. Assume that the microcanonical equilibrium entropy $S(U, \Omega)$ is extensive, that is, $S(U, \Omega) \sim \Omega k s(u)$, as $U \to \infty$ with $U/\Omega \equiv u$ fixed. By imagining as in §§ 6.4.6 and 9.3.3 the volume Ω to be split into two parts $\Omega_1 \equiv \lambda\Omega$ and Ω_2, show that S is a concave function of U and that $\varrho'' \le \varrho'^2/\varrho$, where the derivatives are with respect to the energy. Can the density of states of an extensive system have a minimum?

2. Nothing prevents ϱ_{h} to satisfy $\varrho_{\text{h}}'' > \varrho_{\text{h}}'^2/\varrho_{\text{h}}$, for instance, between two energy values, called E_4 and E_5. Show that under those conditions, in the vicinity of an energy E such that $E_4 < E < E_5$, the number of micro-states where the system is homogeneous is negligible as compared to the number of inhomogeneous micro-states; for the latter one may only take into account the configurations where the system is split into two homogeneous parts, characterized by energies per unit volume E_4/Ω and E_5/Ω. The existence of a region where $\varrho_{\text{h}}'' > \varrho_{\text{h}}'^2/\varrho_{\text{h}}$ thus implies that in equilibrium the system contains, for certain values of its internal energy U, at least two phases which are separate in space. Disregarding all complications connected with the geometry of the regions occupied by the different phases, find the microcanonical

entropy $S(U, \Omega)$ associated with the two-phase density of states ϱ, starting from the entropy S_h which would be associated with the homogeneous states by themselves. How can one find ϱ from ϱ_h? Show that the range of energies $E_2 < E < E_3$ over which the system spontaneously splits into two phases extends beyond the E_4, E_5 range at both ends; show that the energies per unit volume u_2 and u_3 of each phase remain fixed when U traverses the range E_2, E_3. What is the interpretation of the points E_4 and E_5?

3. Calculate the canonical entropy S_C associated with the homogeneous micro-states by themselves. Compare S_C with the two microcanonical entropies S_h and S.

4. Compare, in the region where the two phases coexist, the probability distributions $p(E)$ of the energy in a microcanonical equilibrium of energy U and in a canonical equilibrium of temperature $1/k\beta$.

Hints:

1. For an extensive system boundary effects are negligible; hence, the density of states satisfies the identity

$$\varrho(E, \Omega) = \int dE_1\, \varrho(E_1, \Omega_1)\, \varrho(E - E_1, \Omega_2).$$

Taking into account the relation $\varrho\, \Delta E = e^{S/k}$ and using (5.92) to evaluate the integral over E_1, we find

$$s(u) = \max_x \left[\lambda\, s\left(u + \frac{x}{\lambda}\right) + (1 - \lambda)\, s\left(u - \frac{x}{1 - \lambda}\right)\right].$$

This relation implies the concavity $S'' \leq 0$ of S, whence we find the inequality $\varrho'' \leq \varrho'^2/\varrho$.

2. By choosing $\lambda \equiv (E - E_4)/(E_5 - E_4)$, we have, with $u \equiv E/\Omega$,

$$\delta s \equiv \lambda\, s_h(u_5) + (1 - \lambda)\, s_h(u_4) - s_h(u) > 0,$$

since $s'' > 0$ between u_4 and u_5. The density of heterogeneous states comprising two homogeneous phases with, respectively, volumes and energies equal to Ω_1, E_1 and Ω_2, E_2 equals

$$\begin{aligned} \varrho(E, \Omega) &= \int dE_1\, \varrho_h(E_1, \Omega_1)\, \varrho_h(E - E_1, \Omega_2) \\ &= \frac{\Omega}{\Delta E^2} \int dy\, \exp\left[\Omega\lambda\, s_h\left(u_5 + \frac{y}{\lambda}\right) + \Omega(1 - \lambda) s_h\left(u_4 - \frac{y}{1 - \lambda}\right)\right], \end{aligned}$$

which, according to (5.92) satisfies the inequality

$$\frac{\varrho(E, \Omega)}{\varrho_h(E, \Omega)} > e^{\Omega \delta s}.$$

Even if the macroscopic sample is sufficiently small and the change in entropy δs sufficiently minute so that $\delta S \equiv \Omega k \delta s \simeq 10^{-9}$ J K^{-1}, this ratio exceeds $10^{3 \cdot 10^{13}}$.

The density of inhomogeneous states comprising two homogeneous phases with arbitrary volumes equals, if we assume that the only geometric variable is the volume,

$$\varrho(E,\Omega) = \int_0^1 d\lambda \int dE_1\, \varrho_{\rm h}(E_1,\lambda\Omega)\, \varrho_{\rm h}(E-E_1,\Omega-\lambda\Omega),$$

so that

$$\begin{aligned} S(U,\Omega) &\sim \max_{E_1,\lambda} \left[S_{\rm h}(E_1,\lambda\Omega) + S_{\rm h}(U-E_1,\Omega-\lambda\Omega)\right] \\ &\sim \Omega k \max_{x,\lambda} \left[\lambda\, s_{\rm h}\left(u+\frac{x}{\lambda}\right) + (1-\lambda)\, s_{\rm h}\left(u-\frac{x}{1-\lambda}\right)\right]. \end{aligned}$$

The microcanonical entropy $S(U,\Omega)$, calculated by including two phases, is thus the convex envelope of $S_{\rm h}(U,\Omega)$. Like S_1 in Fig.6.2, its representative curve contains a segment which is bitangent to the curve $S_{\rm h}$. The contact points $u_2 = U_2/\Omega$ and u_3, with $u_2 < u_4 < u_5 < u_3$, are defined by

$$s'_{\rm h}(u_2) = s'_{\rm h}(u_3) = \frac{s_{\rm h}(u_3)-s_{\rm h}(u_2)}{u_3-u_2} \equiv \beta_0.$$

In the interval $u_2 < u < u_3$ we have

$$s(u) = \frac{u-u_2}{u_3-u_2}\, s_{\rm h}(u_3) + \frac{u_3-u}{u_3-u_2}\, s_{\rm h}(u_2),$$

and the stable configuration comprises two phases with energy densities u_2 and u_3.

When $E_2 < U < E_4$ and when $E_5 < E < E_3$, the homogeneous system may be metastable, since its density of states decreases, if one splits it into parts with energy densities close to the original one; a non-infinitesimal energy transfer is then necessary between the two parts to make the density of states increase, whereas an infinitesimal transfer suffices when $E_4 < U < E_5$, in which case metastability is precluded.

3. The canonical entropy is the Legendre transform with respect to β of

$$k \ln Z_{\rm C}(\beta) = k \ln \int dE\, \varrho_{\rm h}(E,\Omega)\, {\rm e}^{-\beta E} \sim \max_E \left[S_{\rm h}(E,\Omega) - k\beta E\right].$$

It is the same as the convex envelope of $S_{\rm h}$, thus as S. Even though we restricted ourselves to homogeneous micro-states, the fact that we are dealing with a canonical ensemble has sufficed to reconstruct the correct thermodynamic functions, the same as if we had included the states where different phases coexist.

4. When β passes through β_0, U jumps from E_3 to E_2. We always have $p(E) \sim \delta(E-U)$ in microcanonical equilibrium, but $p(E)$ is in canonical equilibrium spread over the segment E_2, E_3 when β is close to β_0. By retaining in the exponential of (5.54) or (5.88) only the extensive contributions we would find that $p(E)$ is constant over E_2, E_3, since S is a linear function of the energy with slope $k\beta_0$. However, near $\beta = \beta_0$ the non-extensive contributions, for instance, those from the interfaces, or those from small external influences, such as gravity, govern the shape of $p(E)$ between E_2 and E_3 in canonical equilibrium.

Subject Index for Volume I

absolute entropy, *see* Third Law
absolute temperature 35–36, 125, 197–201, 223, 298–299, 330; *see* Second Law
absolute zero 203
action 81–82, 339
additivity 105, 107–110, 115, 162, 200, 244–246, 248, 357–358
adiabatic compressibility 263, 265
adiabatic demagnetization 40–41, 235
adiabatic expansion 323–325, 345–347, 356
adiabatic invariants, — principle, — theorem 132, 193–194; *see* Born–Oppenheimer
adiabatic transformations 132, 192–194, 286, 387
adiabats 298–299, 323
adsorption 174–175, 216, 319, 345
algebra 52, 76–78; *see* exterior, observables, operators
allotropy 271
alloys 412, 426, 438
analytical mechanics, *see* classical mechanics
angular momentum, — velocity 61, 171, 338–342, 369, 375–376, 379; *see* rotations
anharmonicity 372
anisotropy 44
annealing 438
antilinear 61
antisymmetry, *see* Pauli principle, symmetry
apparatus, *see* measurements
approach to equilibrium 148–149, 331; *see* relaxation
a priori measure or probabilities, *see* prior
argon 393–394
assignment of probabilities, — of macro-state 22, 141–145; *see* indifference, maximum entropy
astrophysics 6; *see* expansion of Universe, ionization, planets, stars
atmosphere 317, 325, 337
atomic nuclei 375; *see* nuclear structures
atomism 5, 127
atoms, *see* chemical equilibrium, monatomic
averages, *see* expectation values
Avogadro number 104, 137, 229, 323, 337–338, 462

balance, macroscopic 275, 277; *see* conservation
balance, microscopic 8, 91, 326–327, 330, 332, 334
ballistic regime 331, 347
balloons 327
barometric equation 325, 337
BBGKY hierarchy 92
Bernoulli 4, 328
bias 142, 144–145, 179
binomial law 98, 464
bit 103–104
black-body radiation 6, 131
black holes 134
blood 278; *see* hemoglobin
Bohr effect 390
Bohr magneton 19, 462
Bohr–van Leeuwen theorem 176
boiling 418, 423, 443; *see* liquid–vapour equilibrium
Boltzmann 5, 123–129, 155–156, 326
Boltzmann constant 26, 36, 104, 112, 200, 229, 323, 337
Boltzmann entropy 126, 129, 138
Boltzmann equation 91, 126, 129

Boltzmann–Gibbs distribution 33, 34, 42, 141–180, 201, 225–226, 227, 229, 313
bootstrap 412
Born–Oppenheimer approximation 366–369, 393–394
Bose–Einstein statistics, *see* bosons, Pauli principle
bosons 61, 375
box potential 311, 350, 394
Boyle 322
bras 50
Bragg–Williams method 392, 412; *see* variational
Brillouin 133
Brillouin curves 39, 44, 47
broken invariances or symmetries 425–426, 429
Brownian motion 9, 224, 227
bubble chamber 423

caloric 124, 189
canonical distribution or ensemble 143, 146, 165–167, 172, 183, 220, 224–226, 229, 234
canonical ensembles 165–172, 220; *see* canonical distribution, equivalence, grand canonical, isobaric-isothermal, microcanonical
canonical partition function 25, 166, 207–209, 212, 214, 220, 314, 351
capillarity, *see* surface tension
Carathéodory 197
carbon dioxide and monoxide 280–281, 378; *see* hemoglobin
Carnot 124–125, 189, 197
Carnot cycle 277, 298
Carnot theorem 276
centrifuge 325–326, 338–340
chain 231, 435–437; *see* polymers
chaotic dynamics 9, 134–135, 149
characteristic function 99; *see* partition functions
characteristic temperature 39, 309–310, 363–365, 367, 369–370, 373–374, 376
charge 205, 244, 285–286, 292
charge carriers 216–217, 331
chemical equilibrium 147–148, 151, 170, 215, 319, 359–363, 381, 387–390
chemical equilibrium, macroscopic 244–246, 279–283
chemical potentials 213, 215–219, 250, 251, 278–283, 306
chemical potentials in classical fluids, in mixtures, or in solutions 296–297, 357, 360–361, 416–418, 442–445
chemical potentials in gases 312, 323, 355–356, 403
Clapeyron relation 273, 418, 445
classical entropy 122–123, 323–324
classical fluids, — gases 307, 311–313, 394–396
classical limit 53–54, 62, 84–89, 351, 370–372, 377
classical mechanics 78–84, 126, 132, 134–153
classical partition functions 171–172, 314, 319, 370–373
classical statistical mechanics, *see* density in phase, reduced densities
Clausius 125, 197–198; *see* virial
closed cycle, *see* cycle
closed systems 187; *see* isolated, open
closure 51
cluster expansion, *see* virial expansion
coarse-graining 130, 135, 138
coding 106–107
coefficient, *see* compressibility, elasticity, expansion, linear expansion
coexistence of phases 269–273, 279, 299, 418, 420–426, 440–442, 446–448
Colding 188–189
collapse 206
collective variables, *see* macroscopic
collisions 91, 307, 310–311, 326, 328–334, 352
communication theory 102–111, 113, 133
commutator 52, 58, 88
commuting or compatible observables 56, 73, 76, 121
complete base 51
composite systems 51, 62, 108, 115–117, 183–188, 244–246, 274, 281; *see* subsystems
compressibility 263, 265–266, 268, 305
compression, *see* condensation
concavity of entropy 106, 117, 139, 161, 200, 264–265, 446
concentration, *see* chemical equilibrium, mixtures, solutions
condensation 391–408
condensed matter 3; *see* liquids, solids
conductivity, *see* thermal
conjugate momenta in classical or quantum mechanics 53, 78, 82, 85–87

conjugate variables in electromagnetism 284–295
conjugate variables in thermostatics 34, 205, 207, 211–212, 214, 253–256, 260
conjugation, *see* Hermitean
connected diagrams 403
conservation laws 7, 60–61, 73, 88, 146–152, 187–188, 244, 382; *see* angular momentum, energy, First Law, momentum, particle number
constants of the motion, *see* conservation laws
convexity 159, 161, 255, 258, 265
cooling 328; *see* adiabatic demagnetization, freezing mixture, Joule–Thomson expansion, refrigerators
cooperative phenomena 392; *see* phase transitions
correlations 31–32, 106, 115–117, 119–120, 159, 162, 202, 302–304, 321, 396, 436
Coulomb forces 6, 205–206, 393; *see* charge
critical opalescence 227, 304
critical phenomena 7, 228, 301–304, 392, 424–426, 433–435; *see* fluctuations, phase equilibria
critical points 245, 268–269, 272, 414–419, 428, 432
cryogenics, *see* cooling
crystallization 425–426, 443
crystal structures 152, 426
Curie 7
Curie constant, — law 17–18, 38, 44, 47, 173–174, 235
Curie principle 284
Curie temperature 301–303, 425, 428
Curie–Weiss law 428, 432
current, *see* electric
current density 335
cycle 276; *see* Carnot, diesel, Otto

Dalton law 357, 361
de Broglie wavelength 312
Debye 7
deformation tensor 244, 250, 283–284
degassing 175, 345
degeneracy 164, 166, 204, 308, 363, 365, 369, 370, 375
degree of disorder, *see* entropy, information
degrees of freedom 370–372; *see* electrons, rotations, translations, vibrations of molecules
demixture, *see* isotope separation, mixture
demon, *see* Maxwell
dense gases 391–408
density in phase 50, 78–93, 101, 122–123, 141–145, 313, 321; *see* reduced densities
density matrix, — operator 50, 63–78, 101, 111–112, 132, 141–145; truncation of –, *see* wavepacket reduction
density of particles, *see* particle density
density of states, *see* level density
depleted uranium, *see* isotope separation
desalination 279
determinism 4, 10, 181, 221
diagonalization, 54, 55, 67, 76
diagrammatic expansion 402–403, 405
diamagnetism 16, 173
diamond 152, 271
diatomic gases, — molecules 309, 366–382
dielectrics 170, 284–292
diesel 277
diffusion, *see* effusion
dilatation, *see* expansion, linear expansion
dilute, *see* ionization, perfect gas, solutions
dimensional analysis 13, 297
dipole moment 170, 244, 285
Dirac constant 462; *see* Planck
Dirac distribution or function 51, 322, 465
direct sum 52
disorder 22–27, 112, 127, 142, 144–145, 248–249, 358; *see* entropy, information
displacement, *see* electric induction, position variables
dissipation 249, 276; *see* irreversibility
dissociation 280, 359–360
dissolving, *see* solutions
distribution, *see* Dirac, probabilities
DNA, *see* genetic code
Doppler profile, — shift 317, 343, 347
Drude 7
dyadics 52, 76–77
dynamical systems 134–135

dynamic equilibrium, *see* stationary states

effective field 411–412, 430, 433; *see* variational methods
effective force, — interaction, — potential 367–368, 393–394, 409
efficiency 276–277, 383–385
effusion 216, 318, 326–328, 331, 344–345, 347–348, 383–385
Ehrenfest 129, 132, 178
Ehrenfest theorem 59, 73, 88, 312
eigenvalues, eigenvectors, *see* diagonalization
Einstein 6, 131, 227
elasticity 231–235, 283–284
electric current 285–286, 292–295
electric dipole, *see* dipole
electric displacement or electric induction 286–287
electric energy, — work 250, 286–292
electric field 170, 285–292
electric potential 216–217, 286–287, 290, 292
electrochemical potential 216; *see* chemical potentials
electromagnetism 131, 245, 250, 284–295
electromotive force 216
electrons in molecules 363–365, 366–367
electrostatic equilibrium 147, 205, 216–217, 245, 286–292; *see* charge, dielectrics, dipole
empiricism 6, 127
endothermic 283
energetics 127, 189
energy, *see* electric, Fermi, free, internal, magnetic
energy conservation 23–24, 125; *see* conservation laws, First Law, internal
energy downgrading 197–198; *see* heat, irreversibility, work
energy eigenstates, *see* excited states, ground state, level density, spectra
energy partition 28–31, 184–186, 222, 244–245; *see* equipartition, probability
energy transfer, *see* current density, flux, transport
energy units 463
enriched uranium, *see* isotope separation
ensembles, *see* canonical, equivalence, grand canonical, isobaric-isothermal, microcanonical, statistical
enthalpy 214, 220, 258–260, 283, 362, 406–407
entropic elasticity 232
entropy, *see* additivity, Boltzmann, classical, Kolmogorov, maximum, mixing, quantum
entropy of gases 323–324, 354, 356–357
entropy, relative or relevant 129–130, 137–139, 247–248
entropy, statistical 26–27, 30, 101–140, 164, 199, 203–204, 225–226, 246–248
entropy, thermodynamic 35–37, 124–125, 132, 197–201, 203–204, 210–211, 220, 242–248
equations of motion, *see* evolution
equations of state 37, 219, 231, 249–250, 253, 286, 296–298, 322–323, 353–354, 398, 404, 414
equilibrium 141, 152–165, 242, 251; *see* approach, chemical, electrostatic, gravitational, hydrostatic, liquid-vapour, magnetostatic, metastability, osmotic, phase, rotational, solutions, thermal, thermostatics
equilibrium constant 361–363
equipartition 370–372, 373–374
equiprobability 24, 104–105, 142, 225–226; *see* indifference principle
equivalence principle, *see* First Law
equivalence of ensembles 34, 165, 186, 207–210, 237, 319–320, 446–448
ergodicity 134–135, 146–147
escape velocity 317
Euler–Lagrange equations 82
Euler–Maclaurin formula 464
evacuation 175, 239, 344–345
evaporation 216, 421; *see* liquid–vapour equilibrium, vaporization heat
evaporation plateau, *see* Maxwell construction
event 102, 113
evolution, macroscopic, *see* approach to equilibrium, irreversibility, magnetic resonance, relaxation
evolution, microscopic 58–60, 73–74, 81–84, 90–92, 118–119
evolution operator 59, 73–74, 118
exchange interaction 427
exchange of particles, *see* indistinguishability, Pauli principle

exchange operator 87–88
exchanges 183–188, 195–197, 243–248, 251; *see* energy, heat, particle
excited states 308–309, 385–386; *see* level density, spectra
exclusive events 64, 67, 139
expansion coefficient 252, 263, 268, 274; *see* linear expansion
expansion of Universe 347
expansions, *see* perturbation
expectation values 21, 76–77, 79, 93, 143–144, 146, 226
expectation values at equilibrium 158–159, 163–164, 166, 168, 226
expectation values, quantum 57, 64–66, 73, 75–77
exponential dominance 29–30, 208, 210; *see* saddle-point method
exponential of operator 54–55
extensive variables 29, 205–207, 212, 244–245, 250
extensivity 49, 134, 205–210, 221, 226, 244, 248, 324, 359, 395, 403, 412, 424, 446–448
exterior algebra 261–263

factorization 161–163, 166, 314, 351
Fermi–Dirac statistics, *see* fermions, Pauli principle
Fermi energy 217
fermions 61, 375, 379
ferroelectricity 426
ferromagnetism 45, 204, 301–304, 392, 425, 427–440, 442
fibre 231–235
field, *see* electric, magnetic, molecular
fine structure 365
finite systems 186, 221–230, 424; *see* fluctuations
First Law 125, 188–197, 244
first-order, *see* phase transition
fluctuations 98–99, 130–131, 221–224, 227–229, 238–239
fluctuations and responses 173–174, 238, 432
fluctuations, critical 227, 302–304, 424, 428
fluctuations, methods of calculation 99, 159, 167, 169, 371
fluctuations, quantum 57–58, 312
fluctuations, smallness 30, 207–209, 226
fluids 165, 253, 263–264, 391–426; *see* gases, interfaces, liquid–vapour equilibrium, solutions
fluid–solid equilibrium, *see* crystallization
flux 326–327, 335
Fock space 52, 93, 144, 167
forces 191, 231–235; *see* pressure, surface tension, viscosity
force variables 191–193, 250, 291, 292
form, *see* exterior algebra
free energy 213–214, 220, 233, 258, 259, 288, 290–291, 294, 322, 403
free enthalpy 214, 220, 232–233, 260, 278, 283
freezing, *see* crystallization
freezing in 40, 309–310, 359, 363–365, 367, 372, 373–374, 377, 393–394
freezing mixture 443
fusion, *see* crystallization, thermonuclear

Galilean invariance 60–61, 170, 177, 341, 350
galvanometer 230
gamma function 464
gas constant 323
gaseous diffusion, *see* effusion
gases, *see* adsorption, centrifuge, classical, condensation, diatomic, ionization, liquid–vapour equilibrium, monatomic, perfect, photon, polyatomic, rare, relativistic
gauge 20, 56, 286
Gaussian distribution 30, 97, 98, 210, 229, 239, 464
generating functions 99; *see* partition functions
generators 61
genetic code, — information 107, 136
g factor 43–44
Gibbs 5, 127, 156
Gibbs–Duhem relation 219, 251, 271, 306
Gibbs ensemble 128, 226; *see* statistical
Gibbs paradox 127–128, 324, 358–359
Gibbs phase rule 272, 283
Gibbs potential, *see* free enthalpy
glasses 7, 152, 204
glycerine 152
grand canonical distribution or ensemble 144, 146, 167–169, 172, 187, 215, 220, 223–224

grand partition function 168, 170, 207–209, 212, 218, 220, 319, 443
grand potential 218–220, 258, 323, 351–352, 397, 402
graph, *see* diagrammatic expansion
graphite 152
Grassmann, *see* exterior algebra
gravitational equilibrium 206, 245, 250, 304–306
gravity 192, 299, 301, 325, 421, 445
ground state 308, 364
Guldberg and Waage, *see* mass action
gyromagnetic ratio, *see* g factor

Hamilton equations 82
Hamiltonian 58, 60–61, 73–74, 82–83, 255, 340, 350, 367–368
Hamilton principle 81
harmonic, *see* oscillator
heat 35–36, 124–125, 190–191, 198–199, 217, 243, 275; *see* enthalpy, latent, reaction, thermal
heat baths, *see* thermostats
heat capacity, *see* specific heats
heat conduction, — death, *see* thermal
heat exchanges 195–197, 259–260; *see* energy partition, thermal contact
heat pumps 276–277
Heisenberg inequalities 58, 312
Heisenberg model 427
Heisenberg picture 60, 74, 77
helium 204, 312
Helmholtz 188–189, 213; — potential, *see* free energy
hemoglobin 387–390
Henry law 443
Hermitean conjugation 52
Hermitean matrix, — operator 54, 56, 66
hidden variables 145, 148, 246, 382; *see* order parameters
Hilbert spaces 50–52, 75, 77
Hooke law 235, 284
H-theorem 126, 129
hydrodynamic regime, *see* local equilibrium
hydrodynamic velocity 335–336, 339, 341
hydrogen 245, 312, 378–382, 462
hydrogen chloride 151, 368–369
hydrostatic equilibrium 147, 305–306, 337
hyperfine splitting 365
hysteresis 271; *see* metastability

ice 443
ideal, *see* measurements, perfect gas, preparations
identical particles, *see* indistinguishability, Pauli principle
impulse 329
inaccessibility principle, *see* Carathéodory
incomplete knowledge, — measurements, — preparations 20, 23, 50, 62, 65, 120–121, 137–138
indicator method 99
indifference principle 142, 155, 226, 248
indistinguishability 56, 206, 342, 352–353; *see* Pauli principle, symmetry of wavefunctions
indistinguishability in the classical limit 80, 85, 87–88, 359, 370, 378; *see* Gibbs paradox
induction, *see* electric, magnetic
inert gases 309, 313, 363–364
information 101–140, 145, 149, 201, 247–249, 358
inhomogeneous systems, *see* coexistence, electrostatic, gravitational, interfaces
insufficient reason, *see* indifference
intensive variables 29, 205, 212, 249–251
interactions 158, 310–311, 352, 391–448
interfaces 245, 299–301, 421, 440–442
internal degrees of freedom 78, 350–351; *see* electrons, rotations, vibrations
internal energy 22, 90, 188–189, 196–197, 213, 220, 244, 288, 294
internal energy of classical fluids 324, 354, 398, 401, 413
internal partition function 350–353, 373–374, 376
International System of Units (SI) 462
invariance laws 7, 60–61, 111, 122, 350, 375, 425–426, 429; *see* conservation laws, Galilean invariance, symmetries
inversion temperature 407
ionization 309, 313, 353, 386–387
ions, *see* chemical equilibrium, solutions
irreversibility 4, 119, 120, 122, 125, 197–198, 201–202, 259, 276; *see* dissipation
irreversibility paradox 128–130, 131, 135, 149

irreversible processes 7, 243, 245, 276–277, 326–336, 358, 385, 406
Ising model 301, 392, 426, 427–437
isobaric-isothermal ensemble 171, 214, 220, 232–233
isobars 419
isolated systems 73, 118, 147, 187, 244–246; *see* closed, open
isothermal transformations, *see* thermostat
isotherms 298–299, 415
isotopes 396
isotope separation by effusion 327–328, 347–348, 383–385
isotope separation by ultracentrifuging 326, 338–342

jacobians 83, 261–263
jet 318–319, 328
Joule 188–189
Joule expansion, — law 263, 324, 346, 354, 401
Joule–Thomson expansion 264, 406–408

Kappler experiment 228–229, 239
kelvin 36, *see* Thomson
kets 50, 57, 62, 66
kinetic energy 318, 334
kinetic theory 5, 125–127, 307–308, 326–336, 345–347
knowledge, *see* information, measurements
Knudsen regime, *see* ballistic
Kolmogorov entropy 135

lack of information, *see* information
Lagrangian 81–82, 255, 339–342
Lagrangian multipliers 152–156, 159, 160, 184, 187–188, 250, 341
Landau 7, 50, 63
Landau theory of phase transitions 299–304, 426
Landé factor, *see* g factor
Langevin paramagnetism 48, 176
Langmuir isotherms 175
language, *see* communication
Laplace 142, 325
large numbers, — systems 8–10, 23, 30, 128, 149, 225–226; *see* extensivity
Larmor frequency, — precession 42, 95–97
latent heat 272–273; *see* melting, vaporization
lattice gas 413, 426
lattice temperature 47, 150–151, 185
Laws of thermodynamics 182; *see* Zeroth, First, Second, Third
least action principle, *see* action
Le Chatelier–Braun principle 267–268
Le Chatelier principle 266–267, 283
Legendre transformation 160–161, 208–209, 211, 213, 218, 253–255
level density 25–26, 166, 169, 206–207, 210, 446–448
Linde 408
linear expansion 232, 235
linear responses, *see* responses
linear transformations 52
Liouville equation 84, 88–89
Liouville representations of quantum mechanics 53–54, 77, 87–88
Liouville theorem 83–84, 91, 122
Liouville–von Neumann equation 73–74
liquefaction, *see* liquid–vapour equilibrium
liquids 312, 393–397; *see* solutions
liquid nitrogen trap 344–345
liquid–vapour equilibrium 269, 299–301, 302, 391–392, 395, 408–426; *see* interfaces, solutions
local equilibrium 331–336, 347; — temperature 201
long-range interactions 205, 206, 433
Lorentz force 82
Loschmidt 128
low temperatures, *see* cooling

macroscopic data or variables 142–144, 146–152, 182, 222, 243
macro-state 20–21, 63–70, 79–81, 113; *see* Boltzmann–Gibbs distribution, density in phase, density operator
magnetic dipole or moment 17–19, 43, 170, 192, 285, 294
magnetic energy, — work 35, 250, 292–295
magnetic field H, — induction B 17, 43, 170, 285–286, 292–295
magnetic resonance 41–43, 150; *see* relaxation
magnetic susceptibility 16, 18, 173, 302–303, 428, 432
magnetism 16, 49, 176, 204, 234, 285–286, 292–295; *see* dia-, para-, and ferromagnetism

magnetization 16, 38–39, 302–303, 285–286, 292–295, 427, 430, 432
magneton, *see* Bohr
magnetostatic equilibrium 147, 170; *see* magnetism
many-body theory 7
marginal energy 217
Mariotte 322
mass action law 362, 381, 442–445
Massieu functions 211–213, 220–221, 227, 256, 296
mathematics 2
matrices, *see* density operator, observables, operators
maximum of Massieu function 256–258; *see* variational methods
maximum of statistical entropy 114, 134, 141–152, 156–158, 171, 186, 222, 225–226, 246–249; *see* disorder
maximum of thermostatic entropy 243–248, 251, 264–273, 274, 279, 305, 361; *see* Second Law
Maxwell 5, 125–126, 326, 335–336
Maxwell construction 272, 298, 423–424
Maxwell demon 130–131, 132–134, 249
Maxwell distribution 155, 313–322, 396–397
Maxwell equations 285–286
Maxwell relations 260–261
Mayer 188–189
Mayer relation 324–325, 354
mean field, *see* effective field, variational methods
mean free path 331–336
mean square, *see* fluctuations
mean value 226; *see* expectation
measurements 57–58, 62, 68, 70–73, 111, 119–122, 132, 230
mechanical equivalents 463
mechanics, *see* classical
melting, *see* crystallization
melting heat 248, 446
message, *see* communication
metastability 150–152, 246, 271, 380–382, 421–423, 437–440, 448
microcanonical distribution or ensemble 22, 28–29, 97, 140, 143, 147, 169, 172, 185–186, 220, 225, 233, 321–322
microcanonical partition function 169, 211, 220; *see* level density
microelectronics 331
micro-state 20–21, 24, 57, 63–64, 78, 113
minimum of thermodynamic potentials 257–258, 409, 411, 421–422; *see* variational methods
missing information, *see* information
mixing entropy 106, 246, 357–359, 383–385
mixtures 81, 127–128, 139, 148, 151, 215, 217, 244–245, 249, 357–363, 381, 426; *see* isotope separation, solutions, statistical
modulus, *see* elasticity
mole 279
molecular beams or jets 318–319, 328
molecular effusion, *see* effusion
molecular field 412; *see* variational methods
molecules 349–390; *see* chemical equilibrium
moment of inertia 171, 342, 374, 376
momentum 56, 61, 170, 177, 315, 340–342; *see* conjugate, conservation
momentum transport 335–336
monatomic fluids 309, 363–365, 393–394
multiplicity, *see* degeneracy
multipliers, *see* Lagrangian
myoglobin 390

natural uranium, *see* isotope separation
natural variables 210–212, 214, 254
negative temperatures 37, 46–47, 185
negentropy 133, 275; *see* information
Nernst 131–132; *see* Third Law
neutron stars 305
Newton 336
nitrogen 344–345, 378
noise 107, 230, 238, *see* random evolution
non-equilibrium, *see* irreversible processes, transport
normal conditions 308, 463
nuclear energy, *see* isotope separation
nuclear magnetism 150–151, 364
nuclear reactions 151, 306
nuclear structures 3, 171, 375, 412
nucleation 423

observables 56–57, 75, 76–77
one-dimensional models 231–235, 435–437
one-particle, *see* reduced densities
Onsager 7, 392
opalescence 227, 304

open systems 127, 156, 167, 187–188, 319; *see* closed, isolated
operators 52–56
orbital magnetism 20, 43, 47
order parameters 301–304, 425–426
orthohydrogen, 379
orthonormality 51, 64
oscillators 131, 140, 177–178, 229, 237, 370–372; *see* anharmonicity, vibrations
osmotic equilibrium 147, 216, 245, 278–279; *see* solutions
osmotic pressure 278, 442, 444
Otto cycle 277
oxygen 312, 378; *see* hemoglobin

paradoxes 127–131
parahydrogen 379
paramagnetism 15–48, 95–97, 175–176, 185, 204, 427, 432
parity 60
partial densities, — pressures 357, 361
particle density 90, 299–301, 316, 320, 325, 337–340, 410, 417, 420, 440–441
particle exchanges 187–188, 215–217, 244–246, 271, 278–283, 327–328, 347–348
particle number 60, 76, 93, 167–168, 357, 359–362; *see* chemical potentials, conservation laws, flux, probabilities
particle physics 3
particles, indistinguishable –, *see* indistinguishability, Pauli principle
partition, *see* energy, particle exchanges
partition functions 158–165, 212, 220–221, 433–435; *see* canonical, classical, grand, internal, microcanonical
Pauli matrices 94
Pauli principle 61, 78, 352–353, 379; *see* bosons, fermions, indistinguishability, symmetry of wavefunctions
perfect gas 80, 296–297, 307–348
Perrin 5, 337–338
perturbation expansions 397–405
petit canonical, *see* canonical
phase density, *see* density in phase
phase diagram 272, 419–420, 422
phase equilibria, — transitions 7, 127, 152, 158, 216, 222–223, 269–273, 283, 301–304, 319, 391–392, 425–448; *see* ferromagnetism, liquid–vapour equilibrium
phase separation, *see* coexistence
phase space 78–84, 89–91, 122–123
photon gas 216, 347
physical constants 462–463
Planck 6, 131
Planck constant 53, 79, 84, 357, 462
Planck law 140, 177–178
Planck length 3
planets 304–306, 317
plasma, *see* ionization
Poincaré 128, 134
point particles 308–309, 350, 393
Poisson bracket 84, 88–89
Poisson distribution 98, 238
Poisson formula 464
polarization 285–291; *see* dielectrics, spin
polyatomic gases, — molecules 309, 366–368, 372–374
polymers 7, 231–235
porous wall, *see* effusion
position variables 191–192, 250, 291, 292
positive operator 54, 67
positivism 127
potential, *see* chemical, electric, Massieu, thermodynamic, vector
power 275–276
precession, *see* Larmor
preparations 57–58, 63; *see* measurements, wavepacket reduction
pressure 171, 213, 219, 305–306, 328–330, 337, 357; *see* equations of state
prior measure 111, 122
probabilities 20–22, 49–50, 68, 72, 75, 102–103, 113, 126–128, 191, 221, 228–230; *see* assignment, statistics
probability distributions: for energy 29–30, 166, 210, 447–448; for macroscopic variables 227–228; for momentum or velocity 315–319; for particle number 93, 98–99, 168; for subsystems 224–226
projection 55, 56, 58, 66
pure states 20, 57–58, 63, 66, 94, 114, 137

quantization 56, 58, 72
quantum entropy 111–122, 131–132, 139, 226

quantum mechanics 49–78, 131–132, 309–310, 312, 324, 351, 375–382; *see* measurements
quasi-equilibrium 320; *see* metastability
quasi-static 198, 245; *see* reversibility
quenching 20, 44; *see* freezing in

radiation, *see* black-body, oscillators, stars
random evolution 62, 74, 95, 118, 128; *see* chaotic dynamics
random variables 68, 76–77, 79; *see* probabilities
Rankine 188
Raoult laws 442–444
rare gases, *see* inert gases
ratio of specific heats 324–325
Rayleigh 6
reaction heat 283, 362–363
reactive mixture or system, *see* chemical equilibrium, nuclear reactions
reactor, *see* isotope separation
recurrence time 128–129
red cells 278; *see* hemoglobin
reduced densities 89–93, 315–316, 320–321, 339, 396
reduction, *see* wavepacket
redundancy 107
refrigerators 248, 276–277; *see* cooling
regime, *see* ballistic, local equilibrium
relative, *see* entropy
relative temperature 33, 183–187; *see* Zeroth Law
relativistic gases 342
relaxation 95–97, 148–150
relevant, *see* entropy
repeatability 72–73
responses in thermodynamics 335–336
responses in thermostatics 173–174, 238, 252–253, 261, 263–264, 266–268, 273–274, 432
reversibility, *see* irreversibility
rotational equilibrium 171, 325, 338–342
rotations 61, 350, 425–426; *see* invariance laws, angular momentum
rotations of molecules 81, 369–370, 373–377, 379
rubber 231–232
Rydberg 462

Sackur–Tetrode formula 323, 356
saddle-point method 209–210, 236–238, 321, 433–435
Saha equation 386–387
saturated vapour 417–421
saturation curve 272, 419–420
saturation of magnetization 38–39, 44, 47
scalar potential, *see* electric
scattering 328
Schottky 365–366
Schrödinger equation 58–59, 367–369, 375–376
Schrödinger picture 59–60, 74
second derivatives 252, 255, 260, 266
Second Law 28, 124, 126, 130, 133, 185, 189, 197–202, 243–246, 385
second-order, *see* critical points, phase transitions
self-consistent potential 411–412, 430, 433; *see* variational methods
self gravitation, *see* gravitational equilibrium
semiconductors 7, 217, 331
semi-permeable membranes, *see* osmotic equilibrium
separation, *see* coexistence of phases, isotope separation
Shannon theorems 106–107; *see* communication
shot noise 230, 238
SI units 462
single-particle, *see* reduced densities
singlet 69, 379–380
small systems, *see* finite
Smoluchowski 130, 227
solidification, *see* crystallization
solids 7, 49, 283–284
solutions 325, 396, 442–446; *see* chemical, mixtures, osmotic
solvent, *see* solutions
sound 316, 325
sources 195–197, 202, 248–249, 274–277, 291; *see* exchanges, thermostats
specific heats 39, 238–239, 252, 263, 266, 268, 274, 302–303, 428, 431
specific heats of gases 296–297, 324–325, 354–355, 363–382
spectra 127, 349, 351, 355, 366–369, 375–376
spectral decomposition 58
spectral lines 317, 343, 355
speed, *see* sound
spherical harmonics 375, 379; *see* angular momentum

spin 19, 65, 69, 94–97, 365, 369, 375, 379–380, 437–438; *see* Ising model, paramagnetism
spin temperature 35–37, 47, 150–151, 185
spontaneous magnetization, *see* broken invariances, ferromagnetism
stability 205–207, 264–268, 305; *see* metastability
standard conditions, *see* normal
standard deviation, *see* fluctuations
stars 205–206, 304–306, 317; *see* ionization
states, *see* density in phase, density operator, macro-state
state vectors, *see* kets, micro-state, pure states
stationary phase, *see* saddle-point
stationary states 146–147, 347–348, 406–408
statistical ensembles 8, 20, 57, 63, 71, 75, 128
statistical entropy, *see* entropy
statistical mixture 63–70, 75–77, 94, 117
statistics 5, 8–10, 57–58, 145, 149, 221, 228–230, 307, 326; *see* Bose–Einstein, Fermi–Dirac, fluctuations, probabilities
steady states, *see* stationary
steam engines 277
steepest descent, *see* saddle-point
Stern–Gerlach experiment 65
Stirling formula 26, 237, 464
stochastic, *see* random
Stokes law 338
strain, *see* deformation
stress tensor 250, 284
structureless particles, *see* point particles
sub-additivity 105, 115–116
subsystems 62, 68–72, 115–117, 137, 147, 224–226; *see* composite systems
sum over states, *see* partition functions
Sun 463; *see* stars
superconductivity 7
supercooling 152, 271
superheating 271, 422
superposition principle 57
supersaturation 422
superselection 57
surface tension 245, 299–301
surprisal 103
susceptibility, *see* magnetic
symmetries 7, 81, 284; *see* broken, conservation, invariance
symmetry number 81, 370, 381
symmetry of wavefunctions 61, 87–88, 352–353, 375, 379; *see* Pauli principle
Szilard 132

temperature, *see* absolute, characteristic, lattice, local, negative, relative, spin, thermometry
tension, *see* elasticity, surface
tensor 44, 283–284
tensor product 51, 54
thermal baths, *see* thermostats
thermal capacity, *see* specific heats
thermal conduction, — conductivity 332–336
thermal contact 28–31, 183–187, 201–202, 222, 245; *see* energy partition, heat exchanges
thermal death 129, 274
thermal engines 171, 197, 235, 248, 274–278
thermal equilibrium 22–33, 147, 183–187
thermal excitation 363–382, 385–386
thermal expansion, *see* expansion, linear expansion
thermal ionization, *see* Saha
thermal length 311–312, 351
thermal noise, *see* noise
thermal pollution 275
thermionic effect 238
thermochemistry, *see* chemical equilibrium
thermodynamic entropy, *see* entropy
thermodynamic equilibrium, *see* equilibrium, thermostatics
thermodynamic identities 249–253
thermodynamic inequalities 265–266
thermodynamic limit, *see* extensivity
thermodynamic potentials 210–221, 256, 288–295, 296; *see* enthalpy, entropy, free energy, free enthalpy, grand potential, internal energy, Massieu functions
thermodynamics 124–125, 241–243; *see* irreversible processes, thermostatics
thermometry 18, 187, 317
thermonuclear fission 151, 306
thermostatics 141, 146–152, 181–235, 241–306

thermostats 32–33, 186–187, 195–197, 214, 224–226, 257, 259
Third Law 36, 132, 203–204, 273, 277, 356, 359, 364, 372, 439
Thomson, Lord Kelvin 125, 128, 197
time reversal 4, 61
trace 55–56, 64, 66, 85, 87
transfer matrix method 435–437
transformations 60–61
transitions, *see* phase
translations 61, 170, 350, 426; *see* invariance laws, momentum
translations of molecules 78, 307, 367–368, 371, 374, 377
transport 331–336; *see* thermal conductivity, viscosity
trap 344–345
trial state, *see* variational methods
triple point 272
triplet 379–380
truncation, *see* wavepacket reduction
two-level system 19, 95
two-particle, *see* reduced densities

ultracentrifuging 325–326, 338–342
unbiased, *see* bias
uncertainty, *see* disorder, entropy, fluctuations, Heisenberg inequalities, information
unification of sciences 1–4, 241–242, 330
unitary operators 54, 59–61, 73–74
units 462–463
universality 426
Universe, *see* expansion, ionization, stars
uranium, *see* isotope separation
urn model 178–179
Ursell–Yvon–Mayer expansion, *see* virial expansion

vacuum, *see* evacuation
van der Waals equation 296–298, 398–401, 405, 414
van't Hoff law 362–363, 443
vaporization, *see* liquid–vapour equilibrium
vaporization heat 418, 421; *see* latent
variance 67; *see* fluctuations
variational methods 156–158, 257, 408–412, 421–422, 424, 427–432, 436, 441
vector potential 56, 82–83, 286, 294
velocity 170, 177, 315–318; *see* hydrodynamic, probability distributions, sound
vibrations of molecules 369–370, 373, 376–377; *see* oscillators
virial coefficents, — expansion 401–405, 442
virial theorem 305–306, 401
viscosity 335–336, 338
Vlasov equation 92
volume 171, 244–245
von Neumann 50, 63, 112, 132, 226;
von Neumann entropy, *see* quantum

wall 311, 328–330, 331
wavepacket reduction 58, 70–73, 119–121, 138
Weiss field 392, 412, 430; *see* variational methods
Wigner representation or transform 53–54, 87–89
Wilson theory 304
wool 231–232
work 125, 191–197, 214, 217, 223, 250, 259, 275, 277, 383–385; *see* electric, magnetic

Young modulus, *see* elasticity

Zermelo 128
zeros of partition functions 424, 433–435
Zeroth Law 183–188, 251
zeta-function 465
Zustandssumme, *see* partition functions

Units and Physical Constants

We use the international system of units, the so-called SI system, which is adopted by most official international organizations. Its fundamental units are the metre (m), the kilogram (kg), the second (s), the ampere (A), the kelvin (K), the mole (mol), and the candela (cd).

Derived SI units with special names are the radian (rad), the steradian (sr), the hertz (Hz $= \mathrm{s}^{-1}$), the newton ($\mathrm{N} = \mathrm{m\ kg\ s}^{-2}$), the pascal ($\mathrm{Pa} = \mathrm{N\,m}^{-2}$), the joule ($\mathrm{J} = \mathrm{N\,m}$), the watt ($\mathrm{W} = \mathrm{J\,s}^{-1}$), the coulomb ($\mathrm{C} = \mathrm{A\,s}$), the volt ($\mathrm{V} = \mathrm{W\,A}^{-1}$), the farad ($\mathrm{F} = \mathrm{C\,V}^{-1}$), the ohm ($\Omega = \mathrm{V\,A}^{-1}$), the siemens ($\mathrm{S} = \mathrm{A\,V}^{-1}$), the weber ($\mathrm{Wb} = \mathrm{V\,s}$), the tesla ($\mathrm{T} = \mathrm{Wb\,m}^{-2}$), the henry ($\mathrm{H} = \mathrm{Wb\ A}^{-1}$), the Celsius temperature (°C), the lumen ($\mathrm{lm} = \mathrm{cd\,sr}$), the lux ($\mathrm{lx} = \mathrm{lm\,m}^{-2}$), the becquerel ($\mathrm{Bq} = \mathrm{s}^{-1}$), the gray ($\mathrm{Gy} = \mathrm{J\,kg}^{-1}$), and the sievert ($\mathrm{Sv} = \mathrm{J\,kg}^{-1}$).

Prefixes used with SI units to indicate powers of 10 as factors are: deca (da = 10); hecto (h = 10^2); kilo (k = 10^3); mega (M = 10^6); giga (G = 10^9); tera (T = 10^{12}); peta (P = 10^{15}); exa (E = 10^{18}); deci (d = 10^{-1}); centi (c = 10^{-2}); milli (m = 10^{-3}); micro ($\mu = 10^{-6}$); nano (n = 10^{-9}); pico (p = 10^{-12}); femto (f = 10^{-15}); atto (a = 10^{-18}).

Constants for electromagnetic units	$\mu_0 = 4\pi \times 10^{-7}\,\mathrm{N\,A}^{-2}$ (definition of the ampere)
	$\varepsilon_0 = \frac{1}{\mu_0 c^2},\ \frac{1}{4\pi\varepsilon_0} \simeq 9 \times 10^9\,\mathrm{N\ m^2\,C^{-2}}$
velocity of light	$c = 299\,792\,458\,\mathrm{m\ s}^{-1}$ (definition of the metre)
	$c \simeq 3 \times 10^8\,\mathrm{m\,s}^{-1}$
Planck's constant	$h = 6.6260755 \times 10^{-34}\,\mathrm{J\,s}$
Dirac's constant	$\hbar = \frac{h}{2\pi} \simeq 1.055 \times 10^{-34}\,\mathrm{J\,s}$
Avogadro's number	$N_\mathrm{A} \simeq 6.022 \times 10^{23}\,\mathrm{mol}^{-1}$ (by definition the mass of one mole of $^{12}\mathrm{C}$ is 12 g)
(Unified) atomic mass unit	$1\,\mathrm{u} = 1\,\mathrm{g}/N_\mathrm{A} \simeq 1.66 \times 10^{-27}\,\mathrm{kg}$ (or dalton or amu)
neutron and proton masses	$m_\mathrm{n} \simeq 1.0014\,m_\mathrm{p} \simeq 1.008\,\mathrm{u}$
electron mass	$m \simeq 1\,\mathrm{u}/1823 \simeq 9.11 \times 10^{-31}\,\mathrm{kg}$
Elementary charge	$e \simeq 1.602 \times 10^{-19}\,\mathrm{C}$
Faraday's constant	$N_\mathrm{A} e \simeq 96\,485\ \mathrm{C\ mol}^{-1}$
Bohr magneton	$\mu_\mathrm{B} = \frac{e\hbar}{2m} \simeq 9.27 \times 10^{-24}\ \mathrm{J\ T}^{-1}$
nuclear magneton	$\frac{e\hbar}{2m_\mathrm{p}} \simeq 5 \times 10^{-27}\ \mathrm{J\ T}^{-1}$
Fine structure constant	$\alpha = \frac{e^2}{4\pi\varepsilon_0 \hbar c} \simeq \frac{1}{137}$
Hydrogen atom:	
Bohr radius	$a_0 = \frac{\hbar}{mc\alpha} = \frac{4\pi\varepsilon_0\hbar^2}{me^2} \simeq 0.53\ \text{Å}$
binding energy	$E_0 = \frac{\hbar^2}{2ma_0^2} = \frac{m}{2\hbar^2}\left(\frac{e^2}{4\pi\varepsilon_0}\right)^2 \simeq 13.6\ \mathrm{eV}$
Rydberg constant	$R_\infty = \frac{E_0}{hc} \simeq 109\,737\ \mathrm{cm}^{-1}$

Boltzmann's constant	$k \simeq 1.381 \times 10^{-23}$ J K^{-1}
molar gas constant	$R = N_A k \simeq 8.32$ J K^{-1} mol^{-1}
Normal conditions: pressure	1 atm = 760 Torr = 1.01325×10^5 Pa
temperature	Triple point of water 273.16 K (definition of the kelvin) or 0.01°C (definition of the Celsius scale)
molar volume	22.4×10^{-3} m^3 mol^{-1}
Gravitational constant	$G \simeq 6.67 \times 10^{-11}$ m^3 kg^{-1} s^{-2}
gravitational acceleration	$g \simeq 9.81$ m s^{-2}
Stefan's constant	$\sigma = \dfrac{\pi^2 k^4}{60\hbar^3 c^2} \simeq 5.67 \times 10^{-8}$ W m^{-2} K^{-4}
Definition of photometric units	A 1 W luminous power, emitted at a frequency of 540 THz, is equivalent to 683 lm
Energy units and equivalents	1 erg = 10^{-7} J (non SI)
	1 kWh = 3.6×10^6 J
electric potential	1 eV $\leftrightarrow$ 1.602×10^{-19} J $\leftrightarrow$ 11600 K
heat	1 cal = 4.184 J (non SI; specific heat of 1 g of water)
chemical binding	23 kcal mol^{-1} $\leftrightarrow$ 1 eV (non SI)
temperature (kT)	290 K $\leftrightarrow$ $\frac{1}{40}$ eV (room temperature)
mass (mc^2)	9.11×10^{-31} kg $\leftrightarrow$ 0.511 MeV (electron rest mass)
wavenumber (hc/λ)	109 700 cm^{-1} $\leftrightarrow$ 13.6 eV (Rydberg)
frequency ($h\nu$)	3.3×10^{15} Hz $\leftrightarrow$ 13.6 eV

It is useful to keep these equivalents handy for quickly finding orders of magnitude.

Various non SI units	1 angstrom (Å) = 10^{-10} m (atomic scale)
	1 fermi (fm) = 10^{-15} m (nuclear scale)
	1 barn (b) = 10^{-28} m^2
	1 bar = 10^5 Pa
	1 gauss (G) = 10^{-4} T
	1 nautical mile = 1852 m
	1 knot = 1 nautical mile per hour = 0.51 m s^{-1}
	1 astronomical unit (AU) $\simeq 1.5 \times 10^{11}$ m (Sun-Earth distance)
	1 parsec (pc) $\simeq 3.1 \times 10^{16}$ m (1 AU/arc sec)
	1 light year (ly) $\simeq 0.95 \times 10^{16}$ m
Solar data	Radius 7×10^8 m = 109 Earth radii
	Mass 2×10^{30} kg
	Average density 1.4 g cm^{-3}
	Luminosity 3.8×10^{26} W

A Few Useful Formulae

Normalization of a Gaussian function:

$$\int_{-\infty}^{+\infty} dx\, \mathrm{e}^{-ax^2} = \sqrt{\frac{\pi}{a}};$$

differentiation of this formula with respect to a gives us the moments of the Gaussian distribution.

Euler's gamma-function:

$$\Gamma(t) \equiv \int_0^{\infty} x^{t-1}\, \mathrm{e}^{-x}\, dx = (t-1)\Gamma(t-1),$$

$$\Gamma(t)\Gamma(1-t) = \frac{\pi}{\sin \pi t}, \qquad \Gamma(\tfrac{1}{2}) = \sqrt{\pi}.$$

Stirling's formula:

$$t! = \Gamma(t+1) \underset{t\to\infty}{\sim} t^t\, \mathrm{e}^{-t} \sqrt{2\pi t}.$$

Binomial series:

$$(1+x)^t = \sum_{n=0}^{\infty} \frac{x^n}{n!} \frac{\Gamma(t+1)}{\Gamma(t+1-n)} = \sum_{n=0}^{\infty} \frac{(-x)^n}{n!} \frac{\Gamma(n-t)}{\Gamma(-t)}, \qquad |x| < 1.$$

Poisson's formula:

$$\sum_{n=-\infty}^{+\infty} f(n) = \sum_{l=-\infty}^{+\infty} \tilde{f}(2\pi l) \equiv \sum_{l=-\infty}^{+\infty} \int_{-\infty}^{+\infty} dx\, f(x)\, \mathrm{e}^{2\pi i l x}.$$

Euler-Maclaurin formula:

$$\frac{1}{\varepsilon} \int_a^{a+\varepsilon} dx\, f(x) \approx \frac{1}{2}\left[f(a) + f(a+\varepsilon)\right] - \frac{\varepsilon}{12} f'(x)\Big|_a^{a+\varepsilon} + \frac{\varepsilon^3}{720} f'''(x)\Big|_a^{a+\varepsilon} + \ldots$$

$$\approx f\left(a + \frac{1}{2}\varepsilon\right) + \frac{\varepsilon}{24} f'(x)\Big|_a^{a+\varepsilon} - \frac{7\varepsilon^3}{5760} f'''(x)\Big|_a^{a+\varepsilon} + \ldots;$$

this formula enables us to calculate the difference between an integral and a sum over n, when we put $a = n\varepsilon$.

Constants:

$$e \simeq 2.718, \qquad \pi \simeq 3.1416,$$

$$\gamma \equiv \lim \left(1 + \cdots + \tfrac{1}{n} - \ln n\right) \simeq 0.577 \qquad \text{Euler's constant.}$$

Riemann's zeta-function:

$$\zeta(t) \equiv \sum_{n=1}^{\infty} \frac{1}{n^t}, \qquad \int_0^{\infty} \frac{x^{t-1}\,dx}{e^x - 1} = \Gamma(t)\,\zeta(t),$$

$$\int_0^{\infty} \frac{x^{t-1}\,dx}{e^x + 1} = \left(1 - 2^{-t+1}\right)\Gamma(t)\,\zeta(t).$$

t	1.5	2	2.5	3	3.5	4	5
ζ	2.612	$\frac{1}{6}\pi^2$	1.341	1.202	1.127	$\frac{1}{90}\pi^4$	1.037

Dirac's δ-function:

$$\frac{1}{2\pi}\int_{-\infty}^{+\infty} dx\, e^{ixy/a} = \delta\left(\frac{y}{a}\right) = |a|\delta(y);$$

$$\lim_{t\to\infty} \frac{\sin tx}{x} = \lim_{t\to\infty} \frac{1 - \cos tx}{tx^2} = \pi\,\delta(x);$$

$$f(x)\,\delta(x) = f(0)\,\delta(x), \qquad f(x)\,\delta'(x) = -f'(0)\,\delta(x) + f(0)\,\delta'(x).$$

If $f(x) = 0$ in the points $x = x_i$, we have

$$\delta[f(x)] = \sum_i \frac{1}{|f'(x_i)|}\,\delta(x - x_i).$$